Elements of Probabilistic Analysis with Applications

Gheorghe Constantin and Ioana Istrăţescu

Elements of Probabilistic Analysis

with Applications

Translated from the Romanian by
Victor Giurgiuţiu, Ph. D.

The National Institute for Scientific and Technical Creation, Bucharest, Romania

EDITURA ACADEMIEI
Bucureşti, Romania

KLUWER ACADEMIC PUBLISHERS
Dordrecht/Boston/London

Library of Congress Cataloging in Publication Data

CIP

Constantin, Gheorghe.
Elements of probabilistic analysis with applications.
(Mathematics and its applications. East European series)
Translation of: Elemente de analiză probabilistă cu aplicații.
Bibliography: p.
Includes indexes.
1. Probabilities. I. Istrățescu, Ioana. II. Title. III. Series: Mathematics and its applications (D. Reidel Publishing Company). East European series.
QA273. C74713 1989 519.2 88–28342
ISBN 90-277-2838-0

Distributors for the U.S.A. and Canada
Kluwer Academic Publishers
101 Philip Drive, Norwell, MA 02061, U.S.A.

Distributors for Albania, Bulgaria, Chinese People's Republic, Cuba, Czechoslovakia, German Democratic Republic, Hungary, Korean People's Republic, Mongolia, Poland, Romania, the U.S.S.R., Vietnam, and Yugoslavia
Editura Academiei, Bucharest, Romania
ISBN 973-27-0009-2 (Romania)

Distributors for all remaining countries
Kluwer Academic Publishers Group.
P. O. Box 322, 3300 AH Dordrecht, The Netherlands

Original title: "*Elemente de analiză probabilistă cu aplicații*".
First edition published in 1981 by Editura Academiei.
Second edition published in 1989 by Editura Academiei and Kluwer Academic Publishers.

Printed in Romania

Table of Contents

Series Editor's Preface

'Et moi, . . ., si j'avait su comment en revenir, je n'y serais point allé.'

Jules Verne

The series is divergent; therefore we may be able to do something with it.

O. Heaviside

One service mathematics has rendered the human race. It has put common sense back where it belongs, on the topmost shelf next to the dusty canister labelled 'discarded nonsense'.

Eric T. Bell

Mathematics is a tool for thought. A highly necessary tool in a world where both feedback and nonlinearities abound. Similarly, all kinds of parts of mathematics serve as tools for other parts and for other sciences.

Applying a simple rewriting rule to the quote on the right above one finds such statements as: 'One service topology has rendered mathematical physics . . .'; 'One service logic has rendered computer science . . .'; 'One service category theory has rendered mathematics . . . '. All arguably true. And all statements obtainable this way form part of the raison d'être of this series.

This series, *Mathematics and Its Applications*, started in 1977. Now that over one hundred volumes have appeared it seems opportune to reexamine its scope... At the time I wrote

> "Growing specialization and diversification have brought a host of monographs and textbooks on increasingly specialized topics. However, the 'tree' of knowledge of mathematics and related fields does not grow only by putting forth new branches. It also happens, quite often in fact, that branches which were thought to be completely disparate are suddenly seen to be related. Further, the kind and level of sophistication of mathematics applied in various sciences have changed drastically in recent years: measure theory is used (non-trivially) in regional and theoretical economics; algebraic geometry interacts with physics; the Minkowsky lemma, coding theory and the structure of water meet one another in packing and covering theory; quantum fields, crystal defects and mathematical programming profit from homotopy theory; Lie algebras are relevant to filtering; and prediction and electrical engi-

> neering can use Stein spaces. And in addition to this there are such new emerging subdisciplines as 'experimental mathematics', 'CFD', 'completely integrable systems', 'chaos, synergetics and large-scale order', which are almost impossible to fit into the existing classification schemes. They draw upon widely different sections of mathematics".

By and large, all this still applies today. It is still true that at first sight mathematics seems rather fragmented and that to find, see, and exploit the deeper underlying interrelations more effort is needed and so are books that can help mathematicians and scientists do so. Accordingly MIA will continue to try to make such books available.

If anything, the description I gave in 1977 is now an understatement. To the examples of interaction areas one should add string theory where Riemann surfaces, algebraic geometry, modular functions, knots, quantum field theory, Kac-Moody algebras, monstrous moonshine (and more) all come together. And to the examples of things which can be usefully applied let me add the topic 'finite geometry'; a combination of words which sounds like it might not even exist, let alone be applicable. And yet it is being applied : to statistics via designs, to radar/sonar detection arrays (via finite projective planes), and to bus connections of VLSI chips (via difference sets). There seems to be no part of (so-called pure) mathematics that is not in immediate danger of being applied. And, accordingly, the applied mathematician needs to be aware of much more. Besides analysis and numerics, the traditional workhorses, he may need all kinds of combinatorics, algebra, probability, and so on.

In addition, the applied scientist needs to cope increasingly with the nonlinear world and the extra mathematical sophistication that this requires. For that is where the rewards are. Linear models are honest and a bit sad and depressing: proportional efforts and results. It is in the nonlinear world that infinitesimal inputs may result in macroscopic outputs (or vice versa). To appreciate what I am hinting at: if electronics were linear we would have no fun with transistors and computers; we would have no TV; in fact you would not be reading these lines.

There is also no safety in ignoring such outlandish things as nonstandard analysis, superspace and anticommuting integration, p-adic and ultrametric space. All three have applications in both electrical engineering and physics. Once, complex numbers were equally outlandish, but they frequently proved the shortest path between 'real' results. Similarly, the first two topics named have already provided a number of 'wormhole' paths. There is no telling where all this is leading — fortunately.

Thus the original scope of the series, which for various (sound) reasons now comprises five subseries: white (Japan), yellow (China), red (USSR), blue (Eastern Europe), and green (everything else), still applies. It has been enlarged a bit to include books treating of the tools from one subdiscipline which are used in others. Thus the series still aims at books dealing with:

- a central concept which plays an important role in several different mathematical and/or scientific specialization areas;
- new applications of the results and ideas from one area of scientific endeavour into another;
- influences which the results, problems and concepts of one field of enquiry have, and have had, on the development of another.

The present volume deals with questions in probabilistic analysis. Trust the mathematicians to generalize anything and to come up with impossible sounding phrases: fractal dimension, finite geometry, ... And now, "random fixed point", the main topic of this book. The concept results from the bringing together of three streams in mathematics: probability, functional analysis and topology (in the form of fixed point theory). Of course, fixed points are a powerful tool in analysis in solving differential equations. Thus, once one has the right concepts, random fixed points ought to constitute a powerful tool in stochastic and integral equations, stochastic processes, stochastic optimization, etc. And so they are. In addition, stochastic operators have a wide range of applications from telephone traffic to pollution, and from random vibrations to chemotherapy.

Perusing the present volume is not guaranteed to turn you into an instant expert, but it will help, though perhaps only in the sense of the last quote on the right below.

The shortest path between two truths in the real domain passes through the complex domain.

J. Hadamard

La physique ne nous donne pas seulement l'occasion de résoudre des problèmes ... elle nous fait pressentir la solution.

H. Poincaré

Never lend books, for no one ever returns them; the only books I have in my library are books that other folk have lent me.

Anatole France

The function of an expert is not to be more right than other people, but to be wrong for more sophisticated reasons.

David Butler

Bussum, October 1988 MICHIEL HAZEWINKEL

Introduction

As is well known, progress in any science is highly dependent upon developments in methodology. Recently, probabilistic analysis has evolved into a very dynamical area of mathematical research that employs and extends concepts and results from functional analysis, operator theory and measure theory, as well as other branches of mathematics. In this way, many attempts have been made to develop and unify the theory of stochastic or random equations using the concepts and methods of probability theory and functional analysis. In the functional analytic formulation of a random equation, account is taken of the correspondence between elements of a given metric or linear space as being defined by a random mapping, and then, utilizing the properties of the mappings and the properties of the space on which the mapping is defined, the problem is to obtain conditions under which there exists a random function with values in the metric or linear space which satisfies the equation with probability one. Numerous recent researches have led to the development of powerful methodological tools for applied mathematics.

Regarding applications, stochastic models for various phenomena in the biological, engineering and physical sciences are obtained. For example, applications to the following areas are given : wave propagation in random media, turbulence theory, random vibration of elastic strings and bars, random methods in stream temperature modelling, stochastic models for stream pollution, chemotherapy, population growth, telephone traffic theory, stochastic nonlinear programming and stochastic optimal control.

For an adequate theoretical framework, the first part of the book gives a comprehensive survey of the most significant recent results concerning the theory of fixed points for operators on probabilistic structures and, respectively, for random operators on metric or linear spaces. The connection between these two important methods of investigation for obtaining fixed points for singlevalued and multivalued mappings is dealt with and the latter are then used for the study of random operator equations and their applications.

Chapter 1

Probabilistic structures and related topics

The extensive analytical and geometrical modelling of present-day physics has yielded new results and interpretations for some natural phenomena from the micro- and macrocosmos. These results have in turn stimulated new ways of tackling space and other fundamental mathematical structures and formulae that model real life aspects. For example, H. Poincaré characterized the difference between the mathematical and the physical continua by showing that the physical continuum is a system S of elements any two of which can be joined by a chain of elements of S such that each element of the chain is indistinguishable from the next element.

The tremendous progress made in the theory of relativity and quantum mechanics during the first part of the twentieth century as well as the development of many new elementary particles have stimulated philosophers, physicists and mathematicians alike to reconsider their previously world-wide accepted ideas regarding the space and time distribution of elementary particles together with other fundamental aspects of mathematics and physics. This theoretical shift prompted new methods and techniques that grew into a fundamentally new mathematical apparatus, a correspondingly new geometry, offering enlarged possibilities for interpreting real life phenomena.

The introduction of the general axiomatic concept of statistical metric spaces is due to K. Menger [290] who dealt with probabilistic geometry [291], with the theory of probabilistic relations and with the mathematical entails of several fundamental ideas due to philosopher-physicists and mathematicians such as H. Poincaré, A. Einstein, etc. He also dealt with the general theory of functions and variables, with the development of general geometry, etc.

The new theory of fundamental probabilistic structures was later developed by numerous authors, as can be realized upon consulting the list of references appended at the end of this book, as well as those in [393] and [115].

The development of probabilistic metric space theory is due to B. Schweizer and A. Sklar and to their coworkers who improved upon Menger's results in several aspects. Recently, their excellent monograph "Probabilistic Metric Spaces" [393] furnished basic results to the mathematical reader. Probabilistic structures are also treated in refs. [323], [380], [236] and [176]. Important contributions were also brought by A. Wald [434], A. N. Šerštnev [398], [399], [401], A. Špaček [414], [415], H. Sherwood [404], [405], O. Onicescu [323], P. Calabrese [85], G. Sâmboan and R. Theodorescu [408], V. Istrăţescu [236] and I. Istrăţescu [220]. Among the related topics, worth mentioning are the studies on probabilistic topology and convergence (B. Schweizer, A. Sklar and E. Thorp [392], E. Thorp [429], Fritsche [151], Sibley [407], Morrel and Nagata [294], Constantin and Radu [111]), continuity of the probabilistic distance (B. Schweizer [381]), measures of compactness (Gh. Bocşan and Gh. Constantin [56], [112], Gh. Bocşan [49], Do Hong Tan [425], G. Cain [83]), completion (Sherwood [404], Muštari [304], Nishiura [312]), entropy (Saleski [374], [375]), product and quotient spaces (V. Istrăţescu and Văduva [231], Schweizer [383], Egbert [130], Xavier [436], Tardiff [428], Radu [347], Alsina [9], I. Istrăţescu [220]), probabilistic metric spaces and Markov chains (Moynihan [298]), copulas (Sklar, Schweizer and Sklar [393]), a.s.o.

1.1. TRIANGULAR NORMS, TRIANGULAR FUNCTIONS AND PROBABILISTIC STRUCTURES

Probabilistic metric spaces were introduced by K. Menger (1942) who generalized the theory of metric spaces, to the development of which he already brought a major contribution. In Menger's theory the concept of distance is considered to be statistical or probabilistic, i.e., he proposed to replace the number $d(p, q)$ by a distribution function $F_{p,q}$ whose value $F_{p,q}(x)$, for any real number x, is interpreted as the probability that the distance between p and q is less than x.

Let us recall that a mapping $F : \mathbb{R} \to \mathbb{R}^+$ is called a *distribution function* if it is nondecreasing, left-continuous, and $\operatorname{Ran} F \subseteq [0,1]$.

Let Δ denote the set of all one-dimensional distribution functions and let Δ^+ be the set $\{F \in \Delta : F(0) = 0\}$. The set of all F in Δ for which $\sup_{x\in\mathbb{R}} F(x) = 1$ and $\inf_{x\in\mathbb{R}} F(x) = 0$ will be denoted by $\mathcal{D}$ and $\mathcal{D}^+ = \{F \in \mathcal{D},\ F(0) = 0\}$, respectively. Let us call a function $F \in \Delta^+$ a *distance distribution function* (d.d.f.). Let Δ^+ be partially ordered *via* $F \leqslant G$ if $F(x) \leqslant G(x)$ for all x. For any $a \geqslant 0$, let $H_a \in \Delta^+$ be defined by $H_a(x) = 0$ for $x \leqslant a$ and $H_a(x) = 1$ for $x > a$.

The *probabilistic distance* on a set S is the mapping $\mathcal{F} : S \times S \to \Delta^+$ whose value $\mathcal{F}(p, q)$ at any pair $(p, q) \in S \times S$, usually denoted by $F_{p,q}$, is assumed to satisfy

(1) $F_{p,q}(x) = 1$ for all $x > 0$, if and only if $p = q$,

(2) $F_{p,q} = F_{q,p}$ for all $p, q \in S$,

(3) the probabilistic triangle inequality.

Different generalizations were proposed for the probabilistic triangle inequality and the study of alternative triangle inequalities has been a central theme in the development of the theory of probabilistic metric spaces.

In his original paper, Menger [290] introduced a triangle inequality using a *triangular norm* (the *Menger t-norm*) T, i.e., a function $T : [0,1] \times [0,1] \to [0,1]$ satisfying

(M_1) $T(a, b) = T(b, a)$.

(M_2) $T(a, b) \leqslant T(c, d)$ whenever $a \leqslant c$, $b \leqslant d$,

(M_3) $T(a, 1) > 0$ whenever $a > 0$ and $T(1, 1) = 1$.

Thus, Menger gave the following generalized triangle inequality

(3.M) $F_{p,r}(x + y) \geqslant T(F_{p,q}(x), F_{q,r}(y))$

for all $p, q, r \in S$ and all real numbers x, y.

A. Wald [434] suggested the substitution of (3.M) by the inequality

$$\text{(3.W)}\quad F_{p,r}(x) \geqslant (F_{p,q} * F_{q,r})(x) = \int_{-\infty}^{\infty} F_{p,q}(x - y)\, \mathrm{d}F_{q,r}(y).$$

One can see that (3.M) has the disadvantage that constructing T is rather cumbersome whereas finding sufficient justification for a particular choice of T is not always easy. On the other hand, (3.W) is rather too strong, i.e., the structure it imposes is too similar to the usual structure of a metric space. In order to attain an appropriate level of generality, Schweizer and Sklar [384], [385] proposed:

(3.S—S) If $F_{p,q}(x) = 1$ and $F_{q,r}(y) = 1$, then $F_{p,r}(x + y) = 1$, which has the advantage that it is implied by both (3.M) and (3.W) and it is weak enough to be valid in a very large class of spaces, yet strong enough to imply that this class includes all ordinary metric spaces. Its principal inconvenience is that it is vacuous in all spaces in which the functions $F_{p,q}$, for $p \neq q$, never attain the value 1. The most fertile way chosen by Schweizer and Sklar [385] was that of using some particular Menger t-norms, thus changing (M_3) into the stronger boundary condition

($S-S_3$) $T(a, 1) = a$ for all $a \in [0, 1]$

and also adding the requirement that T be *associative*, i.e., that

($S-S_4$) $T(T(a, b), c) = T(a, T(b, c))$ for all $a, b, c \in [0, 1]$.

This enabled them to extend (3.M) to a polygonal inequality.

A function $T: [0, 1] \times [0,1] \to [0, 1]$ satisfying (M_1), (M_2), ($S-S_3$) and ($S-S_4$) is called a ***triangular norm** of Schweizer-**Sklar** type* (the $S-S$ t-norm or t-norm).

Examples: $T(x, y) = \text{Min}\{x, y\}$ ($T = \text{Min}$),; $T(x,y) = \text{Max}\{x + y - 1, 0\}$ ($T = T_m$); $T(x,y) = xy$ ($T = \text{Prod}$).

With the rediscovery of Menger's triangle inequality in 1957. Schweizer and Sklar [393] focused their attention on two objectives: (a) to relate the geometric and probabilistic aspects of this inequality, (b) to develop a repertory of t-norms for use in (M_3). The first objective led to the theory of copulas [383], [390], whereas the second came off as the structure and representation theory of semigroups on real intervals.

Let $\bar{\mathbb{R}} = [-\infty, \infty]$ and let j_A, $A \subseteq R$ be the identity function on $A \subseteq R$. It is known [393] that a function g is a *quasi-inverse* of a function f if $g \circ j_{\text{Ran}f}$ (the restriction of g to $\text{Ran}f$) is a right inverse of f, where range is equal to Rang, i.e., $f \circ g \circ j_{\text{Ran}f} = j_{\text{Ran}f}$ and $\text{Ran}\,(g \circ j_{\text{Ran}f}) = \text{Rang}$.

A t-norm T is *archimedean* if it is continuous on $[0, 1] \times [0, 1]$ and $T(x, x) < x$ for all $x \in (0, 1)$.

A t-norm T is *strict* if it is continuous on $[0, 1] \times [0, 1]$ and strictly increasing in each place on $(0,1] \times (0, 1]$. A strict t-norm is archimedean.

A characterization of the archimedean t-norm is given in ref. [285].

THEOREM 1.1.1. *A t-norm T is continuous and archimedean if and only if it admits the representation*

(*) $T(x, y) = f(g(x) + g(y))$,

where

(i) *g is a continuous and strictly decreasing function from* $[0, 1]$ *into* $[0, \infty] = \overline{\mathbb{R}}^+$ *with* $g(1) = 0$;

(ii) *f is a continuous function from* $\overline{\mathbb{R}}^+$ *onto* $[0, 1]$ *that is strictly decreasing on* $[0, g(0)]$ *and such that* $f(x) = 0$ *for all* $x > g(0)$.

(iii) *f is a quasi-inverse of g.*

We say [393] that T is additively generated by the pair (f, g) and we refer to f as an *outer additive generator* and to g as an *inner additive generator*.

A t-norm T is strict if and only if it admits the representation $T(x, y) = g^{-1}(g(x) + g(y))$, where g is a continuous and strictly decreasing function from $[0, 1]$ onto $\overline{\mathbb{R}}^+$, with $g(0) = \infty$ and $g(1) = 0$. g is said to be an *additive generator*. Hence, if T is strict, then the semigroups $([0, 1], T)$ and $(\overline{\mathbb{R}}^+, +)$ are isomorphic and every strict t-norm could be characterized completely in terms of real functions of a single real variable. The representation (*) has been shown to hold even under weaker conditions (see Krause [266]).

We mention the ordinal sum construction, i.e.,

DEFINITION 1.1.1 [393]. Let $(S_\alpha, T_\alpha)_{\alpha \in \Gamma \subset R}$ be a family of binary systems. For $\alpha, \beta, \gamma \in \Gamma$, let these binary systems satisfy the following compatibility conditions: (i) if $\alpha < \beta < \gamma$ and $S_\alpha \cap S_\gamma \neq \emptyset$, then $S_\beta = S_\alpha \cap S_\gamma$; (ii) if $\alpha < \beta$ and $x \in S_\alpha \cap S_\beta$, then x is the unique element in $S_\alpha \cap S_\beta$ and x is the identity of T_α and the null element of T_β.

Finally, let $S = \bigcup_{\alpha \in \Gamma} S_\alpha$ and T be defined on S as follows: for $x \in S_\alpha$ and $y \in S_\beta$,

$$T(x, y) = \begin{cases} x, & \alpha < \beta \\ T_\alpha(x, y), & \alpha = \beta. \\ y, & \alpha > \beta \end{cases}$$

Then the binary system (S, T) is the *ordinal sum* of the family of binary systems $(S_\alpha, T_\alpha)_{\alpha \in \Gamma}$.

THEOREM 1.1.2. *An ordinal sum of associative binary systems is itself associative.*

DEFINITION 1.1.2. Let $[a, e] \subseteq \overline{\mathbb{R}}$ be a closed interval. Then $\mathcal{O}[a, e]$ is the class of all associative binary operations T on $[a, e]$ that satisfy: (i) T is nondecreasing in each place; (ii) the endpoint e is the identity of T (i.e., $T(x, e) = T(e, x) = x$ for all $x \in [a, e]$); (iii) $T(x + 0, x + 0) = x$ for some $x \in [a, e]$

implies that $T(x, x) = x$; (iv) $T(x, x) = x$ implies that $v_x(y) = T(x, y)$ and $h_x(y) = T(y, x)$ are continuous.

The principal result which extends Mostert and Shields' Theorem B [297] on the characterization of continuous topological semigroups on [0,1] to not necessarily continuous associative operations is given by

THEOREM 1.1.3 [393, Th. 5.3.8]. *Let T be in $\mathcal{O}[a, e]$. Then we have:*

(i) *If $T(x, x) < x$ for all $x \in (a, e)$, then T is archimedean on $[a, e]$ and conversely.*

(ii) *If $T(x, x) = x$ for all $x \in [a, e]$, then T is the restriction of* Min *to $[a, e]$ and conversely.*

(iii) *Otherwise, the semigroup $([a, e], T)$ is an ordinal sum of semigroups $\{(S_\alpha, T_\alpha)\}$ where each S_α is a closed subinterval of $[a, e]$, each T_α is in $\mathcal{O}[S_\alpha]$ and T_α is either archimedean on S_α or the restriction of* Min *to S_α.*

COROLLARY 1.1.1. *If a t-norm T is continuous, then*

$$[0,1] \times [0,1] = \left(\bigcup_{k\in K} J_k \times J_k\right) \cup \mathrm{C}\left(\bigcup_{k\in K} J_k \times J_k\right), \quad \mathrm{card}\,(K) \leqslant \aleph_0,$$

for every $k \in K J_k$ is an open subinterval of $[0,1]$ such that $J_k \times J_s = \emptyset$ for every $k \neq s$, and $T_{|J_k \times J_k}$ is an archimedean semigroup for every $k \in K$.

A theorem which leads readily to a construction for t-norms which proceeds by piecing together t-norms is given by

THEOREM 1.1.4 [388]. *Let T_1 and T_2 be t-norms and let λ be a number in the open interval $(0,1)$. Let U_1 be a mapping on the square $[0, \lambda) \times [0, \lambda)$ defined by*

$$U_1(a, b) = \lambda_{T1}(a/\lambda, b/\lambda),$$

and let U_2 be a mapping on the square $[\lambda, 1] \times [\lambda, 1]$ defined by

$$U_2(a, b) = \lambda + (1 - \lambda) T_2\left(\frac{a - \lambda}{1 - \lambda}, \frac{b - \lambda}{1 - \lambda}\right).$$

Then the mapping T defined on the closed unit square $[0,1] \times [0,1]$ by

$$T(a, b) = \begin{cases} U_1(a, b), & a \in [0, \lambda), \; b \in [0, \lambda], \\ a = \mathrm{Min}(a, b), & a \in [0, \lambda), \; b \in [\lambda, 1], \\ b = \mathrm{Min}(a, b), & a \in [\lambda, 1], \; b \in [0, \lambda), \\ U_2(a, b), & a \in [\lambda, 1], \; b \in [\lambda, 1], \end{cases}$$

is a t-norm.

An interesting class of t-norms to be used in Chapter 3 was introduced by O. Hadžić [173] by

DEFINITION 1.1.3. We say that a t-norm T is *$\mathscr{H}$-type* (or $T \in \mathscr{H}$) if $\{T^n(t)\}_{n\in\mathbb{N}}$ is equicontinuous at $t = 1$, where $T^1(t) = T(t, t)$, $T^{n+1}(t) = T(T^n(t), t)$ $n \geqslant 1$, $t \in [0, 1]$.

Let us first give an example of such t-norm T[173], [172]. Suppose that the family of intervals $[c_n^{(1)}, c_n^{(2)}] \subseteq [0, 1]$, $n \in \mathbb{N}$, is such that $c_n^{(1)} < c_n^{(2)} < c_{n+1}^{(1)}$ for every $n \in \mathbb{N}$ and $\lim_{n\to\infty} c_n^{(1)} = 1$. If $\bar{T}$ is an arbitrary continuous t-norm, define a t-norm T in the following way:

$$T(x, y) = \begin{cases} c_n^{(1)} + (c_n^{(2)} - c_n^{(1)})\bar{T}\left(\dfrac{x - c_n^{(1)}}{c_n^{(2)} - c_n^{(1)}}, \dfrac{y - c_n^{(1)}}{c_n^{(2)} - c_n^{(1)}}\right) \\ \qquad\qquad \text{if } x \in \bigcup_{n\in\mathbb{N}} [c_n^{(1)}, c_n^{(2)}] \times [c_n^{(1)}, c_n^{(2)}] \\ \operatorname{Min}\{x, y\} \text{ if } x \notin \bigcup_{n\in\mathbb{N}} [c_n^{(1)}, c_n^{(2)}] \times [c_n^{(1)}, c_n^{(2)}]. \end{cases}$$

Using the above theorem and the fact that $T : [c_n^{(1)}, c_n^{(2)}] \times [c_n^{(1)}, c_n^{(2)}] \to [c_n^{(1)}, c_n^{(2)}]$ and $\lim_{n\to\infty} c_n^{(1)} = 1$, it is easy to see that T is a continuous t-norm such that the family $\{T^n(t)\}_{n\in\mathbb{N}}$ is equicontinuous at the point $t = 1$. It is clear that t-norm $T \neq \operatorname{Min}$.

The following lemma characterizes the *$\mathscr{H}$-type* t-norms in the continuous case.

LEMMA 1.1.1 [35]. *Let T be a continuous t-norm. Then*

(i) *$T \in \mathscr{H}$ if and only if: $\forall a \in [0, 1)$, $\exists b \in [a, 1)$ such that $T(b, b) = b$,*

(ii) *$T \notin \mathscr{H}$ if and only if: $\exists a \in [0, 1)$ such that $T(a, a) = a$, $T(x, x) < x$, $\forall x \in (a, 1)$.*

Proof. Let T be a continuous t-norm, $T \in \mathscr{H}$ and let $a > 0$ be given. Then there exists a $c \in (0, 1)$ such that $T^n(x) > a$, $\forall x \geqslant c$, $\forall n \geqslant 1$. Since clearly $\{T^n(c)\}_{n\in\mathbb{N}}$ is nonincreasing, then it is convergent to some limit $b \geqslant a$. As

$$T^{2n+1}(c) = T(T_n(c), T_n(c)),$$

then $b = T(b, b)$. The converse is obvious.

In order to prove (ii), suppose that $T \notin \mathscr{H}$. Then, according to (i), there exists an $a_0 \in [0, 1)$ such that $T(x, x) < x$ for every $x \in (a_0, 1)$. Let $a = \lim_{n\to\infty} T^n(a_0)$. It is obvious that $T(a, a) = a \leqslant a_0$. If $x \leqslant a_0$ and $T(x, x) = x$, then $x = T^n(x) \leqslant$

$\leqslant T^n(a_0)$, which shows that $x \leqslant a$. Therefore, $T(x, x) < x$ for each $x \in (a, 1)$. The converse implication follows from (i).

From Lemma 1.1.1 and [297], one obtains [355]

LEMMA 1.1.2. *Let T be a continuous t-norm which is not $\mathscr{H}$-type. Then there exists an $a \in [0,1)$ and $h_T : [a,1] \to [0,1]$ such that*

(i) *$h_T(a) = 0$, h_T is continuous and bijective;*

(ii) *$T(a, b) = h_T^{-1}(T_*(h_T(a), h_T(b)))$, $\forall\, a, b \geqslant 0$,*

where $T =$ Prod or $T_* = T_m$, and a is from Lemma 1.1.1. i).

Referring again to the triangle inequality, let us recall that in 1963, A. N. Šerstnev proposed a triangle inequality that included Menger's (3.M), Wald's (3.W) and also (3.S—S) as special cases [399], [401]. It goes as follows:

A two-place function $\tau : \Delta^+ \times \Delta^+ \to \Delta^+$ is called a *triangle function (t-function)* if

(α) $\tau(F, H_0) = F$,

(β) $\tau(F_1, G_1) \geqslant \tau(F_2, G_2)$ for $F_1 \geqslant F_2$, $G_1 \geqslant G_2$,

(γ) $\tau(F, G) = \tau(G, F)$,

(δ) $\tau(\tau(F_1, F_2), F_3) = \tau(F_1, \tau(F_2, F_3))$.

Šerstnev's triangle inequality then reads as

(3.Š) $F_{p,r} \geqslant \tau(F_{p,q}, F_{q,r})$.

When τ is given by $\tau(F, G) = F * G$, Šerstnev's inequality is Wald's inequality; when τ is given by τ_T, where

(4) $\tau_T(F, G; x) = \sup_{u+v=x} T(F(u), G(v))$,

and T is a *left-continuous t-norm* (l.c. t-norm) i.e., T is left-continuous in each place, it reduces to Menger's inequality (see [382, Th.4.3]); and when τ is such that $\tau(H_a, H_b) > H_{a+b}$ for any $a, b > 0$, it implies (3.S—S).

A t-function τ will be called an *l.c. t-function* if it is continuous in each place with respect to nondecreasing sequences.

A class of triangle functions which are, in a certain sense, the duals of the functions τ_T, is given as follows: a *t-conorm* [386] is a function $U : [0,1] \times [0,1] \to [0,1]$ satisfying (M_1), (M_2), (S—S_4) and $S(a, 0) = a$ for all $a \in [0,1]$, i.e., ($[0,1], U$) is a commutative semigroup for which 0, rather than 1, is the unit and 1, rather than 0, the null element.

If T is a t-norm, then the function T^* defined by

(5) $T^*(a, b) = 1 - T(1 - a, 1 - b)$

is the *t-conorm of* T.

In the theory of probabilistic metric spaces, t-conorms arise naturally when one deals with *the tails* of the distance distribution functions rather than with the d.d.f. themselves. For, if for any $F \in \Delta^+$ we denote the tail of F by F^t, so that

$$F^t(x) = 1 - F(x), \text{ for any } x,$$

then using (5) the Menger triangle inequality (3.M) assumes the form

$$(3'.\mathrm{M})\quad F^t_{p,r}(x+y) \leqslant T(F^t_{p,q}(t), F^t_{q,r}(y)).$$

Following Šerstnev, we deduce that, for any $x \in \mathbb{R}$,

$$(3'.\check{\mathrm{S}})\quad F^t_{p,r}(x) \leqslant \inf_{u+v=x} T(F^t_{p,q}(u), F^t_{q,r}(v)) = \inf_{u+v=x} [1 - T(1 - F^t_{p,q}(u),$$

$$1 - F^t_{q,r}(v))] = 1 - \sup_{u+v=x} T(F_{p,q}(u), F_{q,r}(v)) = [\tau_T(F_{p,q}, F_{q,r})]^t.$$

From this it follows that any triangle function τ on Δ^+ induces a corresponding "co-triangle" function $\bar{\tau}$ on the set $\tilde{\Delta}^+$ of tails of elements of Δ^+ which is given by

$$\bar{\tau}(F^t, G^t) = 1 - \tau(F, G) = [\tau(F, G)]^t.$$

Having now at our disposal sufficient information regarding the probabilistic generalization of the triangle inequality on Δ^+, we can define some concrete probabilistic structures :

DEFINITION 1.1.4. A *probabilistic metric space* (PM-space) is a triple $(S, \mathscr{F}, \tau)$, where S is a nonempty set, $\mathscr{F}$ is a probabilistic distance on S, i.e., $\mathscr{F} : S \times S \to \Delta^+$ and the triangle inequality (3.Š) is satisfied for the triangle function τ.

In Chapter 3, PM-space will be used to denote the case when $\mathscr{F} : S \times S \to \mathscr{D}^+$ unless otherwise specified. If, instead of τ, one finds T and if, instead of (3.Š), one verifies (3.M), then one has a *Menger space* and similarly a *Wald space* and a *Šerstnev space* for (3.W) and $(S, \mathscr{F}, \tau)$, respectively.

A *nonarchimedean PM-space* is an ordered pair $(S, \mathscr{F})$, where $\mathscr{F}$ is a nonarchimedean probabilistic distance [220], i.e., $\mathscr{F} : S \times S \to \Delta^+$ satisfies conditions (1), (2) and

$$(3.\mathrm{NP})\quad \text{If } F_{p,q}(x) = 1 \text{ and } F_{q,r}(y) = 1, \text{ then } F_{p,r}(\max\{x, y\}) = 1.$$

If, instead of (3.NP), one has

$$(3.\mathrm{NM})\quad F_{p,r}(\max\{x, y\}) \geqslant T(F_{p,q}(x), F_{q,r}(y)),$$

then the triplet $(S, \mathscr{F}, T)$ is called a *nonarchimedean Menger space*.

An important concept having numerous applications in Chapter 4 is

DEFINITION 1.1.5 [405]. Let $(\Omega, \mathscr{K}, P)$ be a probability space and let (M, d) be a metric space. The ordered pair $(S, \mathscr{F})$ is an *E-space over the metric space* (M, d) if the elements of S are functions from Ω into M such that, for every $p, q \in S$ and every real number x,

$$\{\omega \in \Omega : \mathrm{d}(p(\omega), q(\omega)) < x\} \in \mathscr{K}$$

(i.e., the composite one-place function $d(p, q)$ is P-measurable) and $\mathscr{F}$ is the mapping from $S \times S$ into $\mathscr{D}^+$ defined *via* $\mathscr{F}(p, q) = F_{p,q}$, where

$$F_{p,q}(x) = P(\{\omega \in \Omega : d(p(\omega), q(\omega)) < x\})$$

for every $x \in \mathbb{R}$.

H. Sherwood proved that every E-space is a Menger space under $T = T_m$.

In a PM-space, neighbourhoods and corresponding topological structures may be defined in many nonequivalent ways (see [393, Chaps. 12, 13]).

The first important problems were solved by Schweizer, Sklar and Thorp [392], [393], [381], [294], [382], [429], Fritsche [151], Tardiff [428], etc. A two-parameter family of neighbourhoods of a point $p \in S$ was introduced in [385] by defining the ε, λ-neighbourhood:

$$N_p(\varepsilon, \lambda) = \{q \in S : F_{p,q}(\varepsilon) > 1 - \lambda\} \text{ for } \varepsilon > 0 \text{ and } \lambda > 0.$$

The topological structure induced by the system of ε, λ-neighbourhoods (ε, λ-*topology* or $\mathscr{F}$-topology) was studied extensively. In 1960 it was proved [385] that if T is a t-norm satisfying the condition $\sup\{T(x, x) : 0 < x < 1\} = 1$, then in any Menger space under T, the ε, λ-neighbourhoods induce a topology that is metrizable. Subsequently, it was shown [294] that no condition weaker than the above can guarantee that ε, λ-neighbourhoods induce a *bona fide* topology. A topology which received most attention to date is the *strong topology* given by

DEFINITION 1.1.6 [393]. Let $(S, \mathscr{F})$ be a PM-space. For $p \in S$ and $t > 0$, the strong t-neighbourhood of p is the set $N_p(t) = \{q \in S : F_{p,q}(t) > 1 - t\}$.

The *strong neighbourhood system* at p is the collection $\mathscr{N}_p = \{N_p(t) : t > 0\}$ and the strong neighbourhood system for S is the union $\mathscr{N} = \bigcup_{p \in S} \mathscr{N}_p$.

THEOREM 1.1.5 [393, Th. 12.1.2]. *Let $(S, \mathscr{F}, \tau)$ be a PM-space. If τ is continuous, then the strong neighbourhood system $\mathscr{N}$ determines a Hausdorff topology for S.*

DEFINITION 1.1.7. The Hausdorff topology determined by the strong neighbourhood system $\mathcal{N}$ is the *strong topology* on S.

Let $(S, \mathcal{F})$ be a PM-space. Then, for any $t > 0$, the *strong t-vicinity* is the subset $\mathcal{U}(t)$ of $S \times S$ given by $\mathcal{U}(t) = \{(p, q) \in S \times S : F_{p,q}(t) > 1 - t\}$ and the *strong vicinity system* for S is the union $\mathcal{U} = \bigcup_{t>0} \mathcal{U}(t)$.

THEOREM 1.1.6 [393]. *Let $(S, \mathcal{F}, \tau)$ be a PM-space, where τ is continuous. Then the family $\mathcal{U}$ is a base for a Hausdorff uniformity for S (the so-called strong uniformity). The strong uniformity or, equivalently, the strong topology, is metrizable.*

Since $N_p(t, t) = N_p(t)$ for $t > 0$ and $N_p(\mathrm{Min}\,(\varepsilon, \lambda)) \subseteq N_p(\varepsilon, \lambda)$ for any $\varepsilon, \lambda > 0$, the strong neighbourhood system is equivalent to the system of ε, λ-neighbourhoods. Hence, the strong topology coincides with the ε, λ-topology. Similarly, the strong vicinity system is equivalent to the ε, λ-*vicinity* system defined by $\mathcal{U}(\varepsilon, \lambda) = \{(p, q) \in S \times S : F_{p,q}(\varepsilon) > 1 - \lambda\}$ for $\varepsilon, \lambda > 0$.

Let $(S, \mathcal{F})$ be a PM-space such that, for each $a \in [0, 1)$, $F_{p,q}(x) > a$ and $F_{q,r}(y) > a$ imply that $F_{p,r}(x + y) > a$ for all $p, q, r \in S$; then the family $\{d_a\}_{a \in [0,1)}$ defined from $S \times S$ into $\mathbb{R}^+$ by

$$d_a(p, q) = \inf \{x : F_{p,q}(x) > a\}$$

is a family of pseudometrics given by Nishiura [312]. Then d_a, $0 \leqslant a < 1$ are metrics if and only if $F_{p,q}$ are continuous at zero. For example, in the Menger space $(S, \mathcal{F}, \mathrm{Min})$ the uniformity defined by $\{d_a\}_{a \in [0,1)}$ on S is the same as the ε, λ-uniformity defined by $\mathcal{F}$.

PROPOSITION 1.1.1. *Let $(S, \mathcal{F}, T_m)$ be an E-space over (M, d). Then the following conclusions are equivalent:*

(i) *The sequence $\{p_n\}_{n \in \mathbb{N}} \subset S$ converges to $p \in S$ in the ε, λ-topology;*

(ii) $\lim_{n\to\infty} F_{p_n, p}(t) = H_0(t)$, $t \in \mathbb{R}$ *(pointwise),*

(iii) *$\{p_n\}_{n \in \mathbb{N}} \subset S$ converges in probability to $p \in S$.*

The following proposition is easy to reproduce:

PROPOSITION 1.1.2 [355]. *Let T be a continuous t-norm, $T \in \mathcal{H}$.*

(i) *If $(S, \mathcal{F}, T_m)$ is a Menger space, then $(S, e^{\mathcal{F}-1}$, Prod) is a Menger space with the same ε, λ-uniformity;*

(ii) *If $(S, \mathcal{F}, T)$ is a Menger space, then $(S, h_T^{-1} \circ \mathcal{F}, T)$ is a Menger space with the same ε, λ-uniformity;*

(iii) *If $(S, \mathcal{F}, T)$ is a Menger space, then $(S, h_T \circ \mathcal{F}, T_*)$ is a Menger space with the same ε, λ-uniformity (where $T_* = T_m$ or $T_* =$ Prod).*

Before defining other linear probabilistic structures, we shall mention first those of *diameter* of a set and *distance* between two sets.

DEFINITION 1.1.8 [130]. Let $(S, \mathscr{F}, \tau)$ be a PM-space with τ continuous and let A be a nonempty subset of S. The function D_A defined by

$$D_A(x) = \sup_{t<x} \inf_{p,q\in A} F_{p,q}(t), \ x \in \mathbb{R}, \ (D_A(\infty) = 1)$$

is called the *probabilistic diameter* of A.

It follows immediately that $D_A \in \Delta^+$ for any $A \subseteq S$. The set $A \subseteq S$ is *probabilistic bounded* if $D_A \in \mathscr{D}^+$ i.e., $\sup_{x\in\mathbb{R}} D_A(x) = 1$; A is semi-bounded if $0 < \sup_{x\in\mathbb{R}} D_A(x) \leqslant 1$; A is unbounded if $D_A = H_\infty$.

A probabilistic generalization of the Hausdorff-Pompeiu distance between two nonempty subsets A and B, $d(A, B) = \text{Max}\,\{\sup_{p\in A} \inf_{q\in B} d(p, q), \sup_{q\in B} \inf_{p\in A} d(p, q)\}$ is given by

DEFINITION 1.1.9 [130]. Let $(S, \mathscr{F}, T)$ be a Menger space with a continuous t-norm T and let $A, B \subseteq S$ be nonempty subsets of S. The probabilistic distance between A and B is the function $F_{A,B}$ defined by

$$F_{A,B}(x) = \sup_{t<x} T(\inf_{p\in A} \sup_{q\in B} F_{p,q}(t), \inf_{q\in B} \sup_{p\in A} F_{p,q}(t)).$$

Then, for $A, B \neq \emptyset$, $F_{A,B} \in \Delta^+$; $F_{A,B} = F_{B,A}$; for $A \neq \emptyset$, $F_{A,A} = H_0$; $F_{A,B} = F_{\bar{A},\bar{B}}$ and $F_{A,B} = H_0 \Leftrightarrow \bar{A} = \bar{B}$; $\emptyset \subsetneq A, B, C \subseteq S$ implies that $F_{A,B}(x + y) \geqslant T(F_{A,C}(x), F_{C,B}(y))$ for any $x, y \in R$; if $\mathscr{C}$ is a nonempty collection of nonempty closed subsets of S, then $(\mathscr{C}, \mathscr{F}_{\mathscr{C}}, T)$ is a Menger space, where $\mathscr{F}_{\mathscr{C}}$ is defined for any $A, B \in \mathscr{C}$ by $\mathscr{F}_{\mathscr{C}}(A, B) = F_{A,B}$.

In [190] O. Hadžić introduced

DEFINITION 1.1.10. Let $(S, \mathscr{F}, T)$ be a Menger space. A mapping $W : S\times S\times[0, 1] \to S$ is said to be a *convex structure* if for every $(u, x, y, \lambda) \in S\times S\times S\times(0, 1)$:

$$F_{u,W(x,y,\lambda)}(2t) \geqslant T\left(F_{u,x}\left(\frac{t}{\lambda}\right), F_{u,y}\left(\frac{t}{1-\lambda}\right)\right)$$

for every $t \in R^+$ and $W(x, y, 0) = y$, $W(x, y, 1) = x$.

A Menger space with a convex structure W *satisfies condition* (P II) if for all $(x, y, z, \lambda) \in S\times S\times S\times(0,1)$:

$$F_{W(x,z,\lambda), W(y,z,\lambda)}(\lambda t) \geqslant F_{x,y}(t), \text{ for every } t \in \mathbb{R}^+.$$

Before concluding this paragraph, we shall present the definition of three important classes of probabilistic structures. Let us first consider the following definition due to A. N. Šerstnev in [398]:

DEFINITION 1.1.11. Let S be a linear space, let $\mathscr{F} : S \to \Delta^+$, T a t-norm such that $T(x, y) \geqslant T_m(x, y)$, for every $x, y \in [0, 1]$ and the folowing conditions are satisfied:

(1) $F_p = H_0 \Leftrightarrow p = 0$,

(2) $F_{\lambda p}(x) = F_p\left(\frac{x}{|\lambda|}\right)$ for every $x > 0$, $\lambda \neq 0$ in the scalar field and $p \in S$,

(3) $F_{p+q}(x + y) \geqslant T(F_p(x), F_q(y))$ for every $p, q \in S$ and $x, y > 0$.

Then we say that the triplet $(S, \mathscr{F}, T)$ is a *random normed space* (RN-space).

Remarks. (1) Every RN-space $(S, \mathscr{F}, T)$ is a Menger space under $\mathscr{F}'$ defined by $\mathscr{F}'(p, q) = F_{p-q}$, $p, q \in S$.

(2) Every RN-space is a probabilistic metric space with a convex structure defined by $W(x, y, \lambda) = \lambda x + (1 - \lambda)y$, for $(x, y, \lambda) \in S \times S \times [0, 1]$, and the convex structure W satisfies condition (P II).

(3) Every RN-space $(S, \mathscr{F}, T)$ for which T is continuous and $T \in \mathscr{H}$ is a locally convex space in the ε, λ-topology [173] and the converse [354, Th. 4] is also true; if T is a continuous t-norm such that $T \notin \mathscr{H}$, then there exists an RN-space $(S, \mathscr{F}, T)$ which, when endowed with the ε, λ-topology, is not locally convex.

(4) The first attempt to develop a theory of inner product spaces was made by M. L. Senechal [397]. Some complications appear (arising partly from the fact that the value of an inner product can be negative or even complex). Hence, the probabilistic inner product must contend with functions in Δ rather than in Δ^+. For the definition of a probabilistic inner product space, a weak Menger inner product space or a Menger inner product space, see [397], [393]. We give a modified definition for the PIP-space based on S. Chang's version [99].

DEFINITION 1.1.12. A *probabilistic inner product space* (PIP-space) is a triplet $(S, \mathscr{F}, \tilde{T})$, where S is a real linear space, T is a strong t-norm (i.e., a t-norm T for which $T(a, b) > 0$, $\forall\, a, b \in (0, 1]$), $\mathscr{F}$ is a mapping of $S \times S$ into $\mathscr{D}$ satisfying the following conditions :

(PI$-$1) $F_{p,q} = F_{q,p}$, $\forall\, p, q \in S$;

(PI$-$2) $F_{p,p}(0) = 0$, $\forall\, p \in S$, and $F_{p,p}(t) = H_0(t)$, $\forall\, t \in \mathbb{R}$, if and only if $p = \theta$;

(PI$-$3) $\tilde{T}^*(F_{p,r}(t_1), F_{q,r}(t_2)) \geqslant F_{p+q,r}(t_1 + t_2) \geqslant T(F_{p,r}(t_1), F_{q,r}(t_2))$,

$\forall\, p, q \in S$, $t_1, t_2 \in \mathbb{R}$, where $\tilde{T}^*$ is the t-conorm of $\tilde{T}$.

(PI$-$4) $\sup\{t \in \mathbb{R}, F_{p,q}(t) < 1\} < \infty$, $p, q \in S$;

(PI$-$5) $\sup\{t \in \mathbb{R}, F_{ap,q}(t) < 1\} = a \sup\{t \in \mathbb{R}, F_{p,q}(t) < 1\}$,

$\forall\, a \in \mathbb{R}$, $p, q \in S$

PROPOSITION 1.1.3 [99]. *Let* $(S, F, \tilde{T})$ *be a PIP-space. Then the function* $\sup\{t \in \mathbb{R} : F_{p,q}(t) < 1\} : S \times S \to \mathbb{R}$ *has the following properties* :

(i) $\sup\{t \in \mathbb{R} : F_{p,q}(t) < 1\} = \sup\{t \in \mathbb{R} : F_{q,p}(t) < 1\}$, $\forall\, p, q \in S$;

(ii) $\sup\{t \in \mathbb{R} : F_{p,p}(t) < 1\} \geqslant 0$ *and* $\sup\{t \in \mathbb{R} : F_{p,p}(t) < 1\} \Leftrightarrow$ $\Leftrightarrow p = \theta$;

(iii) $\sup\{t \in \mathbb{R} : F_{ap,q}(t) < 1\} = a \sup\{t \in \mathbb{R}; F_{p,q}(t) < 1\}$,

$\forall\, a \in \mathbb{R}$;

(iv) $\sup\{t \in \mathbb{R} : F_{p+q}(t) < 1\} = \sup\{t \in \mathbb{R} : F_{p,r}(t) < 1\} +$ $+ \sup\{t \in \mathbb{R} : F_{q,r}(t) < 1\}$, $\forall\, p, q \in S$.

PROPOSITION 1.1.4 [99]. *Let* $(S, \mathscr{F}, T)$ *be a PIP-space. Then the following conditions are true* :

(i) $|\sup\{t \in \mathbb{R} : F_{p,q}(t) < 1\}| \leqslant (\sup\{t \in \mathbb{R} : F_{p,p}(t) < 1\})^{1/2}$

$(\sup\{t \in \mathbb{R} : F_{q,q}(t) < 1\})^{1/2}$, $\forall\, p, q \in S$;

(ii) $(\sup\{t \in \mathbb{R} : F_{p+q,p+q}(t) < 1\})^{1/2} \leqslant (\sup\{t \in \mathbb{R} : F_{p,p}(t) <$

$< 1\})^{1/2} + (\sup\{t \in \mathbb{R} : F_{q,q}(t) < 1\})^{1/2}$, $\forall\, p, q \in S$.

It is easy to see that every *PIP*-space $(S, \mathscr{F}, \tilde{T})$ can be normed by $\|p\| = (\sup\{t : F_{p,p}(t) < 1\})^{1/2}$, $\forall\, p \in S$ and, hence, the sequence $\{p_n\}_{n\in\mathbb{N}} \subset S$ is said to be convergent to $p \in S$, if

$$\lim_{n\to\infty} (\sup\{t \in \mathbb{R} : F_{p_n - p, p_n - p}(t) < 1\}^{1/2} = 0.$$

In a similar way are defined the Cauchy sequence and the complete *PIP*-space $(S, \mathscr{F}, T)$.

A self-mapping f on $(S, \mathcal{F}, T)$ is said to be a continuous mapping if $\lim_{n\to\infty} f(p_n) = f(p)$, where $\lim_{n\to\infty} p_n = p$.

A consistent definition having many interesting applications and interpretations of *PIP*-spaces was dealt with by J. Fortuny in his Ph. D. thesis [149].

Another probabilistic structure including the above-mentioned two is given by

DEFINITION 1.1.13. Let S be a linear space over a real or complex field K and let $\mathcal{F} = \{\mathcal{F}^i\}_{i\in I}$ be a family of functions $\mathcal{F}^i : S \to \mathcal{D}^+$. The space S is a *probabilistic locally convex space* (*PLC-space*) if, for every $i \in I$, the following are satisfied :

(i) $F^i_\theta = H_0$,

(ii) $F^i_{\lambda p}(t) = F^i_p\left(\frac{t}{|\lambda|}\right)$, for every $\lambda \in K - 0$, every $p \in S$ and every $t > 0$.

(iii) $F^i_{p+q}(s+t) \geqslant T(F^i_p(s), F^i_q(t))$, for every $p, q \in S$, every $s, t > 0$, where T is a t-norm.

If the t-norm T is continuous and if the family $U^i(\varepsilon, \lambda) = \{p \in S : F^i_p(\varepsilon) > 1 - \lambda\}$, $i \in I$, $\varepsilon > 0$, $\lambda \in (0, 1)$ is taken as the fundamental system of neighbourhoods of zero, then $(S, \mathcal{F}, T)$ becomes a topological vector space.

If T is a continuous t-norm such that $T \in \mathcal{H}$, then $(S, \mathcal{F}, T)$ is a locally convex topological vector space in the ε, λ-topology [173].

1.2. PROBABILISTIC MEASURES OF NONCOMPACTNESS AND APPLICATIONS

The introduction by K. Kuratowski in 1930 [270] of the measure of noncompactness α together with its topological applications, followed later by G. Darbo [122], L. S. Goldenstein, I. Gokhberg and A. S. Markus [161], [162], B. N. Sadovski [370], [371], [372], J. Daneš [121], V. Istrăţescu [233], M. Furi and A. Vignoli [153], [154], R. Nussbaum [339], [342] and others opened new perspectives for the development of several chapters of general topology, abstract space differential equations, general functional equations, optimization theory, etc. The Hausdorff measure of noncompactness χ was first used in the linear operator theory [161], [162], [275]. Important aplications to the theory of differential equations were given in [12], [371], [372], [373], [437], [158], [320], [422]. Applications of the measure of noncompactness in extremum problems were also given in refs. [348], [49].

Various measures of noncompactness for topological locally compact linear spaces can be found in refs. [372], [65], [64] and for multifunctions in ref. [211].

The most important applications of the measure of noncompactness refer to the generalization of fixed point theorems, having in turn numerous applications in functional analysis, integral and differential equations, game theory, etc.

Given the importance of the measures of noncompactness in probabilistic analysis, several such measures were introduced. Of these, mention should be made of the probabilistic Kuratowski and Hausdorff measures of noncompactness [56], [112]. Other more general types of probabilistic measures of noncompactness in *PLC*-spaces were introduced in refs. [23], [423], [425], [427] and [117]. Numerous applications to the theory of fixed points on probabilistic structures are discussed in refs. [45], [49], [50], [62], [57], [348].

1.2.1. *Kuratowski's Probabilistic Measures of Noncompactness*

Let (X, d) be a complete metric space and let A be a bounded subset of X. The Kuratowski number $\alpha(A)$ is the greatest lower bound of the set of positive real numbers ε such that A can be covered by a finite family of subsets of X whose diameters are less than or equal to ε [270]. A natural probabilistic generalization is given by

DEFINITION 1.2.1 [56]. Let $(S, \mathscr{F})$ be a *PM*-space and let $\mathscr{P}_0(S)$ the family of all nonempty subsets of S. The mapping $\alpha : \mathscr{P}_0(S) \to \mathscr{D}$ defined by $\alpha(A) = \alpha_A(x) = \sup\{\varepsilon \geqslant 0 : \exists \{A_j\}_{j\in J}, A \subseteq \bigcup_{j\in J} A_j,\ J = \text{finite and } D_{A_j}(x) \geqslant \varepsilon,\ \forall j \in J\}$ is called the *Kuratowski function* or *Kuratowski probabilistic measure of noncompactness*. The function $\alpha_A(.)$ is called the Kuratowski function of A.

PROPOSITION 1.2.1. *The Kuratowski function $\alpha_A(\cdot)$ has the following properties*:

(i) $\alpha_A \in \Delta^+$; (ii) $\alpha_A(x) \geqslant D_A(x)$; (iii) *if* $\emptyset \neq A \subseteq B \subset S$, *then* $\alpha_A(x) \geqslant \alpha_B(x)$; (iv) $\alpha_{A\cup B}(x) = \min\{\alpha_A(x), \alpha_B(x)\}$; (v) *if* $(S, \mathscr{F}, T)$ *is a Menger space with a continuous t-norm T, then* $\alpha_{\bar{A}}(x) = \alpha_A(x)$, *where $\bar{A}$ is the closure of A in the* ε, λ-*topology*.

Proof. (i) Since $A \in \mathscr{P}_0(S)$ and is probabilistic bounded, then $D_A \in \mathscr{D}^+$ and, hence, $D_A(0) = 0$ and $\alpha_A(0) = 0$. The nondecreasing property for α_A is implied by the same property for every D_{A_j}, $j \in J$.

In order to prove that α_A is left-continuous, we consider $K_{A,x} = \{\varepsilon \geqslant 0, \exists \{A_j\}_{j\in J}, J = \text{finite}, \ A = \bigcup_{j\in J} A_j, \ D_{A_j}(x) \geqslant \varepsilon\}$, $x_0 \in (0, \infty)$, and $\varepsilon \in K_{A,x_0}$. If $\eta > 0$, then by the left-continuity of D_{A_j}, there exists a $\delta_j > 0$ such that $D_{A_j}(x_0) - D_{A_j}(x) < \eta$ if $0 < x_0 - x < \delta_j$. Let $\delta = \min_{j\in J} \delta_j$ and let $\eta \leqslant \varepsilon$. If $0 < x_0 - x < \delta$, then $D_{A_j}(x) \geqslant \varepsilon - \eta$, $\forall j \in J$. Thus, $\varepsilon - \eta \in K_{A,x}$ and, hence, $\varepsilon - \eta \leqslant \alpha_A(x)$. This implies that $\alpha_A(x_0) - \alpha_A(x) \leqslant \eta$ for every $x \in (x_0 - \delta, x_0)$ which proves the left-continuity of α_A on $\mathbb{R}^+$.

It remains to prove that $\lim_{x\to\infty} \alpha_A(x) = 1$. Let us observe that the boundedness of A implies that there exists a partition $\{A_j\}_{j\in J}$ such that $D_{A_j} \in \mathscr{D}^+$. Then, for every $\eta \in (0, 1)$, there exists an $x_j \in D_{A_j}(x_j) \geqslant 1 - \eta$, for every $j \in J$. If $x = \max_{j\in J} x_j$, then $D_{A_j}(x) \geqslant 1 - \eta$, which implies that $\lim_{x\to\infty} \alpha_A(x) = 1$ and (i) is proven.

To prove (ii), consider $x \in \mathbb{R}$ and $\varepsilon \geqslant 0$ such that $D_A(x) \geqslant \varepsilon$. Since $\{A\}$ is a cover of A, then $\varepsilon \in K_{A,x}$ and $\alpha_A(x) \geqslant D_A(x)$.

Statement (iii) is obvious by the inclusion $K_{A,x} \geqslant K_{B,x}$ for every $x \in \mathbb{R}$.

Since (iii) implies that $\alpha_A(x) \geqslant \alpha_{A\cup B}(x)$ and $\alpha_B(x) \geqslant \alpha_{A\cup B}(x)$, then to prove (iv) it suffices to show that $\min \{\alpha_A(x), \alpha_B(x)\} \leqslant \alpha_{A\cup B}(x)$. According to the definition of α_A, for every $\delta > 0$, there exists an $\varepsilon\delta \in K_{A,x}$ and a cover $\{A_j\}_{j\in J}$ of A such that $\alpha_A(x) - \delta < \varepsilon\delta < D_{A_j^\delta}(\varepsilon), j \in J$. Similarly, there exist $\eta\delta \in K_{B,x}$ and $\{B_i^\delta\}$, $i \in I$, $I = \text{finite}$ such that $\alpha_B(x) - \delta < \eta_0 \leqslant D_{B_i^\delta}(x)$. Let us suppose that $\alpha_A(x) \leqslant \alpha_B(x)$. Then $\alpha_A(x) - \delta \leqslant \alpha_B(x) - \delta \leqslant D_{B_i^\delta}(x)$. Thus, we obtain a cover $\{C_m\}_{m\in M}$ of $A \cup B$ such that $D_{C_m}(x) \geqslant \alpha_A(x) - \delta$, $m \in M$, i.e., $\alpha_A(x) - \delta \in K_{A\cup B,x}$, which implies that $\alpha_A(x) \leqslant \alpha_{A\cup B}(x)$ and the proof of property (iv) is complete.

(v) Since $A \subseteq \bar{A}$, it follows that $\alpha_A(x) \geqslant \alpha_{\bar{A}}(x)$ and for the converse inequality, we observe that if $\varepsilon \in K_{A,\varepsilon}$, then there exists $\{A_i\}_{i\in I}$, $I = \text{finite}$, with the property that $A \subseteq \bigcup_{i\in I} A_i$ and $D_{A_i}(x) \geqslant \varepsilon$. Also, since $\bar{A} \subseteq \bigcup \bar{A}_i$ and $D_{\bar{A}_i}(x) =$

$= D_{A_i}(x) \geqslant \varepsilon$ [130], it follows that $\varepsilon \in K_{\overline{A},x}$, which implies that $\alpha_A(x) \leqslant \alpha_{\overline{A}}(x)$ and the proof of the proposition is complete.

Remark. Definition 1.2.1 can be extended to every $A \in \mathscr{P}_0(S)$ and then $\alpha_A \in \Delta^+$. The probabilistic measures of noncompactness offer a characterization of the boundedness of sets in *PM*-spaces [49].

An equivalent definition for the Kuratowski function which can be better used in Šerstnev spaces $(S, \mathscr{F}, \tau)$ springs from [56].

PROPOSITION 1.2.2. *Let $(S, \mathscr{F})$ be a PM-space and, for $\varphi \neq \neq A \subset S$, let K_A be the set of all functions $F \in \mathscr{D}^+$ such that there exists a finite cover $\{A_j\}_{j \in J}$, $J =$ finite, of A with $D_{A_j}(x) \geqslant \geqslant F(x)$ for every $x > 0$ and $j \in J$. Then, for every $x \in \mathbb{R}$, $\alpha_A(x) = = \sup\{F(x) : F \in K_A\}$.*

Proof. Let $F \in K_A$. Then $F(x)$ $K_{A,x}$ and, thus, $F(x) \leqslant \alpha_A(x)$. Hence, $\sup\{F(x) : x \in K_A\} \leqslant \alpha_A(x)$.

For the converse assertion, let $\varepsilon \geqslant 0$, $x > 0$. From the definition of $\alpha_A(x)$, there exists an $\varepsilon' \subset K_{A,x}$ such that $\alpha_A(\varepsilon) - - \varepsilon < \varepsilon'$. Since $\varepsilon' \in K_{A,x}$, there exists a finite cover of A by A_i, $i \in I =$ finite, such that $D_{A_i}(x) \geqslant \varepsilon'$. Let $F_{x,\varepsilon'}(y) = = \inf_{i \in I} D_{A_i}(y)$, $y \in \mathbb{R}$. Clearly, $F_{x,\varepsilon'} \in K_A$ which implies that $F_{x,\varepsilon'}(y) \leqslant \leqslant \sup\{F(y) : F \in K_A\}$. Since $D_{A_i}(x) \geqslant \varepsilon'$, then $F_{x,\varepsilon'}(x) \geqslant \varepsilon'$ which implies that $\alpha_A(x) - \varepsilon < \varepsilon' \leqslant F_{x,\varepsilon'}(x) \leqslant \sup\{F(x) : F \in K_A\}$. Since ε was chosen arbitary, it follows that $\alpha_A(x) \leqslant \leqslant \sup\{F(x) : F \in K_A\}$.

The importance of precompactness is just as great in probabilistic analysis. Hence, we give

DEFINITION 1.2.2. We say that a *PM*-space $(S, \mathscr{F})$ is a *probabilistic precompact space* if, for every $\varepsilon > 0, 1 > \lambda > 0$, there exists a finite cover $\{A_i\}_{i \in I}$ of S such that $D_{A_i}(\varepsilon) > 1 - \lambda$.

PROPOSITION 1.2.3. *Let $(S, \mathscr{F})$ be a PM-space and let $A \subseteq S$ be a probabilistic precompact set. Then, for every $\varepsilon, \lambda > 0$, there exists a finite subset $A_{\varepsilon,\lambda} \subset S$ such that for each $p \in A$ there exists a $q = q(p) \in A_{\varepsilon,\lambda}$ such that $F_{p,q}(\varepsilon) > 1 - \lambda$ (i.e., $A_{\varepsilon,\lambda}$ is a finite ε, λ-net).*

Proof. Let $D'_A(x) = \inf\{F_{p,q}(x) : p,q \in A\}$. Then clearly $D_A(x_0) = = \lim_{\substack{x \to x_0 \\ x < x_0}} D'_A(x) \leqslant D'_A(x_0) \leqslant F_{p,q}(x_0)$. Let us first consider the particular case $\varepsilon = 1/n$, $\lambda = \dfrac{1}{m}$, $n,m \in \mathbb{N}$. Let $\{A_i^{m,n}\}_{i \in I_{m,n}}$ be a finite cover of A such that $D_{A_i^{m,n}}\left(\frac{1}{n}\right) > 1 - \frac{1}{m}, \forall i \in I_{m,n}$, whose existence is implied by the probabilistic precompactness of A. Let $p_i^{m,n} \in A_i^{m,n}$ and let $A_{m,n} = \{p_i^{m,n}, i \in I_{m,n}\}$. Let $p \in A$. Then there exists an $i_0 \in I_{m,n}$ such that $p \in A_{i_0}^{m,n}$ and, therefore, $F_{p,p_{i_0}^{m,n}}\left(\frac{1}{n}\right) > 1 - \frac{1}{m}$. It follows that $A_{m,n} = A_{1/n,1/m}$.

Consider now the general case when $\varepsilon > 0$ and $1 > \lambda > 0$ are arbitary but fixed. Then let $n, m \in N$ such that $1/n < \varepsilon$ and $1/m < \lambda$. It is easy to verify that $A_{\varepsilon,\lambda} = A_{m,n}$, where $A_{m,n}$ is the above set.

In a large class of Menger spaces, we can prove the converse assertion of the above theorem, i.e., a characterization of the probabilistic precompactness.

THEOREM 1.2.1. *Let* $(S, \mathscr{F}, T)$ *be a Menger space with a t-norm* T *such that* $\sup_{x<1} T(x, x) = 1$. *A set* $A \subset S$ *is probabilistic precompact if and only if, for every* $\varepsilon > 0$ *and* $1 > \lambda > 0$, *there exists a finite* ε, λ*-net* $A_{\varepsilon,\lambda}$ *(as in Proposition* 1. 2. 3).

Proof. Let $\varepsilon > 0$ and $1 > \lambda > 0$ be arbitrary but fixed and let $A_{\varepsilon',\lambda'} = \{p_i\}_{i \in I}$, I = finite, an ε, λ-net for A, where $\varepsilon' = \varepsilon/2$ and $\lambda' > 0$ is chosen such that $T(1 - \lambda', 1 - \lambda') > 1 - \lambda$. It is easy to see that $A \subseteq \cup S_{p_i}(\varepsilon', \lambda')$, where $S_{p_i}(\varepsilon', \lambda') = \{q \in S : F_{p_i,q}(\varepsilon') > 1 - \lambda'\}$. If we denote $A = S_{p_i}(\varepsilon', \lambda')$, $i \in I$, then, for every $r, q \in A_i$,

$$F_{r,q}(\varepsilon) \geqslant T(F_{r,p_i}(\varepsilon/2), F_{p_i,q}(\varepsilon/2)) \geqslant T(1 - \lambda', 1 - \lambda') > 1 - \lambda.$$

It follows that $D_{A_i}(\varepsilon) > 1 - \lambda$ for every $i \in I$ and, hence, A is a probabilistic precompact set. The converse assertion follows from Teorem 1.2.1.

THEOREM 1.2.2. *Let* $(S, \mathscr{F}, T)$ *be a Menger space with a t-norm* T *such that* $\sup_{x<1} T(x, x) = 1$. *Then a set* $A \subset S$ *is a probabilistic precompact set if and only if* A *is a precompact set with respect to the* ε, λ*-uniformity induced by* $\mathscr{F}$.

Proof. Note that A is a precompact set with respect to the ε, λ-uniformity if and only if, for every $\varepsilon > 0$ and $\lambda > 0$, there exists an $A_{\varepsilon,\lambda} = \{p_1, \ldots, p_n\}$ such that $A \subseteq \bigcup_{i=1}^{n} S_{p_i}(\varepsilon, \lambda)$.

Let $\varepsilon > 0$ and $1 > \lambda > 0$ be given numbers and let $\lambda' > 0$, $\eta > 0$ be such that $T(1 - \eta, 1 - \eta) > 1 - \lambda' > 1 - \lambda$. If A is a precompact set with respect to the ε, λ-uniformity, then it follows that there exists $\{p_1, \ldots, p_n\}$ such that $A \subseteq \bigcup_{i=1}^{n} S_{p_i}(\varepsilon/4, \eta)$. It is sufficient to show that $D_{S_{p_i}(\varepsilon/4,\eta)}(\varepsilon) > 1 - \lambda'$ for every $i = 1, \ldots, n$. By the definition of $D_{(.)}$, we have that $D_A(\varepsilon) = \sup_{\varepsilon/2<t<\varepsilon} \inf_{p,q\in A} F_{p,q}(t)$. But

$$F_{p,q}(t) \geqslant T\left(F_{p,p_i}\left(\frac{t}{2}\right), F_{p_i,q}\left(\frac{t}{2}\right)\right) \geqslant T(F_{p,p_i}(\varepsilon/4), F_{p_i,q}(\varepsilon/4)) \geqslant$$

$$\geqslant T(1-\eta, 1-\eta) > 1 - \lambda'$$

for every $p,q \in A_i$ and $t \in (\varepsilon/2, \varepsilon)$. Hence, $D_{A_i}(\varepsilon) > 1 - \lambda' > 1 - \lambda$, i.e., A is a probabilistic precompact set.

Conversely, if A is a probabilistic precompact set, then from Theorem 1.2.1, it follows that, for every $\varepsilon > 0$ and $1 > \lambda > 0$, there exists a finite ε, λ-net $A_{\varepsilon,\lambda} = \{p_1, \ldots, p_n\}$, and, hence, $A \subseteq \bigcup_{i=1}^{n} S_{p_i}(\varepsilon, \lambda)$, i.e., A is precompact with respect to the ε, λ-uniformity.

A characterization of probabilistic precompact subsets with the aid of Kuratowski's function is given by

THEOREM 1.2.3. *Let $(S, \mathscr{F})$ be a PM-space. A nonempty set A is a probabilistic precompact set if and only if* $\alpha_A(x) = H_0(x)$.

Proof. Let A be a probabilistic precompact set and let x be a fixed positive number. Choose $\varepsilon = x$ and $\lambda = 1 - \eta$ where η is an arbitrary number in $(0,1)$. Then there exists a finite cover $\{A_i\}_{i\in I}$ for A such that $D_{A_i}(\varepsilon) > 1 - \lambda = 1 - (1-\eta) = \eta$. Hence, $\eta \in K_{A,x}$ which implies that $\alpha_A(x) \geqslant \eta$. Since η is arbitrary in $(0,1)$, we conclude that $\alpha_A(x) = 1$. Also, x is arbitrary in $\mathbb{R}^+$, which implies that $\alpha_A = H_0$.

Conversely, if $\alpha_A(x) = H_0(x)$, we consider $1 > \lambda > 0$ and $\varepsilon > 0$ and we have $\alpha_A(\varepsilon) = 1$. Hence, there exists an $\varepsilon_\lambda \in K_{A,\varepsilon}$ with the property that $\alpha_A(\varepsilon) - \lambda < \varepsilon_\lambda$ and $1 - \lambda < \varepsilon_\lambda$. Since $\varepsilon_\lambda \in K_{A,\varepsilon}$ implies that there exists a finite family $\{A_i\}_{i\in I}$ with $D_{A_i}(\varepsilon) \geqslant \varepsilon_\lambda > 1 - \lambda$, i.e., A is probabilistic precompact.

Remark. From the above theorem, it follows that the Kuratowski functions offer a measure of noncompactness (in the complete case, precompactness coincides with relative compactness).

The relationship between the Kuratowski function $\alpha_A(x)$ in the *PM*-space associated with d (i.e., $\mathscr{F}_d(p, q) = H_0(x - d(p, q))$ and the Kuratowski number α_A in (S, d) is given by

PROPOSITION 1.2.4 [56]. *If* $(S, \mathscr{F}_d)$ *is a PM-space generated by a metric space* (S, d),

$$\alpha_A(x) = H_0(x - \alpha(A)).$$

Hence, $A \subseteq (S, d)$ is precompact if it is a probabilistic precompact set of $(S, \mathscr{F}_d)$.

Let (S, d) be a linear metric space with a metric d invariant to translation, with a locally convex system of neighbourhoods of the origin. Then

$$F_{,}(x) = \begin{cases} \max\left\{0{,}1 - d\left(0, \frac{1}{x}p\right)\right\}, & x > 0 \\ 0, & x \leqslant 0. \end{cases}$$

Hence, $p \in F_p$ is a random norm [304] on S with respect to τ_{Min}, the ε, λ-topology coincides with the topology of linear metric space and

$$\alpha_A(x) = \max\left\{0{,}1 - \alpha\left(\frac{1}{x}A\right)\right\}, \quad x > 0.$$

A hereditary property for $\alpha_A(.)$ is given by

PROPOSITION 1.2.5. *Let* $(S, \mathscr{F})$ *be a PM-space and let* $S^1 \subset S$ *and* $\mathscr{F}^1 = \mathscr{F}|_{S^1 \times S^1}$. *Then, for every* $A \subset S^1$, *we have* $\alpha_A(.) = \alpha_A^1(.)$, *where* $\alpha_A : \mathscr{P}_0(S) \to \Delta^+$ *and* $\alpha_A^1 : \mathscr{P}_0(S^1) \to \Delta^+$ *are Kuratowski functions on* $(S, \mathscr{F})$ *and* $(S^1, \mathscr{F}^1)$, *respectively.*

Let $(S^1, \mathscr{F}^1, T)$ *and* $(S^2, \mathscr{F}^2, T)$ *be Menger spaces under the same left-continuous t-norm* T. *Then their product* $(S^1 \times S^2, \mathscr{F}_T, T)$ *is also a Menger space under* T [130], *where* $\mathscr{F}_T(p,q\,; x) = T(F_{p_1,q_1}(x), F_{p_2,q_2}(x))$ *for any* $p = (p_1, p_2)$, $q = (q_1, q_2)$ *in* $S^1 \times S^2$. We can prove

THEOREM 1.2.4 [112]. *Let* α *be a Kuratowski function on* $(S^1 \times S^2, \mathscr{F}_T, T)$, *where* T *is a continuous t-norm. Then, for every* $A = A_1 \times A_2 \subset S^1 \times S^2$, *we have*

$$\alpha_A(x) = T(\alpha_{A_1}^1(x), \alpha_{A_2}^2(x)), \quad x \in \mathbb{R},$$

where α^1 *and* α^2 *are the Kuratowski functions in* S^1 *and* S^2, *respectively.*

Remark. A well-known topology on Δ is the topology induced by the Lévy metric $d_L(F, G) = \inf\{h : F(x - h) - h \leqslant G(x) \leqslant F(x + h) + h,\ x \in \mathbb{R}\}$. A useful modification of the Lévy metric was introduced by D. A. Sibley [407], and it is a slight modification of this modified Lévy metric that is best suited to our purposes [393].

Let $F, G \in \Delta$, let $h \in [0,1]$ and let $(F, G; h)$ denote the condition

$$F(x-h)-h \leqslant G(x) \leqslant F(x+h)+h \text{ for all } x \in \left(-\frac{1}{h}, \frac{1}{h}\right).$$

The modified Lévy distance is the function $d_{\mathscr{L}} : \Delta \times \Delta \to \mathbb{R}^+$ by

$$d_{\mathscr{L}}(F, G) = \inf\{h : \text{both } (F, G; h) \text{ and } (G, F; h) \text{ hold}\}.$$

It is known that a sequence $\{F_n\} \subset \Delta$ converges weakly to $F \in \Delta$ (i.e., $F_n(x) \to F(x)$ for each continuity point x of F) if and only if $d_{\mathscr{L}}(F_n, F) \to 0$ as $n \to \infty$.

The metric space $(\Delta, d_{\mathscr{L}})$ is compact and, hence, complete.

THEOREM 1.2.5 [49]. *Let $(S, \mathscr{F}, T)$ be a complete Menger space with a continuous t-norm T and let $CB(S)$ be the family of all nonempty, probabilistic bounded and ε, λ-closed subsets of S.*

If $\{A_n\}_{n\in\mathbb{N}} \subseteq CB(S)$ is a decreasing sequence of subsets such that $\alpha_{A_n} \xrightarrow{L} H_0$, then $\{A_n\}_{n\in\mathbb{N}}$ converges in the ε, λ-topology defined by $\mathscr{F}'$ ($\mathscr{F}' : CB(S) \to \Delta^+$, $\mathscr{F}'(A, B) = F_{A,B}$ from Def. 1.1.9) and $\lim_{n\to\infty} A_n = \bigcap_{n\in\mathbb{N}} A_n = \tilde{A}$, where $\tilde{A}$ is a nonempty and compact set with respect to the ε, λ-topology defined by $\mathscr{F}$.

COROLLARY. *Let $(S, \mathscr{F}, T)$ and $\{A_n\}_{n\in\mathbb{N}}$ be as in the Theorem. If $D_{A_n} \xrightarrow{L} H_0$, then A contains exactly one point.*

An interesting result used in the topological degree theory is

THEOREM 1.2.6 [49]. *Let $(S, \mathscr{F}, \text{Min})$ be a Menger space. Then the Kuratowski function α is a continuous mapping from $(CB(S), \mathscr{F}', \text{Min})$ into (Δ^+, L).*

1.2.2. *Hausdorff's Probabilistic Measures of Noncompactness*

Let us give some generalizations on *PM*-spaces of the results of Hausdorff's measures of noncompactness [49], [112].

DEFINITION 1.2.3. *Let $(S, \mathscr{F})$* be a *PM*-space an let $F_{A,B}$ be given by Definition 1.1.9 for $T = \text{Min}$. The mapping $\chi : \mathscr{P}_0(S) \to \Delta^+$, given by

$$\chi_A(x) = \sup\{\varepsilon \geqslant 0 : \exists\, A_\varepsilon \subset S,\ A_\varepsilon = \text{finite},\ F_{A,A_\varepsilon}(x) \geqslant \varepsilon\},$$

$A \in \mathscr{P}_0(S)$, is called *Hausdorff's function of A*.

Let us [49] define $\chi_A^0(x) = \sup_{t<x} \sup\{\varepsilon \geqslant 0 : \exists$ a finite ε, t-net $A_{\varepsilon,t}$ for $A\}$. It can be proved that $\chi_A^0 = \chi_A$. For $(S, \mathscr{F}, T)$ with T continuous, we have $\chi_A(x) \geqslant \alpha_A(x) \geqslant T(\chi_A(x/2),$

$\chi_A(x/2))$, $x \in \mathbb{R}$. In general, $\alpha_A \neq \chi_A$ but properties similar with those above also hold for χ_A.

If T is a continuous t-norm, one can define [49]

$$\chi_A^T(x) = \sup\{F_{A,B}(x) : B \subset S,\ B = \text{finite}\},\ x \in \mathbb{R},$$

which is called Hausdorff's T-function. One has: (i) $\chi_A^T = T(\chi_A(x), \chi_A(x))$, $x \in \mathbb{R}$; (ii) $\chi_A^T \in \Delta^+$ and, for every bounded set $A \subset S$, $\chi_A^T \in \mathscr{D}^+$; (iii) $\chi_A^T(x) \geqslant T(D_A(x), D_A(x))$; (iv) $A \subset B$ implies that $\chi_A^T \geqslant \chi_B^T$; (v) $\chi_{A\cup\{p\}}^T = \chi_A^T$ for all $p \in S$; (vi) $\chi_A^T = \chi_{\bar{A}}^T$; (vii) $A \subseteq S$ is a probabilistic precompact set if and only if $\chi_A^T = H_0$.

It is known that Sadovski [372] and Petryshyn and Fitzpatrick [341] introduced measures of noncompactness on uniform structures. The same measures can be defined [49] for the Nishiura uniformity, i.e., $\xi_A(a) = \alpha_a(A)$ or $\eta_A(a) = \chi_a(A)$ for $a \in [0, 1)$. It can be proved that, for every bounded set $A \subset (S, \mathscr{F}, \text{Min})$, the functions ξ_A and η_A are random variables on $([0, 1), \mathscr{B}_{[0,1)}, P)$ and that the distribution functions of ξ_A and η_A coincide with α_A and χ_A, respectively.

Let $\mathscr{M}$ be a family of nonempty subsets in $(S, \mathscr{F})$ which contains probabilistic precompact subsets. A mapping $\gamma : \mathscr{M} \to \Delta^+$ with the property that $\gamma(A) = H_0$ if and only if A is a probabilistic precompact set is called a *random measure of noncompactness*.

Let $(S, \mathscr{F}, \text{Min})$ be a Menger space and let $\{d_a\}_{a\in[0,1)}$ be the family of Nishiura pseudometrics on S. Let γ_a, $a \in [0,1)$, be a real measure of noncompactness on (S, d_a) (i.e., $\gamma_a(A) = 0 \Leftrightarrow A =$ precompact with respect to d_a and $\gamma_a(A) < \infty$ for $A \in \mathscr{M}$). Let $\theta_A(a) = \gamma_a(A,\ a \in [0, 1))$, $A \in \mathscr{M}$, be such that, for every $A \in \mathscr{M}$, θ_A is a random variable on $([0, 1), \mathscr{B}_{[0,1)}, P)$. Then $\theta = \{\theta_A, A \in \mathscr{M}\}$ is a random process and their 1-dimensional distribution functions induce a random measure of noncompactness.

1.2.3. *Some Remarks on the Probabilistic Measures of Noncompactness on Probabilistic Linear Structures*

Let $(S, \mathscr{F}, \tau)$ be an RN-space and let γ be the Kuratowski function (α) or Hausdorff function (χ) on $\mathscr{P}_0(S)$ with respect to the induced metric $\mathscr{F}_1(\mathscr{F}_1(p, q) = F_{p-q})$. Then, we have [44], [49], [191, p. 233]

(i) $\gamma_A(x) = \gamma_{\bar{A}}(x)$,

(ii) $\chi_A(x) \geqslant \alpha_A(x) \geqslant \tau_{T_m}(\chi_A, \chi_A)(x)$,

(iii) $\gamma_{\lambda A}(x) = \gamma_A\left(\frac{x}{|\lambda|}\right)$, $\lambda \neq 0$,

(iv) $\gamma_{A+B}(x) \geqslant \tau(\gamma_A, \gamma_B)(x)$,

(v) if $\tau = \tau_{\text{Min}}$, $\gamma_{\text{co}A}(x) = \gamma_A(x)$, $x \in \mathbb{R}$, co$A$ = convex hull of A,

(vi) if $\tau = \tau_{\text{Min}}$ and T is a continuous t-norm, then $\chi^T_{\text{co}A} = \chi_A$.

As an interesting application, consider a probabilistic version of Ascoli-Arzela's theorem :

THEOREM 1.2.7 [47]. *Let X be a compact topological space, let $(S, \mathscr{F}, \tau_{\text{Min}})$ be an RN-space and let M be a bounded equicontinuous family of functions from the RN-space $C(X, S)$. Then*

$$\alpha_M(x) = \sup_{t<x} \inf_{p\in X} \alpha_{M(p)}(t),$$

where $M(p) = \{f(p), f \in M\}$.

Some interesting applications of random measures of noncompactness to the study of some topological problems and of some classes of operators on probabilistic structures are given in ref. [49]. A random measure of noncompactness on the set of Banach space-valued random variables having many applications in random operator and random equations theory will be studied in sub-section 2.1.

Do Hong Tan [423] introduced the function $\tilde{F}_{A,B}(t) = \sup_{s<t} \inf_{p\in A} \sup_{q\in B} F^i_{p-q}(s)$, $t \in \mathbb{R}$, $i \in I$, on $B(S)$ (the set of all nonempty probabilistic bounded sets in PLC-space $(S, \mathscr{F}^i, \text{Min})$) and proved that $F_{A,B} \in \mathscr{D}^+$. A set A is said to be *P-precompact* if, for every i, ε, λ, there exists a finite set A_f such that $\tilde{F}^i_{A,A_f}(\varepsilon) > 1 - \lambda$. It was shown that this is equivalent with A being probabilistic precompact. For every $A \subset B(S)$ and $i \in I$, one introduces

$$\beta^i_A(t) = \sup\{\rho \geqslant 0 : \exists\, A_f = \text{finite},\ \tilde{F}^i_{A,A_f}(t) \geqslant \rho\}$$

and proves the same properties as for χ_A and that $\chi_A = \beta_A$.

A generalized probabilistic measure of noncompactness can be introduced [425], [427] as a function $\varphi : B(S) \times I \to \mathscr{D}^+$ which satisfies the following

(i) $\varphi^i_A(t) = H_0(t)$, $\forall\, i \in I \Leftrightarrow A$ is probabilistic precompact,

(ii) $\varphi^i_A(t) = \varphi^i_{\text{co}A}(t)$, $\forall\, i \in I$, $\forall\, A \in B(S)$,

(iii) $\varphi^i_{A\cup B}(t) = \min\{\varphi^i_A(t), \varphi^i_B(t)\}$, $\forall\, i \in I$, $\forall\, A, B \in B(S)$.

It can be proved that $\Phi_{i,\lambda}(A) = \inf\{t: \varphi_A > 1 - \lambda\} = \sup\{t: \varphi_A^i(t) \leqslant 1 - \lambda\}$ for every $i \in I$, $\lambda \in (0,1)$, where $A \subset B(S)$ is a generalized measure of noncompactness in the locally convex space $(S, \|\cdot\|_\lambda^i, i \in I, \lambda \in (0,1))$ associated, where $\|\cdot\|_\lambda^i = \sup\{t: F_{(.)}(t) \leqslant 1 - \lambda$ (i.e., $\Phi_{i,\lambda}(A) = 0$, $\forall\, i \in I$, $\lambda \in (0,1) \Leftrightarrow A$ is precompact; $\Phi_{i,\lambda}(\mathrm{co}A) = \Phi_{i,\lambda}(A)$, $\forall\, i \in I$, $\forall\, A \in B(S)$, $\lambda \in (0,1)$; $\Phi_{i,\lambda}(A \cup B) = \max\{\Phi_{i,\lambda}(A), \Phi_{i,\lambda}(B)\}$, $\forall\, i \in I$, $A, B \in B(S)$, $\lambda \in (0,1)$). Conversely, if $\Phi_{i,\lambda}$ is a generalized measure of noncompactness, then

$$\varphi_A^{i,\lambda}(t) = \begin{cases} 1 & \text{if } \Phi_{i,\lambda}(A) < t \\ 0 & \text{otherwise} \end{cases}$$

is a probabilistic measure of noncompactness on $(S, \mathscr{F}^i, \mathrm{Min})$.

As was suggested in ref. [84], it follows that, for every fixed A, the function $\Phi_{i,\lambda}(A)$ is nonincreasing and left-continuous in λ and then the dependence of λ in $\varphi_A^{i,\lambda}$ can be avoided taking

$$\varphi_A^i(t) = 1 - \sup\{\lambda \in (0,1) : \Phi_{i,\lambda}(A) \geqslant t\},$$

which will be a probabilistic measure of noncompactness, and we have $\Phi_{i,\lambda}(A) = \sup\{t: \varphi_A^i(t) \leqslant 1 - \lambda\}$. The fact that $\Phi_{i,\lambda}$ is exactly the Kuratowski or Hausdorff measure requires an additional condition, namely, the strictness of φ_A^i, i.e., $\varphi_A^i(.)$ are strictly monotone. It is proved that if φ_A^i and $\Phi_{i,\lambda}$ are related to each other as above, then φ_A^i is strict if and only if $\Phi_{i,\lambda}$ is continuous in λ.

A probabilistic inner measure of noncompactness and its properties are investigated in ref. [426].

Chapter 2

Measurable multivalued mappings and randomness

There are numerous situations (for instance, in game theory, mathematical economics and decision theory) which give rise to multivalued mappings defined on an abstract measurable space and having compact values in a separable metric space. The purpose of this chapter is to present some general theorems which arise frequently in applications of such multivalued mappings [86], [87], [214], [215], [433], [132], [136]. Some problems on the existence of random solutions to nonlinear equations involving random multivalued (correspondence) mappings will be also discussed.

2.1. PRELIMINARIES TO THE PROBABILITY THEORY IN BANACH SPACES

Throughout this book, we shall assume that $(\Omega, \mathscr{K}, P)$ is a probability measure space (i.e., the set Ω is a nonempty abstract set whose elements ω are termed elementary events, $\mathscr{K}$ is a σ-algebra of subsets of Ω and P is a probability measure on $\mathscr{K}$; that is, P is a set function, with domain $\mathscr{K}$, which is non-negative, countably additive and such that $P(A) \in [0,1]$ for all $A \in \mathscr{K}$, with $P(\Omega) = 1$). Obviously, assume that P is a complete probability measure unless otherwise specified. In some theorems, P will be replaced by a σ-finite measure. Let $\mathscr{A}$ be some statement about the elements of the set Ω. Then the phrase "the statement $\mathscr{A}$ holds with probability one" or "the statement $\mathscr{A}$ holds almost surely" (briefly, a.s.) means that the set of those $\omega \in \Omega$ for which the statement $\mathscr{A}$ holds belongs to the σ-algebra $\mathscr{K}$ and has its probability measure equal to one.

In this section, we present some of the basic concepts of the probability theory in Banach spaces; hence, we shall be concerned with random variables defined on $(\Omega, \mathscr{K}, P)$ with values in $(X, \mathscr{B}_X)$, where X is a Banach space and $\mathscr{B}_X$ is the σ-algebra of the Borel subsets of X.

DEFINITION 2.1.1. A mapping $x : \Omega \to X$ is said to be a *random variable with values in* X (X-valued r.v.) if $x^{-1}(B) \in \mathscr{K}$ for all $B \in \mathscr{B}_X$.

A mapping $x : \Omega \to X$ is a *P-almost separably valued random variable* if there exists a set $A_0 \in \mathscr{B}_X$ such that $P(A_0) = 0$ and $x(\Omega - A_0)$ is separable.

A mapping $x : \Omega \to X$ is said to be a *strong* (or *Bochner*) *random variable* if there exists a sequence $\{x_n(\omega)\}$ of countably valued random variables which converges to $x(\omega)$ almost surely; that is, there exists a set $A_0 \in \mathscr{K}$, with $P(A_0) = 0$ such that $\lim_{n \to \infty} \| x_n(\omega) - x(\omega) \| = 0$ for every $\omega \in \Omega - A_0$.

A mapping $x : \Omega \to X$ is said to be a *weak* (or *Pettis*) random variable if the functions $x^*(x(\omega))$ are real-valued random variables for each $x^* \in X^*$ (the set of all bounded linear functionals on X).

A mapping $x : \Omega \to X$ is said to a be a *Price random variable* if, for every $x_0 \in X$ and positive real number r, $\{\omega : \| x(\omega) - x_0 \| < r\} \in \mathscr{K}$.

We shall now confine our attention to the important case when X is a separable Banach space. In this situation, we have

THEOREM 2.1.1 [31, p. 16–18]. *Let X be a separable Banach space. Then* (i) *the above four definitions of random variables with values in X are equivalent*; (ii) *the function $\| x(\omega) \|$ is a non-negative real-valued random variable*; (iii) *the set $\mathscr{V}(\Omega, X) = \mathscr{V}_X$ of all X-valued random variables is a linear space* (*i.e.*, $x, y \in \mathscr{V}_X \Rightarrow \alpha x + \beta y \in \mathscr{V}_X$, $\alpha, \beta \in \mathbb{R}$); (iv) *if $f : (X, \mathscr{B}_X) \to (Y, \mathscr{B}_Y)$ is a Borel measurable mapping and x is an X-valued r.v., then $y(\omega) = f(x(\omega))$ is a Y-valued random variable, where Y is an arbitrary Banach space*; (v) *if $\{x_n(\omega)\} \subset \mathscr{V}_X$ is a weakly almost surely convergent sequence to $x(\omega)$, then there exists an X-valued r.v. $x_0 \in \mathscr{V}_X$ such that $x_n \xrightarrow{a.s.} x_0$ and $P(\{\omega ;\, x_0(\omega) = x(\omega)\}) = 1$ and the set of those $\omega \in \Omega$, for which the sequence converges, belongs to $\mathscr{K}$.*

In order to stress the fact that X is a Banach space, we shall sometimes write $\mathscr{V}(\Omega, X) = \mathscr{V}_X$. At other times, we shall consider directly the set $\tilde{\mathscr{V}}_X$ of all equivalence classes (i.e., $\tilde{x} = \{y \in \mathscr{V}_X : P(\{\omega : y(\omega) \in C\} \Delta \{\omega : x(\omega) \in C\}) = 0,\ \forall\, C \in \mathscr{K}\}$; if X is separable, then equivalence means that $P(\{\omega : x(\omega) \neq y(\omega)\}) = 0$).

DEFINITION 2.1.2. An X-valued r.v. is said to be *Pettis integrable* if there exists an element $m_A \in X$ corresponding to each

$A \in \mathscr{K}$ such that

$$x^*(m_A) = (L)\int_A x^*(x(\omega))\,d\mu \quad \text{for all } x^* \in X^*,$$

where the integral is assumed to exist in the sense of Lebesgue. The Pettis integral is defined as

$$m_A = (P)\int_A x(\omega)\,d\mu.$$

$x(.)$ is said to be *Bochner integrable* if there exists a sequence of simple $r.v.\{x_n(\omega)\}_{n\in\mathbb{N}}$ converging a.s. to $x(\omega)$ and such that

$$\lim_{n\to\infty}\int_\Omega \| x_n(\omega) - x(\omega)\|\,d\mu = 0$$

and

$$(B)\int_A x(\omega)\,d\mu = \lim_{n\to\infty}(B)\int_A x_n(\omega)\,d\mu = \lim_{n\to\infty}\sum_{i=1}^{\infty} x_n^i\mu(A\cap A_i),$$

where $x_n(\omega) = x_n^i$ on $A_i \in \mathscr{K}$, $i \in \mathbb{N}$, for every $A \in \mathscr{K}$.

The *expectation* of $x \in \mathscr{V}_X$ is

$$E_w(x(\omega)) = (P)\int_\Omega x(\omega)\,d\mu, \quad \text{resp.} \quad E_s(x(\omega)) = (B)\int_\Omega x(\omega)\,d\mu.$$

As in the case of the Banach space-valued $r.v.$, there exist several ways of defining a random operator, all of which utilize in some fashion the notion of X-valued $r.v.$ We consider

DEFINITION 2.1.3 [31]. Let $(\Omega, \mathscr{K}, P)$ be a complete probability measure space and let $(X, \mathscr{B}_X)$ and $(Y, \mathscr{B}_Y)$ be Banach spaces (not necessarily separable). A mapping $T:\Omega\times X\to Y$ is said to be a *random operator* if $\{\omega : T(\omega, x) = T(\omega)x \in C\} \in \mathscr{K}$ for all $x \in X$, $C \in \mathscr{B}_Y$, i.e., one simply says that $T(\omega)$ is a random operator if $T(\omega)x = y(\omega)$ is a Y-valued $r.v.$

A random operator $T:\Omega\times X\to X$ is called *continuous* (*compact*, etc.) if $T(\omega.,)$ is continuous (compact, etc.) for each $\omega\in\Omega$.

These properties are sometimes not considered pointwise but almost surely.

A random operator $T:\Omega\times X\to X$ is said to be stochastically continuous if, for every $x\in X$ and every $\varepsilon>0$, $\lim_{\|y-x\|\to 0} P(\{\omega : \| T(\omega)y - T(\omega)x\| > \varepsilon\}) = 0.$

Let X be a separable Banach space and let $\mathscr{F}$ be the countable class of all sets of the form $\{x : \|x - x_i\| \leqslant r\}$, $\{x : \|x - x_i\| \geqslant r\}$ and their finite intersections, where $\{x_i\}_{i \in \mathbb{N}}$ is dense in X and r is a rational number.

DEFINITION 2.1.4 [303]. A random operator $T : \Omega \times X \to X$ is said to be *separable* if there exist a countable set S in X and a negligible set $N \in \mathscr{K}$ such that

$$\{\omega : T(\omega)x \in K,\ x \in F\}\ \Delta\{\omega : T(\omega)x \in K,\ x \in F \cap S\} \subset N$$

for every compact set K and every $F \in \mathscr{F}$. The set S is called a *separant* for the random operator T.

LEMMA 2.1.1 [303]. *Let Y be a compact metric space. The random operator $T : \Omega \times X \to Y$ is separable if there exists an N with $P(N) = 0$ such that*

$$T(\omega, x) \in \bigcap_{F : x \in F} \overline{T(\omega, (F \cap S))}, \quad \text{for all } x \in X,\ \omega \notin N,$$

where S is a separant for T.

Let us now state a result which shows that under certain conditions every random operator is equivalent to a separable random operator.

THEOREM 2.1.2 [303], [31]. *Let X be a separable Banach space and let E be a compact subset of X. If $T : \Omega \times X \to E$ is a random operator, then there exists a separable random operator $\widetilde{T} : \Omega \times X \to X$ such that $P(\{\omega : T(\omega, x) = \widetilde{T}(\omega, x)\}) = 1$.*

Then $\| T(\omega) \|$ is a non-negative real-valued r.v. too.

A sufficient condition for separability is given by

THEOREM 2.1.3 [303], [31]. *Let X be a separable Banach space and let $T : \Omega \times X \to X$ be an almost continuous random operator. Then T is separable.*

In several areas of mathematics, as are stochastic processes and control theory, it is of primary interest to impose minimal conditions on a mapping of two variables regarding each variable separately such that measurability with respect to both variables jointly is guaranteed. We give now

THEOREM 2.1.4 [254], [215, Th. 6.1]. *Let $(\Omega, \mathscr{K}, \mu)$ be a measure space, let X be a separable metric space and let Y be a metric space. Assume that $T : \Omega \times X \to Y$ is a continuous random operator. Then T is a measurable mapping from $\Omega \times X$ into Y; in fact, for each closed subset B of Y, $T^{-1}(B)$ is the countable intersection of countable unions of basic rectangles in $\mathscr{K} \times \mathscr{B}_X$.*

Let us now give some results to be used in obtaining some connexions between random operators and operators on PM-spaces.

Let X be a separable Banach space. Then, for every $x \in \mathscr{V}_X$, one can define $F_x(t) = P(\{\omega \in \Omega : \| x(\omega) \| < t\})$, $t \in \mathbb{R}$. The mapping $x \xrightarrow{\mathscr{F}} F_x(.)$ is a random seminorm on $\mathscr{V}_X$ with respect to the t-function τ_{T_m} and the ε, λ-topology on $\mathscr{V}_X$ induced by $\mathscr{F}$ is the topology of the convergence in probability [50], [304].

Let $(\Omega, \mathscr{K}, P)$ be an *atomic probability space*, i.e., $\Omega = \bigcup_{n \in \mathbb{N}} \Omega_n$, Ω_n being the distinct atoms. Then each random variable $x : \Omega \to X$ is constant a.s. on every atom [31]. If S is the set of all random variables which are constant on each atom, then $(S, \mathscr{F}, \tau_{T_m})$ is an RN-space [304]. For every $\omega \in \Omega$, the mapping

$$x \to \| x(\omega) \|_\omega = \| x(\omega) \|, \quad x \in S$$

is a seminorm on S; the locally convex topology on S induced by the family of seminorms $\Gamma = \{\| \cdot \|_\omega, \ \omega \in \Omega\}$ will be called Γ-topology on S.

PROPOSITION 2.1.1 [50]. *The Γ-topology and the ε, λ-topology induced by the random norm $\mathscr{F}$ are identical on S.*

Considering the locally convex space (S, Γ), it is easy to see that the measure of noncompactness of a bounded subset $A \subseteq S$ in the sense of Sadovski-Petryshyn-Fitzpatrick is the function $\omega \to \alpha(A(\omega))$, $\omega \in \Omega$ and where $A(\omega) = \{x(\omega) : x \in A\}$. Then,

PROPOSITION 2.1.2 [50]. *The measure of noncompactness in the sense of Sadovski-Petryshyn-Fitzpatrick of a probabilistic bounded subset A of the random normed space $(S, \mathscr{F}, \tau_{T_m})$ is a non-negative random variable.*

PROPOSITION 2.1.3 [50]. *The mapping $\tilde{\alpha} : A \to \tilde{\alpha}_A(.)$ defined on the class $\mathscr{M}(S)$ of all probabilistic bounded subsets of $(S, \mathscr{F}, \tau_{T_m})$, where*

$$\tilde{\alpha}_A(t) = P(\{\omega \in \Omega : \alpha(A(\omega)) < t\}), \quad t \in \mathbb{R}$$

is a random measure of noncompactness in the RN-space $(S, \mathscr{F}, \tau_{T_m})$(i.e., *$\tilde{\alpha}(A) = H_0 \Leftrightarrow A$ is a probabilistic precompact set); it has the following properties*: (i) $\tilde{\alpha}_{A \cup \{p\}} = \tilde{\alpha}_A$ *for every* $p \in S$; (ii) $\tilde{\alpha}_{\overline{\mathrm{co}} A} = \tilde{\alpha}_A$; (iii) $\tilde{\alpha}_{\overline{A}} = \tilde{\alpha}_A$.

2.2. MEASURABLE MULTIVALUED MAPPINGS AND RANDOM MAPPINGS

The study of measurability of multivalued mappings has largely developed over the last two decades owing to their numerous applications. Important contributions were made by R. Aumann [16], G. Debreu [123], Castaing [86], C. Himmelberg [215], D. Wagner [433], etc. In this section, we shall confine ourselves to giving without proof some oustanding results of use in the following chapters.

Unless stated otherwise, $(\Omega, \mathscr{K}, \mu)$ will denote a σ-finite measure space. X will almost always be a separable metrizable space. Following Bourbaki, we shall call X *Polish*, if X is separable and metrizable by a complete metric, and *Souslin*, if X is metrizable and the continuous image of a Polish space.

A relation (or correspondence, or multifunction, or multivalued mapping) $T : \Omega \to X$ (i..e, $T : \Omega \to \mathscr{P}(X)$) is a subset of $\Omega \times X$. Alternatively, T may be regarded as a mapping from Ω to the set of subsets of X. Also, whenever we want to emphasize the properties of T as a subset of $\Omega \times X$, we shall refer to its graph $\mathrm{Gr}\,T = \{(\omega, x) \in \Omega \times X : x \in T(\omega)\}$. The set $\{\omega \in \Omega : T(\omega) \neq \emptyset\}$ is called the domain of $T : \Omega \to X$. If $B \subset X$, then $T^{-1}(B) = \{\omega \in \Omega : T(\omega) \cap B \neq \emptyset\}$. By 2^X we denote $\{A \subseteq X : A \neq \emptyset \wedge A = \text{closed}\}$, by $CB(X) = \{A \subseteq X : A \in 2^X \wedge \wedge A \text{ bounded}\}$, by $CC(X) = \{A \subseteq X,\ A \in 2^X \wedge A \text{ convex}\}$, by $CK(X) = \{A \subseteq X : A \neq \emptyset$, compact and convex$\}$, by $WK(X) = \{A \subseteq X,\ A \neq \emptyset$, weakly compact$\}$, by $K(X) = \{A \subseteq X : A \neq \emptyset$ and compact$\}$. We say that T is (adjective)-valued if $T(\omega)$ is (adjective) for $\omega \in \Omega$ and we apply operations on sets to operations on set-valued mappings in the obvious way.

A multivalued mapping $T : E \subset X \to 2^X$ is called *demiclosed* if $\{x_n\} \subset E$ and $\{y_n\} \subset X$ with the sequences $y_n \in T(x_n)$ such that if $\{x_n\}$ converges to x_0 and $\{y_n\}$ converges to y_0 in X, then $y_0 \in T(x_0)$.

DEFINITION 2.2.1 [215]. A multivalued mapping $T : \Omega \to \to \mathscr{P}_0(X)$ is called *measurable* ($\mathscr{B}$-measurable, $\mathscr{C}$-measurable, K-measurable) if $T^{-1}(B)$ is measurable (i.e., is in $\mathscr{K}$) for each open (resp. Borel, closed, compact) subset of X. If $T : Y \to \to \mathscr{P}(X)$, where Y is a topological space, then the assertion that T is measurable ($\mathscr{B}$-measurable, etc.) means that T is measurable ($\mathscr{B}$-measurable, etc.) when Y is assigned the σ-algebra $\mathscr{B}_Y$ of Borel subsets of Y. Likewise, if $T : \Omega \times Y \to X$, then the various kinds of measurability of T are always defined in terms of the product σ-algebra $\mathscr{K} \times \mathscr{B}_Y$ on $\Omega \times Y$ generated by the sets $A \times B$ where $A \in \mathscr{K}$ and $B \in \mathscr{B}_Y$.

DEFINITION 2.2.2. A multivalued $T : \Omega \times X \to \mathscr{P}_0(X)$ is called a *multivalued* random operator if, for every $x \in X$, $T(.x)$ is measurable.

The following assertions can be proved [215]:

PROPOSITION 2.2.1. (a) *If* $T : \Omega \to \mathscr{P}_0(X)$ *is measurable or* $\mathscr{C}$*-measurable, then the domain of* T *is measurable.*

(b) $T : \Omega \to \mathscr{P}_0(X)$ *is measurable if and only if* $\bar{T} : \Omega \to \mathscr{P}_0(X)$, *defined by* $\bar{T}(\omega) = \overline{T(\omega)}$, *is measurable.*

The situation for closed-valued mappings can be summarized as in

THEOREM 2.2.1. *Let* X *be separable metric space and let* $T : \Omega \to \mathscr{P}_0(X)$ *have closed values. Consider the following statements*: (a) T *is* $\mathscr{B}$*-measurable*; (b) T *is* $\mathscr{C}$*-measurable*; (c) T *is measurable*; (d) T *is* $\mathscr{K}$*-measurable*; (e) $\omega \to d(\omega, T(\omega))$ *is a measurable function of* ω *for each* $x \in X$; (f) $\operatorname{Gr} T$ *is* $\mathscr{K} \times \mathscr{B}$*-measurable.*

We have (i) (a) $\Rightarrow$ (b) $\Rightarrow$ (c) $\Leftrightarrow$ (e) $\Rightarrow$ (d) *and* (c) $\Rightarrow$ (f);

(i) *If* X *is* σ*-compact, then* (a) $\Rightarrow$ (b) $\Leftrightarrow$ (c) $\Leftrightarrow$ (d) $\Leftrightarrow$ (e) $\Rightarrow$ (f);

(iii) *If* $(\Omega, \mathscr{K}, \mu)$ *is complete and* X *is Souslin, then* (a) $\Leftrightarrow$ (b) $\Leftrightarrow$ (c) $\Leftrightarrow$ (e) $\Leftrightarrow$ (f) $\Rightarrow$ (d).

THEOREM 2.2.2. *Let* X *be a separable metrizable space and let* $T_n : \Omega \to \mathscr{P}_0(X)$ *be a measurable closed-valued mapping for every* $n \in J$. *Assume also that* T_n *is a compact-valued mapping for some* $n \in J$. *Then* $T = \bigcap_{n \in J} T_n$ *is* $\mathscr{C}$*-measurable.*

THEOREM 2.2.3. *Let* X *be a separable Fréchet space and let* $T : \Omega \to \mathscr{P}_0(X)$ *be a closed-valued measurable mapping. Then the multivalued mappings* $\operatorname{co} T$ *and* $\overline{\operatorname{co}}\, T$ (*where* $\overline{\operatorname{co}}\, A$ = *the closed convex hull of* A) *defined by* $\omega \to \operatorname{co} T(\omega)$, $\omega \to \overline{\operatorname{co}}\, T(\omega)$, *are also measurable.* (X *need only be a separable metric locally convex space if every* $T(\omega)$ *is assumed to be complete*).

DEFINITION 2.2.3. A measurable mapping $x : \Omega \to (X, d)$ is called a *measurable selector* of a measurable multivalued mapping $T : \Omega \to \mathscr{P}_0(X)$ if $x(\omega) \in T(\omega)$ for every $\omega \in \Omega$.

THEOREM 2.2.4 (Kuratowski, Ryll-Nardzewski [271, Th. 5.1]). *If* X *is separable metric space and* $T : \Omega \to \mathscr{P}_0(X)$ *is a measurable complete-valued mapping, then* T *has a measurable selector.*

THEOREM 2.2.5. *Let* X *be a separable metric space and let* $T : \Omega \to \mathscr{P}_0(X)$ *be a complete-valued mapping. Then* T *is measurable if and only if there exists a family* U *of measurable selectors for* T *such that* $T(\omega) = \overline{\{u(\omega) : u \in U\}} = \overline{U(\omega)}$ *for all*

$\omega \in \Omega$ (i.e., *T has a Castaing representation* [433]). *If X is also σ-compact, then T need only be closed-valued mappings.*

THEOREM 2.2.6. *If* $(\Omega, \mathscr{K}, \mu)$ *is σ-finite, X is a Souslin space and* $T : \Omega \to \mathscr{P}_0(X)$ *is a multivalued mapping with measurable graph, then there exists a measurable function* $y : \Omega \to X$ *such that* $y(\omega) \in T(\omega)$ *for all ω except those in some set of measure zero.*

Remark 2.2.1. If $(\Omega, \mathscr{K}, \mu)$ is a complete measurable space, then the above theorem with $y(\omega)$ a measurable selector becomes Theorem 3 of M. F. Sainte-Beuve [373].

Corollary 1 [373] confirms that the graph measurability of T implies the existence of a measurable selector for T.

We are now able to recall the following implicit function theorem:

THEOREM 2.2.7 [214]. (i) *Let X be a compact metric space and let Y be a separable metric space. Let* $T : \Omega \to \mathscr{P}_0(X)$ *be a measurable closed-valued mapping, let* $f : \Omega \times X \to Y$ *be a continuous random operator and let* $g : \Omega \to Y$ *be a measurable function such that* $g(\omega) \in f(\{\omega\} + T(\omega))$ *for all* $\omega \in \Omega$. *Then there exists a measurable selector y of T and* $g(\omega) = f(\omega, y(\omega))$ *for all* $\omega \in \Omega$.

(ii) *If we assume in* (i) *that Y is a σ-compact metric space and that measurability is everywhere replaced by K-measurability, then X need only be a complete separable metric space.*

A sufficient condition for the measurability of min $f(\{\omega\} \times T(\omega))$ is given by

THEOREM 2.2.8 [214]. *Let X be a separable metric space, let* $f : \Omega \times X \to \mathbb{R}$ *be a continuous random operator and let* $T : \Omega \to \mathscr{P}_0(X)$ *be a compact-valued measurable mapping. Then the function* $y : \Omega \to \mathbb{R}$, *defined by* $y(\omega) = \min f(\{\omega\} \times T(\omega))$, *is measurable.*

DEFINITION 2.2.4. A multivalued mapping on a metric space (X, d), $T : X \to 2^X$, is called *compact* if $T(X) = \bigcup_{x \in X} T(x)$ is precompact. T is called *upper* (*lower*) *semicontinuous* (briefly usc, resp. lsc) if, for any closed (open) subset C of X, $T^{-1}(C)$ is closed (open). T is called *continuous* if T is both upper and lower semicontinuous.

As the definition of uppersemicontinuity varies in the literature, we state the well-known

LEMMA 2.2.1 [137]. *Let Y be a Banach space,* $K \subseteq 2^Y$, $T : K \to 2^Y$ *such that* $\mathrm{cl}(\bigcup_{x \in K} T(x))$ *is compact. Then the following are equivalent*: (1) $\{(x, y) \in K \times Y : y \in T(x)\}$ *is closed in the product-topology* : (2) *for all* $\{x_n\} \in K^{\mathbb{N}}$, $x \in K$, $y_n \in T(x_n)$, $y \in Y$ *with* $\{x_n\} \to x$ *and* $\{y_n\} \to y$ *we have* $y \in T(x)$; (3) *for all* $x \in K$ *and all open* $G \subseteq Y$ *with* $T(x) \subseteq G$, *there exists a neighbourhood* U *of* x *such that for all* $z \in U \cap K$, $T(x) \subseteq G$; (4) T *is compact in the above sense and* T *is a compact-valued usc mapping.*

A characterization of the continuity is given by

THEOREM 2.2.9 [86, Th. 4.1]. *Let* X *be a metrizable space with the metric* d. *Let* $(K(X), D)$ *be the metric space with the Hausdorff metric* D *induced by* d. *Then a compact-valued mapping* T(*i.e.*, $T(y) \in K(X)$ *for all* $y \in X$) *is continuous if and only if* T *is continuous from* (X, d) *into* $(K(X), D)$.

With the aid of the Hausdorff metric D on $CB(X)$, where X is a Polish space, S. Itoh [237] obtained some results related to measurability and measurable selectors :

PROPOSITION 2.2.2. (a) *Let* $\{T_n\}_{n\in\mathbb{N}}$ *be a sequence of measurable mappings* $T_n : \Omega \to CB(X)$ *and let* $T : \Omega \to CB(X)$ *be a mapping such that for every* $\omega \in \Omega$, $D(T_n(\omega), T(\omega)) \to 0$ *as* $n \to \infty$. *Then* T *is measurable.*

(b) *Let* $T : \Omega \times X \to CB(X)$ *be a random operator such that for every* $x, y \in X$, $D(T(\omega, x), T(\omega, y)) \leqslant k(\omega)\, d(x, y)$ *for every* $\omega \in \Omega$, *where* k *is a real function on* Ω, $k : \Omega \to R^+$. *Let* $u : \Omega \to X$ *be a measurable mapping. Then the mapping* $S : \Omega \to$ $\to CB(X)$, *defined by* $S(\omega) = T(\omega, u(\omega))$, $\omega \in \Omega$, *is measurable.*

PROPOSITION 2.2.3. *Let* T, $S : \Omega \to CB(X)$ *be measurable mappings, let* $u : \Omega \to X$ *be a measurable selector of* S *and let* $r : \Omega \to (0, \infty)$ *be a measurable function. Then there exists a measurable selector* $v : \Omega \to X$ *of* T *such that for any* $\omega \in \Omega$, $d(u(\omega), v(\omega)) \leqslant D(S(\omega), T(\omega)) + r(\omega)$.

Another result was given by N. S. Papageorgiou [336]:

LEMMA 2.2.2. (a) *If* S, $T : \Omega \to CB(X)$ *are measurable multivalued mappings*, $u : \Omega \to X$ *is a measurable selector of* S *and* $r : \Omega \to (1, \infty)$, *then there exists a measurable selector* $v : \Omega \to X$ *for* T *such that* $d(u(\omega), v(\omega)) \leqslant r(\omega) . D(S(\omega), T(\omega))$.

(b) *If* $T : X \to CB(X)$ *is a continuous multivalued mapping, then* $x \to d(x, T(x))$ *is a continuous real-valued function.*

LEMMA 2.2.3 [241]. *Let* X *be a separable metric space, let* $T : \Omega \to 2^X$ *be a measurable multivalued mapping and let* $u : \Omega \to X$ *be a measurable mapping. Then* $d(u(.), T(.))$ *is a measurable function on* Ω.

Remark 2.2.2. A result to be used in Chap. 4, § 1, on the closed-convex-bounded-valued mappings E on an atomic probability measure space (see [50], Prop. 4) implies that the set $\mathscr{S}$ of all measurable selectors of the measurable multivalued mapping $\omega \to E(\omega)$ is nonempty, ε, λ-closed, convex and probabilistic bounded in the RN-space $(S, \mathscr{F}, \tau_{T_m})$.

THEOREM 2.2.10 [251]. *Let X be a separable Banach space and let $T : \Omega \times X \to \mathscr{P}_0(X)$ be a compact-valued mapping such that $T(\omega, .)$ is upper semicontinuous for every $\omega \in \Omega$, and $T(., x)$ is measurable for every $x \in X$. Then T is $\mathscr{K} \times \mathscr{B}_X$-measurable.*

A generalization of Theorem 2.1.3 is given by

THEOREM 2.2.11 [251]. *Let X be a separable Banach space and let $T : \Omega \times X \to \mathscr{P}_0(X)$ be a.s. an upper semicontinuous compact-valued random operator. Then T is separable.*

Let us now present a multivalued analogue to Nashed and Salehi's theorem [309]:

THEOREM 2.2.12 [251]. *Let $(\Omega, \mathscr{K}, P)$ be a complete probability space and let X and Y be separable Banach spaces. Let $T : \Omega \times X \to Y$ be a separable random multivalued operator such that a.s. $T(\omega, .)$ is invertible and $T^{-1}(\omega, .)$ is upper semicontinuous and compact-valued. Then T^{-1} is also a random multivalued operator from $\Omega \times Y$ into X.*

Let X and Y be Banach spaces and let T be a linear operator with dense domain $\mathscr{D}(T)$ in X and range in Y. We assume that the null space of T, $N(T)$, has a topological complement, say M, and the closure of the range of T, $\overline{R(T)}$, has a topological complement, say S; i.e., there exists (continuous) projectors P, Q with $PX = N(T)$ and $QY = \overline{R(T)}$ and $X = N(T) \oplus M$, $Y = \overline{R(T)} \oplus S$. Let $\tilde{T} = T/M$ be the restriction of T to M. Then $\tilde{T} : M \to R(T)$ is one-to-one and onto. We define the *generalized inverse* T^+ of T as the linear extension of $\tilde{T}^{-1}$ to $R(T) + S$ such that $T^+S = \{0\}$. Note that T^+ depends on the topological complements M and S or, equivalently, on the projectors P and Q.

Concerning the joint measurability, we give

PROPOSITION 2.2.4 [241]. *Let X be a separable metric space with $\{x_k\}_{k \in \mathbb{N}}$ a countable dense subset of X and let Y be a separable Banach space. Let $T : \Omega \times X \to CK(X)$ be a mapping having the properties:* (a) *for each $\omega \in \Omega$, $T(\omega, .)$ is usc;* (b) *for each $x \in X$, $T(., x)$ is measurable. Then, the mapping $G : \Omega \times X \to \mathscr{P}(Y)$, defined as*

$$G(\omega, x) = \bigcap_{n=1}^{\infty} \operatorname{cl} \operatorname{co}\left(\cup \left\{ T(\omega, x_k) : \; d(x_k, x) < \frac{1}{n} \right\} \right),$$

$\omega \in \Omega$, $x \in X$ *satisfies the following conditions*: (i) *for each* $\omega \in \Omega$ *and* $x \in X$, $T(\omega, x) \supset G(\omega, x) \neq \emptyset$; (ii) $G(\omega, .)$ *is usc for each* $\omega \in \Omega$; (iii) G *is* $\mathscr{K} \times \mathscr{B}_X$*-measurable.*

Remark 2.2.3 [310]. Let X and Y be separable Banach spaces and let T be a random linear operator from $\Omega \times X$ into Y such that $\bigcap_\omega \mathscr{D}(T(\omega)) = D$ is dense in X. If X^* is separable, then, for each $y^* \in \bigcap_\omega \mathscr{D}(T^*(\omega))$, $T^*(.)y^*$ is a generalized random variable. In particular, if T is a random bounded linear operator, then so is T^*, and conversely.

In the last part of this chapter, we give some results to be used in Chapts. 4 and 5, to obtain some existence theorems of random fixed points for *random operators with stochastic domain* or for *multivalued random operators with stochastic domain* [138] (i.e., mappings $T : \mathrm{Gr}C \to X$ resp. 2^X such that for all open $D \subseteq X$ and all $x \in X$, $\{\omega \in \Omega : x \in C(\omega) \wedge T(\omega, x) \in \in D \neq \emptyset\} \in \mathscr{K}$ resp. $\{\omega \in \Omega : x \in C(\omega) \wedge T(\omega, x) \cap D \neq \emptyset\} \in \mathscr{K}$, where $C : \Omega \to 2^X$ is a measurable multivalued mapping). A mapping $x : \Omega \to X$ will be called a *wide-sense fixed point of* T if, for all $\omega \in \Omega$, $x(\omega) \in C(\omega)$ and $x(\omega) = T(\omega, x(\omega))$ resp. $x(\omega) \in T(\omega, x(\omega))$; it is called *random fixed point* if it fulfills the properties of a wide-sense fixed point for μ-almost all $\omega \in \Omega$ and is measurable ($(\Omega, \mathscr{K}, \mu)$ is supposed to be a σ-finite measure space).

The problem of the existence of a random fixed point of a random operator (or a random solution of a random operator equation) is not equivalent to the non-stochastic problem of the existence of the wide-sense fixed point, as can be seen from the following example (see ref. [138, Ex. 3]):

Let $\Omega = R$, $\mathscr{K} = \sigma$-algebra of all at most countable subsets of $\mathbb{R}$ and their complements, for all $A \in \mathscr{K}$, $\mu(A) = \mathrm{card}(A)$. $(\Omega, \mathscr{K}, \mu)$ is not a (not σ-finite) measure space. Let $T : \Omega \times \times \mathbb{R} \to \mathbb{R}$ be defined by

$$T(\omega, x) = \begin{cases} 1, & (x = 1 \wedge \omega = 1) \vee (x \neq 1 \wedge \omega \neq x), \\ 0, & x = 1 \wedge \omega \neq 1 \\ x, & x \neq 1 \wedge \omega = x. \end{cases}$$

For each $x \in \mathbb{R}$, $T(., x)$ has the form $c_1 + c_2 . i(A)$, where $\mu(A) = 1$ and $c_1, c_2 \in \mathbb{R}$, (i is the indicator function). Therefore, $T(., x)$ is $\mathscr{K}$-measurable, which means that T is a random operator. Obviously, $x : \Omega \to \mathbb{R}$, defined by $x(\omega) = \omega$, is a unique wide-sense fixed point of T (note that $\mu(A) = 0 \Leftrightarrow A = \emptyset$). But x is not measurable, since $\{\omega \in \Omega : x(\omega) < 0\} \notin \mathscr{K}$. So, T has no random fixed point.

PROPOSITION 2.2.5 [59]. *If f is a random continuous function with a random domain C and if $a(\omega) = \inf\{f(\omega, x) : x \in C(\omega)\} < \infty$ for every $\omega \in \Omega$, then $a(.)$ is a random variable.*

Remark 2.2.4. If X, Y, Z are separable Banach spaces, $C : \Omega \to 2^X$ is a multivalued separable mapping, T is a continuous random operator with stochastic domain C and values in Y, U is a random operator from $\Omega \times Z$ into X such that for all $\omega \in \Omega$, $U(\omega)(\mathscr{D}(U(\omega)) \subseteq C(\omega)$, then, for every measurable function $z : \Omega \to X$ such that $x(\omega) \in C(\omega)$ for all $\omega \in \Omega$, we have: (i) $\omega \to T(\omega, z(\omega))$ is measurable; (ii) $T \circ U$ (mapping each (ω, z) into $T(\omega, U(\omega, z))$ is a random operator from $\Omega \times Z$ into Y. Also (see ref. [215, Th. 6.5.]), for every continuous multivalued operator with complete values the multifunction $G(\omega) = T(\omega, g(\omega))$ is measurable for every measurable multifunction $g : \Omega \to \mathscr{P}(X)$.

Since several results of Chap. 6 will be formulated for separable multivalued mappings (i.e., the measurable multivalued mappings $C : \Omega \to 2^X$ for which there exists a countable set $Z \subseteq X$ such that for all $\omega \in \Omega$, $\mathrm{cl}(Z \cap C(\omega)) = C(\omega)$), we recall some results which give sufficient conditions for the separability and the existence of measurable selectors of $\mathrm{int} C$.

LEMMA 2.2.4 [139]. *Let X, U be separable Banach spaces, let $C : \Omega \to 2^X$ and $K : \Omega \to 2^U$ be measurable. Then the multivalued mapping $C \times K : \Omega \to 2^{X \times U}$, defined by $(C \times K)(\omega) = C(\omega) \times K(\omega)$, is measurable. If C and K are separable, then $C \times K$ is separable.*

THEOREM 2.2.13 [136]. *Let X be a real separable reflexive Banach space and let $(\Omega, \mathscr{K}, P)$ be complete, $C : \Omega \to 2^X$ measurable, uniformly bounded (i.e., there exists an $r > 0$ such that for all $\omega \in \Omega$, $C(\omega) \subseteq \{x \in X, \|x\| \leqslant r)$ with closed, convex and solid values (i.e., $\mathrm{int} C(\omega) \neq \emptyset$). Then there exists a measurable selector $x_0 : \Omega \to X$ with $x_0(\omega) \in \mathrm{int} C(\omega)$ for all $\omega \in \Omega$.*

LEMMA 2.2.5 [140]. *Let X be a separable Hilbert space. Let $S : \Omega \to 2^X$ and $P : \Omega \times X \to X$ be such that for all $\omega \in \Omega$, $S(\omega)$ is a closed subspace of X and $P(\omega, .)$ is the orthogonal projector into $S(\omega)$. Then the following statements are equivalent:* (i) *S is measurable*; (ii) *P is a random operator*; (iii) *there exists a sequence of measurable functions $\psi_1, \psi_2, \ldots; \Omega \to X$ such that, for all $\omega \in \Omega$, $\{\psi_1(\omega), \psi_2(\omega), \ldots\}$ is an orthonormal basis of $S(\omega)$ (with the understanding that some $\psi_i(\omega)$ may be* 0); (iv) *there exists a sequence of measurable functions $\Phi_1, \Phi_2, \ldots : \Omega \to X$ such that for all $\omega \in \Omega$,* $\mathrm{cl}(\mathrm{span}\{\psi_1, \psi_2, \ldots\}) = S(\omega)$.

Nowak [315] considered a multivalued mapping $C : \Omega \to \to \mathscr{P}_0(X)$ whose graph $\mathrm{Gr}C \in \mathscr{K} \times B_X$ and $T : \mathrm{Gr}C \to \mathscr{P}_0(X)$ (C is called the stochastic *domain* of T). He gave sufficient conditions for the measurability of the graph of T by

THEOREM 2.2.14. *Let T be a multivalued mapping with stochastic domain C. Assume that C is separable, T is closed-valued, measurable in ω and continuous in x (resp. continuous in the generalized Hausdorff metric); then* $\mathrm{Gr}C \in \mathscr{K} \times \mathscr{B}_X \times \mathscr{B}_X$.

It is not known whether the assumption of continuity in the above theorem can be weakened to lower semicontinuity. If we assume that $T(\omega, x)$ is upper semicontinuous, then the graph of T need not be measurable. Indeed, H. Engl [138] used the supplemental condition "$T : K \to 2^Y$, for $K \in 2^Y$, T is a compact-valued" for a usc multivalued mapping in the sense of Definition 2.2.4 to give

Example 2.2.1 [138, Example 15]. Let $(\Omega, \mathscr{K}, \mu)$ be the interval $[-1, 1]$ with the Borel σ-algebra and the Lebesgue-measure, $X = \mathbb{R}$, $C = [-1, 1]$, X is endowed with the usual topology and the Borel σ-algebra $\mathscr{B}$. Let $S^1 \subseteq \Omega \times C$ be the unit circle and let E be a subset of S^1 which is not in $\mathscr{K} \times \mathscr{B}$. Such a set E exists if we assume the validity of the axiom of choice. Define $T : \Omega \times C \to 2^{\mathbb{R}}$

$$(\omega, x) \to \begin{cases} \{0\} & \text{for } (\omega, x) \notin E, \\ [0, 1] & (\omega, x) \in E. \end{cases}$$

For all $x \in C$ and all open $D \subseteq X$,

$$\{\omega \in \Omega : T(\omega, x) \cap D \neq \emptyset\} = \begin{cases} \theta & \text{if } D \cap [0, 1] = \emptyset \\ & \text{if } 0 \in D \\ B_1(x) & \text{if } 0 \notin D \text{ and } D \cap [0, 1] \neq \emptyset, \end{cases}$$

where $B_1(x) = \{\omega \in \Omega : (\omega, x) \in E\}$. Because of $\mathrm{card} B_1(x) \leqslant 2$, $\{\omega \in \Omega : T(\omega, x) \cap D \neq \emptyset\} \in \mathscr{K}$.

For all $\omega \in \Omega$ and all closed $A \subseteq X$,

$$\{x \in C : T(\omega, x) \cap A \neq \emptyset\} = \begin{cases} \theta & \text{if } A \cap [0, 1] = \theta \\ [-1, 1] & \text{if } 0 \in A \\ B_2(\omega) & \text{if } 0 \notin A \text{ and } A \cap [0, 1] \neq \emptyset, \end{cases}$$

where $B_2(\omega) = \{x \in C : (\omega, x) \in E\}$. Because of $\mathrm{Card}(B_2(\omega)) \leqslant 2$, we have $\{x \in C : T(\omega, x) \cap A \neq \emptyset\} \in \mathscr{K}$ is closed.

So, T is measurable as a function of ω and usc. as a function of x. But because $\left\{(\omega, x) \in \Omega \times X : T(\omega, x) \cap \left(-\frac{1}{2}, 1\right) \neq \emptyset\right\} \notin \notin \mathscr{K} \times \mathscr{B}$, T is not jointly measurable.

In order to prove a random fixed point theorem for *nonexpansive mappings* (i.e. $\| T(\omega, x) - T(\omega, y)\| \leqslant \| x - y \|$ for all $\omega \in \Omega$ and all $x, y \in C(\omega)$), use is made of an idea from the deterministic case which characterizes fixed points as Chebysheff-centers of certain sequences. If X is a real normed linear space, C is a nonempty subset of X, T a mapping $T : C \to X$ and $x_0 \in C$, we call a sequence $\{x_n\} \subset X^{\mathbb{N}}$ a "*normal Mann iteration process for T* (with initial guess x_0)" if for all $n \in \mathbb{N}$, $x_n \in C$ and there exists a sequence $\{c_n\} \in [0, a]^{\mathbb{N}_0}$ $(a < 1)$ with $\sum_{n=0}^{\infty} c_n = \infty$ such that for all $n \in \mathbb{N}_0$, $x_{n+1} = (1 - c_n)x_n + c_n T(x_n)$. X will be called the "*Opial space*" if, for all sequences $\{z_n\}_{n \in \mathbb{N}} \subset X$ converging weakly to $x \in X$ and all $y \neq x$, we have $\liminf \| z_n - y \| > \liminf \| z_n - x \|$.

THEOREM 2.2.15 [134]. *Let X be an Opial space and let C be weakly compact, $T : C \to X$ nonexpansive, $x_0 \in C$. If there exists a normal Mann iteration process $\{x_n\} \subset X$ for T with initial guess x_0, then T has a fixed point. Furthermore, $\{x_n\}$ converges weakly to a fixed point $\bar{x}$: $\bar{x}$ which can be characterized as the unique $\bar{x} \in C$ with* $\limsup \| x_n - \bar{x} \| = \inf_{y \in C} \limsup \| x_n - y \|$.

Hence, $\bar{x}$ is the Chebysheff-center of x_n with respect to C which, in general, is the set $\check{C}\ (C, \{x_n\}) = \{x \in C : \limsup \| x_n - x \| = \inf_{y \in C} \limsup \| x_n - y \|\}$.

COROLLARY 2.2.1. *Let X be an Opial space, let C be convex and weakly compact and let $T : C \to C$ be nonexpansive. Then T has a fixed point and, for all $0 < \lambda < 1$ and $x_0 \in C$, the sequence $\{x_n\} = \{(1 - \lambda)\, id + \lambda T)^n(x_0)$ converges weakly to a fixed point which can be characterized as in Theorem* 2.2.15.

PROPOSITION 2.2.6 [135]. *Let $(\Omega, \mathscr{K}, \mu)$ be complete, let X be a reflexive Banach space, let $C : \Omega \to 2^X$ be measurable with $C(\omega)$ uniformly bounded and weakly compact for all $\omega \in \Omega$, let $x_n : \Omega \to X$ be measurable with $x_n(\omega) \in \{x : \| x \| \leqslant r, r > 0$ fixed$\}$ for all $n \in \mathbb{N}$, $\omega \in \Omega$. Then the mapping $\check{C} : \Omega \to 2^X$, $\omega \to \check{C}(C(\omega)$, $\{x_n(\omega)\})$, is measurable.*

Before concluding this section, we recall some results of interest in applications.

LEMMA 2.2.6 [136]. (a) *Let $C_n[0, 1] = \{x : [0, 1] \to \mathbb{R}^n$, $x =$ continuous$\}$ and $\mathbb{R}^n$ be endowed with an arbitrary norm $\| \ \|_n$ and $C_n[0, 1]$ with the generated sup-norm $\| \ \|$. Let $x : \Omega \times [0, 1] \to \mathbb{R}^n$ be such that $x(\omega, .)$ is continuous for all $\omega \in \Omega$. Then $x(., t)$ is measurable for all $t \in [0, 1]$ iff $\tilde{x} : \Omega \to C_n[0, 1]$, defined by $\tilde{x}(\omega)(t) = x(\omega, t)$, is measurable (we shall identify x and $\tilde{x}$).*

(b) *Let* $f : [0,1] \times \mathbb{R}^n \times \Omega \to \mathbb{R}^n$ *be such that, for all* $\omega \in \Omega$ *and* $x \in C_n[0,1]$, $f(0, x(.), \omega)$ *is Riemann-integrable and, for all* $(t, x) \in [0,1] \to \mathbb{R}^n$, $f(t, x, .)$ *is measurable. Let* $T : \Omega \times$

$$\times\, C_n[0,1] \to C_n[0,1], \quad (\omega, x) \to \left(t \to \int_0^t f(s, x(s), \omega)\, \mathrm{d}s\right).$$ *Then*

T is a random operator with $C(\omega) = C_n[0,1]$ *for all* $\omega \in \Omega$.

Likewise, let us recall some deterministic fixed point theorems.

THEOREM 2.2.16 [274, Th. 3.2]. *Let X be a Banach space which satisfies Opial's condition. If C is a nonempty convex weakly compact subset of X and* $T : C \to K(C)$ *is a compact-valued nonexpansive mapping, then there exists a fixed point* $x_0 \in Tx_0$.

THEOREM 2.2.17 [15, Th. 1]. *Let H be a complete and convex metric space, let K be a nonempty, closed subset and let T be a nonexpansive multivalued operator on K with closed and bounded values. If* $T(x) \subseteq K$ *for each* $x \in \partial K$, *then there exists an* $x_0 \in K$ *such that* $x_0 \in T(x_0)$.

THEOREM 2.2.18 (Kakutani-Bohnenblust-Karlin). *Let Y be a Banach space, let* $K \in C \subset (Y)$ *and* $T : K \to C \subset (Y)$ *be compact such that, for all* $x \in K$, $T(x) \subseteq K$. *Then T has a fixed point.*

LEMMA 2.2.7 [337]. (a) *If T is a nonexpansive operator on a Hilbert space, then* $\mathrm{cl}(R(I-T))$ *is convex.*

(b) *If T is a nonexpansive operator on a closed and convex subset* $C \subset H$, *then T has a fixed point in C if and only if there exists an element* $x_0 \in C$ *such that the sequence* $\{\| T^n x_0 \|\}$ *is bounded.*

Chapter 3

Fixed point theory in probabilistic structures

A fixed point of a mapping defined on an arbitrary set is an invariant point of this set with respect to this mapping, provided that such a point exists.

In a series of papers on curves described by differential equations, H. Poincaré (1885) considered a continuous vector field over a closed surface, assigning an index to each isolated singular point. He proved that if the surface is orientable and of genus $\neq 1$, then it has at least one singular point.

Mention should be made here of the famous results of L.E.J. Brouwer, who established the degree of a continuous mapping from an n-manifold into another extending Poincaré's idea of index from 2 to n dimensions. Brouwer also gave and proved his fixed point theorems regarding such mappings for n-cells, n-spheres and projective planes.

Extensions of the Brouwer theorem to functional spaces were given by Birkhoff and Kellogg (1922), Lefschetz (1926, 1942), Hoph (1929), Schauder (1927, 1930), Tychonoff (1935), Kakutani (1941), Bohnenblust and Karlin (1950), Fan (1952), Glicksberg (1952) and others.

The theory of topological degree was extended by Leray-Schauder (1934), Browder (1960) and others.

Fixed point theorems are generally of two kinds : (a) topological and (b) algebraic.

Topological fixed point theorems are strictly existence theorems, i.e., they establish the conditions under which a fixed point exists but do not provide methods for finding it. Algebraic fixed point theorems give the method for finding a fixed point. The prototype of most algebraic fixed point theorems is the well-known contraction mapping principle due to Banach (1932). This is an abstract formulation of the classical method of successive approximation due to Picard. (This theorem was extended in various ways, and was given many applications in analysis and in the theory of operator equations by

Nemytskiĭ (1936), Brodskiĭ and Milman (1948), M. A. Krasnosel'skiĭ (1950), G. Darbo (1955), B. N. Sadovski (1967), and others).

In probabilistic metric spaces, random normed spaces and probabilistic locally convex spaces, some fixed point theorems were obtained by V. M. Sehgal [394], [396], H. Sherwood [403], [406], A. T. Bharucha-Reid [32], [396], V. Istrăţescu [234], [236], I. Istrăţescu [218], [221], [222], Gh. Constantin [109], [110], G. L. Cain and R. H. Kasriel [84], O. Hadžić [167], [169], [191], Gh. Bocşan [45], [48], [49], V. Radu [348], [352], [355], Shih-sen Chang [95], [98], [100], L. Ćirić [105], T. L. Hicks [210] and others.

In the following section, we shall present several outstanding results regarding the existence and the uniqueness of the fixed points for classes of mappings on probabilistic metric spaces, random normed spaces or probabilistic locally convex spaces.

3.1. FIXED POINT THEOREMS FOR MAPPINGS ON *PM*-SPACES

The study of contraction mapping theorems for *PM*-spaces was initiated by V. M. Sehgal [394], [396], H. Sherwood [403], [406], and A. T. Bharucha-Reid [32], [396]. More recently, many papers and books dealing with numerous generalizations and applications have emerged [234], [235], [236], [105], [45], [48], [49], [56], [84], [109], [110], [112], [113], [114], [115], [166], [176], [177], [178], [180], [182], [183], [191], [217], [218], [215], [221], [222], [97], [99], [101], [353], [354], [356], [357].

Throughout this chapter, we shall assume that for a *PM*-space the condition $\mathscr{F}(S \times S) \subseteq \mathscr{D}^+$ is satisfied; otherwise, we accept that $(S, \mathscr{F})$ is a generalized *PM*-space.

3.1.1. *Contraction Mappings on Complete PM-Spaces*

First, we shall briefly note that an analogue of the Banach contraction mapping theorem was obtained by H. Sherwood [403] using

DEFINITION 3.1.1. Let $(S, \mathscr{F}, T)$ be a Menger space with a strict t-norm T; let g be an additive generator in the sense defined in § 1, Chap. 1, and let $k \in (0, 1)$. A mapping $f: S \to S$ is a *contraction mapping* in S with respect to g and k if, for every $p, q \in S$,

$$g(F_{f(p),f(q)}(kx)) \leqslant kg(F_{p,q}(x)).$$

The probabilistic version of Banach's theorem due to H. Sherwood [403] is

THEOREM 3.1.1. *Let* $(S, \mathscr{F}, T)$ *be a complete Menger space under the strict t-norm* T *with additive generator* g. *Let* f *be a contraction mapping on* S *with respect to* g *and let* $k \in (0, 1)$. *Then there exists a unique point* p *in* S *such that* $f(p) = p$.

The proof is easy to reproduce from Theorem 3.1.55.

V. M. Sehgal introduced [394] a very natural definition of the contraction mapping:

DEFINITION 3.1.2. Let $(S, \mathscr{F})$ be a *PM*-space. A mapping $f: S \to S$ is a *contraction mapping* on $(S, \mathscr{F})$ if and only if there exists a $k \in (0,1)$ such that

$$(*)\ F_{f(p),f(q)}(kx) \geqslant F_{p,q}(x)$$

for every p, $q \in S$ and $x \in \mathbb{R}^+$.

Denoting by j the identical mapping of $\mathbb{R}^+$ on $\mathbb{R}^+$, one can write the inequality $(*)$ as

$$F_{f(p),f(q)} \geqslant F_{p,q}\left(\frac{j}{k}\right),$$

for all p, $q \in S$.

V. M. Sehgal also proved

THEOREM 3.1.2. *Every contraction mapping* f *on a complete Menger space* $(S, \mathscr{F}, \mathrm{Min})$ *has a unique fixed point* p. *Moreover,* $f^n(q) \to p$ *for each* $q \in S$.

Proof. If there exist p, $q \in S$ with $f(p) = p$ and $f(q) = q$, then, for every $n \geqslant 1$ and every $x > 0$,

$$F_{p,q}(x) = F_{f^n(p)f^n(q)}(x) \geqslant F_{p,q}\left(\frac{x}{k^n}\right).$$

Since $\sup\limits_{n \in \mathbb{N}} F_{p,q}\left(\frac{x}{k^n}\right) = 1$, it follows that $F_{p,q} = H_0$, i.e., $p = q$.

To prove the existence of a fixed point, consider an arbitrary $p_0 \in S$, and define the sequence of iterates $p_n = f(p_{n-1})$, $n \in \mathbb{N}$, of p_0 under f. Let us show that the sequence $\{p_n\}_{n \in \mathbb{N}}$ is fundamental in S. Thus, for every positive integer m, n, $n < m$, and $k \in (0, 1)$, we have

$$(1 - k)(k + \ldots + k^{m-n-1}) = k - k^{m-n} < 1,$$

whence

$$F_{p_n,p_m}(x) \geqslant F_{p_n,p_m}((1-k)(1+k+\ldots+k^{m-n-1})x) \geqslant$$
$$\geqslant \mathrm{Min}(F_{p_n,p_{n+1}}((1-k)x),\ F_{p_{n+1},p_m}((1-k)(k+\ldots+k^{m-n-1})x)) \geqslant$$
$$\geqslant \mathrm{Min}\left(F_{p_0,p_1}\left(\frac{(1-k)x}{k^n}\right), \ldots, F_{p_0,p_1}\left(\frac{(1-k)x}{k^n}\right)\right) =$$
$$= F_{p_0,p_1}\left(\frac{(1-k)x}{k^n}\right).$$

Thus, since $F_{p_0,p_1} \in \mathscr{D}^+$

$$\lim_{n\to\infty} F_{p_n,p_m}(x) = 1$$

for all $x > 0$ independent of m, $m > n$, i.e. $\{p_n\}_{n\in\mathbb{N}}$ is a fundamental sequence in $(S, \mathscr{F}, \mathrm{Min})$. Since $(S, \mathscr{F}, \mathrm{Min})$ is a complete Menger space, there exists a $p \in S$ such that $p_n \to p$, that is $f^n(p_0) \to p_0$. To prove that $f(p) = p$, it is sufficient to prove that $f^n(p_0) \to f(p)$. Thus, for all $x > 0$,

$$F_{f(p_n),f(p)}(x) \geqslant F_{p_n,p}\left(\frac{x}{k}\right) \geqslant F_{p_n,p}(x).$$

Since $(S, \mathscr{F}, \mathrm{Min})$ is a Hausdorff space [385, Th. 7.2], we conclude that $f(p) = p$, and thus the existence is proven.

The first general result on the existence of fixed points for contraction mappings on complete PM-spaces was given by H. Sherwood in [406].

THEOREM 3.1.3. *Let $(S, \mathscr{F}, \tau)$ be a complete PM-space, where τ is a continuous triangle function in the point (H_0, H_0), and suppose that $\mathscr{D}^+$ is closed under τ.*
Let f be a contraction mapping on $(S, \mathscr{F}, \tau)$. Then either

(i) *f has a unique fixed point,*

or

(ii) *for every $p \in S$, $\sup_{x\in R} D_{O_f(p,0,\infty)}(x) < 1$, where $O_f(p, 0, \infty) = \{f^n(p) : n \in \mathbb{N} \cup \{0\}\}$.*

Proof. Suppose there exists a $p_0 \in S$ such that $O_f(p_0, 0, \infty)$ is bounded. Then, if $\{p_n\}_{n\in\mathbb{N}}$ is the sequence of iterates of p_0 under f,

$$F_{p_n,p_{n+m}} \geqslant F_{p_0,p_m}\left(\frac{j}{k^n}\right) \geqslant D_{O_f(p_0,0,\infty)}\left(\frac{j}{k^n}\right).$$

Since $D_{O_f(p_0,0,\infty)} \in \mathscr{D}^+$, it follows that $\{p_n\}_{n\in \mathbb{N}\cup\{0\}}$ is a Cauchy sequence in $(S, \mathscr{F}, \tau)$ which is a complete space. Hence, there exists a point $q \in S$ such that $\lim_{n\to\infty} p_n = q$. Let us prove that $f(q) = q$. Indeed,

$$F_{f(q),q} \geqslant \tau(F_{q,p_n}, F_{p_n\ f(q)}) \geqslant \tau\left(F_{q,p_n}, F_{p_{n-1},q}\left(\frac{j}{k}\right)\right).$$

Following the lines of Theorem 3.1.2, one can conclude that q is the unique fixed point of the mapping f.

Remark. Condition (ii) is satisfied by a point if and only if it is satisfied by every point, i.e., for every $p,\ q \in S$, $O_f\ (p, 0, \infty)$ is bounded if and only if $O_f(q, 0, \infty)$ is bounded.

Indeed, if $O_f(p, 0, \infty)$ is bounded, it follows that, for every $m,\ n \in \mathbb{N} \cup \{0\}$,

$$F_{f^m(q),f^n(q)} \geqslant \tau(F_{f^m(q),f^m(p)}, \tau(F_{f^m(p),f^n(p)}, F_{f^n(p),f^n(q)})) \geqslant$$
$$\geqslant \tau(F_{p,q}, \tau(D_{O_f(p,0,\infty)}, F_{p,q})),$$

and hence

$$D_{O_f(q,0,\infty)} \geqslant \tau(F_{p,q}, \tau(D_{O_f(p,0,\infty)}, F_{p,q})).$$

This implies that $D_{O_f(q,0,\infty)} \in \mathscr{D}^+$, i.e., $O_f(q, 0, \infty)$ is bounded.

Similarly, from $D_{O_f(q,0,\infty)} \in \mathscr{D}^+$ it follows that $D_{O_f(p,0,\infty)} \in \mathscr{D}^+$.

On the other hand, H. Sherwood [406] showed that the above results are an exception to rather than a rule for *PM*-spaces. He proved that it is possible to construct complete *PM*-spaces together with contraction mappings which have no fixed points.

DEFINITION 3.1.3. Let $\tau : \Delta^+ \times \Delta^+ \to \Delta^+$ be a triangle function, $k \in (0, 1)$ and $G \in \Delta^+$. Then $G^{\tau,k}$ is the function in Δ^+ given by

$$G^{\tau,k}(x) = \lim_{t<x} \lim_{n\to\infty} G_n^{\tau,k}(t),$$

where $G_n^{\tau,k} = \tau^n(F_0, F_1, \ldots, F_n) = \tau(\tau^{n-1}(F_0, \ldots, F_{n-1}), F_n)$, $\tau^1 = \tau(F_0, F_1)$, $F_i = G\left(\frac{j}{k^i}\right)$.

Since τ is associative, then $G_n^{\tau,k}$ is well defined for all $n \in \mathbb{N}$, and since $G_n^{\tau,k} \geqslant G_{n+1}^{\tau,k}$ the limit $G^{\tau,k}$ always exists.

THEOREM 3.1.4 [406]. *Let T be an l.c. t-norm and let τ_T the l.c. t-function induced by T. Then there exist a complete PM-space $(S, \mathscr{F}, \tau_T)$ and a contraction mapping f on $(S, \mathscr{F}, \tau_T)$ which has no fixed point if and only if there exist $G \in \mathscr{D}^+$ and a number $k \in (0, 1)$ such that $G^{\tau_T, k} \notin \mathscr{D}^+$.*

Proof. Let $G \in \mathscr{D}^+$ and $k \in (0,1)$ be such that $G^{\tau_T k} \notin \mathscr{D}^+$. Let also $S = \mathbb{N}$ and let $\mathscr{F}$ be defined by

$$\begin{cases} F_{n,n+m} = F_{n+m,n} = \tau_T^m\left(G\left(\dfrac{j}{k^{n-1}}\right), \ldots, G\left(\dfrac{j}{k^{n+m-1}}\right)\right) = \\ = G^{\tau_T, k}\left(\dfrac{j}{k^{n-1}}\right) \\ F_{n,n} = H_0. \end{cases}$$

It is easy to show that $F_{n,m} \in \mathscr{D}^+$ for all $n, m \in \mathbb{N}$. In order to prove that $(S, \mathscr{F}, \tau_T)$ satisfies Serstnev's inequality, let us consider the following three cases:

Case 1. $F_{n,n+m+s} \geqslant \tau_T(F_{n,n+m}, F_{n+m,n+m+s})$.

As τ_T is associative, nondecreasing and commutative, then

$$\begin{aligned} F_{n,n+m+s} &= \tau_T^m\left(G\left(\frac{j}{k^{n-1}}\right), \ldots, G\left(\frac{j}{k^{n+m+s-1}}\right)\right) = \\ &= \tau_T\left(\tau_T^{m+s}\left(G\left(\frac{j}{k^{n-1}}\right), \ldots, G\left(\frac{j}{k^{n+m+s-1}}\right)\right), H_0\right) = \\ &= \tau_T\left(\tau_T^m\left(G\left(\frac{j}{k^{n-1}}\right), \ldots, G\left(\frac{j}{k^{n+m-1}}\right)\right), \tau_T^s\left(G\left(\frac{j}{k^{n+m}}\right), \ldots\right.\right. \\ &\quad \left.\left.\ldots, G\left(\frac{j}{k^{n+m+s-1}}\right), H_0\right)\right) \geqslant \\ &\geqslant \tau_T(F_{n,n+m}, \tau_T^s\left(G\left(\frac{j}{k^{n+m-1}}\right), \ldots, G\left(\frac{j}{k^{n+m+s-1}}\right), G\left(\frac{j}{k^{n+m-1}}\right)\right) = \\ &= \tau(F_{n,n+m}, F_{n+m,n+m+s}). \end{aligned}$$

Case 2. $F_{n,n+m} \geqslant \tau_T(F_{n,n+m+s}, F_{n+m+s,n+m})$.

As in Case 1, one has

$$F_{n,n+m} = \tau_T^m\left(G\left(\frac{j}{k^{n-1}}\right), \ldots, G\left(\frac{j}{k^{n+m-1}}\right)\right) =$$

$$= \tau_T\left(\tau_T^m\left(G\left(\frac{j}{k^{n-1}}\right), \ldots, G\left(\frac{j}{k^{n+m-1}}\right), H_0\right.\right. \geqslant$$

$$\geqslant \tau_T\left(\tau_T^m\left(G\left(\frac{j}{k^{n-1}}\right), \ldots, G\left(\frac{j}{k^{n+m-1}}\right), \tau_T^{s-1}G\left(\frac{j}{k^{n+m}}\right), \ldots,\right.\right.$$

$$\left.\left.\ldots, G\left(\frac{j}{k^{n+m+s-1}}\right)\right)\right) =$$

$$= \tau_T^{m+s}\left(G\left(\frac{j}{k^{n-1}}\right), \ldots, G\left(\frac{j}{k^{n+m+s-1}}\right)\right) = F_{n,n+m+s} \geqslant$$

$$\geqslant \tau_T(F_{n,n+m}, F_{n+m,n+m+s}) \geqslant \tau_T(F_{n,n+m+s}, F_{n+m+s,n+m}).$$

Case 3. $F_{n+m,n+m+s} \geqslant \tau_T(F_{n+m,n}), F_{n,m+n+s})$, and hence, as in case 2, one can prove the triangle inequality.

To prove that $(S, \mathscr{F}, \tau_T)$ is complete, let us show that the only possible Cauchy sequences are those which are eventually constant.

Indeed, let $\{p_n\}_{n\in\mathbb{N}}$ be a Cauchy sequence in S which is not eventually constant. Then two cases are possible:

(i) for every positive integer s, there exists a positive integer n_s such that $p_{n_s} > p_{n_{s-1}}$, where $n_0 = 1$,

or

(ii) there exists a positive integer m such that $p_n \leqslant m$ for all n.

In the first case, let $\{p_n\}_{n\in N}$ be a Cauchy sequence in $(S, \mathscr{F}, \tau_T)$. Then, for the isometric image of $\{p_n\}_{n\in\mathbb{N}}$ in the completion of $(S, \mathscr{F}, \tau_T)$ (see Th. 21), one writes

$$\lim_{s\to\infty} F_{1,p_{n_s}} \in \mathscr{D}^+.$$

On the other hand, for any $x \in R$,

$$\lim_{s\to\infty} F_{1,p_{n_s}}(x) = \lim_{s\to\infty} \tau_T^{p_{n_s}-1}\left(G(j),\ G\left(\frac{j}{k}\right), \ldots, G\left(\frac{j}{p_{s_n}-1}\right)\right)(x) =$$

$$= \limsup_{s\to\infty} T^{p_{n_s}-1}\left(G(\beta_0 x),\ G\left(\frac{\beta_1 x}{k}\right), \ldots, G\left(\frac{\beta_{p_{n_s}-1}x}{k^{p_{n_s}-1}}\right)\right);$$

$$\left\{\sum_{i=0}^{p_{n_s}-1} \beta_i = 1,\ 0 \leqslant \beta_i \leqslant 1\right\} = G^{\tau_T,k}(x) \notin \mathscr{D}^+$$

which is in contradiction with the above.

In the second case, let us see that since $\{p_n\}_{n\in\mathbb{N}}$ is not eventually constant, it follows that for every $n_0 \in \mathbb{N}$, there exist $i, j \in \mathbb{N}$, $i, j > n_0$ such that $p_i \neq p_j$. Hence, there exists an $x > 0$ such that:

$$F_{p_i,p_j}(x) \leqslant \max\{F_{r,s}(x) : 0 \leqslant r, s \leqslant m, r \neq s, r, s \in \mathbb{N}\} < H_0(x).$$

Then $\{p_n\}_{n\in\mathbb{N}}$ is not a Cauchy sequence, thus contradicting again the hypothesis. By default, it follows that $(S, \mathscr{F}, \tau_T)$ is complete.

Let $f : \mathbb{N} = S \to S$ be the mapping defined by $f(n) = n + 1$ for all $n \in S$. This mapping has obviously no fixed point. To see that the mapping f is a contraction mapping, it is sufficient to note that, for $n > m$, n, $m \in \mathbb{N}$ and $x > 0$,

$$F_{m+1,n+1}(x) = \tau_T^{n-m}\left(G\left(\frac{j}{k^m}\right), \ldots, G\left(\frac{j}{k^n}\right)\right)(x) =$$

$$= \sup\left\{T^{n-m}\left(G\left(\frac{\beta_0 x}{k^m}\right), G\left(\frac{\beta_1 x}{k^{m+1}}\right), \ldots, G\left(\frac{\beta_{n-m} x}{k^n}\right)\right);\right.$$

$$\left.\sum_{i=0}^{n-m} \beta_i = 1,\ 0 \leqslant \beta_i \leqslant 1\right\} =$$

$$= \tau_T^{n-m}\left(G\left(\frac{j}{k^{m-1}}\right), \ldots, G\left(\frac{j}{k^{n-1}}\right)\right)\left(\frac{x}{k}\right) = F_{m,n}\left(\frac{x}{k}\right).$$

To prove the converse of this theorem, let us suppose that for every $G \in \mathscr{D}^+$ and for every $k \in (0, 1)$, $G^{\tau_T,k} \in \mathscr{D}^+$. Suppose also that f is a contraction mapping on the complete PM-space $(S, \mathscr{F}, \tau_T)$. For an arbitrary point $p_0 \in S$, let us now show that $O_f(p_0, 0, \infty)$ is bounded. Indeed, let $p_n = f^n(p_0)$ and for some $k \in (0, 1)$,

$$F_{p_0,p_n} \geqslant \tau_T^{n-2}(F_{p_0,p_1}, F_{p_1,p_2}, \ldots, F_{p_{n-1},p_n}) \geqslant$$

$$\geqslant \tau_T^{n-2}\left(F_{p_0,p_1}, F_{p_0,p_1}\left(\frac{j}{k}\right), \ldots, F_{p_0,p_1}\left(\frac{j}{k^{n-1}}\right)\right) \geqslant F^{\tau_T,k}_{p_0,f(p_0)}.$$

Then, for any m, $n \in \mathbb{N}$,

$$F_{p_n,p_{n+m}} \geqslant F_{p_0,p_m}\left(\frac{j}{k^n}\right) \geqslant F^{\tau_T,k}_{p_0,f(p_0)}.$$

Since $F^{\tau_T,k}_{p,f(p)} \in \mathscr{D}^+$ implies that $D_{O_f(p_0,0,\infty)} \in \mathscr{D}^+$, it follows from Theorem 3.1.3 that f has a fixed point. Thus, the proof is complete.

Remarks. (1) If τ is a triangle function such that $\mathscr{D}^+$ is closed under τ and if $G^{\tau,k}$ is in $\mathscr{D}^+$ for any $G \in \mathscr{D}^+$ and any $k \in (0, 1)$, then any contraction mapping on any complete PM-space $(S, \mathscr{F}, \tau)$ has a unique fixed point. For example, the triangle function $\tau = \tau_{\mathrm{Min}}$ gives

$$G^{\tau_{\mathrm{Min}},k} = G((1 - k)j) \in \mathscr{D}^+,$$

and thus Theorem 3.1.2 is a corollary of Theorem 3.1.4.

(2) For a large class of triangle functions, the assumptions of the above remark are also necessary. Indeed, one can show that there exists τ such that $\mathscr{D}^+$ is closed under τ and

$$(**) \quad \tau(F, G)(aj) = \tau(F(aj), G(aj))$$

or all $a \in \mathbb{R}^+$ and all F, $G \in \mathscr{D}^+$; however, if for some $a \in \mathbb{R}^+$ and some $G \in \mathscr{D}^+$ the function $G^{\tau,k} \notin \mathscr{D}^+$ then there exist a complete PM-space $(S, \mathscr{F}, \tau)$ and a contraction mapping f on $(S, \mathscr{F}, \tau)$ which has no fixed point [191].

The class of triangle functions having the property (**) includes the convolutions.

(3) Since any PM-space $(S, \mathscr{F}, \tau)$ such that τ is an l.c. t-function has a completion $(S^*, \mathscr{F}^*, \tau)$ which is unique up to isometry, it follows that if $(S, \mathscr{F}, T)$ is a Menger space such that T is an l.c. t-norm, and f is a contraction mapping on S, then there exist a completion $(S^*, \mathscr{F}^*, T)$ of $(S, \mathscr{F}, T)$ and a unique extension f^* of f such that f^* is a contraction on S, with the same constant k as f.

COROLLARY. *There exists a complete PM-space $(S, \mathscr{F}, \tau_T)$ with $T = T_m$ or $T =$ Prod and a contraction mapping on $(S, \mathscr{F}, \tau_T)$ having no fixed point.*

Proof. Let k be any number such that $k \in (0, 1)$ and let G be defined as

$$G(x) = \begin{cases} 0, & x \leqslant \dfrac{1}{k^2} \\[2ex] 1 - \dfrac{1}{n}, & \dfrac{1}{k^n} < x \leqslant \dfrac{1}{k^{n+1}}, \quad n > 1, \ n \in \mathbb{N}. \end{cases}$$

It can be easily verified that the conditions of Theorem 3.1.4 are satisfied.

THEOREM 3.1.5 [406]. *If T is any Archimedean t-norm and τ_T is the l.c. t-function corresponding to T, then there exist a complete PM-space $(S, \mathscr{F}, \tau_T)$ and a contraction mapping f on $(S, \mathscr{F}, \tau_T)$ with no fixed point.*

Proof. If one can define $K \in \mathscr{D}^+$ such that $K^{\tau_{T},k} \notin \mathscr{D}^+$, then the result follows from Theorem 3.1.4. Let us choose $G \in \mathscr{D}^+$ as the function defined in the proof of the above corollary and let $k \in (0, 1)$. Define K by $K(x) = h^{[-1]}(G(x))$ for every $x \in \mathbb{R}$, where h is the multiplicative generator for the Archimedean t-norm T. Note that $K \in \mathscr{D}^+$. Let x be any number such that $G(x) > h(0)$ and let $n \in \mathbb{N}$ such that $\frac{1}{k^n} < x \leqslant \frac{1}{k^{n+1}}$. Then, for $i \geqslant 1$, $G\left(\frac{x}{k^i}\right) > h(0)$,

$$K^{\tau_{T},k}(x) = \lim_{m\to\infty} \tau_T^m \left(K, K\left(\frac{j}{k}\right), \ldots, K\left(\frac{j}{k^m}\right)\right)(x) =$$

$$= \lim_{m\to\infty} \sup \left\{T^m \left(K(\beta_0 x), K\left(\frac{\beta_1 x}{k}\right), \ldots, K\left(\frac{\beta_m x}{k^m}\right)\right) :\right.$$

$$\left.\sum_{i=0}^{m} \beta_i = 1,\ 0 \leqslant \beta_i \leqslant 1\right\} \leqslant \lim_{m\to\infty} T^m \left(K(x), \ldots, K\left(\frac{x}{k^m}\right)\right) =$$

$$= \lim_{m\to\infty} h^{[-1]}\left[\prod_{i=0}^{m} hK\left(\frac{x}{k^i}\right)\right] = \lim_{m\to\infty} h^{[-1]}\left[\prod_{i=0}^{m} hh^{[-1]}G\left(\frac{x}{k^i}\right)\right] =$$

$$= h^{[-1]}\left[\lim_{m\to\infty} \prod_{i=0}^{m} G\left(\frac{x}{k^i}\right)\right] = h^{[-1]}\left[\lim_{n\to\infty} \prod_{i=0}^{m} \left(1 - \frac{1}{n+i}\right)\right] =$$

$$= h^{[-1]}\left[\lim_{m\to\infty} \prod_{i=0}^{m} \frac{n+i-1}{n+i}\right] = h^{[-1]}\left[\lim_{m\to\infty} \frac{n-1}{n+m}\right] =$$

$$= h^{[-1]}(0) = 0$$

and the proof is complete.

The above results show that the existence of fixed points for probabilistic contraction mappings on PM-spaces is strongly related to the triangle inequality. However, the t-norm $T = \text{Min}$ in Sehgal's theorem (Theorem 3.1.2) can be replaced by weaker t-norms. In this way, new (generalized or refined) classes of contraction mappings can be defined.

For example, Sehgal and Bharucha-Reid [396] introduced:

DEFINITION 3.1.4. Let ε and λ be positive reals. A mapping f of a PM-space into itself will be called an (ε, λ)-*local contraction* if there exists a constant $k \in (0, 1)$, such that, if $p \in S$ and $q \in U_p(\varepsilon, \lambda)$, then

$$F_{f(p),f(q)}(kx) \geqslant F_{p,q}(x) \text{ for } x > 0.$$

In probabilistic terms, this definition states that whenever the probability of the distance between p and q being less than

ε is greater than $1-\lambda$, then f acts as a contraction mapping for the pair p and q in the sense of Definition 3.1.2.

DEFINITION 3.1.5. Let $(S, \mathcal{F})$ be a PM-space and let ε, λ be positive reals. The space $(S, \mathcal{F})$ is called (ε, λ)-*chainable* if for each $p, q \in S$ there exists a finite sequence $p=p_0, p_1, \ldots, p_n=q$ with elements in E such that $p_{i+1} \in U(\varepsilon, \lambda)$, i.e. $F_{p_{i+1}, p_i}(\varepsilon) > 1-\lambda$ for $i=0,1, \ldots, n-1$.

It is known that a t-norm T of $\mathscr{H}$-type is different from $T = \text{Min}$ [173], [182], [191].

THEOREM 3.1.6. *Let $(S, \mathcal{F}, T)$ be a complete (ε, λ)-chainable Menger space, where T is a continuous t-norm such that the family $\{T^n(t)\}_{n \in \mathbb{N}}$ is equicontinuous at the point $t=1$, i.e., T is of type $\mathscr{H}$. If $f: S \to S$ is an (ε, λ)-contraction, then f has a unique fixed point p and for all $q \in S$, $\lim_{n \to \infty} f^n(q) = p$.*

Proof. Let us first prove that for each $p \in S$ and for any positive real x, there exists a positive integer $n(p, x)$ such that $F_{f^m(p), f^{m+1}(p)}(x) > 1-\lambda$ for all $m \geqslant n(p, x)$. The reason for this is similar to that presented in [406] and [191].

Let $p=p_0, p_1, \ldots, p_n=f(p)$ be a finite sequence such that $F_{p_{i+1}, p_i}(\varepsilon) > 1-\lambda$, $i=0,1, \ldots, n-1$. Since f is an (ε, λ)-local contraction, it follows that $F_{f(p_{i+1}), f(p_i)}(\varepsilon) \geqslant F_{p_{i+1}, p_i}\left(\frac{\varepsilon}{k}\right) > 1-\lambda$ and that the sequence of elements $f(p_0), f(p_1), \ldots, f(p_n)$ is an (ε, λ)-chain for $f(p)$ and $f^2(p)$. Hence, by induction, $f^r(p_0)$, $f^r(p_1), \ldots, f^r(p_n)$ is an (ε, λ)-chain for $f^r(p)$ and $f^{r+1}(p)$, whatever the positive integer r. Therefore, for $x>0$ and for every integer $r>0$,

$$F_{f^r(p_{i+1}), f^r(p_i)}(x) \geqslant F_{p_{i+1}, p_i}\left(\frac{x}{k^r}\right)$$

and

$$F_{f^r(p_0), f^r(p_n)}(x) \geqslant T\left(F_{p_0, p_1}\left(\frac{x}{2k^r}\right), F_{p_1, p_n}\left(\frac{x}{2k^r}\right)\right) \geqslant$$

$$\geqslant T\left(F_{p_0, p_1}\left(\frac{x}{2k^r}\right), T\left(F_{p_1, p_2}\left(\frac{x}{2^2k^r}\right), F_{p_2, p_n}\left(\frac{x}{2^2x^r}\right)\right)\right) \geqslant \ldots \geqslant$$

$$\geqslant T\left(F_{p_0, p_1}\left(\frac{x}{2k^r}\right), T\left(F_{p_1, p_2}\left(\frac{x}{2^2k^r}\right), T\left(F_{p_2, p_3}\left(\frac{x}{2^3k^r}\right), \ldots,\right.\right.\right.$$

$$\left.\left.\left.\ldots, T\left(F_{p_{n-2}, p_{n-1}}\left(\frac{x}{2^{n-1}k^r}\right), F_{p_{n-1}, p_n}\left(\frac{x}{2^{n-1}k^r}\right)\right)\right) \cdots\right).$$

Since n is a finite integer, then, for all $\delta > 0$, there exists an integer $m_i(x) > 0$ such that $F_{p_i, p_{i+1}}\left(\frac{x}{2^{i+1}k^r}\right) > \delta$ for each $r \geqslant m_i(x)$, $i = 0, 1, \ldots, n-1$. Let $n(p, x) = \max\{m_0(x), \ldots, \ldots, m_{n-1}(x)\}$. Then

$$F_{f^r(p), f^{r+1}(p)}(x) \geqslant T^{n-1}(\delta)$$

for all $r \geqslant n(p, x)$.

Since the family $\{T^n(t)\}_{n \in \mathbb{N}}$ is equicontinuous at the point $x = 1$, then, for every $\lambda \in (0, 1)$, there exists $\delta(\lambda)$ such that

$$t > \delta(\lambda) \Rightarrow T^s(t) > 1 - \lambda, \text{ for every } s \in \mathbb{N}.$$

It follows that for every $\delta > \delta(\lambda)$ and $r \geqslant n(p, x)$,

$$F_{f^r(p), f^{r+1}(p)}(x) \geqslant T^{n-1}(\delta) > 1 - \lambda.$$

Let $p \in S$ be arbitrary. From the above argument it springs that, for every $\varepsilon > 0$, there exists an integer $n(p, \varepsilon)$ such that

$$F_{f^n(p), f^{n+1}(p)}(\varepsilon) > 1 - \lambda \text{ for all } n \geqslant n(p, \varepsilon) = n_0.$$

Let $f^{n_0} p = q$ and hence $F_{f^m(q), f^{m+1}(q)} > 1 - \lambda$ for all integer $m \geqslant 0$. According to the definition of the (ε, λ)-local contraction,

$$F_{f^m(q), f^{m+1}(q)}(x) \geqslant F_{q, f(q)}\left(\frac{x}{k^m}\right), \quad m = 0, 1, \ldots, \ \forall\, x > 0.$$

Let us now prove that the sequence $q_n = f^n(q)$ is fundamental in S. Let ε, λ be positive reals and $m > n$; then

$$F_{q_n, q_m}(\varepsilon) \geqslant T(F_{q_n, q_{n+1}}(\varepsilon - k\varepsilon), F_{q_{n+1}, q_m}(k\varepsilon)) \geqslant$$

$$\geqslant T\left(F_{q, q_1}\left(\frac{\varepsilon - k\varepsilon}{k^n}\right), T(F_{q_{n+1}, q_{n+2}}(k\varepsilon - k^2\varepsilon), F_{q_{n+2}, q_m}(k^2\varepsilon))\right) \geqslant \ldots$$

$$\ldots \geqslant T\left(F_{q, q_1}\left(\frac{\varepsilon - \alpha\varepsilon}{k^n}\right), T\left(F_{q, q_1}\left(\frac{\varepsilon - k\varepsilon}{k^n}\right), \ldots\right.\right.$$

$$\left.\left.\ldots, T\left(F_{q, q_1}\left(\frac{\varepsilon - k\varepsilon}{k^n}\right), F_{q, q_1}\left(\frac{\varepsilon}{k^n}\right)\right)\right) \ldots\right) \geqslant$$

$$\geqslant T^{m-n-1}\left(\frac{\varepsilon}{k^n}\right) > 1 - \lambda$$

for all $n > n(\lambda)$ and $m \in \mathbb{N}$. Hence, $\{q_n\}_{n \in \mathbb{N}}$ is fundamental in S. Note that $f^n(p)$ is fundamental in S too.

If $f^n(p) \to p_* \in S$, for $n \to \infty$, it can be easily seen that $f^n(p) \to \to f(p)$ for $n \to \infty$. Indeed, for any neighborhood $U_{f(p^*)}(\delta, \mu)$ of $f(p_*)$, the assertion $f^n(p) \to p_*$ for $n \to \infty$ implies that there exists an integer $m \geqslant 0$ such that $F_{f^n(p), p_*}(\delta) > 1 - \mu$ and $F_{f^n(p), p_*}(\varepsilon) > 1 - \lambda$ for all $n \geqslant m$. Then

$$F_{f^{n+1}(p), f(p_*)}(\delta) \geqslant F_{f^n(p), p_*}\left(\frac{\delta}{k}\right) \geqslant F_{f^n(p), p_*}(\delta) > 1 - \mu, n \geqslant m.$$

Therefore, $f^n(p) \to f(p_*)$ and, since $(S, \mathscr{F}, T)$ is a Hausdorff space, one concludes that $f(p_*) = p_*$. The existence of the fixed point is thus proven.

To prove the uniqueness, let us first observe that for $f(p) = p$ and $f(q) = q$ there exists an (ε, λ)-chain for p and q (for example, $p_0 = p, p_1, \ldots, p_n = q$). We have already shown that, for any positive integer m, the functions $f^m(p), f^m(p_1), \ldots, f^m(q)$ form an (ε, λ)-chain for $f^m(p)$ and $f^m(q)$. Since $\sup_{m \in \mathbb{N}} F_{p,q}\left(\frac{x}{k^m}\right) = 1$, it follows that, for any $m \geqslant 1$ and any $x > 0$,

$$F_{p,q}(x) = F_{f^m(p), f^m(q)}(x) \geqslant F_{p,q}\left(\frac{x}{k^m}\right) \to H_0(x) \text{ when } m \to \infty,$$

thus implying that $p = q$. The proof of Theorem 3.1.6 is thus complete.

Let us note that Theorem 3.1.2 can be considered a consequence of Theorem 3.1.6, and so is also the following

COROLLARY [406, Th. 3 and Th. 7]. *Let $(S, \mathscr{F}, \mathrm{Min})$ be a complete Menger space. If f is any contraction mapping of S into itself, then there exists a unique $p \in S$ such that $f(p) = p$. Moreover, $\lim_{n \to \infty} f^n(q) = p$ for each $q \in S$.*

If $(S, \mathscr{F}, \mathrm{Min})$ is a complete (ε, λ)-chainable Menger space and if $f : S \to S$ is an (ε, λ)-contraction, then f has a unique fixed point $p \in S$ and $\lim_{n \to \infty} f^n(q) = p$ for each $q \in S$.

We shall show that the condition of continuity imposed upon T in Theorem 3.1.6 plays no role. Moreover, we shall prove that the class of t-norms of $\mathscr{H}$-type is essentially the largest class for which the contraction principle holds.

DEFINITION 3.1.6. We say that the t-norm T has the Fixed Point Property (shortly FPP) if every contraction mapping on a complete Menger space $(S, \mathscr{F}, T)$ has a fixed point (which is clearly unique).

Remark. A t-norm T has FPP if for every contraction mapping f on a Menger space $(S, \mathscr{F}, T)$ and for each fixed p_0 in S, the sequence $p_n = f^n(p_0)$ is $\mathscr{F}$-Cauchy. Moreover, it suffices to consider contraction mappings having the constant k such that $k \in \left(0, \frac{1}{2}\right]$. The converse holds for l.c. t-norms [356].

THEOREM 3.1.7 [356, Th. 1]. *Every t-norm $\mathscr{H}$-type has the fixed point property.*

Proof. Let $(S, \mathscr{F}, T)$ be a Menger space such that T is $\mathscr{H}$-type and consider a mapping $f : S \to S$ which verifies Definition 3.1.2 with $k \in \left(0, \frac{1}{2}\right]$. Let $p_0 \in S$ and $x \in (0, \infty)$ be fixed. If m is a positive integer, then

$$F_{p_0, f^{m+1}(p_0)}(2x) \geqslant T(F_{p_0, f(p_0)}(x),\ F_{f(p_0), f^{m+1}(p_0)}(x)) \geqslant$$

$$\geqslant T(F_{p_0, f(p_0)}(x),\ F_{p_0, f^{m}(p_0)}(2x)),$$

and, therefore,

$$F_{p_0, f^{m}(p_0)}(2x) \geqslant T^m(F_{p_0, f(p_0)}(x)),\ \forall\, m \geqslant 1.$$

Thus, for any positive integer n, m,

$$F_{f^n(p_0), f^{n+m}(p_0)}(2x) \geqslant T^m(F_{p_0, f(p_0)}(xk^{-n})).$$

Since T is $\mathscr{H}$-type and $F_{p_0, f(p_0)} \in \mathscr{D}^+$, then

$$\lim_{n \to \infty} F_{f^n(p_0), f^{n+m}(p_0)}(2x) = 1, \text{ uniformly in } m, \text{ for any } x \in (0, \infty).$$

By definition, this means that $\{f^k(p_0)\}_{k \in \mathbb{N}}$ is $\mathscr{F}$-Cauchy. The proof of Theorem 3.1.7 follows from the above remark.

A partial converse of this theorem is given by

THEOREM 3.1.8 [356, Th. 2]. *If T is an l.c. t-norm with fixed point property, then T is $\mathscr{H}$-type.*

Proof. Suppose that T is not $\mathscr{H}$-type. Then there exists an $a \in (0,1)$ such that, for every $b > a$, there exists an $m \geqslant 1$ with $T^{m_b}(b) < a$. Let now $b_n \in (0, 1)$, and increasing to 1. Then there exists an $m_n \geqslant 1$ such that

$$T^{m_n}(b_n) < a, \quad n = 1, 2, \ldots .$$

Obviously, we can suppose that m_n is also increasing. Consider now $G \in \mathscr{D}^+$ defined as

$$G(x) = \begin{cases} 0 & \text{if} \quad x \leqslant 1, \\ b_1 & \text{if} \quad x(1, 2^{2+m_1}], \\ b_{n+1} & \text{if} \quad x \in (2^{2n+m_n}, 2^{2n+2+m_{n+1}}], \quad n \geqslant 1. \end{cases}$$

For the Menger space $S = \{1,2,3,\ldots\}$ and for the probabilistic metric on S, defined as

$$\begin{cases} F_{n,n+m}(x) = T^m(G(2^n x), G(2^{n+1}x), \ldots, G(2^{n+m}x)) \\ F_{n,n} = 0, \end{cases}$$

it follows that $(S, \mathscr{F}, T)$ is a Menger space and that the mapping $f : S \to S$, $n \xrightarrow{f} n + 1$ is a contraction with the constant $k = \frac{1}{2}\cdot$

Also, we have successively

$$F_{n,n+m_n}(1) \leqslant F_{n,n+m_n}(2^n) = T^{m_n}(G(2^{2n}), G(2^{2n+1}), \ldots, G(2^{2n+m_n})) \leqslant$$
$$\leqslant T^{m_n}(b_n) < a.$$

Therefore, the sequence $\{f^n(1)\}_{n \in \mathbb{N}}$ is not $\mathscr{F}$-Cauchy. From the afore-mentioned remark it follows that T does not have the fixed point property. The proof is complete.

Considering the above results together with the characterization of the t-norms of $\mathscr{H}$-type, we get :

THEOREM 3.1.9 [356, Th. 3]. *Let T be a continuous t-norm. Then the following statements are equivalent:*

(i) *T has the fixed point property;*

(ii) *T is $\mathscr{H}$-type;*

(iii) *for each $a \in (0, 1)$, there exists a $b \geqslant a$ such that $T(b, b) = b < 1$.*

COROLLARY. *If T is an Archimedean t-norm, then T does not have the fixed point property.*

Proof. By definition, T is continuous and has no idempotents. On the other hand, we shall investigate further the fixed point property of t-norms, by use of the structure of continuous t-norms as given in [353, Lemma 1, Th. 1, Lemma 2]. Essentially, we shall prove that a t-norm does not have the fixed point property if, in a neighbourhood of 1, its behaviour is similar to that of T_m. Thus, the counter-example of H. Sherwood [406] can be generally used for all t-norms which are not of $\mathscr{H}$-type. To this purpose, let us consider

THEOREM 3.1.10. *Let T be an arbitrary continuous but fixed t-norm such that T is not $\mathscr{H}$-type. Then the following are equivalent:*

(i) *T does not have the fixed point property;*
(ii) *$T =$ Prod does not have the fixed point property;*
(iii) *$T = T_m =$ Max(Sum-1,0) does not have the fixed point property.*

Proof. Observe that the construction of new equivalent Menger spaces given in Proposition 1.1.2 does not change, the property of f being a probabilistic contraction. Therefore, the equivalence (ii) $\Leftrightarrow$ (iii) results from Proposition 1.1.2 (i), for Prod $\geqslant T_m$.

The implications (i) $\Rightarrow$ (ii) and (i) $\Rightarrow$ (iii) are corollaries of Proposition 1.1.2 (iii). The implications (ii) $\Rightarrow$ (i) and (iii) $\Rightarrow$ (i) are corollaries of Proposition 1.1.2 (ii). The proof is complete.

Remark. From the above theorem and the corollary of Theorem 3.1.4, we conclude that if T is a continuous t-norm which is not $\mathscr{H}$-type, then it does not have the fixed point property.

The above results can be extended to contraction maps on $G.PM$-spaces given by

DEFINITION 3.1.7. A complete *generalized PM-space* (briefly, $G.PM$-space) is an ordered pair $(S, \mathscr{F})$, where S is an abstract set and $\mathscr{F} : S\times S \to \Delta$ with $\mathscr{F}(p, q) = F_{p,q}$. The functions $F_{p,q}$ are assumed to satisfy the following conditions:

(I) $F_{p,q}(x) = 1$ for all $x > 0$ if and only if $p = q$,
(II) $F_{p,q}(0) = 0$ for all $p, q \in S$,
(III) $F_{p,q} = F_{q,p}$ for all $p, q \in S$,
(IV) if $F_{p,q}(x) = 1$ and $F_{q,r}(y) = 1$, then $F_{p,r}(x + y) = 1$,
(V) every $\mathscr{F}$-Cauchy sequence in $(S, \mathscr{F})$ is $\mathscr{F}$-convergent, i.e. $F_{p_n,p_m} \to H_0$ as $n, m \to \infty$ implies the existence of a $p \in S$ with $F_{p,p_n} \to H_0$.

If for some t-norm T the inequality

(VI) $F_{p,r}(x + y) \geqslant T(F_{p,q}(x), F_{q,r}(y))$

holds for all triplets p, q and r in S and all $x,y \in R^+$, then $(S, \mathscr{F}, T)$ is a generalized Menger space.

It is known [130] that the relation ρ:

$$p\rho q \Leftrightarrow \lim_{x\to\infty} F_{p,q}(x) = 1 \quad (\text{i.e. } p,q \in \mathscr{D}^+)$$

is a relation of equivalence on S. Let us denote by $\tilde{p}$ the equivalence class determined in S by p and ρ. It can be also shown [130] that $\tilde{p}$ is a closed subset in the (ε, λ)-topology. If T is a continuous t-norm $(S/\rho, \mathscr{F}_{S/\rho}, T)$ was shown to be a Menger space in which the distance between distinct equivalence classes is perhaps infinite (i.e., $\lim_{x\to\infty} F_{\tilde{p},\tilde{q}}(x) < 1$), where $S/\rho = \{\bar{p} = \rho(p), p \in S\}$.

It is easy to verify that every generalized complete metric space, as defined by Luxemburg [286], may be regarded as a generalized complete Menger space with $T = \text{Min}$ and

$$F_{p,q}(x) = \begin{cases} H_0(x - d(p, q)) & \text{for } x \in R, \text{ if } d(p, q) < \infty \\ a \in [0, 1) & \text{for } x > 0, \text{ if } d(p, q) = \infty \end{cases}$$

for every $p, q \in S$.

Let $(S^i, \mathscr{F}^i)_{i\in I}$ be a family of disjoint PM-spaces. Then there exists a natural way of obtaining a $G.PM$-space $(S, \mathscr{F})$, taking $(S, \mathscr{F}) = \bigcup_{i\in I} (S^i, \mathscr{F}^i)$ and defining, for every $p, q \in S$,

$$F_{p,q}(x) = \begin{cases} F_{p,q}(x) & \text{if } p, q \in S^i \text{ for some } i \in I,\ x \in \mathbb{R}, \\ a_{ij} \in [0, 1) & \text{for } x > 0, \text{ if } p \in S^i,\ q \in S^j \text{ for some } \\ & i, j \in I,\ i \neq j. \end{cases}$$

It is easy to verify that $(S, \mathscr{F})$ is a $G.PM$-space and if $(S^i, \mathscr{F}^i)$ is complete for each $i \in I$, then $(S, \mathscr{F})$ is a complete $G.PM$-space.

Let us now prove that the above procedure is the only way to obtain $G.PM$-spaces or complete $G.PM$-spaces. Indeed, for $(S, \mathscr{F})$ a $G.PM$-space and for relation ρ defined as before, S can be uniquely decomposed into equivalence classes S^i, $i \in I$. This decomposition of S will be called *canonical decomposition*. A result similar to that for ordinary generalized metric spaces defined by Luxemburg [286] can be obtained.

PROPOSITION 3.1.1. *Let $(S, \mathscr{F})$ be a $G.PM$-space, let $S = \bigcup_{i\in I} S^i$ be the canonical decomposition and let*

$$\mathscr{F}^i = \mathscr{F}_{/S^i\times S^i},$$

for each $i \in I$. Then:

(i) *for each $i \in I$, $(S^i, \mathscr{F}^i)$ is a PM-space,*

(ii) *for any $i, j \in I$, with $i \neq j$, $\lim_{n\to\infty} F_{p,q}(x) = a_{ij} \in [0, 1)$ for any $p \in S^i$ and $q \in S^j$,*

(iii) *$(S, \mathscr{F})$ is a complete $G.PM$-space if and only if $(S^i, \mathscr{F}^i)$ is a complete PM-space for each $i \in I$.*

Proof. Considering the developments in [130], (i) and (ii) become obvious. Therefore let us now prove only (iii). Suppose that $(S, \mathscr{F})$ is a complete $G.PM$-space. For any $i \in I$, let $\{p_n\}_{n\in N}$ be a $\mathscr{F}^i$-Cauchy sequence in S^i. Then $\{p_n\}_{n\in\mathbb{N}}$ is a $\mathscr{F}$-Cauchy sequence in S and hence there exists a $p \in S$ such that $F_{p,p_n}(x) \to \to H_0(x)$, as $n \to \infty$. Since, for $x > 0$ and every $\varepsilon > 0$, there exists an integer $n(\varepsilon, x)$ such that $F_{p,p_n}(x) > 1 - \varepsilon$ whenever $n \geqslant n(\varepsilon, x)$, it follows that $F_{p,p_n}(y) \geqslant F_{p,p_n}(x) > 1 - \varepsilon$ for all p_n with $n \geqslant n(\varepsilon, x)$ and $y > x$. Thus, there exists an $i_0 \in I$ such that $p \in S^{i_0}$ and the sequence $\{p_n\}_{n \in \mathbb{N}}$ $\mathscr{F}^{i_0}$ converges to p.

Conversely, if $(S^i, \mathscr{F}^i)$ is a complete PM-space for each $i \in I$ and $\{p_n\}_{n \in \mathbb{N}}$ is a $\mathscr{F}$-Cauchy in S, then it follows that $F_{p_n,p_m}(x) \to H_0(x)$ as $n, m \to \infty$. Hence, for every $\varepsilon > 0$, there exists a positive integer $n(\varepsilon, x)$ such that $F_{p_m,p_n}(x) > 1 - \varepsilon$ for $n, m \geqslant n(\varepsilon, x)$. This implies that there exists an $i_0 \in I$ such that $p_n \in S^{i_0}$ for $n \geqslant n(\varepsilon, x)$ and the sequence $\{p_n\}_{n \geqslant n(\varepsilon,x)}$ is $\mathscr{F}^{i_0}$-Cauchy in S^{i_0}. Thus, $\{p_n\}_{n \geqslant n(\varepsilon,x)}$ is $\mathscr{F}^{i_0}$-convergent to $p \in \mathscr{F}^{i_0}$. Since $\{p_n\}_{n \in \mathbb{N}}$ is clearly $\mathscr{F}$-convergent to p in S, the proof is complete.

Therefore, the natural way considered above is the only way to obtain a $G.PM$-space.

We give without proof the following proposition which can be easily verified :

PROPOSITION 3.1.2. *Let $(S, \mathscr{F})$ be a $G.PM$-space, let*

$$S = \bigcup_{i \in I} S^i$$

be its canonical decomposition and let $f : S \to S$ be a mapping such that $\lim_{x \to \infty} F_{f(p),f(q)}(x) = 1$ whenever $p, q \in S$ and $\lim_{x \to \infty} F_{p,q}(x) = 1$. Then f has a fixed point if and only if

$$f_i = f|_{S^i} : S^i \to S^i$$

has a fixed point for some $i \in I$.

PROPOSITION 3.1.3. *Let $(S, \mathscr{F})$ be a complete $G.PM$-space and let $S = \bigcup_{i \in I} S^i$ be its canonical decomposition. Let also $f : S \to S$ be a contraction mapping, i.e., there exists a constant $k \in (0,1)$ such that, for each $p, q \in S$, $F_{f(p),f(q)}(kx) \geqslant F_{p,q}(x)$ for all $x > 0$. If there exists a $p_* \in S$ such that $\lim_{x \to \infty} F_{p_*,f(p_*)}(x) = 1$ then, for some $i \in I$, the restriction*

$$f_i = f|_{S^i} : S^i \to S^i$$

is a contraction mapping.

Proof. Since $\lim_{x \to \infty} F_{p_*,f(p_*)}(x) = 1$, it follows that p_* and $f(p_*)$ belong to the same S^{i_0} for some $i_0 \in I$. Since f is a contraction mapping, we have that $f(S^{i_0}) \subseteq S^{i_0}$ and the restriction $f_{i_0} = f_{S^{i_0}}$ is a contraction mapping from S^{i_0} into S^{i_0}. The proof is complete.

THEOREM 3.1.11 [217]. *Let $(S, \mathscr{F}, T)$ be a complete generalized Menger space, where T is a continuous t-norm such that the family $\{T^n(t)\}_{n\in N}$ is equicontinuous at the point $t = 1$. For $f: S \to S$ an arbitrary contraction mapping and for p a point of S, let us consider the "sequence of successive approximations with initial element p": $p, f(p), f^2(p), \ldots, f^n(p), \ldots = O_f(p, 0, \infty)$. The following alternatives hold:*

A) *for every $n \in \mathbb{N} \cup \{0\}$, one has*

$$F_{f^n(p), f^{n+1}(p)}(x) = a \in [0, 1) \text{ for } x > 0$$

or

B) *the sequence of successive approximations $O_f(p, 0, \infty)$ is $\mathscr{F}$-convergent to a fixed point of f.*

Proof. Consider the sequence of distribution functions

$$F_{p,f(p)}(x), F_{f(p),f^2(p)}(x), \ldots, F_{f^n(p),f^{n+1}(p)}(x), \ldots,$$

which is the sequence of distances between consecutive neighbours of the sequence of successive approximations with initial element p. There exist two mutually exclusive possibilities:

(a) for every $n \in \mathbb{N} \cup \{0\}$, one has

$$\lim_{x\to\infty} F_{f^n(p),f^{n+1}(p)}(x) = a \in [0, 1)$$

or

(b) for some $n \in \mathbb{N} \cup \{0\}$, one has

$$\lim_{x\to\infty} F_{f^n(p),f^{n+1}(p)}(x) = 1.$$

Case (a) is precisely alternative (A). Hence, to complete the proof, it only remains to show that (b) implies alternative (B).

Let $m = n(p)$ be the smallest integer for which

$$\lim_{x\to\infty} F_{f^m(p),f^{m+1}(p)}(x) = 1.$$

Then, by the definition of contraction mappings, there exists a $k \in (0,1)$ such that

$$F_{f^{m+1}(p),f^{m+2}(p)}(kx) \geqslant F_{f^m(p),f^{m+1}(p)}(x) \text{ for all } x > 0.$$

Therefore,

$$\lim_{x\to\infty} F_{f^{m+1}(p),f^{m+2}(p)}(x) = 1.$$

The mathematical induction process yields

$$F_{f^{m+r}(p),f^{m+r+1}(p)}(x) \geqslant F_{f^m(p),f^{m+1}(p)}(x/k^r) \text{ for all } x > 0,$$

i.e.,

$$\lim_{x\to\infty} F_{f^{m+r}(p),f^{m+r+1}(p)}(x) = 1$$

for every $r = 0, 1, 2, \ldots$.

As in Theorem 3.1.6, let us prove that the sequence of successive approximations $O_f(p, 0, \infty)$ is an $\mathscr{F}$-Cauchy sequence. Since $(S, \mathscr{F}, T)$ is a complete generalized Menger space, there exists a p_* S such that

$$F_{f^n(p),p_*}(x) \to H_0(x) \text{ as } n \to \infty.$$

We can also prove that $f^n(p) \to f(p_*)$ since, for $F_{f^n(p),p_*}(x) \to$ $\to H_0(x)$ as $n \to \infty$, one has

$$F_{f(f^n(p)),f(p_*)}(\varepsilon) \geqslant F_{f^n(p),p_*}(\varepsilon/k) \geqslant F_{f^n(p),p_*}(\varepsilon) > 1 - \lambda$$

for every $\varepsilon > 0$, $\lambda > 0$ and $n > n(\varepsilon, \lambda)$. As T is a continuous t-norm, then $(S, \mathscr{F}, T)$ is a Hausdorff space in the (ε, λ)-topology. We conclude that $f(p_*) = p_*$, thus proving the existence in S of a fixed point of f.

Remarks. (1) Another proof of Theorem 3.1.11 comes as a corollary to Proposition 3.1.3 since if (A) does not hold, then, for some $m \in \mathbb{N}$, $\lim_{x\to\infty} F_{f^m(p),f^{m+1}(p)}(x) = 1$. Letting $q = f^m(p)$, it follows from Proposition 3.1.3 that

$$f_i = f\,|_{S^i} : S^i \to S^i$$

is a contraction mapping, where S^i is the complete PM-space which contains q. According to Theorem 3.1.6, the sequence $\{f^r(p)\}_{r\geqslant m}$ is $\mathscr{F}^i$-convergent to a fixed point p_* of f_i. This implies that (B) holds.

(2) For the uniqueness of the fixed point, we give two additional conditions:

(c_1) for every sequence of successive approximations with initial element $p \in S$, $\{f^n(p)\}_{n\in\mathbb{N}\cup\{0\}}$, there exists an index $m = n(p)$ such that $\lim_{x\to\infty} F_{f^m(p),f^{m+r}(p)}(x) = 1$ for all $r \in \mathbb{N}$.

(c_2) if p and q are two fixed points of f, then $\lim_{x\to\infty} F_{p,q}(x) = 1$.

Indeed, alternative (A) is excluded by hypothesis (c_1). Condition (c_2) implies that f can have at most one fixed point because if $p \neq q$ and $f(p) = p$, $f(q) = q$, then by (I) there exist an $x > 0$ and an $a \in [0, 1)$ such that $F_{p,q}(x) = a$. However, for each $n \in \mathbb{N}$, we have

$$a = F_{p,q}(x) = F_{f^n(p), f^n(q)} \geq F_{f^n(p), f^n(q)}(x/k^n) \to 1 \text{ as } n \to \infty$$

which implies that $a = 1$. We obtain a contradiction with the selection of a.

Therefore, if (c_1) and (c_2) are not satisfied, then every two of its fixed points must be at an infinite distance.

3.1.2. *Relation between Fixed Point Theorems on Uniform Spaces, Metric Spaces and PM-Spaces*

It is known [393], [396], [84], [191], [210], [357] that connections between *PM*-spaces, uniform spaces and metric spaces can be established using the probabilistic metric $\mathscr{F}$. Thus, some results regarding the existence and uniqueness of fixed points for contractions on *PM*-spaces can be transposed into metric spaces and conversely.

The connection between metric spaces and Menger spaces is given by

PROPOSITION 3.1.4 [396, Th. 2]. *If (S, d) is a metric space, then the metric d induces a mapping $\mathscr{F} : S \times S \to \mathscr{D}^+$, where $\mathscr{F}(p, q)$ is defined by*

$$\mathscr{F}(p, q)(x) = H_0(x - d(p, q)), \; x \in R, \; p, q \in S.$$

Further, if $T = $ Min, *then $(S, \mathscr{F}, T)$ is a Menger space. It is complete if the metric d is complete.*

The space $(S, \mathscr{F}, T)$ thus obtained will be called the *induced Menger space.*

We shall now prove the well-known Banach contraction mapping theorem using a probabilistic metric.

THEOREM 3.1.12 [396, Th. 4]. *Let $(S, \mathscr{F}, d)$ be a complete metric space and let $f : S \to S$ satisfy the following condition: there exists a constant $k \in (0, 1)$ such that $d(f(p), f(q)) \leq kd(p, q)$ for all $p, q \in S$. Then f has a unique fixed point $p_* \in S$ and $f^n(q) \to p_*$ for each $q \in S$, as $n \to \infty$.*

Proof. If $\mathscr{F} : S \times S \to \mathscr{D}^+$ is the mapping induced by the metric d, then from Proposition 3.1.4 it follows that $(S, \mathscr{F}, T)$ is a complete Menger space, where $T = $ Min. Since, for each $x > 0$,

$$F_{f(p), f(q)}(kx) = H_0(kx - d(f(p), f(q))) \geq H_0(kx - kd(p, q)) =$$
$$= H_0(x - d(p, q)) = F_{p,q}(x),$$

and hence f is a contraction of S into itself. The conclusion follows from Theorem 3.1.2.

It is known [128], [396] that a metric space (S, d) is called ε-chainable for a fixed positive number ε, if, for every $p, q \in S$, there exists an ε-chain $p = p_0, p_1, \ldots, p_n = q \in S$ joining p and q, i.e., $d(p_i, p_{i+1}) < \varepsilon$, $i = 0, 1, \ldots, n-1$.

PROPOSITION 3.1.5 [396, Th. 6]. *If (S, d) is an ε-chainable metric space, then the induced Menger space $(S, \mathscr{F}, \mathrm{Min})$ is (ε, λ)-chainable in the sense of Definition* 3.1.5.

Proof. Let $p, q \in S$ and let $p = p_0, p_1, \ldots, p_n = q$ be the ε-chain joining p and q. Then $d(p_i, p_{i+1}) < \varepsilon$, $i = 0, 1, \ldots, n-1$, implies that $F_{p_i, p_{i+1}}(\varepsilon) = H_0(\varepsilon - d(p_i, p_{i+1})) = 1 > 1 - \lambda$ for all $\lambda > 0$. The proof is complete.

COROLLARY [128, 396]. *Let (S, d) be a complete ε-chainable metric space and let $f: S \to S$ satisfy the condition that $d(p, q) < \varepsilon$ implies $d(f(p), f(q)) \leqslant kd(p, q)$ for some $k \in [0, 1)$ and for all $p, q \in S$. Then f has a unique fixed point $p_* \in S$ and $f^n(p) \to p_*$ for each $p \in S$ as $n \to \infty$.*

Proof. Let $(S, \mathscr{F}, \mathrm{Min})$ be the induced Menger space. Since $(S, \mathscr{F}, \mathrm{Min})$ is an (ε, λ)-chainable space for each $\lambda > 0$, let us choose $\lambda < 1$ and then conclude that f is an (ε, λ)-contraction. Indeed, for $q \in U_p(\varepsilon, \lambda)$, one has $H_0(\varepsilon - d(p, q)) > 1 - \lambda > 0$, and, therefore, $d(p, q) < \varepsilon$, $d(f(p), f(q)) \leqslant kd(p, q)$. Hence, for $x > 0$,

$$F_{f(p), f(q)}(kx) = H_0(kx - d(f(p), f(q))) \geqslant H_0(kx - kd(p, q)) = F_{p,q}(x)$$

and the rest follows from Theorem 3.1.6.

Some fixed point theorems in probabilistic structures are obtained from similar results on clasical uniform structures generated by a collection of pseudometric [84], [178], [194] or metric structures [210], [352], [353], [354], [356].

In what follows, we shall give some results on this converse problem.

PROPOSITION 3.1.6 [84]. *Let $(S, \mathscr{F}, \mathrm{Min})$ be a Menger space and let $d_\alpha: S \times S \to R$ be the family of functions defined by*

$$d_\alpha(p, q) = \sup\{t : F_{p,q}(t) \leqslant 1 - \alpha\}, \quad \alpha \in (0, 1).$$

Then the family $\{d_\alpha\}_{\alpha \in (0,1)}$ is a family of pseudometrics associated with the probabilistic metric $\mathscr{F}$ which satisfies the following properties:

(i) *$d_\alpha(p, q)$ is finite and $d_\alpha(p, q) = 0$ for all $\alpha \in (0, 1)$ if and only if $p = q$;*

(ii) *for each* $(p, q) \in S \times S$, $d_\alpha(p, q)$ *is a left-continuous nonincreasing function of* α *such that* $F_{p,q}(d_\alpha(p, q)) \leqslant 1 - \alpha$;

(iii) *the topology on* S *generated by* $\{d_\alpha\}_{\alpha \in (0,1)}$ *is the same as the topology induced by* $\mathscr{F}$.

Proof. (i) It is easy to see that $d_\alpha(p, q)$ is finite, since $F_{p,q} \in \mathscr{D}^+$. Then $F_{p,q}(0) = 0$ and $F_{p,q}$ nondecreasing implies that $d_\alpha(p, q) \geqslant 0$. $F_{p,q} = F_{q,p}$ implies that $d_\alpha(p, q) = d_\alpha(q, p)$ and $F_{p,q}(t) = 1$ for all $t > 0$ if and only if $p = q$ implies that $d_\alpha(p, p) = 0$ and $d_\alpha(p, q) = 0$ if and only if $p = q$.

To prove the triangle inequality, let us suppose that

$$d_\alpha(p, q) + d_\alpha(q, r) < d_\alpha(p, r)$$

for some $p,q,r \in S$. One can choose $t > d_\alpha(p, q)$ and $s > d_\alpha(q, r)$ so that

$$t + s < d_\alpha(p, r).$$

Therefore,

$$F_{p,q}(t) = 1 - \alpha_1 > 1 - \alpha$$

and

$$F_{p,q}(s) = 1 - \alpha_2 > 1 - \alpha.$$

If we choose $\bar{\alpha} = \max\{\alpha_1, \alpha_2\}$, then we have

$$1 - \bar{\alpha} \leqslant \operatorname{Min}(1 - \bar{\alpha}, 1 - \bar{\alpha}) \leqslant \operatorname{Min}(F_{p,q}(t), F_{q,r}(s)) \leqslant$$
$$\leqslant \operatorname{Min} F_{p,r}(t + s) \leqslant 1 - \alpha,$$

which is a contradiction. The proof of the triangle inequality is complete.

(ii) Since $F_{p,q}$ is nondecreasing, d_α is clearly nonincreasing. By the same reason and from the left continuity of $F_{p,q}$, it follows that $F_{p,q}(d_\alpha(p, q)) \leqslant 1 - \alpha$.

To see that d_α is left-continuous, we consider a nondecreasing sequence $\{\alpha_n\}$, $\alpha_n \to \alpha$ as $n \to \infty$, for $\alpha \in (0, 1)$ fixed. It follows that $\{d_{\alpha_n}\}_{n \in \mathbb{N}}$ is a nonincreasing sequence bounded from below by zero and, hence, it converges to a number $\beta \in [0, 1)$, $\beta \geqslant d_\alpha$ since $d_{\alpha_n} \geqslant d_\alpha$. Moreover, $\beta = d_\alpha$ since, for $\beta > d_\alpha$, $F_{p,q}(\beta) > 1 - \alpha$ and choosing n large enough (i.e. such that $F_{p,q}(\beta) > 1 - \alpha_n \geqslant 1 - \alpha$), one gets a contradiction to the condition $\beta \leqslant d_{\alpha_n}$ and $F_{p,q}(\beta) \leqslant F_{p,q}(d_{\alpha_n}) \leqslant 1 - \alpha_n$.

(iii) From the definition of d_α and the fact that $F_{p,q}(d_\alpha(p, q)) \leqslant \leqslant 1 - \alpha$, we conclude that

$$F_{p,q}(\varepsilon) > 1 - \alpha \text{ if and only if } d_\alpha(p,q) < \varepsilon.$$

From this relation, it follows that the (ε, α)-neighbourhood $U_p(\varepsilon, \alpha)$ of $p \in (S, \mathscr{F}, \mathrm{Min})$ in the $\mathscr{F}$-topology is a d_α spherical neighbourhood of p of radius ε, which implies (iii). The proof of the proposition is complete.

Remark. Since, for fixed $p, q \in S$, $d_\alpha(p, q)$ is a nonincreasing function of α, it follows that, for any $\delta > 0$, the family $\{d_\alpha\}_{\alpha \in (0, \delta)}$ generates the same topology as that generated by the entire family $\{d_\alpha\}_{\alpha \in (0, 1)}$.

For a given set S and a family of pseudometrics on S, $\{d_\alpha\}_{\alpha \in (0, 1)}$ with property (ii) of the above proposition, it is natural to ask whether there exists a probabilistic metric $\mathscr{F}$ on S such that $\{d_\alpha\}_{\alpha \in (0, 1)}$ is the family of pseudometrics associated with $\mathscr{F}$. An affirmative answer was given by Cain and Kasriel [84]:

THEOREM 3.1.13. *Suppose that S is a Hausdorff space with a topology generated by a family of pseudometrics $\{d_\alpha\}_{\alpha \in (0, 1)}$ such that, for $p, q \in S$, $d_\alpha(p, q)$ is a nonincreasing left-continuous function of α with $d_\alpha(p, q) = 0$ for all $\alpha \in (0, 1)$ if and only if $p = q$. Then there exists a probabilistic metric $\mathscr{F}$ on S such that $\{d_\alpha\}_{\alpha \in (0, 1)}$ is the family of pseudometrics associated with it.*

Proof. Let p, q be two fixed points in S and define $\mathscr{F}(p, q)$ as follows:

$$F_{p,q}(t) =$$

$$= \begin{cases} 1 - \sup\{\alpha \in (0, 1) : d_\alpha(p, q) \geq t\}, & \text{if } \{\alpha \in (0,1) : d_\alpha(p, q) \geq t\} \neq \emptyset, \\ 1, & \text{if } d_\alpha(p, q) < t \text{ for each } \alpha \in (0, 1). \end{cases}$$

Since $F_{p,q}(t) = 0$ for $t \leqslant 0$, $\inf_{t \in \mathbb{R}} F_{p,q}(t) = 0$, $\sup_{t \in \mathbb{R}} F_{p,q}(t) = 1$ and $F_{p,q}$ is nondecreasing, it remains to prove only the left continuity of $F_{p,q}$. This property follows from the left continuity of the function

$$f_{p,q}(t) =$$

$$= \begin{cases} \sup\{\alpha \in (0, 1) : d_\alpha(p, q) \geqslant t\}, & \text{if } \{\alpha \in (0, 1), d_\alpha(p, q) \geqslant t\} \neq \emptyset, \\ 0, & \text{if } d_\alpha(p, q) < t \text{ for each } \alpha \in (0, 1). \end{cases}$$

Indeed, $f_{p,q}(t) = 1$ for $t \leqslant 0$ and $f_{p,q}(t) = 0$ for $t > \sup d_\alpha(p, q)$. For other values of t, we have that for every nondecreasing sequence $t_n \to t$, as $n \to \infty$, the sequence $\{f_{p,q}(t_n)\}_{n \in \mathbb{N}}$ is nonincreasing and is bounded from below by $f_{p,q}(t)$, i.e., $f_{p,q}(t_n) \geqslant f_{p,q}(t)$. It follows that $\lim_{n \to \infty} f_{p,q}(t_n) \geqslant f_{p,q}(t)$. If $\lim_{n \to \infty} f_{p,q}(t_n) > f_{p,q}(t)$, then $\sup\{\alpha \in (0, 1) : d_\alpha(p, q) \geqslant t_n\} \geqslant \lim_{n \to \infty} f_{p,q}(t_n) = \beta \geqslant \sup\{\alpha \in$

$\in (0,1) : d_\alpha(p, q) \geqslant t\}$. Hence, $d_\beta < t$, but $d_\beta = \sup\{t' : F_{p,q}(t') \leqslant \leqslant 1 - \beta\}$ implies that $F_{p,q}(t) > 1 - \beta$. On the other hand, from $\sup\{\alpha \in (0,1) : d_\alpha(p, q) \geqslant t_n\} \geqslant \beta$, it follows that $d_\beta \geqslant t_n$, i.e., $\sup\{t'' : F_{p,q}(t'') \leqslant 1 - \beta\} \geqslant t_n$ and, hence, $F_{p,q}(t_n) \leqslant 1 - \beta$, which implies that $\lim_{n\to\infty} F_{p,q}(t_n) \leqslant 1 - \beta$. This contradiction implies that $\lim_{n\to\infty} f_{p,q}(t_n) = f_{p,q}(t)$.

That $\mathscr{F}$ is a probabilistic metric follows from the observations below:

(i) $F_{p,q}(t) = 1$ if and only if $t > d_\alpha(p, q)$ for every $\alpha \in (0,1)$.
(ii) $F_{p,q}(t) = 0$ if and only if $t \leqslant d_\alpha(p, q)$ for every $\alpha \in (0,1)$.
(iii) $F_{p,q}(t) > 1 - \alpha$ if and only if $d_\alpha(p, q) < t$.

One must show that the family of pseudometrics associated with the probabilistic metric $\mathscr{F}$ is precisely the family $\{d_\alpha\}_{\alpha\in(0,1)}$, i.e.,

$$d_\alpha(p, q) = \sup\{t : F_{p,q}(t) \leqslant 1 - \alpha\}.$$

Let us first show that $d_\alpha(p, q)$ is both a member of and an upper bound for the set $\{t : F_{p,q}(t) \leqslant 1 - \alpha\}$.

Indeed, one has

$$\alpha \in \{\beta \in (0,1) : d_\beta(p, q) \geqslant d_\alpha(p, q)\},$$

which implies that

$$\alpha \leqslant \sup\{\beta \in (0,1) : d_\beta(p, q) \geqslant d_\alpha(p, q)\}$$

and, consequently,

$$\alpha \leqslant f_{p,q}(d_\alpha(p, q)) = 1 - F_{p,q}(d_\alpha(p, q))$$

or

$$F_{p,q}(d_\alpha(p,q)) \leqslant 1 - \alpha.$$

Thus,

$$d_\alpha(p, q) \in \{t : F_{p,q}(t) \leqslant 1 - \alpha\}.$$

To see that $d_\alpha(p, q)$ is an upper bound for $\{t : F_{p,q}(t) \leqslant 1 - \alpha\}$, we observe that (iii) implies that $d_\alpha(p, q) \geqslant t$ if $F_{p,q}(t) \leqslant 1 - \alpha$.

Remark. It is easy to see that $(S, \mathscr{F}, \mathrm{Min})$ is a Menger space.

Let us now apply some results derived for fixed and periodic points in uniform spaces to obtain fixed-point-type theorems on PM-spaces [84].

We shall assume that S is a set and $\mathscr{P} = \{d_\alpha\}$ is a family of pseudometrics on S. The uniformity generated by $\mathscr{P}$ is constructed by taking as a sub-base all sets in $S \times S$ of the form $U_{\alpha,\varepsilon} = \{(p, q) : d_\alpha(p, q) < \varepsilon\}$, where $d_\alpha \in \mathscr{P}$ and $\varepsilon > 0$. The topology $\mathscr{T}$ determined by this uniformity has all d_α-spheres as a sub-base. Also, we shall assume that, for $p, q \in S$, $p \neq q$, there exists at least one $d_\alpha \in \mathscr{P}$ for which $d_\alpha(p, q) > 0$, so that the space $(S, \mathscr{T})$ is necessarily Hausdorff.

A uniform version of metric chainability is given by

DEFINITION 3.1.8. *S is said to be (d_λ, ε)-chainable if, for $d_\lambda \in \mathscr{P}$ and $\varepsilon > 0$, and any points $p, q \in S$, there exists a finite set $\{p = p_0, p_1, \ldots, p_n = q\}$ such that $d_\lambda(p_{i-1}, p_i) < \varepsilon$ for all $i = 1, 2, \ldots, n$.*

THEOREM 3.1.14. *Let $(S, \mathscr{F}, \text{Min})$ be a complete Menger space, let f be a mapping $f: S \to S$ and suppose there exists a $\delta \in (0, 1)$ such that, for each $\alpha \in (0, \delta)$, there exists a $k_\alpha \in (0, 1)$ such that for $p, q \in S$,*

$$F_{f(p),f(q)}(k_\alpha t) \geqslant F_{p,q}(t)$$

whenever $F_{p,q}(t) > 1 - \alpha$. Then f has a unique fixed point $r \in S$ and $f^n(p) \to r$, as $n \to \infty$, for every $p \in S$.

Proof. Let us first recall a direct extension of Banach's theorem.

LEMMA 3.1.1 [84, Th. 21]. *Suppose that $(S, \mathscr{P})$ is sequentially complete and f is a mapping of S into S having the property that, for every $d_\alpha \in \mathscr{P}$, there exists a constant $k_\alpha \in (0, 1)$ such that $d_\alpha(f(p), f(q)) \leqslant k_\alpha \cdot d_\alpha(p, q)$ for all $p, q \in S$. Then there exists a unique point $r \in S$ such that $f(r) = r$ and $f^n(p) \to r$, as $n \to \infty$, for any $p \in S$.*

In order to prove the theorem, let us consider $\{d_\alpha\}$ to be the family of pseudometrics associated with $\mathscr{F}$; $\alpha \in (0, \delta)$ and $p, q \in S$. Let $t = d_\alpha(f(p), f(q))/k_\alpha$ and suppose that $f_{p,q}(t) > 1 - \alpha$. Then

$$F_{f(p),f(q)}(k_\alpha t) \geqslant F_{p,q}(t) > 1 - \alpha,$$

which is a contradiction since, according to Proposition 3.1.6,

$$F_{f(p),f(q)}(k_\alpha t) = F_{f(p),f(q)}(d_\alpha(f(p), f(q))) \leqslant 1 - \alpha.$$

Thus, $F_{p,q}(t) \leqslant 1 - \alpha$, so $t \leqslant d_\alpha(p, q)$. Hence, $d_\alpha(f(p), f(q)) \leqslant$ $\leqslant k_\alpha d_\alpha(p, q)$. The proof of the theorem follows from the above lemma since $\{d_\alpha\}_{\alpha \in (0, \delta)}$ generates the topology of S.

THEOREM 3.1.15. *Let $(S, \mathscr{F}, \text{Min})$ be a complete (ε, λ)-chainable Menger space for some λ and ε and $f: S \to S$. Suppose that there exists a $\delta \in (0, 1)$ so that, for each $\alpha \in (0, \delta)$, there exists a constant $k_\alpha(0, 1)$ such that $F_{p,q}(\varepsilon) > 1 - \lambda$ implies that*

$$F_{f(p),f(q)}(k_\alpha t) \geqslant F_{p,q}(t)$$

whenever $F_{p,q}(t) > 1-\alpha$. *Then.* f *has a unique fixed point* $r \in S$ *and* $f^n(p) \to r$, *as* $n \to \infty$, *for all* $p \in S$.

Proof. Let $\{d_\alpha\}$ be the family of pseudometrics associated with $\mathscr{F}$. Suppose that $d_\lambda(p, q) < \varepsilon$. Then $F_{p,q}(\varepsilon) > 1-\lambda$, so $F_{f(p), f(q)}(k_\alpha t) \geqslant F_{p,q}(t)$ when $F_{p,q}(t) > 1 - \alpha$. As in the proof of the above theorem, this implies that $d_\alpha(f(p), f(q)) \leqslant k_\alpha d_\alpha(p, q)$ for $\alpha \in (0, \delta)$.

For any $p, q \in S$, let $p = p_0, p_1, \ldots, p_n = q$ be an (ε, λ)-chain, i.e.,

$$F_{p_{i-1}, p_i}(\varepsilon) > 1 - \lambda \text{ for } i = 1, 2, \ldots, n.$$

From Proposition 3.1.6, $d_\lambda(p_{i-1}, p_i) < \varepsilon$. The proof of the theorem follows from

LEMMA 3.1.2 [84, Th. 2.2]. *Let* $(S, \mathscr{P})$ *be sequentially complete and* $f: S \to S$. *Suppose that there exist a* $d_\lambda \in \mathscr{P}$ *and an* $\varepsilon > 0$ *such that* S *is* (d_α, ε)*-chainable. If, for each* $d_\alpha \in P$, *there exists a constant* $k_\alpha \in (0, 1)$ *such that* $d_\alpha(f(p), f(q)) \leqslant k_\alpha d_\alpha(p, q)$ *whenever* $d_\lambda(p, q) < \varepsilon$, *then there exists a unique* $r \in S$ *for which* $f(r) = r$, *and* $f^n(p) \to r$, *as* $n \to \infty$, *for any* $p \in S$.

Regarding the existence of periodic points of a mapping on uniform structures, let us recall

LEMMA 3.1.3 [84, Th. 2.4, Th. 2.5]. *Let* $(S, \mathscr{P})$ *be sequentially compact and suppose that* $f: S \to f(S) \subset S$ *is a mapping with the property that there exist a* $d_\lambda \in \mathscr{P}$ *and an* $\varepsilon > 0$ *such that* $d_\alpha(f(p), f(q)) < d_\alpha(p, q)$ *for all* $d_\alpha \in \mathscr{P}$ *whenever* $d_\alpha(p, q) < \varepsilon$ *and* $d_\alpha(p, q) > 0$. *Then the set* $A = \{r \in S; \exists\, m \in \mathbb{N}, f^m(r) = r\}$ *of periodic points of* f *is finite and nonempty. Moreover, for every* $p \in S$, $\lim_{n\to\infty} f^m(p) = r$ *for some* $r \in A$.

If, in addition, S *is* (d_λ, ε)*-chained, then* A *consists of exactly one point* r, *the unique fixed point of* f, *and* $f^n(p) \to r$, *as* $n \to \infty$, *for every* $p \in S$.

THEOREM 3.1.16. *Suppose that* f *is a mapping of the compact Menger space* $(S, \mathscr{F}, \mathrm{Min})$ *into itself. Suppose, further, that there exist the numbers* $\varepsilon > 0$ *and* $\lambda \in (0, 1)$ *such that, whenever* $F_{p,q}(\varepsilon) > 1 - \lambda$, *there exists a* $\delta > 0$ *for which*

$$F_{f(p), f(q)}(t - \delta) \geqslant F_{p,q}(t)$$

for all t. *Then the set* A *of periodic points of* f *is finite and nonempty. Moreover, if* m *is a period of the points of* A, *then* $\lim_{n\to\infty} f^{mn}(p) = r_p \in A$, *for every* $p \in S$.

Proof. Let $\{d_\alpha\}$ be the family of pseudometrics associated with $\mathscr{F}$, and suppose that $d_\lambda(p, q) < \varepsilon$. Then, according to Proposition 3.1.6, $F_{p,q}(\varepsilon) > 1 - \lambda$, which implies the existence of a $\delta > 0$ such that

$$F_{f(p),f(q)}(t - \delta) \geqslant F_{p,q}(t).$$

Let d_α be such that $d_\alpha(p, q) > 0$ and let $\bar{t} = d_\alpha(p, q)$. Then, we have $F_{p,q}(\bar{t}) \leqslant 1 - \alpha$ and $F_{p,q}(t) > 1 - \alpha$ for all $t > \bar{t}$. If we suppose that

$$d_\alpha(f(p), f(q)) \geqslant d_\alpha(p, q) = \bar{t},$$

then, from the definition of $d_\alpha(p, q)$ (Proposition 3.1.6), it follows that

$$F_{f(p),f(q)}(\bar{t}) \leqslant F_{f(p),f(q)}(d_\alpha(f(p), f(q)) \leqslant 1 - \alpha$$

and

$$F_{p,q}(\bar{t} + \delta) > 1 - \alpha.$$

On the other hand,

$$F_{f(p),f(q)}(\bar{t} + \delta - \delta) \geqslant F_{p,q}(\bar{t} + \delta) > 1 - \alpha$$

and, from Proposition 3.1.6, one concludes that $d_\alpha(f(p), f(q)) < \bar{t}$, which contradicts the assumption that $d_\alpha(f(p), f(q)) \geqslant \bar{t}$. It follows that

$$d_\alpha(f(p), f(q)) < d_\alpha(p, q)$$

and the proof of the theorem follows from Lemma 3.1.3.

COROLLARY. *If $(S, \mathscr{P})$ is (d_λ, ε)-chainable, then A consists of exactly one point $r \in S$, the unique fixed point of f, and, for each $p \in S$, $\lim_{n\to\infty} f^n(p) = r$.*

In ref. [194], Hadžić and Budinčević proved that t-norm Min in Proposition 3.1.6 can be replaced by weaker t-norms.

It is known (Coroll. 1.1.1) that every continuous t-norm admits a denumerable decomposition with the aid of the archimedean semigroups $\{t_m\}_{m\in K}$, $\mathrm{card}(K) \leqslant \aleph_0$ and the t-norm Min. Then, one can state:

PROPOSITION 3.1.7. *Let $(S, \mathscr{F}, T)$ be a Menger space and let T be a continuous t-norm such that every semigroup t_m is strict. If T is of $\mathscr{H}$-type, then there exists a denumerable family of pseudometrics $\{d_n\}_{n\in\mathbb{N}}$ such that the (ε, λ)-topology and the topology induced by the family $\{d_n\}_{n\in\mathbb{N}}$ coincide. Here, $d_n(p, q) = \sup\{t : F_{p,q}(t) \leqslant a_n\}$, $(\forall\, n \in \mathbb{N})$ for some sequence $\{a_n\} \subset (0, 1)$ such that $\lim_{n\in\infty} a_n = 1$.*

Proof. Let $J_m = (a_m, b_m)_{m \in K}$ be the family of open subintervals of $[0,1]$ from the structure theorem. First, suppose that $\sup_{m \in K} b_m = b < 1$. Then $T(u, v) = \text{Min}(u, v)$ for every $(u, v) \in \in [b, 1] \times [b, 1]$ and the family of pseudometrics is defined by

$$d_n(p, q) = \sup\{t : F_{p,q}(t) < c_n\}, \quad n \in N,$$

where $\{c_n\}_{n \in N}$ is an arbitrary sequence with $\lim_{n \to \infty} c_n = 1$ and $c_n > b$, for every $n \in N$.

If $b = 1$, we consider a nondecreasing sequence $\{b_{m_k}\}_{k \in \mathbb{N}}$ with $\lim_{k \to \infty} b_{m_k} = 1$ and b_{m_k} corresponding to $J_{m_k} = (a_{m_k}, b_{m_k})$. Then, for every $k \in \mathbb{N}$, let us define d_k as follows:

$$d_k(p, q) = \sup\{t : F_{p,q}(t) \leqslant a_{m_k}\}.$$

This is a family of pseudometrics since, if $k \in \mathbb{N}$ and $x, y, z \in S$ exist such that

$$d_k(p, q) + d_k(q, r) < d_k(p, r),$$

then one can choose the real numbers x and y such that

$$d_k(p, q) < x, \; d_k(q, r) < y, \text{ and } x + y < d_k(p, r).$$

From the definition of d_k, it follows that

$$F_{p,q}(x) = u_1 > a_{m_k}$$

$$F_{q,r}(y) = u_2 > a_{m_k}$$
$$F_{p,r}(x + y) \leqslant a_{m_k}.$$

Let $\bar{u}$, $\bar{\bar{u}}$ be such that

$$a_{m_k} < \bar{\bar{u}} < \bar{u} < \min\{u_1, u_2, b_{m_k}\},$$

Then, since $t_{m_k} = T|_{J_{m_k} \times J_{m_k}}$ is a strict semigroup, one has

$$a_{m_k} = T(a_{m_k}, a_{m_k}) \leqslant T(\bar{\bar{u}}, \bar{\bar{u}}) < T(\bar{u}, \bar{u} T(\bar{u}, \bar{u}) \leqslant$$
$$\leqslant T(F_{p,q}(x), F_{q,r}(y)) \leqslant F_{p,r}(x + y) \leqslant a_{m_k},$$

which is a contradiction. Hence, $\{d_k\}_{k \in N}$ is a family of pseudometrics for which

$$F_{p,q}(\varepsilon) > a_{m_k} \text{ if and only if } d_k(p, q) < \varepsilon.$$

Since $\lim_{k \to \infty} a_{m_k} = 1$, it follows that the family

$$\{U_q(\varepsilon, 1 - a_{m_k}\}_{k \in \mathbb{N}}$$

forms a fundamental system of neighbourhoods of $p \in S$ and the proof of the theorem is complete.

An example of a t-norm having the properties stated in the above proposition is given in (Sub-sec. 1.1). This fact justifies the following condition :

$$(*) \quad \text{There exists a sequence } \{(a_m, b_m)\}_{m \in \mathbb{N}} \text{ such that } (a_i, b_i) \cap (a_j, b_j) \neq \emptyset \text{ for } i \neq j \text{ and } \lim_{m\to\infty} a_m = 1, \text{ where } T\big|_{(a_m, b_m)\times(a_m, b_m)}$$

is a strict semigroup, for every $m \in \mathbb{N}$.

PROPOSITION 3.1.8. *Suppose that* $(S, \mathcal{F}, T)$ *is a complete Menger space,* $f: S \to S$ *and* T *is a continuous t-norm of* $\mathcal{H}$*-type which satisfies condition* $(*)$*. If there exists a* $\delta \in (0, 1)$ *such that, for every* $m \in \mathbb{N}$ *with* $a_m > \delta$*, there exists a* $k_m \in (0, 1)$ *such that*

$$F_{p,q}(x) > a_m \Rightarrow F_{f(p),f(q)}(k_m x) \geqslant F_{p,q}(x),$$

then $\{p \in S : f(p) = p\} \neq \emptyset$.

Proof. From Proposition 3.1.7, it follows that

$$d_m(p, q) = \sup \{t : F_{p,q}(t) \leqslant a_m\}$$

is a pseudometric for every $m \in \mathbb{N}$. Suppose that $m \in \mathbb{N}$ is such that $a_m > \delta$ and

$$t = \frac{d_m(f(p), f(q))}{k_m};$$

then $F_{p,q}(t) \leqslant a_m$. Indeed, if one supposes that $F_{p,q}(t) > a_m$, then

$$F_{f(p),f(q)}(k_m t) \geqslant F_{p,q}(t),$$

whence

$$F_{f(p),f(q)}(k_m t) = F_{f(p),f(q)}(d_m(f(p), f(q)) \leqslant a_m,$$

which is a contradiction. So, one has $F_{p,q}(t) \leqslant a_m$ which implies that $d_m(p, q) \geqslant t$ and then

$$d_m(f(p), f(q)) \leqslant k_m d_m(p, q)$$

for every $m \in \mathbb{N}$ such that $a_m > \delta$. Since $\lim_{m\to\infty} a_m = 1$, there exists an $m_0 \in \mathbb{N}$ such that $a_m > \delta$ for every $m \geqslant m_0$. Since $\{d_m\}_{m : a_m < \delta}$ induces the same topology as the whole family $\{d_m\}_{m \in \mathbb{N}}$, it is easy to see that there exists a $p \in S$ such that $f(p) = p$.

The close relationship between some classes of PM-spaces and metric spaces suggested the introduction of other contractions from metric spaces to PM-spaces as generalized contractions or quasi-contractions [104], [105]. A mapping $f: (S, d) \to (S, d)$

is said to be a generalized contraction iff for every $p,q \in S$ there exist the nonnegative numbers a, b, c and e, which may depend on both p and q such that $\sup\{a + b + c + 2e : p,q \in S\} < 1$ and

$$(1)\quad d(f(p), f(q)) \leqslant ad(p, q) + bd(p, f(p)) + cd(q, f(q)) + \\ + e[d(p, f(q)) + d(q, f(p))].$$

A mapping $f : (S, d) \to (S, d)$ is said to be a quasi-contraction iff there exists a number $k \in (0,1)$ such that

$$(2)\quad d(f(p), f(q)) \leqslant k\{\max d(p,q),\ d(p, f(p)),\ d(q, f(q)),\ d(p, f(q)), \\ d(q, f(p))\}$$ holds for every $p, q \in S$.

It is clear that every generalized contraction is a quasi-contraction and there exist quasi-contractions which are not generalized contractions. Every quasi-contraction on a complete metric space has a unique fixed point. For PM-spaces, let us consider [178]

THEOREM 3.1.17. *Let $(S, \mathscr{F}, T)$ be a complete Menger space with the continuous t-norm T of $\mathscr{H}$-type. Suppose that, for every $m \in \mathbb{N}$, the semigroup t_m is strict. If $f : S \to S$ is a mapping such that, for every $r_i > 0$, $i = 1, 2, 3, 4, 5$, and every $p, q \in S$,*

$$F_{f(p),f(q)}\left(\sum_{i=1}^{5} r_i\right) \geqslant T\left(T\left(T\left(T\left(F_{p,f(q)}\left(\frac{r_5}{a}\right), F_{q,f(p)}\left(\frac{r_4}{b}\right)\right),\right.\right.\right.$$

$$\left.\left.\left.F_{q,f(q)}\left(\frac{r_3}{c}\right)\right), F_{p,f(p)}\left(\frac{r_2}{d}\right)\right), F_{p,q}\left(\frac{r_1}{e}\right)\right),$$

where $a, b, c, d, e \in R^+$ and $a + b + c + d + e < 1$, then this mapping has a unique fixed point.

Proof. As in Proposition 3.1.7, we shall prove that there exists a sequence $\{a_n\}_{n\in\mathbb{N}} \subset (0,1)$ such that $\lim_{n\to\infty} a_n = 1$ and that the family $\{d_n\}_{n\in\mathbb{N}}$ of pseudometrics defines the (ε, λ)-topology, where

$$d_n(p, q) = \sup\{t : F_{p,q}(t) \leqslant a_n\},\ n \in \mathbb{N},\ p, q \in S.$$

Additionally, it will be proved that, for every $n \in \mathbb{N}$ and every $p, q \in S$, one has

$$d_n(f(p), f(q)) \leqslant ed_n(p, q) + dd_n(p, f(p)) + cd_n(q, f(q)) + \\ + bd_n(q, f(p)) + ad_n(p, f(q)).$$

Indeed, if one supposes the contrary, then there exist $r_1, r_2, r_3, r_4, r_5 \in \mathbb{R}^+$ such that

$$ed_n(p, q) < r_1,\ dd_n(p, f(p)) < r_2,\ cd_n(q, f(q)) < r_3,\ bd_n(q, f(p)) < r_4,\ ad_n(p, f(q)) < r_5 \text{ and } d_n(f(p), f(q)) > \sum_{i=1}^{5} r_i.$$

Then it follows that

$$(3)\quad F_{p,q}\left(\frac{r_1}{e}\right) > a_n;\ F_{p,f(p)}\left(\frac{r_2}{d}\right) > a_n;\ F_{q,f(q)}\left(\frac{r_3}{c}\right) > a_n;$$

$$F_{q,f(p)}\left(\frac{r_4}{b}\right) > a_n;\ F_{p,f(q)}\left(\frac{r_5}{a}\right) > a_n \text{ and } F_{f(p),f(q)}\left(\sum_{i=1}^{5} r_i\right) \leqslant a_n.$$

On the other hand, for every $x \in (a_n, 1)$, it follows that $T^s(x) > a_n$ for every $s \in \mathbb{N}$. Indeed, suppose that $s = 1$. Then, if $T = \text{Min}$ or if there exists a $b \in (0,1)$ such that $T|_{[b,1]} = \text{Min}$, then, from $a_n > b$, it follows that $T(x, x) = x > a_n$. If there exist no $b \in (0,1)$ such that $T|_{[b,1]} = \text{Min}$, then $a_n = a_{m_n} = T_{(a_{m_n}, a_{m_n}}$ and choosing $\bar{u}, \bar{\bar{u}} \in (a_{m_n}, b_{m_n})$ such that $a_n < \bar{u} < \bar{\bar{u}} < x$, one concludes that

$$a_n = T(a_n, a_n) \leqslant T(\bar{u}, \bar{u}) < T(\bar{u}, \bar{\bar{u}}) \leqslant T(x, x).$$

Hence, $T(x, x) > a_n$. By induction, supposing that, for some $s_0 \in N$, $T^{s_0}(x) > a_n$ yields

$$(4)\quad a_n < T(T^{s_0}(x), T^{s_0}(x)),$$

since it has been proved that $y > a_n$ implies that $T(y, y) > a_n$ Also, since $T^{s_0}(x) \leqslant \text{Min}^{s_0}(x) = x$, one has

$$a_n < T(T^{s_0}(x), T^{s_0}(x)) \leqslant T(T^{s_0}(x), x) = T^{s_0+1}(x).$$

Now, it follows from (3) that

$$a_n < \min\left\{F_{p,q}\left(\frac{r_1}{e}\right), F_{p,f(p)}\left(\frac{r_2}{d}\right), F_{q,f(q)}\left(\frac{r_3}{c}\right), F_{q,f(p)}\left(\frac{r_4}{b}\right), F_{p,f(q)}\left(\frac{r_5}{a}\right)\right\},$$

and so

$$a < T^4\left(\min\left\{F_{p,q}\left(\frac{r_1}{e}\right), F_{p,f(p)}\left(\frac{r_2}{d}\right), F_{q,f(q)}\left(\frac{r_3}{c}\right), F_{q,f(p)}\left(\frac{r_4}{b}\right), F_{p,f(q)}\left(\frac{r_5}{a}\right)\right\}\right) < T\left(T\left(T\left(T\left(F_{p,f(q)}\left(\frac{r_5}{a}\right)\right), F_{q,f(p)}\left(\frac{r_4}{b}\right)\right)\right),\right.$$

$$F_{q,f(q)}\left(\frac{r_3}{c}\right)\Bigg), F_{p,f(p)}\left(\frac{r_2}{d}\right)\Bigg); F_{p,q}\left(\frac{r_1}{e}\right)\Bigg) \leqslant$$
$$\leqslant F_{f(p),f(q)}\left(\sum_{i=1}^{5} r_i\right) \leqslant a_n,$$

which is a contradiction.

Hence, one has

$$d_n(f(p), f(q)) \leqslant ed_n(p, q) + dd_n(p, f(p)) + cd_n(q, f(q)) + bd_n(q, f(p)) + \\ + ad_n(p, f(q)) \leqslant (a + b + c + d + e)\max\{d_n(p, q), \\ d_n(p, f(p)), d_n(q, f(q)), d_n(q, f(p)), d_n(p, f(q))\}.$$

Since T is continuous, it follows that the (ε, λ)-topology is metrizable. Using the results for quasi-contractions on metric spaces [104, Th. 1], one concludes that there exists a unique fixed point $p_* \in S$ of f and $\lim_{n\to\infty} f^n(r) = p_*$ for every $r \in S$.

L. Čirić introduced the generalized contraction on PM-spaces using

DEFINITION 3.1.9. A mapping $f: (S, \mathscr{F}) \to (S, \mathscr{F})$ will be called a *generalized contraction* if there exists a constant $k \in (0, 1)$ such that, for every $p, q \in S$,

(5) $F_{f(p),f(q)}(kx) \geqslant \min\{F_{p,q}(x), F_{p,f(p)}(x), F_{q,f(q)}(x),$
$F_{p,f(q)}(2x), F_{q,f(p)}(2x)\}$

for all $x > 0$.

The same paper [105, Th. 1] proves that every generalized contraction on a complete Menger space $(S, \mathscr{F}, \mathrm{Min})$ has a unique fixed point $p \in S$ and $\lim_{n\to\infty} f^n(q) = p$ for every $q \in S$.

Since (5) implies (4), a short proof of this theorem can be given using the above technique.

Indeed, if there exist an $a \in (0, 1)$ and $p, q \in S$ such that

$$d_a(f(p), f(q)) > k \max\left\{d_a(p, q),\ d_a(p, f(p)),\ d_a(q, f(q)),\right.$$
$$\left.\frac{1}{2} d_a(p, f(q)),\ \frac{1}{2} d_a(q, f(p))\right\},$$

then there exists a $b > 0$ such that

$$d_a(f(p), f(q)) > b > k \left\{\max d_a(p, q), d_a(p, f(p)), d_a(q, f(q)),\right.$$
$$\left.\frac{1}{2} d_a(p, f(q)),\ \frac{1}{2} d_a(q, f(p))\right\}.$$

But from the definition of d_a, one has

$$F_{f(p),f(q)}(b) \leqslant 1-a \text{ and } \min\left\{F_{p,q}\left(\frac{b}{k}\right), F_{p,f(p)}\left(\frac{b}{k}\right), F_{q,f(p)}\left(\frac{b}{k}\right),\right.$$

$$\left. F_{p,f(q)}\left(\frac{2b}{k}\right), F_{q,f(p)}\left(\frac{2b}{k}\right)\right\} > 1 - a,$$

which is a contradiction. The rest follows from the corresponding theorem on metric spaces.

The same result also implies the existence and uniqueness of a fixed point for the mapping $f: S \to S$ with the following weaker condition :

$$F_{f(p),f(q)}(kx) \geqslant \min\{F_{p,q}(x), F_{p,f(p)}(x), F_{q,f(q)}(x),$$
$$F_{p,f(q)}(x), F_{q,f(p)}(x)\}$$

for every $p, q \in S$ and $k \in (0, 1)$ fixed. This is a *probabilistic quasi-contraction*.

Remark. A result similar to that in Theorem 3.1.16 can be proved [197] for a continuous t-norm T of $\mathscr{H}$-type.

Many of the above results can be simplified if one knowns the metric which metrizes the (ε, λ)-topology and hence the uniformity generated by $\{d_\alpha\}_{\alpha\in(0,1)}$ on S.

A first result in this direction is given by

THEOREM 3.1.18 [406]. *Let $(S, \mathscr{F}, \tau)$ be a complete PM-space under an l.c. t-function τ. If, for every pair $p,q \in S$, there exists a real number x such that $F_{p,q}(x) = 1$, then every contraction $f: S \to S$ has a unique fixed point.*

Proof. For each $p, q \in S$,

$$d(p, q) = \inf\{t : F_{p,q}(t) = 1\}$$

and f is a contraction map on (S, d). Since for every $p_0 \in S$ the sequence $p_n = f^n(p_0)$, $n \in \mathbb{N}$, is a Cauchy sequence in (S, d), then it is also a Cauchy sequence in $(S, \mathscr{F}, \tau)$. As $(S, \mathscr{F}, \tau)$ is complete, there exists a point $p \in S$ such that $p = \lim_{n\to\infty} p_n$. The point p is the unique fixed point of f in S.

For a continuous t-norm T, one can obtain Theorems 3.1.2 and 3.1.7 using the classical Banach principle on generalized complete metric spaces. Thus, one can state

THEOREM 3.1.19 [352]. *Let $(S, \mathscr{F}, \text{Min})$ be a complete Menger space and let F be a fixed distribution function, $F \in \mathscr{D}^+$. Then the* *two-place function* $d_F : S \times S \to \mathbb{R}^+$, *defined as*

$$d_F(p, q) = \inf\{a > 0 : F_{p,q}(ax) \geqslant F(x), \quad \forall\, x \in \mathbb{R}\},$$

is a Luxemburg metric on S ($\inf \emptyset = +\infty$).

Moreover, if $(S, \mathscr{F})$ *is complete, then* (S, d_F) *is complete. Generally, the* d_F*-topology is not weaker than the* (ε, λ)*-topology (called also* $\mathscr{F}$*-topology).*

Proof. It is clear that d_F is symmetric and $d_F(p, p) = 0$ for all $p \in S$. If $d_F(p, q) = 0$, then, for each $a > 0$, $F_{p,q}(ax) \geqslant F(x)$ for all $x \in \mathbb{R}$. Now, if $y = ax$, and $x \to \infty$, then $F_{p,q}(x) = 1$, that is $p = q$.

Suppose now that $d_F(p, r) < \infty$, $d_F(r, q) < \infty$.

If $d_F(p, r) < a' < a$, $d_F(r, q) < b' < b$, then

$$F_{p,q}[(a' + b')x] \geqslant \mathrm{Min}\{F_{p,r}(a'x), F_{r,q}(b'x)\} \geqslant F(x),$$

which shows that $d_{F(p,q)} \leqslant a' + b' < a + b$. Therefore, $d_{F(p,q)} \leqslant d_F(p, r) + d_F(r, q)$, and it follows that d_F is a Luxemburg metric.

Let $\{p_n\}_{n \in \mathbb{N}}$ be a sequence d_F-convergent to p. If $\varepsilon > 0$ and $\lambda \in (0, 1)$, then there exists an x_0 such that $F(x_0) > 1 - \lambda$. Let $a < \dfrac{\varepsilon}{x_0}$ and choose n_0 such that $d_F(p_n, p) < \varepsilon$ for all $n \geqslant n_0$. Then $F_{p_n,p}(\varepsilon) \geqslant F_{p_n,p}(ax_0) \geqslant F(x_0) > 1 - \lambda$, that is $\{p_n\}_{n \in \mathbb{N}}$ is $\mathscr{F}$-convergent to p.

Suppose that $\{p_n\}_{n \in \mathbb{N}}$ is a d_F-Cauchy sequence and $(S, \mathscr{F})$ is complete. Then, one obtains that $\{p_n\}_{n \in \mathbb{N}}$ is an $\mathscr{F}$-Cauchy sequence and thus there exists a $p \in S$ such that $\{p_n\}_{n \in \mathbb{N}}$ is $\mathscr{F}$-convergent to p.

If $a > 0$, then there exists an $n_0 \in \mathbb{N}$ such that $F_{p_n,p_{n+m}}(ax) \geqslant F(x)$ for all $n \geqslant n_0$, all $m \geqslant 1$ and each real x. If $m \to +\infty$, then

$$F_{p_n,p}(ax) \geqslant \lim_{m \to \infty} F_{p_n,p_{n+m}}(ax) \geqslant F(x), \quad \forall\, n \geqslant n_0,\ \forall\, x \in R,$$

which shows that $d_F(p_n, p) \leqslant a$, $\forall\, n \geqslant n_0$, and, hence, p_n is d_F-convergent.

Remark. The left continuity of F is not needed in the above theorem.

Example. If $F(x) = H_1(x) = \begin{cases} 0, & x \leqslant 1 \\ 1, & x > 1 \end{cases}$, then

$$d_F(p, q) = \inf\{x : F_{p,q}(x) = 1\},$$

which shows that d_F need not be nontrivial.

Moreover, if (S, d) is a metric space, considered as a Menger space, with $F_{p,q}(x) = H_0(x - d(p, q))$, then $d_F(p, q) = d(p, q)$.

LEMMA 3.1.4. *If $f: S \to S$ is a contraction on the Menger space $(S, \mathscr{F}, \mathrm{Min})$, then f is a contraction in (S, d_F).*

Proof. Since $F_{f(p),f(q)}(kx) \geqslant F_{p,q}(x)$ for some $k \in (0, 1)$ and every real x, then it is easy to see that $d_F(f(p), f(q)) \leqslant kd_F(p, q)$ and the proof of the lemma follows immediately.

THEOREM 3.1.20 [352]. *Let $(S, \mathscr{F}, \mathrm{Min})$ be a complete Menger space and let $f: S \to S$ be an $\mathscr{F}$-contraction. Then f has a unique fixed point p_0, and, for each $p \in S$, $p_0 = \lim_{n\to\infty} f^n(p)$ in the $\mathscr{F}$-topology.*

Proof. Let $p \in S$ and put $F = F_{p,f(p)}$. From Lemma 3.1.4, f is a d_F-contraction. Moreover, $d_F(p, f(p)) = 1 < \infty$ and, for each $n \geqslant 1$, one has

$$d_F(f^n(p), f^{n+1}(p)) = k^n < \infty.$$

From the fixed point alternative in generalized metric spaces (see [124, 217, 288]), the sequence $f^n(p)$ converges to a fixed point p_0 of f, in the metric d_F, and so in the $\mathscr{F}$-topology. Clearly, p_0 is unique.

Remark. Let F be as defined in the theorem. It is well known that $S_p = \{q : d_F(p, q) < \infty\}$ is a d_F-complete metric space and that $f^n(p) \in S_p$ for all $n \geqslant 1$. Therefore, f has a unique fixed point in S_p. Since f has a unique fixed point in S, the proof of the theorem follows immediately.

THEOREM 3.1.21 [352]. *Let $(S, \mathscr{F}, \mathrm{Min})$ be a complete Menger space and suppose that for some $F \in \mathscr{D}^+$ the d_F-topology and the $\mathscr{F}$-topology ((ε, λ)-topology) are identical. Then, for every $\mathscr{F}$-contraction f on S, one has:*

(i) *for each $p \in S$, $f^n(p)$ is convergent to the unique fixed point of f;*
(ii) *for each $p \in S$, there exists an $n \geqslant 0$ such that*

$$F_{f^n(p), f^{n+1}(p)}(x) \geqslant F(x).$$

Proof. The first assertion follows from Theorem 3.1.20 above. The second assertion follows from the fixed point alternative, since (i) is always true.

It is clear that the above argument gives a whole family of generalized metric topologies on Menger spaces $(S, \mathscr{F}, \mathrm{Min})$. The following example indicates how the value of $\mathscr{F}$ behaves.

Example. Let

$$F(x) = \begin{cases} 0, & x \leqslant 1 \\ 1 - \dfrac{1}{x}, & x > 1 \end{cases}.$$

It is easy to see that

$$d_F(p, q) = \sup \alpha(d_\alpha(p, q)) = \sup\{\alpha \inf\{t : F_{p,q}(t) \geqslant 1 - \alpha\}\}.$$

Suppose that d_F induces an (ε, λ)-topology on S. Therefore, for each $p \in S$, there exists a $b \geqslant 0$ such that

$$F_{f^n(p), f^{n+1}(p)}(x) \geqslant 1 - \frac{1}{x}, \quad \forall\, x \geqslant 1,\ \forall\, n \geqslant b.$$

Similarly, one can take

$$F_1(x) = \begin{cases} 0, & x \leqslant 1 \\ 1 - \dfrac{1}{x^\beta}, & x > 1,\ \beta > 0. \end{cases}$$

Then

$$d_{F_1}(p, q) = \sup \alpha^{1/\beta} d_\alpha(p, q)$$

and

$$F_{f^n(p), f^{n+1}(p)}(x) \geqslant 1 - \frac{1}{x^\beta}, \quad \forall\, x \geqslant 1.$$

Generally, from the fixed point alternative one obtains the following result.

THEOREM 3.1.22. *If $(S, \mathscr{F}, \mathrm{Min})$ is a complete Menger space, $f : S \to S$ is an $\mathscr{F}$-contraction and $F \in \mathscr{D}^+$, then, for each $p \in S$, either*

(i) *$f^n(p)$ is d_F-convergent to the unique fixed point of f;*

or

(ii) *for each $n \geqslant 0$ and each $a > 0$, there exists an $x_{n_a} > 0$ such that*

$$F_{f^n(p), f^{n+1}(p)}(a x_{n_a}) < F(x_{n_a}).$$

The following lemma shows how one can construct generalized metrics on a Menger space under an $\mathscr{H}$-type t-norm.

LEMMA 3.1.5 [353]. *Let T be an $\mathscr{H}$-type t-norm. Take*

$$0 < a_1 < a_2 < \ldots < a_n \to \infty, \text{ as } n \to \infty,$$

$$0 < b_1 < b_2 < \ldots < b_n \to \infty, \text{ as } n \to \infty,$$

$$T(b_n, b_n) = b_n,$$

and set

$$F(x) = \begin{cases} 0 & \text{for } x \leqslant a_1 \\ b_n & \text{for } x \in (a_n, a_{n+1}],\ n = 1, 2, \ldots . \end{cases}$$

Consider a Menger space $(S, \mathscr{F}, T)$ *and define*

$$d(p, q) = \inf\{a > 0, F_{p,q}(ax) \geqslant F(x)\}, \quad \forall\, x \in \mathbb{R}.$$

Then:

(i) *d is a generalized metric on S*;
(ii) *If S is $\mathscr{F}$-complete, then S is d-complete*;
(iii) *The d-topology is not weaker than the $\mathscr{F}$-topology.*

Proof. (i) We only need to prove the triangle inequality. Thus, if $d(p, q) < a' < a$, $d(q, r) < b' < b$ and $x \in (a_n, a_{n+1}]$, then

$$F_{p,r}(a'x + b'x) \geqslant T(F_{p,q}(a'x), F_{q,r}(b'x)) \geqslant T(F(x), F(x)) =$$
$$= T(b_n, b_n) = b_n = F(x).$$

Therefore, $d(p, q) \leqslant a' + b' < a + b$, and the triangle inequality follows shortly.

(ii) and (iii). Let $\{p_n\}_{n \in \mathbb{N}}$ be a Cauchy sequence and let $a > 0$ be fixed. According to the definition of d, there exists an $n_m \geqslant 1$ such that $F_{p_n, p_{n+m}}(ax) \geqslant F(x)$, $\forall\, n > n_m$, $\forall\, m \geqslant 1$, $\forall\, x \in \mathbb{R}$.

If $\varepsilon > 0$ and $\lambda \in (0, 1)$ are given, then let $a > 0$ and $x_0 \in \mathbb{R}$ be such hat $F(x_0) > 1 - \lambda$ and $ax_0 \leqslant \varepsilon$.

If $n \geqslant n_a$, $m \geqslant 1$ then $F_{p_n, p_{n+m}}(\varepsilon) > 1 - \lambda$, which shows that $\{p_n\}_{n \in \mathbb{N}}$ is $\mathscr{F}$-Cauchy. Assuming that S is $\mathscr{F}$-complete, then $\{p_n\}_{n \in \mathbb{N}}$ is $\mathscr{F}$-convergent to some limit p. Therefore,

$$F(x) \leqslant \lim_{m \to \infty} F_{p_n, p_{n+m}}(ax) = F_{p_n, p}(ax)$$

for any real number x and all $n \geqslant n_a$, i.e., $d(p_n, p) \leqslant a$, $\forall\, n \geqslant n$. Thus, p_n is d-convergent and the proof of the lemma is complete.

Remark. For any given $p_0, q_0 \in S$, the value of a_n in the above lemma can be chosen such that $F_{p_0, q_0}(a_n) \geqslant b_n$; in this case, the metric d is nontrivial.

A generalization of Theorem 3.1.20 with respect to the utility of the classical Banach principle in PM-spaces is given by

THEOREM 3.1.23 [353]. *The classical form of Banach's fixed point principle implies that any probabilistic contraction on a complete Menger space $(S, \mathscr{F}, T)$ under a t-norm T of $\mathscr{H}$-type has a unique fixed point which is the limit of successive approximations.*

Proof. Let $(S', \mathscr{F}, T)$ be a complete Menger space with the $\mathscr{H}$-type t-norm T and let $f: S \to S$ be a probabilistic contraction. If p_0 is given in S', then let $\{a_n\}_{n \in \mathbb{N}}$ and $\{b_n\}_{n \in \mathbb{N}}$ be as in

Lemma 3.1.5 and such that $F_{p_0,f(p_0)}(a_n) \geqslant b_n$. Consider the generalized metric d as in Lemma 3.1.5. It is easy to see that $d(p_0, f(p_0)) < \infty$, and, therefore, we conclude (see [288]) that $S_0 = \{q_0 \in S', d(p_0, q_0) < \infty\}$ is a complete metric space and f is a contraction on S_0. Therefore, $p_n = f^n(p_0)$ d-converges to the (evidently unique) fixed point of f and the proof is complete.

The initial idea of using fixed point theorems developed for certain metric-space mappings to obtain similar results for PM-space mappings was later extended in two directions [210]. On the one hand, one looked for convenient metrics in the (ε, λ)-topology such that completitude, fixed point properties and a good correspondance of the mappings are preserved.

On the other hand, T. L. Hicks [210] introduced a class of mappings f on PM-spaces with the following contraction-type property :

(H) If $x > 0$ and $F_{p,q}(x) > 1 - x$, then $F_{f(p),f(q)}(kx) > 1 - kx$

for $k \in (0, 1)$ fixed, and every $p, q \in S$. In precise terms, this leads to

DEFINITION 3.1.10 [210]. Let $(S, \mathscr{F}, T)$ be a Menger space. A mapping $f: S \to S$ satisfying condition (H) is called an *H-contraction*.

The consistence of this new class of probabilistic contractions stems from the fact that it contains the contractions corresponding to the classical contractions from (S, d) induced into $(S, \mathscr{F})$ by $\mathscr{F}(p, q) = F_{p,q}(x) = H_0(x - d(p, q))$. More precisely, if (S, d) is a bounded metric space (i.e., there exists a real number $M > 0$ such that $d(p, q) < M$ for any $p, q \in S$) and if $d(f(p), f(q)) \leqslant k\ d(p, q)$ for $k \in (0, 1)$ fixed and for any $p, q \in S$, then condition (H) is satisfied for f in $(S, \mathscr{F})$.

Indeed,

$$F_{f(p),f(q)}(kx) = H_0(kx - d(f(p), f(q))) \geqslant H_0(kx - k\ d(p, q)) =$$
$$= H_0(x - d(p, q)) = F_{p,q}(x).$$

Since

$$F_{p,q}(x) = H_0(x - d(p, q)) > 1 - x$$

if and only if $F_{p,q}(x) = 1$, i.e., if and only if $F_{p,q}(x) > 1 - kx$, then condition (H) follows.

We now intend to show that, for a wide class of PM-spaces, there exists a compatible metric d such that

$$d(f(p), f(q)) \leqslant kd(p, q) \text{ if (H) holds,}$$

and that the Banach principle for H-contraction is obtained as a consequence of Banach's principle for metric spaces.

In this respect, an important result was obtained by T. L. Hicks in ref. [210].

THEOREM 3.1.24. *Let* $(S, \mathscr{F}, \mathrm{Min})$ *be a Menger space and let*

$$d(x, y) = \begin{cases} \sup \{x : q \notin U_p(x, x)\}, \ 0 < x < 1 \\ 0 \ \textit{if} \ q \in U_p(x, x) \ \textit{for all} \ x > 0. \end{cases}$$

Then

(i) $d(p, q) < x$ *if and only if* $F_{p,q}(x) > 1 - x$.
(ii) d *is a compatible metric for the* (ε, λ)*-topology.*
(iii) *If* $f : S \to S$ *and* $0 < k < 1$ (H) *holds if and only if* $d(f(p), f(q)) \leqslant k\ d(p, q)$.
(iv) $(S, \mathscr{F})$ *is complete if and only if* (S, d) *is complete.*

Proof. Let us first observe that if $x < y$, $U_p(x, x) \subset U_p(y, y)$, where $U_p(x, x) = \{q : F_{p,q}(x) > 1 - x\}$, $x > 0$. Also, $\cap \{U_p(x, x) : 0 < x < 1\} = \{x\}$ since, if $p \neq q$, $F_{p,q} \neq H_0$. Thus, there exists an $x > 0$ such that $F_{p,q}(x) = a$, $0 < a < 1$. Set $a = 1 - b$ and let $a_1 = \min\{x, b\}$. Then $F_{p,q}(a_1) \leqslant F_{p,q}(x) = a = 1 - b \leqslant \leqslant 1 - a_1$ gives $q \in U_p(a_1, a_1)$.

(i) If $1 < x$, $d(p, q) \leqslant 1 < x$ and also $F_{p,q}(x) \geqslant 0 > 1 - x$. Suppose that $d(p, q) < x \leqslant 1$. Choose a such that $d(p,q) < < a < x \leqslant 1$. Then, $q \in U_p(a, a)$ and $F_{p,q}(x) \geqslant F_{p,q}(a) > 1 - a > > 1 - x$ since, if one assumes that $q \notin U_p(a, a)$, then $d(p, q) = = \sup\{x : q \notin U_p(x, x), 0 < x < 1\} \geqslant a$, which is a contradiction.

Conversely, suppose that $F_{p,q}(x) > 1 - x$ where $0 < x \leqslant 1$. Then $q \in U_p(x, x)$. Indeed, if $q \notin U_p(a, a)$ for all $a < x$, $F_{p,q}(x) = \lim\limits_{\substack{a \to x \\ a < x}} F_{p,q}(a) \leqslant \lim\limits_{\substack{a \to x \\ a < x}} (1 - a) = 1 - x$, which is a contradiction. Thus, there exists a $0 < a < x$ such that $q \in U_p(a, a)$. Hence, $d(p, q) \leqslant a < x$.

(ii) It is obvious that if d satisfies the triangle inequality, then d is a metric. But (i) implies that d is compatible with the (ε, λ)-topology. To prove that d satisfies the triangle inequality, observe that $d(p, q) < a$ and $d(q, r) < b$ imply that $d(p, r) < < a + b$. Indeed, suppose that $F_{p,q}(a) > 1 - a$ and $F_{q,r}(b) > > 1 - b$. Then, for $F_{p,q}(a) \leqslant F_{q,r}(b)$, it follows that $F_{p,r}(a + b) \geqslant \geqslant \mathrm{Min}\ \{F_{p,q}(a), F_{q,r}(b)\} > 1 - a > 1 - (a + b)$, which gives $d(p, r) < a + b$ and the triangle inequality follows.

(iii) Suppose that $d(f(p), f(q)) \leqslant kd(p, q)$ and $F_{p,q}(x) > 1 - x$. Then, $d(p, q) < x$ and $d(f(p), f(q)) < kx$. Thus, $F_{f(p),f(q)}(kx) > > 1 - kx$. Conversely, if condition (H) holds, let $a > 0$ be given. Set $x = d(p, q) + a$, $d(p, q) = x - a < x$ and, hence,

$$F_{p,q}(x) > 1 - x; \quad \text{then} \quad F_{f(p),f(q)}(kx) > 1 - kx$$

follows from (H). Thus, $d(f(p), f(q)) < kx = k(d(p, q) + a) = kd(p, q) + ka$. Since $a > 0$ was chosen arbitrarily, it follows that $d(f(p), f(q)) \leqslant kd(p, q)$.

COROLLARY. *Let $(S, \mathcal{F}, \mathrm{Min})$ be a complete Menger space. Let $f: s \to S$ be an H-contraction. Then f has a unique fixed point p. If $p_0 \in S$ and $p_n = f^n(p_0)$, then*

(i) $p = \lim_{n\to\infty} p_n$, *and*

(ii) *for* $x > \dfrac{k^{n-1}}{1-k} d(p, f(p)) = a_n$,

$$1 - F_{p_n,p}(x) \leqslant \frac{k^{n-1}}{1-k} d(p, f(p)).$$

Proof. The theorem gives a compatible metric d such that $d(f(p), f(q)) \leqslant k\, d(p, q)$. According to the classical Banach principle, f has a unique fixed point p satisfying (i). Also,

$$d(p_n, p) \leqslant \frac{k^n}{1-k} d(p, f(p)) \leqslant \frac{k^{n-1}}{1-k} d(p, f(p)) = a_n.$$

From condition (i) of the Theorem, it follows that

$F_{p_n,p}(a_n) > 1 - a_n$.

For $x \geqslant a_n$

$F_{p_n,p}(x) \geqslant F_{p_n,p}(a_n) > 1 - a_n$.

Remark 1. The error bound is usable in the sense that for given $\varepsilon > 0$, one can choose $0 < \varepsilon_0 < 1$ and p such that $d(p, f(p)) < \varepsilon_0$. For $x \geqslant b = \dfrac{\varepsilon_0}{1-k} > a_n$,

$$1 - F_{p_n,p}(x) \leqslant \frac{k^{n-1}}{1-k} d(p, f(p)) < \frac{k^{n-1}}{1-k} \varepsilon_0.$$

If $\dfrac{k^{n_0}}{1-k} \varepsilon_0 < \varepsilon$, then $1 - F_{p_n,p}(x) < \varepsilon$ for all $n \geqslant n_0$ and all $x \geqslant b$.

Remark 2. A large number of results regarding the existence of fixed points for certain classes of contraction on metric spaces (S, d) can be directly transposed into PM-spaces $(S, \mathcal{F}, \mathrm{Min})$. More precisely, if $f: (S, d) \to (S, d)$ and if there exists a function $G(p, q)$ such that $d(p, q) \leqslant G(p, q)$ for any $p, q \in S$, then one can show that

(H*) $F_{f(p),f(q)}(kx) > 1 - kx$ whenever $F_{p,q}(x) > 1 - x$ and $x > G(p, q)$ is equivalent to $d(f(p), f(q)) \leqslant k\ G(p, q)$.

A result of this type was previously derived on page 82, where

$$G(p, q) = \max\{d(p, q), d(p, f(p)), d(q, f(q)), d(p, f(q)), d(q, f(p))\}.$$

One can treat similarly Theorem 3.1.15, in which the condition of (ε, λ)-local contraction becomes

$$F_{f(p),f(q)}(\lambda t) > 1 - \lambda t \text{ whenever } F_{p,q}(a) > 1 - a,$$

where $a = \min\{\varepsilon, t\}$ and the condition of (ε, λ)-chainability becomes $(\varepsilon, \varepsilon)$-chainability.

Another example is given by

THEOREM 3.1.25 [210]. *Let* $(S, \mathscr{F}, \mathrm{Min})$ *be a complete Menger space and let* f *be a self-mapping on* S. *Suppose there exists a* $p \in S$ *such that*

(A) *for* $x > 0$, $F_{f(p),f^2(q)}(kx) > 1 - kx$ *whenever*

$$F_{p,f(q)}(x) > 1 - x \text{ and } q \in O_f(p, 0, \infty).$$

Then:

(i) $\lim_{n\to\infty} f^n(p) = q$ *exists*

(ii) *If* f *is continuous at* q, *then* $f(q) = q$.

(iii) *For* $x \geqslant \dfrac{k^{n-1}}{1-k}\, d(p, f(p))$, *one has*

$$1 - F_{f^n(p),p}(x) \leqslant \frac{k^{n-1}}{1-k}\, d(p, f(p)).$$

The proof of this theorem follows directly from the similar results obtained on metric spaces (see [210], Th. 2.2).

Some improvements on Hicks' results are given by V. Radu [356], [357]. He proved that Hicks' two-place function d is a metric in any Menger space $(S, \mathscr{F}, T)$ with $T \geqslant T_m$. Using a slightly modified form of d, we can state

THEOREM 3.1.26. *Let* $(S, \mathscr{F}, T_m)$ *be a Menger space and define*

$$d(p, q) = \sup\{x : F_{p,q}(x) \leqslant 1 - x\}.$$

Then

(i) d *is a metric on* S, *which generates the* (ε, λ)-*uniformity.*

(ii) S *is* $\mathscr{F}$-*complete if* S *is* d-*complete.*

(iii) $f: S \to S$ *is an* H-*contraction iff* f *is a* d-*contraction.*

Proof. The reasoning is as in the above theorem. We shall prove only the triangle inequality.

If $d(p, q) \leqslant a$ and $d(q, r) < b$, then it is easy to see that $F_{p,q}(a) > 1 - a$, $F_{q,r}(b) > 1 - b$.

Therefore,

$$F_{p,r}(a + b) \geqslant T_m(F_{p,q}(a), F_{q,r}(b)) \geqslant F_{p,q}(a) + F_{q,r}(b) - 1 > 1 - (a + b),$$

which shows that $d(p, r) < a + b$. This implies that $d(p, r) \leqslant d(p, q) + d(q, r)$, and the proof is complete.

Remarks. 1) If (S, d_1) is a given metric space considered as a Menger space through $\mathscr{F}(p, q) = F_{p,q}(x) = H_0(x - d(p, q))$, then it is easy to see that $d(p, q) = \min\{1, d_1(p, q)\}$. Thus, if d_1 is $[0, 1)$-valued, then $f: S \to S$ is a d_1-contraction if it is an H-contraction.

2) Another way to prove the above theorem is to observe ([see [354]) that, for $T \geqslant T_m$, the following two-place function

$$d'(p, q) = \inf_{x>0} \{x + 1 - F_{p,q}(x)\}$$

is a metric on S, and, hence, it induces the (ε, λ)-uniformity. Moreover, if $\varepsilon, \lambda \in \left(0, \frac{1}{2}\right)$, then

$$(*) \begin{cases} U(\varepsilon, \lambda) \subset \{(p, q) \in S \times S,\ d'(p, q) < \varepsilon + \lambda\} \\ \{(p, q) \in S \times S,\ d'(p, q) < \varepsilon\} \subset U(\varepsilon, \varepsilon), \end{cases}$$

which shows, in particular, that $(S, \mathscr{F}, T)$ is complete if and only if (S, d') is complete.

More precisely, we shall show that, for an H-contraction, its existence, uniquenes and the expression of its fixed points as the limit of successive approximations stem directly from the classical Banach principle in (S, d').

Indeed, if $d'(p, q) < a$, then it is easy to see that $F_{p,q}(a) > 1 - a$ and hence $F_{f(p),f(q)}(ka) > 1 - ka$. Therefore,

$$ka + 1 - F_{f(p),f(q)}(ka) < 2ka,$$

which shows that $d'(f(p), f(q)) < 2ka$. This implies that $d'(f(p), f(q)) \leqslant 2kd'(p, q)$. Next, it is easy to see that for n sufficiently large, such that $k^n < \frac{1}{2}$,

$$d'(f^n(p), f^n(q)) \leqslant 2k^n d'(p, q)$$

i.e., f^n is a classical contraction. Hence, according to the Banach principle, f^n has a unique fixed point. It follows that

f has a unique fixed point which, clearly, is the limit of successive approximations.

An extension of Theorem 3.1.24 can be obtained by showing that its conclusions are valid in any complete Menger space $(S, \mathscr{F}, T)$ for which $\sup_{a<1} T(a, a) = 1$ [356]. It should be observed that this condition imposed on T is generally the weakest condition which ensures the existence of the (ε, λ)-uniformity on S, i.e., that the (ε, λ)-topology exists. This result implies a situation entirely different from that occurring for probabilistic contractions.

LEMMA 3.1.6. *If f is an H-contraction, then, for every $\varepsilon > 0$ and $\lambda \in (0,1)$, there exists a positive integer $n(\varepsilon, \lambda)$ such that*

$$F_{f^n(p), f^n(q)}(\varepsilon) > 1 - \lambda, \ \forall p, q \in S, \ \forall n \geqslant n(\varepsilon, \lambda).$$

Proof. Since, clearly, $F_{p,q}(1+\varepsilon) > 1 - (1+\varepsilon)$, condition (H) implies that

$$F_{f^n(p), f^n(q)}[k^n(1+\varepsilon)] > 1 - k^n(1+\varepsilon), \ \forall\, n \geqslant 1, \ \forall\, p, q \in S.$$

Since $k \in (0,1)$, there exists an $n(\varepsilon, \lambda)$ such that $k^n(1+\varepsilon) \leqslant \min\{\varepsilon, \lambda\}$. If $n \geqslant n(\varepsilon, \lambda)$, then, taking into account that every distribution function is nondecreasing, it follows that

$$F_{f^n(p), f^n(q)}(\varepsilon) \geqslant F_{f^n(p), f^n(q)}[k^n(1+\varepsilon)] > 1 - k^n(1+\varepsilon) \geqslant 1 - \lambda,$$

and the proof of the lemma is complete.

LEMMA 3.1.7 [356]. *Each H-contraction has at most one fixed point.*

Proof. Let us suppose that $p, q \in S$ are two different fixed points for f. Then $f^n(p) = p$ and $f^n(q) = q$ for each $n \geqslant 1$. Hence, the above lemma implies that

$$F_{p,q}(\varepsilon) > 1 - \lambda, \quad \forall\, \varepsilon > 0, \ \forall\, \lambda \in (0,1),$$

and hence $p = q$.

LEMMA 3.1.8. *Every H-contraction is (uniformly) continuous.*

Proof. Let $\varepsilon > 0$ and $\lambda \in (0,1)$ be fixed and choose $\delta > 0$ such that $k\delta < \min\{\varepsilon, \lambda\}$. If $(p, q) \in U(\delta, \delta)$, then $F_{p,q}(\delta) > 1 - \delta$. Since f is an H-contraction, it follows that $F_{f(p), f(q)}(k\delta) > 1 - k\delta$. Therefore, $F_{f(p), f(q)}(\varepsilon) \geqslant F_{f(p), f(q)}(k\delta) > 1 - k\delta > 1 - \lambda$, that is $(f(p), f(q)) \in U(\varepsilon, \lambda)$ and the proof of the lemma follows shortly.

THEOREM 3.1.27 [356]. *Let $(S, \mathscr{F}, T)$ be a complete Menger space such that $\sup_{a<1} T(a, a) = 1$. Then every H-contraction*

f on S has a unique fixed point which is the limit of the sequence $\{f^n(p_0)\}_{n\in\mathbb{N}}$ for any arbitrary $p_0 \in S$.

Proof. Let $\varepsilon > 0$ and $\lambda \in (0,1)$ be fixed. Then, according to Lemma 3.1.6, there exists an $n(\varepsilon, \lambda)$ such that

$$F_{f^n(p)f^n(q)}(\varepsilon) > 1 - \lambda, \quad \forall\, p, q \in S, \quad \forall\, n \geqslant n(\varepsilon, \lambda).$$

If $p = p_m = f^m(p)$ and $q = p_0$, then

$$F_{p_{n+m}, p_n}(\varepsilon) = F_{f^n(p_m), f^n(p_0)}(\varepsilon) > 1 - \lambda, \quad \forall\, n \geqslant n(\varepsilon, \lambda), \quad \forall\, m \geqslant 1$$

Therefore, $\{p_n\}_{n\in\mathbb{N}}$ is a Cauchy sequence. Since S is complete, there exists a $q_0 \in S$ such that $\{p_n\}$ converges to q_0. From the definition of p_n and Lemma 3.1.8, it follows that $f(q_0) = q_0$, that is q_0 is a fixed point for f. According to Lemma 3.1.7, q_0 is unique and the proof of the theorem is complete.

Remark. It is easy to see that the classical Banach principle is a direct consequence of Theorem 3.1.27.

The idea of introducing classes of improved probabilistic contractions able to solve the fixed-point existence problem even for the t-norms T_m was first formulated by H. Sherwood [406], who introduced the class of strict contractions on the E-space $(S, \mathscr{F})$. Sherwood starts from the fact that every E-space is a PM-space satisfying the Menger inequality under T_m. Given the consistence of these spaces and their special importance, the author wondered whether their particular structure implies the existence of fixed points for probabilistic contractions. Not finding an answer, he suggested a refinement of the definition of probabilistic contractions such that the metric d of the metric-space (M, d) is defined over the values of the elements $p \in S$. This idea led to

DEFINITION 3.1.11 [406]. Let $(S, \mathscr{F})$ be an E-space over the metric space (M, d) and let $(\Omega, \mathscr{K}, P)$ be the associated probability space. The mapping $f : S \to S$ is a *strict contraction* mapping on $(S, \mathscr{F})$ if and only if there exists a $k \in (0,1)$ such that, for every $p, q \in S$ and for every real x,

$$\{\omega \in \Omega : d(p(\omega), q(\omega)) < x\} \subseteq \{\omega \in \Omega : d(f(p(\omega)), f(q(\omega))) < kx\}.$$

Notice that every strict contraction mapping is a contraction mapping.

THEOREM 3.1.28 [406]. *Every strict contraction mapping on a complete E-space has a unique fixed point.*

Proof. Let $(S, \mathscr{F})$ be an E-space over the metric space (M, d) and let $(\Omega, \mathscr{K}, P)$ be the associated probability space.

Let f be a strict contraction mapping on $(S, \mathscr{F})$, let $p_0 \in S$ and let $\{f^n(p_0)\}_{n\in\mathbb{N}} = \{p_n\}_{n\in\mathbb{N}^*}$. If $x > 0$ is fixed and if m is a positive integer, then

$$\{\omega \in \Omega : d(p_0(\omega), p_m(\omega)) < x\} \supseteq \{\omega \in \Omega : d(p_0(\omega), p_1(\omega)) + \ldots$$
$$\ldots + d(p_{m-1}(\omega), p_m(\omega)) < x\} \supseteq \{\omega \in \Omega : (1 + k + \ldots +$$
$$+k^{m-1})d(p_0(\omega), p_1(\omega)) < x\} = \{\omega \in \Omega : (1 -$$
$$-k^m)d(p_0(\omega), p_1(\omega)) < (1 - k)x\} \supseteq \{\omega \in \Omega : d(p_0(\omega), p_1(\omega)) <$$
$$(1 - k)x\}.$$

Thus, $F_{p_0,p_m} \geqslant F_{p_0,p_1}((1 - k)/j)$. Therefore,

$$\inf \{F_{p_0,p_m} : m \in \mathbb{N}\} \geqslant F_{p_0,p_1}((1 - k)/j)$$

and the proof follows from Theorem 3.1.3.

Remark. The above result seems natural if one observes that the strict contractions considered in the pseudo-metrically generated PM-spaces (equivalent to E-spaces) come down to a contraction mapping in the generating family.

It follows that the strict contractions are subjected to more restrictive conditions than the probabilistic contractions given by Definition 3.1.2.

3.1.3. *Some Classes of Generalized Contraction Mappings*

The large number of existing generalized contraction mappings for metric spaces led to the idea of extending such mappings into the PM-space. The so-called generalized contractions and probabilistic quasi-contractions [104], [105], [178] previously encountered in Theorem 3.1.17, are only a few examples in this respect.

A fixed point theorem for a self-mapping on a metric space which has a contractive iterate (or the so-called n-local contraction) at each point, given by V. M. Sehgal [395], led to several useful results for PM-spaces, [222], [224], [225], [186], [95], [97], [98], [83], [115], a few of which are presented in the sequel.

THEOREM 3.1.29 [222]. *Let $(S, \mathscr{F}, \mathrm{Min})$ be a complete Menger space. If $f : S \to S$ is a mapping such that for every $p \in S$ there exists an $n(p) \in \mathbb{N}$ such that, for all $q \in S$ and any $k \in (0, 1)$,*

$$F_{f^{n(p)}(p), f^{n(p)}(q)}(kt) \geqslant F_{p,q}(t) \quad \textit{for all } t > 0,$$

then f has a unique fixed point s and $\lim\limits_{n\to\infty} f^n(r) = s$, for each $r \in S$.

Proof. Let $s_0 \in S$ be arbitrary. Let $m_0 = n(s_0)$, $s_1 = f^{m_0}(s_0)$ and set inductively $m_i = n(s_i)$, $s_{i+1} = f^{m_i}(s_i)$. One has to show that the sequence $\{s_n\}_{n\in\mathbb{N}}$ is convergent. To this purpose, it is sufficient to prove that $\{s_n\}_{n\in\mathbb{N}}$ is a Cauchy sequence. Thus, for $m > n$, we have

$$F_{s_n,s_m}(t) = F_{f^{m_{n-1}}(s_{n-1}),f^{m_{m-1}}(s_{m-1})}(t) =$$

$$= F_{f^{m_{n-1}}(s_{n-1}),f^{m_{m-1}+m_{m-2}+\dots m_n+m_{n-1}}(s_{n-1})}(t) \geqslant$$

$$\geqslant F_{s_{n-1},f^{m_{m-1}+m_{m-2}+\dots+m_n}}(k^{-1}t) =$$

$$= F_{f^{m_{n-2}}(s_{n-2}),f^{m_{m-1}+\dots+m_n+m_{n-2}}(s_{n-2})}(k^{-1}t) \geqslant$$

$$\geqslant F_{s_{n-2},f^{m_{m-1}+\dots+m_n}(s_{n-2})}(k^{-2}t) \geqslant \dots \geqslant F_{s_0,f^{m_{m-1}+\dots+m_n}(s_0)}(k^{-n}t).$$

Let $m_{m-1} + m_{m-2} + \dots + m_n = m'n(s_0) + n'$, $0 < m'$ and $0 \leqslant n' < n(s_0)$. From the properties of f and from $T = \mathrm{Min}$, it follows that

$$F_{s_0,f^{m_{m-1}+\dots+m_n}(s_0)}(k^{-n}t) \geqslant$$

$$\geqslant T(F_{s_0,f^{n(s_0)}(s_0)}(k^{-n}t - k^{-n+1}t), F_{f^{n(s_0)}(s_0),f^{m'n(s_0)+n'}(s_0)}(k^{-n+1}t)) =$$

$$= T(F_{s_0,f^{n(s_0)}(s_0)}(k^{-n}t + k^{-n+1}t), F_{s_0,f^{(m'-1)n(s_0)+n'}(s_0)}(k^{-n}t)) \geqslant$$

$$\geqslant T(F_{s_0,f^{n(s_0)}(s_0)}(k^{-n}t - k^{-n+1}t), T(F_{s_0,f^{n(s_0)}(s_0)}(k^{-n}t -$$

$$- k^{-n+1}t), F_{f^{n(s_0)}(s_0),f^{(m'-1)n(s_0)+n'}}(k^{-n+1}t))) \geqslant$$

$$\geqslant T(F_{s_0,f^{n(s_0)}(s_0)}(k^{-n}t - k^{-n+1}t), F_{s_0,f^{(m'-2)n(s_0)+n'}(s_0)}(k^{-n}t) \geqslant \dots$$

$$\dots \geqslant T(F_{s_0,f^{n(s_0)}(s_0)}(k^{-n}t - k^{-n+1}t), F_{s_0,f^{n'}(s_0)}(k^{-n}t)) \to 1$$

for any $t > 0$ as $n \to \infty$, and hence $\{s_n\}_{n\in\mathbb{N}}$ is a Cauchy sequence.

Let $s = \lim\limits_{n\to\infty} s_n$ and let us show that

$$\lim_{n\to\infty} f^{n(s)}(s_n) = f^{n(s)}(s).$$

Indeed, there exists a positive integer n_0 such that

$$F_{f^{n(s)}(s_n),f^{n(s)}(s)}(t) \geq F_{s_n,s}(k^{-1}t) > 1 - \lambda$$

for all $t > 0$, $\lambda > 0$ and $n \geq n_0$.

Let us now state and prove that $f^{n(s)}(s) = s$. Indeed,

$$F_{f^{n(s)}(s),s}(t) \geq T(F_{f^{n(s)}(s),f^{n(s)}(s_n)}(t - kt), F_{f^{n(s)}(s_n),s}(kt)) \geq$$

$$\geq T(F_{s,s_n}(k^{-1}t - t), F_{f^{n(s)}(s_n),s}(kt)) \geq$$

$$\geq T(F_{s,s_n}(k^{-1}t - t),\ T(F_{f^{n(s)}(s_n),s_n}(kt - \varepsilon), F_{s_n,s}(\varepsilon)))$$

for all $0 < \varepsilon < kt$. On the other hand,

$$F_{f^{n(s)}(s_n),s_n}(kt - \varepsilon) = F_{f^{n(s)}f^{n(s_{n-1})}(s_{n-1}),f^{n(s_{n-1})}(s_{n-1})}(kt - \varepsilon) \geq$$

$$\geq F_{f^{n(s)}(s_{n-1}),s_{n-1}}(t - k^{-1}\varepsilon) \geq \ldots \geq F_{f^{n(s)}(s_0),s_0}(k^{-n+1}t - k^{-n}\varepsilon).$$

From the above argument, one can conclude that there exists an n_0' such that

$$F_{f^{n(s)}(s),s}(t) \geq T(F_{s,s_n}(k^{-1}t - t),\ T(F_{f^{n(s)}(s_0),s_0}(k^{-n+1}t - k^{-n}\varepsilon),$$

$$F_{s_n,s}(\varepsilon))) > 1 - \lambda$$

for all $t > 0$, $0 < \varepsilon < kt$ and $\lambda > 0$. Hence, $f^{n(s)}(s) = s$.

In order to prove the uniqueness of the fixed point s for $f^{n(s)}$, let r, $r \neq s$, be another fixed point of $f^{n(s)}$. Then there exists a real $t > 0$ such that $F_{s,r}(t) = a$ for any given $a, 0 \leq \leq a < 1$. But

$$a = F_{s,r}(t) = F_{f^{n(s)}(s),f^{n(s)}(r)}(t) = F_{s,r}(k^{-1}t) = \ldots = F_{s,r}(k^{-n}t).$$

Since $F_{s,r}(k^{-n}t) \to 1$ as $n \to \infty$, it follows that $a = 1$. This contradicts the selection of a and therefore the fixed point is unique.

From the uniqueness of the fixed point s of $f^{n(s)}$, it follows that

$$f(s) = f(f^{n(s)}(s)) = f^{n(s)+1}(s) = f^{n(s)}(f(s))$$

implies $f(s) = s$. Thus, f has a fixed point, and every fixed point of f is necessarily a fixed point of f^n for all n and hence it is unique.

To show that $f^n(r) \to s$ for each $r \in S$ as $n \to \infty$, let us observe that if n is a sufficiently large integer, then $n = m'n(s) + n'$, $0 \leqslant n' < n(s)$ and

$$F_{f^n(r),s}(t) = F_{f^{m'n(s)+n'}(r), f^{n(s)}(s)}(t) \geqslant$$

$$\geqslant F_{f^{(m'-1)n(s)+n'}(r),s}(k^{-1}t) \geqslant F_{f^{(m'-2)n(s)+n'}(r),s}(k^{-2}t) \geqslant \ldots \geqslant$$

$$\geqslant F_{f^{n'}(r),s}(k^{-m'}t).$$

Since $n \to \infty$ implies that $m' \to \infty$, it follows that $F_{f^n(r),s}(t) \to 1$ as $n \to \infty$. Thus, the proof of the theorem is complete.

A generalization of the above result is given by

THEOREM 3.1.30. *Let* $(S, \mathscr{F}, \mathrm{Min})$ *be a complete Menger space and let* f *be a mapping of* S *onto itself. Let* $\delta, 0 < \delta < 1$, *be such that for each* $\alpha \in (0, \delta)$ *there exists a* $k\alpha < 1$ *such that for every* $p \in S$ *there exists a corresponding positive integer* $n(p; \alpha)$ *for which, letting* $r = f^{n(p;\alpha)}(p)$ *and* $s = f^{n(p;\alpha)}(q)$,

$$F_{r,s}(k\alpha t) \geqslant F_{p,q}(t)$$

for all $t > 0$. *Then* f *has a unique fixed point* u *and* $f^n(p) \to u$ *for all* $p \in S$ *as* $n \to \infty$.

Proof. Let us recall that if $\mathscr{P}$ is a family of pseudometrics on S, the uniformity on S generated by $\mathscr{P}$ is separating if, for every $p, q \in S$, $p \neq q$, there exists a $d \in \mathscr{P}$ for which $d(p, q) > 0$. Let us consider this case.

A mapping $f: (S, \mathscr{P}) \to (S, \mathscr{P})$ is said to have a *d-contractive iterate* at $p \in S$ if there exists a constant $k_d \in (0, 1)$ and an integer $n(p, d)$ such that

$$d(f^{n(p,d)}(p), f^{n(p,d)}(q)) \leqslant k_d d(p, q)$$

for all $q \in S$.

One can prove in a manner similar to that in [222, Th. 1; 82, Th. 1] that if $(S, \mathscr{P})$ is sequentially complete and f is a mapping of S into S having, for each $d \in \mathscr{P}$, a d-contractive iterate at every $p \in S$, then f has exactly one fixed point u, and $f^n(p) \to u$ for any $p \in S$.

Now, for each $\alpha \in (0, \delta)$, let

$$d_\alpha(p, q) = \sup \{t : F_{p,q}(t) \leqslant 1 - \alpha\}$$

and $\mathscr{P} = \{d_\alpha, \alpha \in (0, \delta)\}$. From Proposition 3.1.6, it follows that each d_α is a pseudometric and the collection $\mathscr{P}$ generates the topology induced by the probabilistic metric $\mathscr{F}$. The proof of the theorem can be now obtained by showing that for each d_α, f has a d_α-contractive iterate at every $p \in S$.

To this end, let $p \in S$ and $\alpha \in (0, \delta)$. For $q \in S$, choose

$t_0 = d_\alpha(f^{n(p,\alpha)}(p), f^{n(p,\alpha)}(q))/k_\alpha$.

If $F_{p,q}(t_0) > 1 - \alpha$, then

$F_{r,s}(k_\alpha t_0) \geqslant F_{p,q}(t_0) > 1 - \alpha$

or

$F_{r,s}(d_\alpha(r, s)) > 1 - \alpha$

which contradicts the definition of d_α. Thus

$F_{p,q}(t_0) \leqslant 1 - \alpha$

i.e., $t_0 \leqslant d_\alpha(p, q)$ and $d_\alpha(r, s) \leqslant k_\alpha d_\alpha(p, q)$, thus completing the proof.

In order to ilustrate the consistence of mappings over uniform structures, consider the following example [82]:

Example. Let $X = [0, 1]$. For $r \in (0, 1)$, define d_r

$$d_r(x, y) = \begin{cases} |x - y| \text{ for } x, y > r \\ 0, \text{ for } x, y \leqslant r \\ x - r, \text{ for } y \leqslant r < x \\ y - r, \text{ for } x \leqslant r < y. \end{cases}$$

Each d_r is a pseudometric. The topology generated by the family $\mathscr{P} = \{d_r, r \in (0, 1)\}$ is the usual topology.

Let f be defined by

$$f(x) = \begin{cases} 0, & x = 0, \\ 2(x - 2^{-n}) + 2^{-n-1}, & \text{for } x \in \left[2^{-n}, \frac{5}{4} 2^{-n}\right] \\ 2^{-n}, & \text{for } x \in \left[\frac{5}{4} 2^{-n}, 2^{-n+1}\right] \end{cases}$$

for $n = 1, 2, 3, \ldots$.

Note that $f([2^{-n}, 2^{-n+1}]) = [2^{-n-1}, 2^{-n}]$, and on each interval $[2^{-n}, 2^{-n+1}]$, there exist points x, y such that $|f(x) - f(y)| = 2|x - y|$. Thus, we can see that, with respect to the usual metric, f is not contractive, is not locally contractive, does not have a contractive iterate, and also does not have a contractive iterate at all points. For each $x \in X$, however, f does have a d-contractive iterate at x for each $d \in \mathscr{P}$.

Another result using also uniform structures is obtained by extending the notion of weakly contractive mapping, first considered by Furi and Vignoli [152]. We use the Kuratowski measure of compactness to give some fixed theorems and to apply them to PM-spaces.

DEFINITION 3.1.12. A continuous mapping $f:(S,\mathscr{P})\to(S,\mathscr{P})$ is called α-densifying if, for every bounded set $A\subset S$ such that $k_\alpha(A)>0$, we have

$$k_\alpha(f(A))<k_\alpha(A),$$

where $k_\alpha(A)=\inf\{\varepsilon>0:\exists\, A_i,\ i\in I=\text{finite},\ \bigcup_{i\in I}A_i\supset A,$
$\sup_{p,q\in A}d_\alpha(p,q)<\varepsilon\}$.

DEFINITION 3.1.13. Let G be a real lower-semicontinuous function defined on $S\times S$. The mapping f is said to be *weakly G-contractive* if the condition

$$G(f(p),f(q))<G(p,q)$$

holds for all $p,q\in S$, $p\neq q$.

If we take $G=d_\alpha$, then f is called weakly contractive with respect to d_α.

PROPOSITION 3.1.9. *Suppose that* $(S,\mathscr{P})$ *is sequentially complete and* f *is a mapping of* S *into* S *having the following properties:*

i) *it is weakly G-contractive,*

ii) *it is* α*-densifying for all* $\alpha\in I$, $(\mathscr{P}=\{d_\alpha\}_{\alpha\in I})$,

iii) *for some* $p\in S$, $O_f(p,0,\infty)$ *is a bounded set. Then* f *has a unique fixed point in* S.

Proof. Start with

$$O_f(p,0,\infty)=f(O_f(p,0,\infty))\cup\{p\}.$$

If, for an $\alpha_0\in I$, $k_{\alpha_0}(O_f(p,0,\infty))>0$, then

$$k_{\alpha_0}(O_f(p,0,\infty))=k_{\alpha_0}(f(O_f(p,0\ \infty)))<k_{\alpha_0}(O_f(p,0,\infty)),$$

which is a contradiction. Hence, $k_\alpha(O_f(p,0,\infty))=0$ for all $\alpha\in I$ and $O_f(p,0,\infty)$ is a compact set. The function

$$g(p)=G(p,f(p)),$$

defined on the closure of $O_f(p,0,\infty)$, is lower semicontinuous and since $O_f(p,0,\infty)$ is a compact set it attains its minimum at, say, q. If $q\neq f(q)$, then

$$g(f(q))=G(f(q),f^2(q))<G(q,f(q))=g(q),$$

which is a contradiction. This proves that q is a fixed point.

Using a similar argument (G), one can show that q is unique.

Remarks. (1) The case when

$$d_\alpha(f(p),f(q)) \leqslant k_\alpha d_\alpha(p, q)$$

for some $k_\alpha \in (0,1)$ and $\alpha \in I$ is clearly contained in the above theorem.

(2) Suppose that, for a given right-continuous, nondecreasing function $\varphi : \mathbb{R}^+ \to \mathbb{R}^+ = [0, \infty)$ with $\varphi(t) < t$, the mapping $f : S \to S$ satisfies the following conditions:

(i) $d_\alpha(f(p), f(q)) \leqslant \varphi(d_\alpha(p, q))$,

(ii) S is bounded and sequentially complete.

Then there exists a unique fixed point of f in S. Indeed, if $A \subset S$ and if, for $\alpha \in I$, $k_\alpha(A) > 0$, then it is clear that (i) and $\varphi(t) < t$ imply an obvious contradiction. Thus, $\bar{A}$ is a compact set. For $A = O_f(p, 0, \infty)$, the existence of the fixed point for f is obtained.

THEOREM 3.1.31. *Let* $(S, \mathscr{F}, \mathrm{Min})$ *be a complete Menger space and suppose that there exists a* φ *as defined in remark* (2) *above such that, for* $p, q \in S$,

$$F_{f(p),f(q)}(\varphi(t)) \geqslant F_{p,q}(t).$$

Then $f : S \to S$ *has a unique fixed point.*

Proof. Let $d_\alpha(p, q) = \sup\{t,\ F_{p,q}(t) \leqslant 1 - \alpha\}$, $\alpha \in (0,1)$ be the family of pseudometrics associated with $\mathscr{F}$. The hypotheses on φ and the definition of the pseudometrics give

$$d_\alpha(p, q) > d_\alpha(f(p), f(q)).$$

The proof of the theorem follows shortly.

PROPOSITION 3.1.10. *Let* $(S, \mathscr{F}, \mathrm{Min})$ *be a complete Menger space and let* $f : S \to S$ *be such that* $f(S)$ *is a compact set and* f *is weakly G-contractive. Then* f *has a unique fixed point.*

Proof. The uniqueness is clear from the property of f to be weakly G-contractive and the existence follows from the compactness of $f(S)$ (relative compactness).

New classes of probabilistic contractions can be introduced using the Kuratowski function:

DEFINITION 3.1.14 [45], [48]. A continuous mapping $f : S \to S$ is called a *probabilistic densifying mapping* iff for every probabilistic bounded subset $A \subset S$:

i) $f(A)$ is probabilistic bounded;

ii) $\alpha_{f(A)} > \alpha_A$, if $\alpha_A < H_0$.

If f verifies the condition

$$\Phi(f(p), f(q)) > \Phi(p, q),$$

for every $p, q \in S$, $p \neq q$, where $\Phi : S \times S \to \mathscr{D}$ is upper-semicontinuous in the sense of Birkhoff (i.e., the topology on $(\mathscr{D}, \leqslant)$ has as sub-basis of the closed sets the family $S_F = \{G \in \mathscr{D} : G \geqslant F\}$, $F \in \mathscr{D}$, where $F \leqslant G$ iff $F(x) \leqslant G(x)$ for any real x) then f is called a *probabilistic weakly Φ-contractive mapping*.

PROPOSITION 3.1.11. *Let $(S, \mathscr{F}, T)$ be a probabilistic bounded complete Menger space such that* $\sup_{a<1} T(a, a) = 1$. *If f is probabilistic densifying, then there exists an (ε, λ)-compact proper subset $K \subset S$ such that $f(K) = K$.*

Proof. If $p \in S$, then

$$f(O_f(p, 0, \infty)) \cup \{p\} = O_f(p, 0, \infty).$$

Thus $K = \bar{O}_f(p, 0, \infty) \neq \emptyset$. Since $f(O_f(p, 0, \infty)) \subseteq O_f(p, 0, \infty)$ and since f is continuous, one has $f(\bar{O}_f(p, 0, \infty)) \subseteq \bar{O}_f(p, 0, \infty)$, i.e., $f(\bar{K}) \subset K$.

Let us now show that K is a compact set. It suffices to prove that $O_f(p, 0, \infty)$ is probabilistic precompact, that is $\alpha_{O_f(p,0,\infty)} = H_0$, since the (ε, λ)-uniformity on S is metrizable [392]. If $\alpha_{O_f(p,0,\infty)} < H_0$, then

$$\alpha_{O_f(p,0,\infty)} = \alpha_{O_f(p,0,\infty) \cup \{p\}} = \min \{\alpha_{O_f(p,0,\infty)}, \alpha_{\{p\}}\} = \alpha_{O_f(p,0,\infty)},$$

which contradicts the fact that f is densifying.

Hence, $K \subseteq f(K)$. Consider $q \in K$; there exists a sequence $\{q_n\}_{n \in \mathbb{N}} \subset O_{f(p,0,\infty)}$ such that $q_n \to q$ as $n \to \infty$. If $q_n = f^{m_n}(p)$, then the sequence $\{f^{m_n - 1}(p)\}_{n \in \mathbb{N}} \subset K$. Since K is a compact set, it has a convergent subsequence $\{f^{m_{n_k} - 1}(p)\}_{k \in \mathbb{N}} \subset \{f^{m_n - 1}\}_{n \in \mathbb{N}}$. From the continuity of f, it follows that

$$f(\lim_{k \to \infty} f^{m_{n_k} - 1}(p)) = \lim_{k \to \infty} f^{m_{n_k}}(p) = q,$$

i.e., $q \in f(K)$ and the proposition is proved.

Remarks. 1) The above result is valid for a general *probabilistic γ-condensing*, i.e., in Definition 3.1.14 $\alpha = \gamma$ is a random measure of noncompactness [49].

2) If $(S, \mathscr{F}, \tau)$ is a Serstnev complete space with a left-continuous t-function τ and $\{p_n\}_{n \in \mathbb{N}}$ is a probabilistic bounded sequence such that $F_{p_n, f(p_n)} \xrightarrow{\mathscr{L}} H_0$, then the set $\{p_n\}_{n \in \mathbb{N}}$ is an (ε, λ)-compact set and the accumulation points are fixed points for every probabilistic α-densifying mapping f [49, Th. 1.3.]

THEOREM 3.1.32 [48]. *Let $(S, \mathscr{F}, T)$ be a complete Menger space with the t-norm T such that $\sup_{a<1} T(a, a) = 1$. Let f be a probabilistic weakly Φ-contractive mapping of S into S.*

If there exists a $p \in S$ such that $\sup_{t\in\mathbb{R}} G_p(t) = \sup_{t\in\mathbb{R}} \{\inf_{n\in\mathbb{N}} F_{p,f^n(p)}(t)\} = 1$, then f has a unique fixed point.

Proof. Consider the relation

$$D'_{O_f(p,0,\infty)}(t) = \inf_{m,n\in\mathbb{N}} F_{f^m(p),f^n(p)}(t) \geqslant$$

$$\geqslant \inf_{m,n\in\mathbb{N}} T\left(F_{p,f^m(p)}\left(\frac{t}{2}\right), F_{p,f^n(p)}\left(\frac{t}{2}\right)\right) \geqslant$$

$$\geqslant T\left(G_p\left(\frac{t}{2}\right), G_p\left(\frac{t}{2}\right)\right).$$

Hence, $\sup_{t\in\mathbb{R}} D'_{O_f(p,0,\infty)}(t) = 1$. Since $D_{O_f(p,0,\infty)}(t) = D'_{O_f(p,0,\infty)}(t-0)$, it follows that $O_f(p, 0, \infty)$ is a probabilistic bounded subset of S.

From $f(O_f(p, 0, \infty)) \subseteq O_f(p, 0, \infty)$ and from the continuity of f, it follows that $f(\bar{O}_f(p, 0, \infty)) \subseteq f(O_f(p, 0, \infty)) \subseteq \bar{O}_f(p, 0, \infty)$ which shows that the mapping $\tilde{f}(p) = (p, f(p))$ from $\bar{O}_f(p, 0, \infty)$ into $\bar{O}_f(p, 0, \infty) \times \bar{O}_f(p, 0, \infty)$ is well defined and continuous.

Moreover, $\bar{O}_f(p, 0, \infty)$ is a compact set. In order to show this, it is sufficient to show that $O_f(p, 0, \infty)$ is a precompact set, i.e., $\alpha_{O_f(p,0,\infty)} = H$. Indeed, since the uniformity generated by $\mathscr{F}$ is metrizable and the corresponding metric space is complete, the conclusion is immediate.

If $\alpha_{O_f(p,0,\infty)} < H$, then $\alpha_{O_f(p,0,\infty)} = \alpha_{\{p\}\cup f(O_f(p,0,\infty))} = \min\{\alpha_{\{p\}}, \alpha_{f(O_f(p,0,\infty))}\} = \alpha_{f(O_f(p,0,\infty))}$, which is a contradiction.

Let φ be a mapping from $\bar{O}_f(p, 0, \infty)$ into $\mathscr{D}$ defined as

$$\varphi(p) = \Phi(p, f(p))$$

and let $\tilde{\Phi}$ be the restriction of Φ to $\bar{O}_f(p, 0, \infty) \times \bar{O}_f(p, 0, \infty)$. Then $\tilde{\varphi} = \tilde{\Phi} \circ f$ and φ is semicontinuous in the sense of Birkhoff. On the other hand, since $\bar{O}_f(p, 0, \infty)$ is a compact set, there exists a $q \in \bar{O}_f(p, 0, \infty)$ such that the relation $\varphi(q) < < \varphi(p)$ is false for every $p \in \bar{O}_f(p, 0, \infty)$ (see [42, Th. 16]).

Moreover, q is the fixed point of f. Indeed, if $f(q) \neq q$, then $\varphi(f(q)) = \Phi(f(q), f^2(q)) > \Phi(q, f(q))$, which is a contradiction.

The uniqueness of the fixed point follows from the fact that $\Phi(q, r) = \Phi(f(q), f(r)) > \Phi(q, r)$ for any two fixed points $r \neq q$.

COROLLARY. *Let (S, d) be a complete metric space and let f be a weakly φ-contractive mapping in the sense of Furi and Vignoli* [152]. *If there exists a $p \in S$ such that $O_f(p, 0, \infty)$ is a bounded set, then f has a unique fixed point.*

THEOREM 3.1.33 [48]. *Let $(S, \mathscr{F}, T)$ be a complete Menger space with the t-norm T such that $\sup_{a<1} T(a, a) = 1$. Let f be a probabilistic densifying mapping of S into S such that $F_{f(p),f(q)} > F_{p,q}$ for every $p, q \in S$, $p \neq q$.*

If $\mathscr{F}$ is an upper semicontinuous mapping in the sense of Birkhoff, then f has a unique fixed point.

Proof. Is suffices to observe that f is probabilistic weakly $\mathscr{F}$-contractive.

Remark. The above results (Th. 3.1.32 and Th. 3.1.34) hold for a nonsingular random measure of noncompactness γ which is invariant to closure.

By relaxing even further the restrictions imposed upon the t-norm T, newer generalizations of Theorem 3.1.29 were obtained. For example, consider the results obtained by O. Hadžić [186]:

THEOREM 3.1.34. *Let $(S, \mathscr{F}, T)$ be a complete Menger space with continuous t-norm T, and let $f : S \to S$ be a mapping such that, for every $p \in S$, there exists an $n(p) \in \mathbb{N}$ such that, for any $q \in S$,*

$$F_{f^{n(p)}(p), f^{n(p)}(q)}(kt) \geqslant F_{p,q}(t), \text{ for all } t > 0,$$

where $k \in (0, 1)$. If there exists an $s_0 \in S$ such that $\sup_{t \in \mathbb{R}} G_{s_0}(t) = 1$, with

$$G_{s_0}(t) = \inf \{F_{f^m(s_0), s_0}(t),\ m \in \mathbb{N}\},$$

then f has a unique fixed point $s \in S$ and $\lim_{n\to\infty} f^n(r) = s$ for every $r \in S$.

Proof. Let us define the sequence $\{s_n\}_{n \in \mathbb{N} \cup \{0\}}$ as in Theorem 3.1.29, and then, for $m > n$,

$$F_{s_n, s_m}(t) \geqslant F_{s_0, f^{m_{m-1}+\dots+m_{n(s_0)}}}(k^{-n}t) \geqslant G_{s_0}(k^{-n}t)$$

for all $t > 0$.

Since $\sup_{t \in \mathbb{R}} G_{s_0}(t) = 1$, it follows that $\{s_n\}_{n \in \mathbb{N}}$ is a Cauchy sequence. Let $s = \lim_{n\to\infty} s_n$.

Since the inequality

$$F_{f^{n(s)}(s_n), f^{n(s)}(s)}(kt) \geqslant F_{s_n, s}(t)$$

is true for any $t > 0$ and any $n \in \mathbb{N}$, it follows that $\lim_{n\to\infty} f^{n(s)}(s_n) = f^{n(s)}(s)$. The relation $f^{n(s)}(s) = s$ follows from

$$F_{s,f^{n(s)}(s)}(t) \geqslant T(F_{s,s_n}(k^{-1}t-t),\ T(F_{f^{n(s)}(s_0),s_0}(k^{-n+1}t-k^{-n}\varepsilon),\ F_{s_n,s}(\varepsilon)))$$

for $0 < \varepsilon < kt$. It can be easily proved that s is the unique fixed point of the mapping $f^{n(s)}$ and, hence, the unique fixed point of the mapping f. It is also true that

$$\lim_{n\to\infty} f^n(r) = s \text{ for any } r \in S$$

and the proof of the theorem is complete.

A generalization of Theorem 3.1.29 is given by the following

COROLLARY [186]. *Let* $(S, \mathscr{F}, T)$ *be a complete Menger space with t-norm* T *of* $\mathscr{H}$*-type. If* $f : S \to S$ *is constructed in such a way that, for every* $p \in S$, *there exists an* $n(p) \in N$ *such that for any* $q \in S$ *and any* $t > 0$,

$$F_{f^{n(p)}(p),f^{n(p)}(q)}(kt) \geqslant F_{p,q}(t),$$

with $k \in (0, 1)$, *then* f *has a unique fixed point* $s \in S$, *and, for any* $r \in S$, $\lim_{n\to\infty} f^n(r) = s$.

Proof. We shall prove that for every $s_0 \in S$, $\sup_{t\in\mathbb{R}} G_{s_0}(t) = 1$. Let $m \in \mathbb{N}$ and let $m'n(s_0) < m \leqslant (m'+1)n(s_0)$. Then, for every $t > 0$,

$$\begin{aligned}
F_{f^m(s_0),s_0}(t) &\geqslant T(F_{f^m(s_0),f^{n(s_0)}(s_0)}(kt),\ F_{f^{n(s_0)}(s_0),s_0}(t-kt)) \geqslant \\
&\geqslant T(F_{f^{m-n(s_0)}(s_0),s_0}(t),\ F_{f^{n(s_0)}(s_0),s_0}(t-kt)) \geqslant \ldots \geqslant \\
&\geqslant T(T(\ldots T(F_{f^{m-m'n(s_0)}(s_0),s_0}(t),\ F_{f^{n(s_0)}(s_0),s_0}(t-kt)),\ldots \\
&\ldots,\ F_{f^{n(s_0)}(s_0),s_0}(t-kt)) \geqslant T^{m'}(\min_{1\leqslant j\leqslant n(s_0)} \{F_{f^j(s_0),s_0}(t-kt)\}).
\end{aligned}$$

Since $\lim_{t\to\infty} \min_{1\leqslant j\leqslant n(s_0)} \{F_{f^j(s_0),s_0}(t-kt)\} = 1$ and T is $\mathscr{H}$-type, one concludes that $\sup_{t\in\mathbb{R}} G_{s_0}(t) = 1$.

We have shown that another way of generalizing some existing fixed point theorems was that of relaxing the contraction condition. To the condition of n-local contraction, Chang Shihsen [100] adds the generalized-contraction condition as defined by L. B. Cirić [105], using a function φ with the following properties:

(φ) i) $\varphi : R^+ \to R^+$, ii) φ is nondecreasing, iii) $\varphi(0) = 0$ and $\varphi(t) > kt$, $\forall\, t > 0$ for some constant $k > 1$.

THEOREM 3.1.35. *Let* $(S, \mathscr{F}, T = \mathrm{Min})$ *be a complete Menger space and let* $f : S \to S$ *be a mapping satisfying the following conditions* :

i) *For every* $p \in S$, *there exists an integer* $n(p) \geqslant 1$ *such that, for any* $t > 0$ *and any* $q \in S$,

$$F_{f^{n(p)}(p),f^{n(p)}(q)}(t) \geqslant$$

$$\geqslant \min \{F_{p,q}(\varphi(t)),\ F_{p,f^{n(p)}(p)}(\varphi(t)),\ F_{p,f^{n(p)}(q)}(\varphi(t)),$$

$$F_{q,f^{n(p)}(p)}(\varphi_2(t)),\ F_{q,f^{n(p)}(q)}(\varphi(2t))\},$$

where the function $\varphi(t)$ *satisfies the condition* (φ).

ii) *There exists a* $p \in S$ *such that*

$$\sup_{t\in\mathbb{R}} G_p(t) = \sup_{t\in\mathbb{R}} \{\inf_{q\in O_f(p,0,\infty)} F_{p,q}(t)\} = 1.$$

Then f *has a unique fixed point* $r \in S$, *and, for every* $s \in S$, $f^n(s) \to r$ *as* $n \to \infty$.

Proof. First, let us prove that, for every $p_0 \in S$, the sequence $p_m\}_{m\in\mathbb{N}\cup\{0\}}$, defined by

$$p_{m+1} = f^{n(p_m)}(p_m),\ m \in \mathbb{N} \cup \{0\},$$

is a Cauchy sequence in S.

If $n_i = n(p_i)$, $i = 0, 1, 2, \ldots$, then for any positive integer m, i and $t > 0$, it follows from condition (i) that

$$F_{p_m,p_{m+i}}(t) \geqslant F_{f^{n_{m-1}}(p_{m-1}),f^{n_{m+i-1}+\cdots+n_m+n_{m-1}}(p_{m-1})}(t) \geqslant$$

$$\geqslant \min \{F_{p_{m-1},f^{n_{m+i-1}+\cdots+n_m}(p_{m-1})}(\varphi(t)),\ F_{p_{m-1},f^{n_{m-1}}p_{(m-1)}}(\varphi(t)),$$

$$F_{p_{m-1},f^{n_{m+i-1}+\cdots+n_m+n_{m-1}}(p_{m-1})}\varphi(t)),$$

$$F_{f^{n_{m+i-1}+\cdots+n_m}(p_{m-1}),f^{n_{m-1}}(p_{m-1})}(\varphi(2t)),$$

$$F_{f^{n_{m+i-1}+\cdots+n_m}(p_{m-1}),f^{n_{m+i-1}+\cdots+n_m+n_{m-1}}(p_{m-1})}(\varphi(2t))\}.$$

However,

$$F_{f^{n_{m+i-1}+\dots+n_m}(p_{m-1}),f^{n_{m-1}}(p_{m-1})}(\varphi(2t)) \geqslant$$

$$\geqslant T(F_{f^{n_{m+i-1}+\dots+n_m}(p_{m-1}),p_{m-1}}(\varphi(2t)-kt),\ F_{f^{n_{m-1}}(p_{m-1}),p_{m-1}}(kt)) \geqslant$$

$$\geqslant T(F_{f^{n_{m+i-1}+\dots+n_m}(p_{m-1}),p_{m-1}}(kt),\ F_{f^{n_{m-1}}(p_{m-1}),p_{m-1}}(kt)).$$

Similarly,

$$F_{f^{n_{m+i-1}+\dots+n_m}(p_{m-1}),f^{n_{m+i-1}+\dots+n_m+n_{m-1}}(p_{m-1})}(\varphi(2t)) \geqslant$$

$$\geqslant T(F_{f^{n_{m+i-1}+\dots+n_m}(p_{m-1}),p_{m-1}}(kt),$$

$$F_{f^{n_{m+i-1}+\dots+n_m+n_{m-1}}(p_{m-1})p_{m-1}}(kt)).$$

From the above relations, it follows that

$$F_{p_m,p_{m+i}}(t) \geqslant \inf_{q\in O_f(p_{m-1},0,\infty)} F_{p_{m-1},q}(kt), \quad \forall\ t>0.$$

Repeating this procedure, one gets

$$F_{p_m,p_{m+i}}(t) \geqslant \inf_{q\in O_f(p_m,0,\infty)} F_{p_{m-1},q}(kt) \geqslant \inf_{q\in O_f(p_{m-2},0,\infty)} F_{p_{m-2}}(k^2t) \geqslant$$

$$\geqslant \dots \geqslant \inf_{q\in O_f(p_0,0,\infty)} F_{p_0,q}(k^m t) \geqslant \sup_{u<k^m t} \inf_{q\in O_f(p_0,0,\infty)} (u)$$

and, hence,

$$\lim_{n\to\infty} F_{p_m,p_{m+i}}(t) \geqslant \sup_{u>0} \inf_{q\in O_f(p_0,0,\infty)} F_{p_0,q}(u) = 1,\ \forall\, t\in \mathbb{R}^+.$$

This shows that $\{p_m\}_{m\in\mathbb{N}}$ is a Cauchy sequence in S. Let $r=\lim_{m\to\infty} p_m$ and $n_*=n(r)$. We prove that $f^{n_*}(r)=r$. Indeed, for any $i\in\mathbb{N}$ and any $t\in\mathbb{R}^+$,

$$F_{p_i,f^{n_*}(p_i)}(t) = F_{f^{n_{i-1}}(p_{i-1}),f^{n_*+n_{i-1}}(p_{i-1})}(t) \geqslant \sup_{u<k^i t} \inf_{q\in O_f(p_0,0,\infty)} F_{p_0,q}(u)$$

and, hence,

$$\lim_{i\to\infty} F_{p_i,f^{n_*}(p_i)} \geqslant \sup_{u>0} \inf_{q\in O_f(p_0,0,\infty)} F_{p_0,q}(u) = 1,\ \forall\, t\in\mathbb{R}^+.$$

On the other hand, since $p_i\to r$ as $i\to\infty$ it follows that

$$\lim_{i\to\infty} F_{r,f^{n_*}(p_i)}(t) \geqslant \lim_{i\to\infty} T\left(F_{r,p_i}\left(\frac{t}{2}\right),\ F_{p_i,f^{n_*}(p_i)}\left(\frac{t}{2}\right)\right) = 1,\ \forall t\in\mathbb{R}^+.$$

Taking $h \in (1, k)$ one gets

$$F_{r,f^{n}*(r)} \geqslant T\left(F_{r,f^{n}*(p_i)}\left(t - \frac{t}{h}\right), F_{f^{n}*(r),f^{n}*(p_i)}\left(\frac{t}{h}\right)\right) \geqslant$$

$$\geqslant T\left(F_{r,f^{n}*(p_i)}\left(t - \frac{t}{h}\right), \min\left\{F_{r,p_i}\left(\varphi\left(\frac{t}{h}\right)\right),\right.\right.$$

$$F_{r,f^{n}*(r)}\left(\varphi\left(\frac{t}{h}\right)\right), F_{r,f^{n}*(p_i)}\left(\varphi\left(\frac{t}{h}\right)\right), F_{p_i,f^{n}*(r)}\left(\varphi\left(\frac{2t}{h}\right)\right),$$

$$\left.\left.F_{p_i,f^{n}*(p_i)}\left(\varphi\left(\frac{2t}{h}\right)\right)\right\}\right) \geqslant T\left(F_{r,f^{n}*(p_i)}\left(t - \frac{t}{h}\right),\right.$$

$$\min\left\{F_{r,p_i}\left(\frac{kt}{h}\right), F_{r,f^{n}*(r)}\left(\frac{kt}{h}\right), F_{r,f^{n}*(p_i)}\left(\frac{kt}{h}\right), T\left(F_{p_i,r}\left(\frac{kt}{h}\right),\right.\right.$$

$$\left.\left.\left.F_{r,f^{n}*(r)}\left(\frac{kt}{h}\right)\right), T\left(F_{p_i,r}\left(\frac{kt}{h}\right), F_{r,f^{n}*(p_i)}\left(\frac{kt}{h}\right)\right)\right\}\right), \quad \forall\, t \in \mathbb{R}^{+}.$$

Since $\dfrac{k}{h} > 1$, it follows that

$$F_{r,f^{n}*(r)}(t) \geqslant F_{r,f^{n}*(r)}\left(\frac{kt}{h}\right), \quad \forall\, t \in \mathbb{R}^{+},$$

and, inductively, one can get

$$F_{r,f^{n}*(r)}(t) \geqslant F_{r,f^{n}*(r)}\left(\left(\frac{k}{h}\right)^{i} t\right), \quad i = 1, 2, \ldots \quad \forall\, t \in \mathbb{R}^{+}.$$

Hence,

$$F_{r,f^{n}*(r)}(t) = 1, \quad \forall\, t \in \mathbb{R}^{+}, \quad \text{i.e., } r = f^{n*}(r).$$

Let us now prove that f^{n*} has a unique fixed point in S. Indeed, if there exists another $r' \in S$, $r' \neq r$, $f^{n*}(r') = r'$ then, from condition (i),

$$F_{r,r'}(t) = F_{f^{n}*(r),f^{n}*(r')}(t) \geqslant \min\{F_{r,r'}(\varphi(t)), F_{r,r}(\varphi(t)),$$

$$F_{r,r'}(\varphi(t)), F_{r',r}(\varphi(2t)), F_{r',r'}(\varphi(2t))\} \geqslant F_{r,r'}(\varphi(t)) \geqslant \ldots \geqslant$$

$$\geqslant F_{r,r'}(\varphi^{i}(t)) \geqslant F_{r,r'}(k^{i}t), \quad \forall\, t \in \mathbb{R}^{+}.$$

Letting $i \to \infty$ yields $F_{r,r'}(t) = 1$, $\forall\, t \in \mathbb{R}^{+}$, i.e., $r = r'$. It is easy to prove that r is the unique fixed point of f.

In order to prove that $f^n(p_0) \to r$ as $n \to \infty$, let us observe that, for any integer $n > n_*$, $n = m'n_* + n'$, $0 \leqslant n' < n_*$, and, hence,

$$F_{r,f^n(p_0)}(t) = F_{f^{n_*}(r),f^{m'n_*+n'}(p_0)}(t) \geqslant$$

$$\geqslant \min\{F_{r,f^{(m'-1)n_*+n'}(p_0)}(\varphi(t)), F_{r,r}(\varphi)t)), F_{r,f^n(p_0)}(\varphi(t)),$$

$$F_{f^{(m'-1)n_*+n}(p_0),r}(\varphi(2t)), F_{f^{(m'-1)n_*+n'}(p_0),f^n(p_0)}(\varphi(2t))\} \geqslant$$

$$\geqslant \min\{F_{r,f^{(m'-1)n_*+n'}(p_0)}(kt), F_{r,f^n(p_0)}(kt), T(F_{f^{(m'-1)n_*+n'}(p_0),r}(kt),$$

$$F_{r,f^n(p_0)}(kt)\} = \min\{F_{r,f^{(m'-1)n_*+n'}(p_0)}(kt), F_{r,f^n(p_0)}(kt)\}.$$

Repeating this argument, one can prove that

$$F_{r,f^n(p_0)}(t) \geqslant \min\{F_{r,f^{(m'-1)n_*+n'}(p)}(kt), F_{r,f^n(p_0)}(k^i t)\}, \ \forall\, t \in \mathbb{R}^+.$$

Then for $i \to \infty$, one has

$$F_{r,f^n(p_0)}(t) \geqslant F_{r,f^{(m'-1)n_*+n'}(p_0)}(kt) \geqslant F_{r,f^{n'}(p_0)}(k\ t), \ \forall\, t \in \mathbb{R}^+$$

and for $m \to \infty$ (hence $n \to \infty$),

$$F_{r,f^n(p_0)}(t) \geqslant \lim_{m\to\infty} F_{r,f^{n'}(p_0)}(k^m t) = 1, \ \forall\, t \in \mathbb{R}^+,$$

which implies that $f^n(p_0) \to r$ as $n \to \infty$.

Remarks. (1) If (i) is substituted by

(iii) for every $p \in S$, there exists an integer $n(p) \geqslant 1$ such that, for any $q \in S$ and any $t \in \mathbb{R}^+$,

$$F_{f^{n(p)}(p),f^{n(p)}(q)}(t) \geqslant \min\{F_{p,q}(\varphi(t)), F_{p,f^{n(p)}(p)}(\varphi(t)), F_{p,f^{n(p)}(q)}(\varphi(t))\},$$

then the condition $T = \text{Min}$ in the above theorem can be substituted by the more general condition of continuity for T.

(2) The conclusion of Theorem 3.1.35 still holds if for every $p \in S$ there exists an $n(p) \in \mathbb{N}$ such that, for any $q \in S$ and any $t \in \mathbb{R}^+$,

(iv) $F_{f^{n(p)}(p),f^{n(p)}(q)}(t) \geqslant \min\{F_{p,q}(ht), F_{p,f^{n(p)}(p)}(ht), F_{p,f^{n(p)}(q)}(ht)\}$

or, equivalently,

(v) $F_{f^{n(p)}(p),f^{n(p)}(q)}(t) \geqslant$

$$\geqslant \min\{F_{p,q}(ht), F_{p,f^{n(p)}(p)}(ht), F_{p,f^{n(p)}(q)}(ht), F_{q,f^{n(p)}(q)}(2ht),$$

$$F_{q,f^{n(p)}(p)}(2ht)\},$$

where h is a constant greater than unity [100, Th. 3; 97, Th. 6].

COROLLARY. *Let* (S, d) *be a complete metric space and let* $f: S \to S$ *be a mapping satisfying the following condition: for every* $p \in S$, *there exists a positive integer* $n(p)$ *such that, for any* $q \in S$,

(1) $d(f^{n(p)}(p), f^{n(p)}(q)) \leqslant \psi(\max\{d(p,q), d(p, f^{n(p)}(p)), d(p, f^{n(p)}(q)\})$

or

(2) $$d(f^{n(p)}(p), f^{n(p)}(q)) \leqslant \psi(\max\{d(p,q), d(p, f^{n(p)}(p)), d(p, f^{n(p)}(q)), \frac{1}{2} d(q, f^{n(p)}(p)), \frac{1}{2} d(q, f^{n(p)}(q))\}),$$

where the function $\psi: \mathbb{R}^+ \to \mathbb{R}^+$ *is strictly increasing,* $\psi(0) = 0$ *and* $\psi(t) \leqslant kt$, $t > 0$, *with* $k \in (0,1)$ *a constant.*

Then f *has a unique fixed point* $r \in S$, *and, for every* $q \in S$, $f^n(q) \to r$ *as* $n \to \infty$.

PROPOSITION 3.1.12 [97]. *Let* $(S, \mathscr{F}, \mathrm{Min})$ *be a complete Menger space and let* $f: S \to S$ *satisfy the following condition: for any* $p, q \in S$ *and any* $t \geqslant 0$,

$$F_{f(p),f(q)}(t) \geqslant \min\{F_{p,q}(t), F_{p,f(q)}(t), F_{q,f(q)}, \frac{1}{2}[F_{p,f(q)}(t) + F_{q,f(p)}(t)]\}.$$

Then f *has a fixed point* $r \in S$, *and, for any* $p \in S$, $f^n(p) \to r$, *as* $n \to \infty$, *if and only if there exists a closed set* $M \subset S$ *such that*

(a) $F_{f(p),s}(t) \geqslant F_{p,s}(t), \quad \forall\ p \in S,\ s \in M,\ t \geqslant 0;$

(b) $\limsup_{n\to\infty} \{\sup_{s \in M} F_{f^n(p),s}(t)\} = H_0(t), \ \forall\ t \geqslant 0.$

Proof. To prove the necessity, let f have a fixed point and let us denote by $\mathrm{Fix}(f)$ the set of all the fixed points of f. Taking $M = \mathrm{Fix}(f)$, we have $M \neq \emptyset$, and, for any $p \in S$, $f^n(p) \to u \in M$ and, for any $p \in S$, $q \in M$, $t \geqslant 0$, one has

$$F_{f(p),q} = F_{f(p),f(q)}(t) \geqslant \min\{F_{p,q}(t), F_{p,q}(t), F_{q,q}(t), \frac{1}{2}[F_{p,q}(t) + F_{q,f(p)}(t)]\} = \min\{F_{p,q}(t), H_0(t), \frac{1}{2}[F_{p,q}(t) + F_{q,f(p)}(t)]\}.$$

Now, let us prove that $F_{f(p),q}(t) \geqslant F_{p,q}(t)$, $\forall\ p \in S$, $q \in M$, $t > 0$. If this were not the case, there would be $p' \in S$, $q' \in M$, $t' > 0$ such that $F_{f(p'),q'}(t') < F_{p',q'}(t')$. Hence,

$$F_{f(p'),q'}(t') \geqslant \min\{F_{p',q'}(t'), \frac{1}{2}[F_{p',q'}(t') + F_{q',f(p')}(t')]\} > F_{f(p'),q'}(t').$$

This would lead to a contradiction and, hence, one obtains condition (a). To prove that M is a closed set in S, let $\{q_n\}_{n\in\mathbb{N}} \subset M$ be such that $q_n \to \bar{q}$. Then

$$F_{f(\bar{q}),\bar{q}}(t) \geqslant \mathrm{Min}\left(F_{f(\bar{q}),q_n}\left(\frac{t}{2}\right), F_{q_n,\bar{q}}\left(\frac{t}{2}\right)\right) \geqslant$$

$$\geqslant \mathrm{Min}\left(F_{\bar{q},q_n}\left(\frac{t}{2}\right), F_{\bar{q},q_n}\left(\frac{t}{2}\right)\right) \geqslant F_{\bar{q},q_n}\left(\frac{t}{2}\right) \to H_0(t),$$

as $n \to \infty$, $\forall\, t \geqslant 0$,

i.e., $\bar{q} \in M$.

Moreover, from

$$F_{f^n(p),q} = F_{f(f^{n-1}(p)),q}(t) \geqslant F_{f^{n-1}(p),q}(t), \ \forall\, t \geqslant 0,\ q \in M,$$

it follows that

$$\sup_{q\in M} F_{f^n(p),q}(t) \geqslant \sup_{q\in M} F_{f^{n-1}(p),q}(t), \ \forall\, t \geqslant 0.$$

Hence, $\{\sup_{q\in M} F_{f^n(p),q}(t)\}_{n\in\mathbb{N}}$ is a nondecreasing sequence. Since $u \in M$, one has

$$\lim_{n\to\infty} \sup \{\sup_{q\in M} F_{f^n(p),q}(t)\} \geqslant \lim_{n\to\infty} \sup \{F_{f^n(p),u}(t)\} = \lim_{n\to\infty} F_{f^n(p),u}(t) =$$

$$= H_0(t), \ \forall\, t \geqslant 0.$$

Thus, the proof of necessity is complete.

To prove sufficiency, let us observe that if

$$F_{f^n(p),q}(t) = F_{f(f^{n-1}(p)),q}(t) \geqslant F_{f^{n-1}(p),q}(t), \ \forall\, q \in M,\ t \geqslant 0, \text{ for any}$$

$p \in S$, then

$$\sup_{q\in M} F_{f^n(p),q}(t) \geqslant \sup_{q\in M} F_{f^{n-1}(p),q}(t), \ \forall\, t \geqslant 0.$$

It follows that

$$(*) \quad \lim_{n\to\infty} \{\sup_{q\in M} F_{f^n(p),q}(t)\} = \lim_{n\to\infty} \sup \{\sup_{q\in M} F_{f^n(p),q}(t)\} = H_0(t), \ \forall\, t \geqslant 0$$

and, hence, for any $\varepsilon > 0$, $\lambda > 0$ there exists an $n(\varepsilon, \lambda) \in \mathbb{N}$ such that $\sup_{q\in M} F_{f^n(p),q}(\varepsilon) > 1 - \lambda$, $\forall\, n \geqslant n(\varepsilon, \lambda) = n'$.

Consequently, for any $q \in M$ and any positive integer,

$$F_{f^m(p),f^n(q)}(2\varepsilon) \geqslant \mathrm{Min}\,(F_{f^m(p),q}(\varepsilon), F_{q,f^n(p)}(\varepsilon)) \geqslant$$

$$\geqslant \mathrm{Min}\,(F_{f^{m-1}(p),q}(\varepsilon), F_{f^{n-1}(p),q}(\varepsilon)) \geqslant \ldots \geqslant$$

$$\geqslant \mathrm{Min}\,\{F_{f^{n'}(p),q}(\varepsilon), F_{f^{n'}(p),q}(\varepsilon)\} = F_{f^{n'}(p),q}(\varepsilon) \text{ since, } m, n > n(\varepsilon, \lambda).$$

It follows that $\{f^n(p)\}_{n\in\mathbb{N}}$ is a Cauchy sequence in S. Let $n = \lim_{n\to\infty} f^n(p)$; since M is a closed-set, relation $(*)$ implies that $u \in M$. From condition (a), one has

$$F_{f(u),u}(t) \geqslant F_{u,u}(t) = H_0(t), \ \forall\ t \geqslant 0.$$

i.e., $u = f(u)$ and the proof of the theorem is complete.

COROLLARY. *Let* $(S, \mathscr{F}, \mathrm{Min})$ *be a complete Menger space. Suppose that* $f : S \to S$ *is nonexpansive (i.e.,* $F_{f(p),f(q)}(t) \geqslant F_{p,q}(t)$, $\forall\ p, q \in S,\ t \geqslant 0$). *Then* f *has a fixed point* $r \in S$, *and, for any* $p \in S$, $f^n(p) \to r$ *as* $n \to \infty$, *if and only if there exists a closed set* $M \subset S$ *such that conditions (a) and (b) of the above theorem hold.*

Some new results generalizing and unifying those given by I. Istrăţescu [222] and V. M. Sehgal and A. T. Bharucha-Reid [396] were obtained by Shi-sen Chang and Shi-kun Kang [101].

THEOREM 3.1.36. *Let* $(S, \mathscr{F}, T)$ *be a Menger space with a continuous t-norm* T. *Let* $f : S \to S$ *be a mapping with the property that, for every* $p \in S$, *there exists an* $n(p) \in \mathbb{N}$ *such that, for any* $q \in S$ *and any* $t \geqslant 0$,

$$\text{(I)}\ F_{f^{n(p)}(p),f^{n(p)}(q)}(t) \geqslant F_{p,q}(\Phi(t)),$$

where Φ *is a function satisfying the following condition :*

(Φ_1) $\Phi : \mathbb{R}^+ \to \mathbb{R}^+$, $\Phi(0) = 0$, *is strictly increasing and* $\forall\ t > 0$ $\lim_{n\to\infty} \Phi^n(t) = \infty$.

If, for every $p \in S$, $O_f(p, 0, \infty)$ *is probabilistic bounded (or equivalently, see the Remark of Theorem 3.1.3, for a point* $p \in S$), *then* f *has a unique fixed point* $r \in S$, *and for every* $s \in S$, $\lim_{n\to\infty} f^n(s) = r$.

The main steps of the proof are similar to those outlined in the proof of Theorem 3.1.61 and hence we shall not reiterate them here.

COROLLARY 1 [101, Th. 2]. *Let* $(S, \mathscr{F}, \mathrm{Min} = T)$ *be a complete Menger space and let* $f : S \to S$ *be a mapping satisfying condition (I) of Theorem 3.1.36 with a function* Φ *satisfying condition* (Φ_1) *of the same theorem as well as the following condition :*

$$(\Phi_2)\ \lim_{t\to\infty} (\Phi(t) - t) = \infty.$$

Then, the conclusion of Theorem 3.1.36 remains true for this function too.

Proof. It is sufficient to prove that, for a $p \in S$, the orbit $O_f(p, 0, \infty)$ is probabilistic bounded.

Let $p \in S$ and $\lambda \in (0, 1)$; there exists a $t_\lambda > 0$ such that

$$\min_{0 \leqslant m \leqslant n(p)} F_{p f^{m}(p)}(\Phi(t)) \geqslant 1 - \lambda, \ \forall \, t \geqslant t_\lambda .$$

Let $n > n(p)$ and put $n = m'n(p) + n'$, $0 \leqslant m'$ and $0 \leqslant n' < n(p)$. Then

$$F_{p,f^{n}(p)}(\Phi(t)) = F_{p,f^{m'n(p)+n'}(p)}(\Phi(t)) \geqslant T(F_{p,f^{n(p)}(p)}(\Phi(t) - t),$$

$$F_{f^{m'n(p)+n'}(p), f^{n(p)}(p)}(t) \geqslant$$

$$\geqslant T(F_{p,f^{n(p)}(p)}(\Phi(t) - t), F_{p,f^{(m'-1)n(p)+n'}(p)}(\Phi(t)) \geqslant \ldots \geqslant$$

$$\geqslant T(F_{p,f^{n(p)}(p)}(\Phi(t) - t), F_{p,f^{n'}(p)}(\Phi(t)).$$

Considering condition (Φ_2^1), it follows that there exists a $t_1 > 0$ such that as $t \geqslant t_1$, $\Phi(t) - t > \Phi(t_\lambda)$. For $t \geqslant t' = \max\{t_1, t_\lambda\}$, one has

$$F_{p,f^{n}(p)}(\Phi(t)) > 1 - \lambda, \ n = 0, 1, 2, \ldots .$$

Hence,

$$\inf_{q \in O_f(p,0,\infty)} F_{p,q}(\Phi(t)) \geqslant 1 - \lambda, \ \forall \, t > t',$$

and thus

$$\sup_{t>0} \inf_{q \in O_f(p,0,\infty)} F_{p,q}(t) \geqslant 1 - \lambda,$$

i.e.,

$$\sup_{t>0} \inf_{q \in O_f(p,0,\infty)} F_{p,q}(t) = 1.$$

Moreover, $O_f(p, 0, \infty)$ is probabilistic bounded. Indeed, for any $r, s \in O_f(p, 0, \infty)$,

$$F_{r,s}(t) \geqslant T\left(F_{r,p}\left(\frac{t}{2}\right), F_{p,s}\left(\frac{t}{2}\right)\right) \geqslant \inf_{q \in O_f(p,0,\infty)} F_{p,q}\left(\frac{t}{2}\right), \ \forall \, t \geqslant 0.$$

Hence,

$$\inf_{r,s \in O_f(p,0,\infty)} F_{r,s}(t) \geqslant \inf_{q \in O_f(p,0,\infty)} F_{p,q}\left(\frac{t}{2}\right), \ \forall t \geqslant 0 ,$$

and thus

$$\sup_{t>0} \inf_{r,s \in O_f(p,0,\infty)} F_{r,s}(t) \geqslant \sup_{t>0} \inf_{q \in O_f(p,0,\infty)} F_{p,q}\left(\frac{t}{2}\right) = 1,$$

i.e., $O_f(p, 0, \infty)$ is probabilistic bounded.

COROLLARY 2 [101]. *Let (S, d) be a complete metric space, let $f : S \to S$, and let $\psi : \mathbb{R}^+ \to \mathbb{R}^+$ be a function which is strictly increasing with respect to t, has $\psi(0) = 0$, and has an inverse function ψ^{-1} satisfying the conditions :*

$$\lim_{n\to\infty} \psi^{-n}(t) = \infty,\ \forall\, t > 0 \text{ and } \lim_{t\to\infty} (\psi^{-1}(t) - t) = \infty.$$

If, for every $p \in S$, there exists an $n(p) \in \mathbb{N}$ such that, for any $q \in S$,

$$d(f^{n(p)}(p), f^{n(p)}(q)) \leqslant \psi(d(p, q)),$$

then there exists a unique fixed point $r \in S$, and $f^n(s) \to r$ for every $s \in S$.

Next, we shall give a generalization of Hick's result (see Definition 3.1.10 and Theorem 3.1.24) for mappings with a probabilistic local contractive iterate and $H\varphi$-local contraction-type mappings.

DEFINITION 3.1.15. $f : S \to S$ is said to have an *H-contractive iterate* at $p \in S$ if there exists a positive integer $n(p)$ such that, for all $q \in S$, and $k \in (0,1)$ fixed,

$$(\mathrm{H}_1) \text{ if } t > 0 \text{ and } F_{p,q}(t) > 1 - t, \text{ then } F_{f^{n(p)}(p), f^{n(p)}(q)}(kt) > 1 - kt.$$

Remark. If the metric space (S, d), with a bounded metric d, is made into a PM-space (in the natural way, that is $F_{p,q}(t) = H_0(t - d(p,q))$ and if, for each $p \in S$, there exists a positive integer $n(p)$ such that

$$d(f^{n(p)}(p), f^{n(p)}(q)) \leqslant kd(p, q),\ \forall\, q \in S,$$

then condition (H_1) holds.

Proof. Let f have the above-mentioned properties and let $\mathscr{F}(p, q)$ be defined as above. Then

$$F_{f^{n(p)}(p), f^{n(p)}(q)}(kt) = H_0(kt - d(f^{n(p)}(p), f^{n(q)}(q)) \geqslant H_0(kt - kd(p, q)) = H_0(t - d(p, q)) = F_{p,q}(t).$$

Since $d : S \times S \to R$ is a bounded metric, it follows that

$$F_{p,q}(t) = H_0(t - d(p, q)) > 1 - t \text{ if and only if } F_{p,q}(t) = 1$$

and

$$F_{p,q}(t) = 1 \text{ if and only if } F_{p,q}(t) > 1 - kt,$$

which implies condition (H_1).

THEOREM 3.1.37. *Let $(S, \mathscr{F}, T)$ be a complete Menger space such that $\sup_{a<1} T(a, a) = 1$. Suppose that the mapping $f : S \to S$ has an H-contractive iterate at every $p \in S$. Then f has exactly one fixed point r and $f^n(s) \to r$ for each $s \in S$ and $n \to \infty$.*

Proof. To prove the existence of the fixed point, consider an arbitrary $p_0 \in S$ and define the sequence $\{p_n\}_{n\in \mathbb{N}}$ in the following manner :

$$p_n = f^{n(p_{n-1})}(p_{n-1}) \text{ for every } n \in \mathbb{N}.$$

Observe that, for every $t > 0$, $p_0 \in S$, $n_0 \in N$, one has

$$F_{f^{n_0}(p_0),p_0}(1 + t) > 1 - (1 + t).$$

Then from condition (H_1), one obtains

$$F_{f^{n(p_0)}f^{n_0}(p_0),f^{n(p_0)}(p_0)}(k(1 + t)) = F_{f^{n_0}(p_1),p_1}(k(1 + t)) > 1 - k(1+t)$$

$$F_{f^{n(p_1)}f^{n_0}(p_1),f^{n(p_1)}(p_1)}(k^2(1 + t)) = F_{f^{n_0}(p_2),p_2}(k^2 (1+t)) > 1-k^2(1+t)$$

. . . .

$$F_{f^{n(p_{n-1})}f^{n_0}(p_{n-1}),f^{n(p_{n-1})}(p_{n-1})}(k^n(1 + t)) = F_{f^{n_0}(p_n),p_n}(k^n(1 + t)) >$$
$$> 1 - k^n(1 + t).$$

For every $n \in \mathbb{N}$, put $n_0 = n(p_{n+m-1}) + \ldots + n(p_n)$ and, hence,

$$F_{p_{n+m},p_n}(k^n(1 + t)) > 1 - k^n(1 + t).$$

Since $k \in (0,1)$, there exists a positive integer $n(t)$ such that $k^n(1 + t) \leqslant t$ and, thus,

$$(*)\ F_{p_{n+m},p_n}(t) \geqslant F_{p_{n+m},p_n}(k^n(1 + t)) > 1 - k^n(1 + t) > 1 - t$$

for all $n \geqslant n(t)$ and $p_0 \in S$. Hence, $\{p_n\}_{n\in\mathbb{N}}$ is a fundamental sequence. Since $(S, \mathscr{F}, T)$ is a complete Menger space, there exists a value $r = \lim_{n\to\infty} p_n$, i.e., for every $t > 0$ there exists an $n_1(t)$ such that

$$(**)\ F_{f^{n(r)}(p_n),f^{n(r)}(r)}(kt) > 1 - kt$$

for all $n \geqslant n_1(t)$. Hence,

$$\lim_{n\to\infty} f^{n(r)}(p_n) = f^{n(r)}(r).$$

On the other hand, it follows that, for every $t > 0$ and $n_0 \in \mathbb{N}$, there exists a positive integer $n_2(t, n_0)$ such that

$$(***)\ F_{f^{n_0}(p_n),p_n}(t) > 1 - t$$

for all $n \geqslant n_2(t, n_0)$ and $p_0 \in S$.

Let us now prove that $f^{n(r)}(r) = r$. Indeed, taking into account the relations (**) and (***) and the fact that $T(x, x) \leqslant x$, for all $x \in [0,1]$, it follows that there exists $n_3(t) = \max\{n_1(t),$

$n_2(t, n(r))\}$ such that

$$F_{f^{n(r)}(r),r}(t) \geqslant F_{f^{n(r)}(r),r}(kt) \geqslant T(F_{f^{n(r)}(r),f^{n(r)}(p_n)}(kt - k^2t),$$
$$F_{f^{n(r)}(p_n),r}(k^2t)) \geqslant T(1 - kt + k^2t, T(F_{f^{n(r)}(p_n),p_n}(k^2t - k^3t),$$
$$F_{p_n,r}(k^3t))) \geqslant T(1 - kt, T(1 - k^2t + k^3t, 1 - k^3t)) \geqslant$$
$$\geqslant T(1 - kt, T(1 - kt, 1 - kt)) \geqslant T(T(1 - kt, 1 - kt),$$
$$T(1 - kt, 1 - kt)).$$

Since $\sup\limits_{a<1} T(a,a) = 1$, it follows that

$$F_{f^{n(r)}(r),r}(t) = 1 \text{ for every } t > 0,$$

and, hence,

$$f^{n(r)}(r) = r.$$

To prove the uniqueness of the fixed point r of $f^{n(r)}$, let $q \in S$, $q \neq r$, and let $f^{n(r)}(r) = r$, $f^{n(r)}(q) = q$. Then there exists a real $t > 0$ such that $F_{q,r}(t) = a$ for a given $a \in [0,1)$. Since, clearly, $F_{p,q}(2 + a) > 1 - (2 + a)$, it follows that

$$F_{q,r}(k + (2 + a)) = F_{f^{n(r)}(q),f^{n(r)}(r)}(k(2 + a)) > 1 - k(2 + a),$$

$$\cdots\cdots\cdots\cdots\cdots\cdots\cdots$$

$$F_{q,r}(k^n(2 + a)) = F_{f^{n(r)}(q),f^{n(r)}(r)}(k^n(2 + a)) > 1 - k^n(2 + a)$$

for all $n \in \mathbb{N}$, and hence there exists an $n(t, a)$ such that $k^n(2+a) \leqslant \leqslant \min\{t, 1 - a\}$. If $n \geqslant n(t, a)$, one gets $a = F_{r,q}(t) \geqslant \geqslant F_{f^{n(r)}(r),f^{n(r)}(q)}(k^n(2 + a)) > 1 - (1 - a) = a$ and thus $r = q$.

Since r is the unique fixed point of $f^{n(r)}$ it follows that r is the unique fixed point of f.

To prove that $f^n(s) \to r$ as $n \to \infty$ for each $s \in S$, let us observe that, for all $t > 0$ and $m \in \{1, 2, \ldots, n(r)\}$, one gets

$$F_{f^m(s),r}(1 + t) > 1 - (1 + t)$$

and then conclude that, for $n \geqslant n(t)$, $n = m'n(r) + m$,

$$F_{f^n(s),r}(t) = F_{f^{m'n(r)+m}(s),f^{m'n(r)}(r)}(t) > 1 - t.$$

Since $n \to \infty$ implies that $m' \to \infty$, the proof of the theorem is complete.

Remark. Theorem 3.1.27 can be considered as a corollary of the above theorem.

Using a slightly modified form of Hick's [210] metric d on S which generates the (ε, λ)-uniformity, one can introduce a new class of probabilistic contractions. We shall prove that this class is a class of contractions with respect to that metric. This result will be then applied to obtain a new fixed point theorem for probabilistic metric spaces using the metric spaces obtained in [287], [369].

Let $\varphi : \mathbb{R}^+ \to \mathbb{R}^+$ be a function which satisfies the following conditions:

i) φ is strictly increasing,

ii) $\lim_{n\to\infty} \varphi^n(t) = 0$ for all $t > 0$.

iii) φ is right-continuous.

DEFINITION 3.1.16. Let $(S, \mathscr{F}, T)$ be a Menger space. A mapping $f : S \to S$ which satisfies the condition

(H_φ) for $t > 0$, $F_{f(p),f(q)}(\varphi(t)) > 1 - \varphi(t)$ whenever $F_{p,q}(t) > 1 - t$,

for all $p, q \in S$, is called an *H_φ-probabilistic contraction.*

THEOREM 3.1.38. *Every H_φ-probabilistic contraction on a complete Menger space $(S, \mathscr{F}, T)$ with $T \geqslant T_m$ has a unique fixed point which is the limit of the successive approximations.*

Proof. Let $T = T_m = \mathrm{Max}(\mathrm{Sum} - 1, 0)$ and define

$$d(p, q) = \sup \{t : F_{p,q}(t) \leqslant 1 - t\};$$

then (see Theorem 3.1.26) d is a metric on S which generates the (ε, λ)-uniformity and S is $\mathscr{F}$-complete if S is d-complete, and, for every $t > 0$, $F_{p,q}(t) > 1 - t$ iff $d(p, q) < t$. Using the properties of φ, we can conclude that f is an H_φ-probabilistic contraction on $(S, \mathscr{F}, T_m)$ iff f is a φ-contraction on (S, d), i.e., $d(f(p), f(q)) \leqslant \varphi(d(p, q))$ for all $p, q \in S$.

Indeed, for every $t > 0$ such that $F_{p,q}(t) > 1 - t$ and $p, q \in S$ with $d(f(p), f(q)) \leqslant \varphi(d(p, q))$, it follows that $d(p, q) < t$; hence, $\varphi(d(p, q)) < \varphi(t)$ and $F_{f(p),f(q)}(\varphi(t)) > 1 - \varphi(t)$, which is equivalent with relation (H_φ).

Let f be a mapping satisfying condition (H_φ) and let $t = d(p, q) + \varepsilon$ for arbitrary $\varepsilon > 0$ and for $p, q \in S$. Since $d(p, q) = t - \varepsilon < t$, one concludes that $F_{p,q}(t) > 1 - t$. Hence,

$$F_{f(p),f(q)}(\varphi(t)) > 1 - \varphi(t)$$

and

$$d(f(p), f(q)) < \varphi(t) = \varphi(d(p, q) + \varepsilon).$$

Since ε is arbitrary, it follows that

$$d(f(p), f(q)) \leqslant \varphi(d(p, q)).$$

It is known [369, pp. 33] that one can find an $r \in S$ such that $f(r) = r$ and $f^n(s) \to r$ (uniformly) for all $s \in S$, and the theorem is proven. Hence, the proof is complete.

Following a reasoning similar to that in the above theorem and of the fixed point theorem of Maia [287], one finds

PROPOSITION 3.1.13 [110]. *Let S be a nonempty set, let $\mathscr{F}^1$ and $\mathscr{F}^2$ be two probabilistic metrics on S and let $f : S \to S$ be a mapping. Suppose that*

i) *there exists a $c > 0$ such that*

$$\mathscr{F}^1_{f(p),f(q)}(ct) \geqslant \mathscr{F}^2_{p,q}(t) \text{ for all } p, q \in S \text{ and } t > 0,$$

ii) *$(S, \mathscr{F}^1, T_m)$ is a complete Menger space,*

iii) *$f : (S, \mathscr{F}^1) \to (S, \mathscr{F}^1)$ is continuous,*

iv) *$f : (S, \mathscr{F}^2) \to (S, \mathscr{F}^2)$ is an H_φ-probabilistic contraction.*

Then the mapping $f : (S, \mathscr{F}^1) \to (S, \mathscr{F}^1)$ has a unique fixed point which is the limit of successive approximations.

We shall present now several results regarding the comparison method, using some ideas from [273] for the *PM-spaces of* Šerstnev type with "co-triangle" function, Chap. 1. To this purpose, the comparison method will be used to deduce the existence, uniqueness and the corresponding approximations formulae for the fixed points of $f : S \to S$ using the fixed points of some convenient mapping on the topological semi-group $(\tilde{\Delta}^+, \tau_T)$, T continuous.

THEOREM 3.1.39. *Let $(S, \mathscr{F}, \tau_T)$ be a complete PM-space of Šerstnev type. Let $F \in (\tilde{\Delta}^+, \tau_T)$, $p_0 \in S$ and $S(p_0, F) = \{q \in S : F_{q,p_0} \leqslant F\}$. Assume that $f : S(p_0, F) \to S(p_0, F) \subset S$ is a φ-contraction mapping, i.e., for any $p, q \in S(p_0, F)$ with $F_{p,q} \leqslant F$,*

$$F_{f(p),f(q)} \leqslant \varphi(F_{p,q}),$$

where φ is a continuous and isotone mapping. $\varphi : [\varepsilon_0, F] \to$ $\to [\varepsilon_0, F] \subset (\tilde{\Delta}^+, \tau_T)$, $\varphi^n(F)$ converges for $n \to \infty$ and

$$\forall G \in [\varepsilon_0, F] \text{ such that } \varphi(G) = G \Rightarrow G = \varepsilon_0.$$

Then the mapping f has a unique fixed point $p_* \in S(p_0, F)$ such that:

$$f^n(p_0) \to p_* \text{ and } F_{p_*, f^n(p_0)} \leqslant \varphi^n(F).$$

Proof. For $m \in N$, $p_m = f^m(p_0) \in S(p_0, F)$ and, by induction,

$$F_{p_n, p_{n+m}} \leqslant \varphi^n(F), \quad \forall n, m \in \mathbb{N}.$$

Considering the properties of the mapping φ, it follows that $\varphi^n(F) \searrow \varepsilon_0$. This implies that there exists a $p_* \in S$ such that $\lim_{n\to\infty} p_n = p_*$ and, obviously, $f(p_*) = p_*$. To prove that p_*

is unique, observe that for any other fixed point of f, say p'_*,

$$F_{p'_*, f^n(p_0)} \leqslant \varphi^n(F), \quad \forall n \in \mathbb{N},$$

which implies that $p'_* = p_*$.

Since the ball $S(p, F) = \{q \in S : F_{p,q} \leqslant F\}$ is assumed to be closed in $(S, \mathscr{F}, \tau_T)$, one obtains an error evaluation of p_* as

$$F_{p_*, f^n(p_0)} \leqslant \varphi^n(F).$$

Remarks. 1) If $f : S(p_0, F) \to S$, the above theorem holds provided there exists an isotone mapping $\Phi : [\varepsilon_0, F] \to (\tilde{\Delta}^+, \tau_T)$ such that

$$F_{f(p_0), f(p)} \leqslant \Phi(F_{p_0, p}) \text{ for } p \in S(p_0, F)$$

and

$$\tau_T(G, \Phi(F)) \leqslant F, \quad G \geqslant F_{p_0, f(p_0)}$$

for some $G \in \tilde{\Delta}^+$.

2) The local-property feature used in the above theorem can be avoided either by using some convenient abstract inequalities or by imposing some chainability upon S.

More precisely, one should consider the properties of the semigroup $(\tilde{\Delta}^+, \tau_T)$ and should assume that the mapping $\varphi : (\tilde{\Delta}^+, \tau_T) \to (\tilde{\Delta}^+, \tau_T)$ has the following properties:

(i) φ is isotone and continuous,

(ii) for any decreasing sequence $\{F_n\} \subset \tilde{\Delta}^+$, the sequence $\{\varphi(F_n)\}_{n \in \mathbb{N}}$ converges to some element of $\tilde{\Delta}^+$,

(iii) for any $G \in \tilde{\Delta}^+$, there exists the globally maximal solution $m(\varphi, G)$ of the equation $F = \tau_T(\varphi(F), G)$.

Then, for every $U \in \tilde{\Delta}^+$ such that $U \leqslant \tau_T(\varphi(U), G)$, it follows that $U \leqslant m(\varphi, G)$.

Indeed, if F_0 is the globally maximal solution of the equation $F = \tau_T(\tau_T(U, G), \varphi(F))$ and

$$F_{n+1} = \tau_T(G, \varphi(F_n)), \quad n \in \mathbb{N},$$

then, by induction,

$$F_{n+1} \leqslant F_n, \quad U \leqslant F_n, \quad n \in \mathbb{N}.$$

From the properties of φ, one gets that $\{\varphi(F_n)\}_{n \in \mathbb{N}}$ converges and that $\{F_n\}_{n \in \mathbb{N}}$ converges too. If $\tilde{F} = \lim_{n \to \infty} F_n$, then $\tilde{F}$ is the solution of the equation $F = \tau_T(G, \varphi(F))$ and $U \leqslant \tilde{F}$ Since $\tilde{F} \leqslant m(\varphi, G)$, it follows that $U \leqslant m(\varphi, G)$.

3) Moreover, if ψ has the same properties as φ and if $\psi(F) \leqslant \varphi(F)$, $\forall F \in \tilde{\Delta}^+$, then the globally maximal solution of the equation $F = \tau_T(\psi(F), G)$ exists for any $G \in \tilde{\Delta}^+$ and $m(\psi, G) \leqslant m(\varphi, G)$.

THEOREM 3.1.40. *Let* $f : (S, \mathscr{F}, \tau_T) \to (S, \mathscr{F}, \tau_T)$ *and let* $\varphi : (\tilde{\Delta}^+, \tau_T) \to (\tilde{\Delta}^+, \tau_T)$ *be a continuous isotone mapping with the property that, for any decreasing sequence* $\{F_n\}_{n \in \mathbb{N}} \subset \tilde{\Delta}^+$, *the sequence* $\{\varphi(F_n)\}_{n \in \mathbb{N}}$ *converges to some element of* $\tilde{\Delta}^+$.

If, for any $G \in \tilde{\Delta}^+$, *there exists a globally maximal solution* $m(\varphi, G)$ *of the equation* $F = \tau_T(\varphi(F), G)$, $m(\varphi, \varepsilon_0) = \varepsilon_0$ *and, for any* $p, q \in S$,

$$F_{f(p), f(q)} \leqslant \varphi(F_{p,q}),$$

then there exists in S *a unique fixed point* p_* *of* f, $f^n(p_0) \to p_*$ *as* $n \to \infty$ *for any* $p_0 \in S$ *and*

$$F_{p_*, f^n(p_0)} \leqslant \varphi^n(F_0),$$

where

$$F_0 = m(\varphi, G_0), \quad G_0 = F_{p_0, f(p_0)}.$$

Proof. Let $p_0 \in S$ and let p'_* be any fixed point of f. Then, we have

$$F_{p'_*, p_0} \leqslant \tau_T(F_{f(p'_*), f(p_0)}, F_{f(p_0), p_0}) \leqslant \tau_T(\varphi(F_{p'_*, p_0}), G_0).$$

By the above reason,

$$F_{p'_*, p_0} \leqslant m(\varphi, G_0) = F_0$$

i.e., all fixed points of f lie in $S(p_0, F_0)$. Also, it is easy to verify that $f(S(p_0, F_0)) \subseteq S(p_0, F_0)$ and the rest follows by the above theorem.

DEFINITION 3.1.17. Let $(S, \mathscr{F}, T)$ be a PM-space. A mapping $f : S \to S$ is *a locally probabilistic contraction* on $(S, \mathscr{F})$ if, for every $p \in S$, there exists a $k(p) \in (0, 1)$ such that

$$F_{f(p), f(q)}(k(p)\, t) \geqslant F_{p,q}(t)$$

for all $q \in S$ and $t > 0$.

PROPOSITION 3.1.14. *Let* $f : S \to S$ *be a locally probabilistic contraction on the* PM*-space* $(S, \mathscr{F})$. *If* q *is a fixed point for* f, *then* q *is unique and* $f^n(p) \to q$ *for every* $p \in S$ *as* $n \to \infty$.

The proof mimics that for probabilistic contraction.

DEFINITION 3.1.18. A mapping $f : (S, \mathscr{F}) \to (S, \mathscr{F})$ is said to have *expanding probabilistic orbital diameters* if, for each $p \in S$,

$$\delta_p(t) = \sup_{n \in N \cup \{0\}} D_{O_f(p,n,\infty)}(t) > D_{O_f(p,0,\infty)}(t)$$

for every $t > 0$.

Remark. Some classes of local-contraction mappings and the class corresponding to the class of diminishing orbital diameter in the PM-space associated in the natural way for metric space are included in the class of expanding orbital diameters. The class of mappings for which M. Stojaković [418, Th. 1] or Hadžić [185] proved the existence of fixed point is also contained in the above class.

Other classes of mappings formulated with corresponding properties for probabilistic diameters were introduced by O. Hadžić [183], Shih-sen Chang [95], [97], [98] and M. Stojaković [418].

DEFINITION 3.1.19. Let $(S, \mathscr{F})$ be a PM-space and let $f : S \to S$. A point $p \in S$ is *regular for* f iff $\sup_t D_{O_f(p,0,\infty)}(t) = 1$.

DEFINITION 3.1.20. Let $(S, \mathscr{F})$ be a PM-space and let $f : S \to S$. One says that two points $p, q \in S$ are *asymptotic* under f iff $\lim_{n\to\infty} F_{f^n(p), f^n(q)}(t) = 1$, for every $t > 0$.

THEOREM 3.1.41 [183]. *Let $(S, \mathscr{F})$ be a complete probabilistic metric space and let $f : S \to S$ be a continuous mapping such that each point of S is regular for f and every pair of points in S is asymptotic under f. If there exists a constant $k \in (0, 1)$ such that, for every $p \in S$,*

$$D_{O_f(p,1,\infty)}(kt) \geqslant D_{O_f(p,0,\infty)}(t), \text{ for any } t > 0,$$

then there exists one and only one fixed point q of the mapping f and $q = \lim_{n\to\infty} f^n(p)$ for every $p \in S$.

Proof. For every $t > 0$ and every $n \in N$,

$$D_{O_f(p,n,\infty)}(t) \geqslant D_{O_f(p,n-1,\infty)}\left(\frac{t}{k}\right) \geqslant \ldots > D_{O_f(p,0,\infty)}\left(\frac{t}{k^n}\right).$$

Let us prove that the sequence $\{f^n(p)\}_{n \in \mathbb{N}}$ is a Cauchy sequence.

Since $\sup_t D_{O_f(p,0,\infty)}(t) = 1$, it follows that, for every $\lambda > 0$, there exists a $t(\lambda)$ such that

$$D_{O_f(p,0,\infty)}(t(\lambda)) > 1 - \frac{\lambda}{2}.$$

Let $n_0 \in N$ be such that $\frac{t}{k^{n_0}} \geq t(\lambda)$. It follows that

$$D_{O_f(p,n,\infty)}(t) \geq D_{O_f(p,0,\infty)}\left(\frac{t}{k^{n_0}}\right) \geq D_{O_f(p,0,\infty)}(t(\lambda)) > 1 - \lambda$$

for any $n \geq n_0$. This means that

$$\sup_{u<t} \inf_{q,r \in O_f(p,n,\infty)} F_{q,r}(u) > 1 - \frac{\lambda}{2}, \text{ for any } n \geq n_0.$$

Since $F_{q,r}(t) \geq F_{q,r}(u)$ for any $u < t$ and every $q, r \in S$, it follows that

$$\inf_{q,r \in O_f(p,n,\infty)} F_{q,r}(t) \geq \sup_{u<t} \inf_{q,r \in O_f(p,n,\infty)} F_{q,r}(u) > 1 - \frac{\lambda}{2}$$

for any $n \geq n_0$, and then

$$F_{q,r}(t) \geq 1 - \frac{\lambda}{2} \text{ for every } q, r \in O_f(p, n, \infty) \text{ and any } n \geq n_0.$$

This inequality shows that

$$F_{f^m(p), f^{m+i}(p)}(t) > 1 - \frac{\lambda}{2} \text{ for any } i \in \mathbb{N} \text{ and } m \geq n_0.$$

Since S is complete, there exists a q such that $q = \lim_{n\to\infty} f^n(p)$. The continuity of f implies that q is a fixed point of f. Suppose that there exists an $r \in S$ with $f(r) = r$. Then

$$F_{q,r}(t) = F_{f^n(q), f^n(r)}(t),$$

and since q and r are asymptotic under f, it follows that

$$F_{q,r}(t) = \lim_{n\to\infty} F_{f^n(q), f^n(r)}(t) = 1 \text{ for any } t > 0$$

i.e., $q = r$.

COROLLARY. *Let $(S, \mathcal{F}, T)$ be a complete Menger space with a continuous $\mathcal{H}$-type t-norm T. If $f: S \to S$ is a probabilistic contraction mapping, then f has a unique fixed point.*

It is easy to see that all the conditions of the above theorem are satisfied.

Remarks. The above notions of regular point and asymptotic pair are defined in metric spaces as in topological spaces with the topology defined by a family $\{d_n\}_{n\in\mathbb{N}}$ of pseudometrics. For example, if $(S, \mathcal{F}, T)$ is a Menger space as in Proposition 3.1.7 [187], it is proved that the regular points and the asymptotic pairs in the above sense and in the pseudometrics d_n, for every $n \in \mathbb{N}$, are equivalent.

Moreover, the condition

$$D_{O_f(p,1,\infty)}(kt) > D_{O_f(p,0,\infty)}(t)$$

implies, for every $p \in S$ and $t > 0$, the condition

$$D_n(O_f(p, 1, \infty)) \leqslant kD_n(O_f(p, 0, \infty)), \quad n \in \mathbb{N},$$

where $D_n(A) = \sup\{d_n(p, q),\ p, q \in A \subseteq S\}$.

A result similar with that of the above theorem holds for $(S, \mathcal{P})$, $\mathcal{P} = \{d_n\}_{n\in\mathbb{N}}$. Hence,

PROPOSITION 3.1.15. *Let $(S, \mathcal{F}, T)$ be a Menger space as defined in Proposition 3.1.7, let $f: S \to S$, let every point $p \in S$ be regular for f, let every two points $p, q \in S$ be asymptotic under f and, for every $t > 0$ and every $p \in S$, let*

$$D_{O_f(p,1,\infty)}(kt) > D_{O_f(p,0,\infty)}(t), \quad k \in (0, 1).$$

Then there exists one and only one fixed point q of f and $q = \lim_{n\to\infty} f^n(p)$ for every $p \in S$.

Some generalizations of the above results are given in [97], [420] or [191].

THEOREM 3.1.42 [97]. *Let f be a continuous self-mapping on a complete Menger space $(S, \mathcal{F}, T)$. Suppose that, for a point $p \in S$, $D_{O_f(p,0,\infty)} \in \mathcal{D}^+$ and that, for any $p \in S$, there exists an $n(p) \in \mathbb{N}$ such that, for all $t \geqslant 0$,*

$$(*) \qquad \inf_{q, r \in O_f(p, n(p), \infty)} F_{q,r}(t) = D'_{O_f(p, n(p), \infty)}(t) \geqslant$$

$$\geqslant D'_{O_f(p,0,\infty)}(\Phi t)),$$

where Φ is a function satisfying condition (Φ_1) of Theorem 3.1.36.

Then f has a fixed point $q \in S$ and $\lim_{n\to\infty} f^n(p) = q$ for any $p \in S$.

Proof. Let $\{n(j)\}_{j\in\mathbb{N}\cup\{0\}}$ be a sequence of positive integers defined by

$$n(0) = n(p),\ n(j+1) = n(j) + n(p_{n(j)}),\quad j = 0, 1, 2, \ldots,$$

where $p_{n(j)} = f^{n(j)}(p)$.

One has

$$D'_{O_f(p,\,n(j+1),\,\infty)}(t) \geqslant D'_{O_f(p,\,n(j),\,\infty)}(\Phi(t)) \geqslant \ldots \geqslant$$

$$\geqslant D'_{O_f(p,\,n(0),\,\infty)}(\Phi^{j+1}(t)) \geqslant D'_{O_f(p,\,0,\,\infty)}(\Phi^{j+2}(t)) \geqslant$$

$$\geqslant \sup_{s<\Phi^{j+2}(t)} D'_{O_f(p,\,0,\,\infty)}(s).$$

According to the properties of Φ,

$$\lim_{j\to\infty} D'_{O_f(p,\,n(j+1)\,\infty)}(t) = H_0(t), \quad \forall\, t > 0$$

i.e., $\{f^n(p)\}_{n\in\mathbb{N}}$ is a Cauchy sequence in S. The point $q = \lim_{n\to\infty} f^n(p)$ is the fixed point of f and the proof of the theorem is complete.

Remarks. If $n(p) = n \in \mathbb{N}$, for any $p\in S$, the conclusion of Theorem 3.1.42 still holds.

The conclusion $(*)$ can be replaced by

$$F_{f^n(p),\,f^{n+k}(p)}(t) \geqslant D'_{O_f(p,\,0,\,\infty)}(\Phi(t)), \quad \forall\, t \geqslant 0,$$

where $n(p) = n$ for any $p \in S$. It can also be replaced by the condition : $\exists\, m, n \in \mathbb{N}$ such that, for any $r, s \in S$, and any $t \geqslant 0$,

$$F_{f^m(r),\,f^n(s)}(t) \geqslant D'_{O_f(r,\,s,\,0,\,\infty)}(\Phi(t)),$$

where $O_f(r, s, 0, \infty) = O_f(r, 0, \infty) \cup O_f(s, 0, \infty)$ [see ref. 97, Th. 4].

We shall now conclude the presentation of generalized contraction classes with the results obtained by S. Heikkilä and S. Seikkala [209]. These authors dealt with the family of pseudometrics induced by the probabilistic metrics $\mathscr{F}$ taken as a mapping from $S \times S$ into $R_+^{(0,1)} = \{u : (0,1) \to \mathbb{R}^+,\ u(\alpha) = d(p, q)(\alpha) = \sup\{t : F_{p,q}(t) \leqslant 1 - \alpha,\ \alpha \in (0,1)\},\ p, q \in S\}$. Consider the topology of pointwise convergence on $R_+^{(0,1)}$ and the partial ordering "$\leqslant$" defined by $u \leqslant v$ iff $u(\alpha) \leqslant v(\alpha)$ for all $\alpha \in (0,1)$. It is easy to see that

$$d(p, q) = d(q, p)$$

and

$$d(p, q) = \theta \text{ iff } p = q.$$

The relation $F_{p,q}(\varepsilon) > 1 - \alpha$ iff $d(p, q)(\alpha) < \varepsilon$ implies that

$$p_n \xrightarrow{\mathscr{F}} p \text{ iff } d(p_n, p) \to \theta \text{ in } R_+^{(0,1)}$$

and $(S, \mathscr{F})$ is complete iff a sequence $\{p_n\}_{n\in\mathbb{N}}$ converges in S whenever

$$d(p_{n+m}, p_n) \leqslant u_n, \quad m, n = 1, 2, \ldots$$

is satisfied for a sequence $\{u_n\}_{n\in\mathbb{N}}$ of $R_+^{(0,1)}$ which converges to θ.

Remark. The convergence $p_n \to p$ is equivalent to $F_{p_n, p}(t) \to 1$ for all $t > 0$. Additionally,

$$F_{p,q}(t) = P(\{\alpha : d(p, q)(\alpha) < t, \quad t \in \mathbb{R}\}),$$

where P denotes the Lebesgue measure on $(0,1)$. Hence, $d(p, q)$ is a random variable on the probability measure space $((0,1), \mathscr{B}(0,1), P)$. Then $p_n \to p \Leftrightarrow d(p_n, p) \to \theta$ in the sense of distributions and probability since θ is degenerated.

Let $b > \theta$ be given. Denote

$$[\theta, b] = \{u \in R_+^{(0,1)} : \theta \leqslant u \leqslant b\}$$

and, for $p_0 \in S$,

$$B_b(p_0) = \{p \in S : d(p, p_0) < b\} = \{p \in S : F_{p, p_0}(b(\alpha)) > 1 - \alpha\}$$

for all $\alpha \in (0,1)$.

THEOREM 3.1.43. *Let $(S, \mathscr{F})$ be a PM-space and let $f : S \to S$ satisfy*

(i) *for all $\alpha \in (0,1)$, $u \in [\theta, b]$ and $p, q \in S$, $F_{f(p), f(q)}(Qu(\alpha)) > 1 - \alpha$ whenever $F_{p,q}(u(\alpha)) > 1 - \alpha$, where $Q : [\theta, b] \to [\theta, b]$ satisfies $\lim_{m\to\infty} Q^n b = \theta$.*

If f has a fixed point p, then this point is unique in $B_b(p)$. If f also satisfies

(ii) *there exists a $p_0 \in S$ such that, for all $\alpha \in (0,1)$, $F_{p_0, f(p)}(b(\alpha)) > 1 - \alpha$ whenever $F_{p_0, p}(b(\alpha)) > 1 - \alpha$,*

and if S is complete and Hausdorff, then $p = \lim_{n\to\infty} f^n(p_0)$ exists and is a fixed point of f.

Proof. From condition (i), it follows that, for all $\alpha \in (0,1)$, $u \in [\theta, b]$ and $p, q \in S$

$$(*) \quad d(f(p), f(q))(\alpha) < Q(u(\alpha)) \text{ whenever } d(p, q)(\alpha) < u(\alpha).$$

If p is a fixed point of f, then, for any fixed point q of f in $B_b(p)$,

$$d(p, q) = d(f^n(p), f^n(q)) < Q^n b, \quad n = 0, 1, \ldots$$

Since $\lim_{n\to\infty} Q^n b = \theta$, we obtain $d(p, q) = \theta$, i.e., $p = q$.

Let us suppose now that S is complete and Hausdorff. Condition (ii) implies that $f(B_b(p_0)) \subseteq B_b(p_0)$ and then, for any $m \in \mathbb{N}$,

$$d(f^m(p_0), p_0) < b.$$

Hence, for any $m, n \in \mathbb{N} \cup \{0\}$,

$$d(f^{m+n}(p_0), f^n(p_0)) < Q^n b.$$

Since S is complete and $\lim\limits_{n\to\infty} Q^n b = \theta$, it follows that there exists a $p \in B_b(p_0)$ such that $p = \lim\limits_{n\to\infty} f^n(p_0)$.

To prove that p is a fixed point of f, it suffices to show that $f(x) = \lim\limits_{n\to\infty} f^n(p_0)$. Let $\varepsilon > 0$ and $\alpha \in (0, 1)$. Then there exists an $n_0 \in \mathbb{N}$ such that

$$Q^n b(\alpha) < \varepsilon \quad \text{for } n \geqslant n_0$$

since $Q^n b(\alpha)$ is a sequence of positive numbers converging to zero. Further, $\lim\limits_{n\to\infty} d(p, f^n(p_0))(\alpha) = 0$ implies that there exists an $m_0(\alpha, \varepsilon) \in \mathbb{N}$ such that

$$d(x, f^{m-1}(p_0))(\alpha) < Q^{n_0} b(\alpha) \quad \text{for } m \geqslant m_0(\alpha, \varepsilon).$$

Using $(*)$, it follows that

$$d(f(p), f^m(p_0))(\alpha) < Q^{n_0+1} b(\alpha) < \varepsilon, \text{ for } m \geqslant m_0(\alpha, \varepsilon)$$

and then $f^n(p_0) \to f(p)$, which implies that p is the unique fixed point of the mapping f in $B_b(p)$.

THEOREM 3.1.44. *Let $(S, \mathcal{F}, \mathrm{Min})$ be a complete Menger space and let us assume that $f : S \to S$ satisfies (i) with $Qb < b$ and has a b-Qb-approximate fixed point p_0*, i.e., $d(p_0, f(p_0)) < b - Qb$. *Then, for any $r \in S$ possessing a b-chain between p_0 and r, we have $f^n(r) \to s$, where s is the unique fixed point of f in $C_b(p_0) = \{q \in S :$ possessing a b-chain between q and $p_0\}$.*

Proof. It is said that, for $T = \mathrm{Min}$, d is a pseudometric. From $d(p_0, f(p)) < b - Qb$ and from $(*)$, it follows that, for every $p \in B_b(p_0)$,

$$d(f(p), p_0) \leqslant d(f(p), f(p_0)) + d(f(p_0), p_0) < Qb + b - Qb = b,$$

such that f maps the ball $B_b(p_0)$ into itself. Thus, f also satisfies (ii) of Theorem 3.1.43 and, hence, $f^n(p_0) \to p$, i.e., p is a fixed point of f. Also, one gets

$$d(p, f(p_0)) \leqslant Qb < b$$

such that $\{p_0, f(p_0), p\}$ is a b-chain, i.e., $p \in C_b(p_0)$.

If $r \in C_b(p_0)$, then there exists a b-chain $\{r, p_1, \ldots, p_m = p\}$ between r and p. From the triangle inequality and condition (*), one gets

$$d(f^m(r), p) = d(f^m(r), f^n(p)) \leqslant d(f^m(r), f^n(p_1)) + d(f^m(p_1), f^n(p_2)) + \\ + \ldots + d(f^n(p_{m-1}), f^n(p)) < mQ^n b, \quad n = 0, 1, 2, \ldots,$$

and, hence, $\lim_{n\to\infty} f^n(r) = p$.

If s is another fixed point of f in $C_b(p_0)$, one can choose $r = s$ and from the above, one can obtain $p = \lim_{n\to\infty} f^n(s) = s$. Thus, p is the unique fixed point of f in $C_b(p_0)$.

COROLLARY. *The fixed point p of f from the above theorem satisfies the following estimation :*

$$d(p, p_0) \leqslant u,$$

where u is any positive solution of the inequality

$$d(p_0, f(p_0)) + Qu \leqslant u$$

and p is the fixed point of f.

PROPOSITION 3.1.16. *Let $(S, \mathcal{F}, \mathrm{Min})$ be a complete Menger space and let f be a mapping from S into itself. Assume that, for every $v \in R_+^{(0,1)}$, there exists a $b \in R_+^{(0,1)}$ and $Q : [\theta, b] \to [\theta, b]$ such that (i) from Theorem 3.1.43 holds and $v + Q(b) \leqslant b$. Then, for every $p_0 \in S$, $\lim_{n\to\infty} f^n(p_0) = q$, where q is the unique fixed point of f in S.*

Proof. For $p_0 \in S$, there exist $v, b \in R_+^{(0,1)}$ and $Q : [\theta, b] \to [\theta, b]$ such that $d(p_0, f(p_0)) < v \leqslant b - Qb$, and such that (i) holds. Hence, according to Theorem 3.1.44, $p = \lim_{n\to\infty} f^n(p_0)$ is a fixed point of f. If r is any fixed point of f, choose $b \in R_+^{(0,1)}$ and $Q : [\theta, b] \to [\theta, b]$ such that $d(p, r) < b$ and such that (i) holds. These, together with condition (*), imply that

$$d(p, r) = d(f^n(p), f^n(r)) < Q^n b \to \theta \text{ as } n \to \infty,$$

whence $p = r$.

Remarks. (1) A mapping $Q : [\theta, b] \to [\theta, b]$ has the property $\lim_{n\to\infty} Q^n b = \theta$, for example, if Q is upper-semicontinuous from the right, i.e., if $Q(\lim_{n\to\infty} u_n) \geqslant \lim_{n\to\infty} Qu_n$ whenever $\{u_n\}_{n\in\mathbb{N}}$ is a decreasing sequence in $[\theta, b]$ such that also $\{Qu_n\}_{n\in N}$ decreases, and if either Q is increasing and has θ as the only fixed point or $Q(u(\alpha)) < u(\alpha)$ whenever $0 < u(\alpha) \leqslant b(\alpha)$.

3.1.4. *Common Fixed Point Theorems for a Pair of Mappings*

We shall now present several results regarding the existence and unicity of common fixed points for pairs of probabilistic contractions.

Let $(S, \mathscr{F})$ be a PM-space and let $f, g : S \to S$ be two mappings. This pair of mappings is a *contraction pair* on $(S, \mathscr{F})$ iff there exists a $k \in (0, 1)$ such that

$$F_{f(p), g(q)}(kt) \geqslant F_{p,q}(t), \text{ for every } p, q \in S \text{ and } t \geqslant 0.$$

If (f, g) is a contraction pair and if $p_0 \in S$, then the sequence of iterates of p_0 under the pair (f, g) is the sequence $\{p_n\}_{n \in \mathbb{N}}$ defined as

$$(*) \quad p_{2i+1} = f(p_{2i}), \; p_{2j} = g(p_{2j-1})$$

for all $i = 0, 1, 2, \ldots;\ j = 1, 2, \ldots$.

Define the function G_{p_0} as

$$G_{p_0}(x) = \inf_{n \in \mathbb{N}} \{F_{p_0, p_n}(x)\}, \quad x \in \mathbb{R}^+,$$

and $\{p_n\}_{n \in N}$ defined by the relation $(*)$.

THEOREM 3.1.45. *Let $(S, \mathscr{F}, \tau_T)$ be a complete Menger or Šerstnev space with a t-norm T for which $\sup_{a<1} T(a, a) = 1$ or τ_T is a continuous t-function in (H_0, H_0) and $\mathscr{D}^+$ is closed under τ_T.*

If (f, g) is a contraction pair on $(S, \mathscr{F}, \tau_T)$, then either
(i) *there exists a common unique fixed point for f and g, or*
(ii) *for every $p_0 \in S$, $\sup_{x \in \mathbb{R}} G_{p_0}(x) < 1$.*

Proof. Suppose that there exists a $p_0 \in S$ such that

$$\sup_{x \in \mathbb{R}} G_{p_0}(x) = 1$$

and let $\{p_n\}_{n \in \mathbb{N}}$ be the sequence of iterates. Then

$$F_{p_{2i+1}, p_{2j}}(t) = F_{f(p_{2i}), g(p_{2j-1})}(t) \geqslant F_{p_{2i}, p_{2j-1}}\left(\frac{t}{k}\right) \geqslant \ldots \geqslant$$

$$\geqslant F_{p_0, p_{2j-2i-1}}\left(\frac{t}{k^{2i+1}}\right) \geqslant G_{p_0}\left(\frac{t}{k^{2i+1}}\right),$$

and

$$F_{p_{2i}, p_{2i+2j}}(t) \geqslant T(F_{p_{2i}, p_{2i+1}}(t - \varepsilon), F_{p_{2i+1}, p_{2i+2j}}(\varepsilon))$$

$$F_{p_{2i+1}, p_{2i+2j+1}}(t)) \geqslant T(F_{p_{2i+1}, p_{2i+2j}}(t - \varepsilon), F_{p_{2i+2j}, p_{2i+2j+1}}(\varepsilon))$$

for every $\varepsilon \in (0, t)$.

Hence, $\{p_n\}_{n\in\mathbb{N}}$ is a Cauchy sequence in $(S, \mathcal{F}, T)$ and thus there exists a q such that

$$q = \lim_{n\to\infty} p_n.$$

Since

$$F_{f(q),q}(t) \geqslant T(F_{f(q),p_{2n}}(t-\varepsilon)\, F_{p_{2n},q}(\varepsilon)) \geqslant$$

$$\leqslant T(F_{q,p_{2n-1}}(t-\varepsilon),\ F_{p_{2n},q}(\varepsilon))$$

for every $\varepsilon \in (0, t)$ and $n \to \infty$, we obtain that $f(q) = q$ and similarly $g(q) = q$.

The proof of uniqueness mimics the proof of Theorem 3.1.2.

The following corollary comes as a generalization of Theorem 3.1.2.

COROLLARY. *Every contraction pair on a complete Menger space* $(S, \mathcal{F}, \mathrm{Min})$ *has a unique common fixed point.*

THEOREM 3.1.46 [187]. *Let* $(S, \mathcal{F}, T)$ *be a complete Menger space with continuous t-norm* T *and let* g *and* h *be continuous mappings of* S *into* S. *Then* g *and* h *have a common fixed point in* S *if and only if there exists a continuous mapping* f *of* S *into* $g(S) \cap h(S)$ *which commutes with* g *and* h *and satisfies the following two conditions:*

(i) *for every* $p, q \in S$, $F_{f(p),f(q)}(kt) \geqslant F_{g(p),h(q)}(t)$ *for every* $t > 0$, *where* $k \in (0, 1)$,

(ii) *there exists a* $p_0 \in S$ *such that* $\sup_t \inf_{n\in\mathbb{N}} F_{f(p_n),f(p_0)}(t) = 1$, *where* $\{p_n\}_{n\in\mathbb{N}}$ *is such that* $f(p_{2n-2}) = g(p_{2n-1})$, $f(p_{2n-1}) = h(p_{2n})$ *for every* $n \in \mathbb{N}$. *Then* f, g *and* h *have a unique common fixed point.*

Proof. To prove the necessity of conditions (i) and (ii), let $r \in S$ be such that $g(r) = r = h(r)$. Then

$$f(p) = r \text{ for every } p \in S$$

and f commutes with g and h. Also

$$F_{f(p),f(q)}(kt) = 1 \geqslant F_{g(p),h(q)}(t)$$

for every $p, q \in S$ and any arbitrary $k \in (0, 1)$. Condition (ii) is satisfied since if $p_0 \in S$, then $f(p_0) = r$ and, thus, from $g(r) = r$, it follows that one can choose $p_1 = r$. Since $p_n = r$ for every $n \in \mathbb{N}$, it follows that $f(p_{2n-2}) = g(p_{2n-1})$, $f(p_{2n-1}) = h(p_{2n})$, for every $n \in \mathbb{N}$, and $F_{f(p_n),f(p_0)}(t) = 1$, for every $t > 0$, i.e. condition (ii).

To prove the sufficiency of conditions (i) and (ii), consider $p_0 \in S$. Then $f(p_0) \in g(S) \cap h(S)$, i.e., there exists a $p_1 \in S$ such that $f(p_0) = g(p_1)$ and $f(p_1) = h(p_2)$. Hence, one has

$$f(p_{2n-2}) = g(p_{2n-1}),\ f(p_{2n-1}) = h(p_{2n}),$$

for every $n \in \mathbb{N}$. On the other hand,

$$F_{f(p_n), f(p_{n+1})}(t) \geqslant F_{g(p_n), h(p_{n+1})}\left(\frac{t}{k}\right) \geqslant \ldots \geqslant F_{f(p_0), f(p_1)}\left(\frac{t}{k^n}\right)$$

for every $t > 0$.

To prove that the sequence $\{f(p_n)\}_{n \in \mathbb{N}}$ is Cauchy, let us observe that

$$F_{f(p_{2i}), f(p_{2j+1})}(t) \geqslant F_{g(p_{2j+1}), h(p_{2i})} = F_{f(p_{2j}), f(p_{2i-1})}\left(\frac{t}{k}\right)$$

for every $i, j \in \mathbb{N}$.

For $2i > 2j + 1$, one has

$$F_{f(p_{2i}), f(p_{2j+1})}(t) \geqslant F_{f(p_{2i-1}), f(p_{2j})}\left(\frac{t}{k}\right) \geqslant \ldots$$

$$\ldots \geqslant F_{f(p_{2i-2j-1}), f(p_0)}\left(\frac{t}{k^{2j+1}}\right)$$

and for $j \geqslant i$,

$$F_{f(p_{2i}), f(p_{2j+1})}(t) \geqslant F_{f(p_{2i-1}), f(p_{2j})}\left(\frac{t}{k}\right) \geqslant \ldots$$

$$\ldots \geqslant F_{f(p_0), f(p_{2j+1-2i})}\left(\frac{t}{k^{2i}}\right).$$

From these relations, one can conclude that

$$(*) \quad F_{f(p_n), f(p_{n+m})}(t) \geqslant F_{f(p_0), f(p_m)}\left(\frac{t}{k^n}\right)$$

if $n = 2i$, $m = 2j + 1$ or $n = 2i + 1$, $m = 2j + 1$, and

$$(**) \quad F_{f(p_n), f(p_{n+m})}(t) \geqslant T\left(F_{f(p_0), f(p_1)}\left(\frac{t}{2k^n}\right), F_{f(p_0), f(p_{m-1})}\left(\frac{\varepsilon}{2k^{n+1}}\right)\right)$$

if $n = 2i$, $m = 2j$ or $n = 2i + 1$, $m = 2j$.

On the other hand, it follows from (ii) that, for every $\delta \in (0, 1)$, there exists a $t_\delta > 0$ such that

$$\inf_{n \in N} F_{f(p_n), f(p_0)}(t_\delta) > \delta.$$

Since T is continuous and since $T(1,1) = 1$, it follows that, for every $\lambda \in (0,1)$, there exists a $\delta \in (0,1)$ such that

$$T(u, v) > 1 - \lambda \text{ for } u \geqslant \delta \text{ and } v \geqslant \delta.$$

Let $\delta > 1 - \lambda$ and let $n(\delta) \in \mathbb{N}$ be such that

$$\frac{t}{2k^n} > t_\delta \text{ for } n \geqslant n(\delta),$$

and then

$$F_{f(p_0), f(p_m)}\left(\frac{t}{2k^n}\right) \geqslant F_{f(p_0), f(p_m)}(t_\delta) > \delta$$

for every $m \in \mathbb{N}$ and every $n \geqslant n(\delta)$.

From (*) and (**), it follows that $\{f(p)_n\}_{n \in \mathbb{N}}$ is a Cauchy sequence.

Let $r = \lim_{n\to\infty} f(p_n)$. Then

$$r = \lim_{n\to\infty} g(p_{2n+1}) = \lim_{n\to\infty} h(p_{2n})$$

and

$$f(r) = g(r) = h(r).$$

Since

$$F_{f(r), f^2(r)}(t) \geqslant F_{g(r), h(f(r))}\left(\frac{t}{k}\right) = F_{g(r), f(h(r))}\left(\frac{t}{k}\right) =$$

$$= F_{f(r), f^2(r)}\left(\frac{t}{k}\right) \geqslant \ldots \geqslant F_{f(r), f^2(r)}\left(\frac{t}{k^n}\right),$$

it follows that $F_{f(r), f^2(r)}(t) = 1$ for every $t > 0$, and so $f(r) = f^2(r)$, i.e., $f(r) = r_1$ is a fixed point of g, h and f. Then r_1 is the unique common fixed point. Indeed, if r_2 is another common fixed point, then

$$F_{r_1, r_2}(t) = F_{f^2(r_1), f(r_2)}(t) \geqslant F_{g(f(r_1)), h(r_2)}\left(\frac{t}{k}\right) = F_{f(r_1), f(r_2)}\left(\frac{t}{k}\right) =$$

$$= F_{r_1, r_2}\left(\frac{t}{k}\right) \geqslant \ldots \geqslant F_{r_1, r_2}\left(\frac{t}{k^n}\right)$$

for every $n \in \mathbb{N}$ and every $t > 0$, and so $r_2 = r_1$.

COROLLARY. *Let $(S, \mathscr{F}, T)$ be a complete Menger space with a continuous $\mathscr{H}$-type t-norm T and let g and h be continuous mappings of S into S. Then g and h have a comon fixed point in S if and only if there exists a continuous mapping $f : S \to g(S) \cap$*

$\cap h(S)$ which commutes with g and h and satisfies the following condition :

$$F_{f(p),f(q)}(t) \geqslant F_{g(p),h(q)}\left(\frac{t}{k}\right)$$

for every $t>0$, $k \in (0,1)$ and every $p, q \in S$. Then g, h and f have a unique common fixed point.

Proof. It is sufficient to prove that condition (ii) of the theorem holds. To do this, observe that

$$t = t(k^{m-1} + k^{m-2}(1-k) + \ldots + k(1-k) + (1-k)),$$

and then, for every $j \in \mathbb{N}$ and every $n \in \mathbb{N}$,

$$F_{f(p_0),f(p_j)}\left(\frac{t}{k^n}\right) \geqslant$$

$$\underbrace{T(T(\ldots T(}_{j-1 \text{ times}}T(F_{f(p_j),f(p_{j-1})}\left(\frac{t}{k^{n+1-j}}\right), F_{f(p_{j-1}),f(p_{j-2})}\left(\frac{t(1-k)}{k^{n+2-j}}\right)\Bigg), \ldots$$

$$\ldots, F_{f(p_0),f(p_1)}\left(\frac{t(1-k)}{k}\right)\Bigg) \geqslant T(T(\ldots T(T(F_{f(p_1),f(p_0)}\left(\frac{t}{k^n}\right),$$

$$F_{f(p_1),f(p_0)}\left(\frac{t(1-k)}{k^n}\right)\Bigg), \ldots, F_{f(p_0),f(p_1)}\left(\frac{t(1-k)}{k^n}\right)\Bigg).$$

Since $\dfrac{t}{k^n} \geqslant \dfrac{t(1-k)}{k^n}$ and T is of $\mathscr{H}$-type, we conclude that, for every $\lambda \in (0,1)$, there exists an $n(t, \lambda) \in \mathbb{N}$ such that

$$T^{j-1}(F_{f(p_0),f(p_1)}\left(\frac{t(1-k)}{k^n}\right) > 1 - \lambda$$

for every $n > n(t, \lambda)$ and every $j \in \mathbb{N}$. It also follows that

$$F_{f(p_0),f(p_j)}\left(\frac{\varepsilon}{k^n}\right) > 1 - \lambda$$

for every $n \geqslant n(t, \lambda)$ and every $j \in \mathbb{N}$. Hence,

$$\sup_{t \in \mathbb{R}} \inf_{j \in \mathbb{N}} F_{f(p_0),f(p_j)}(t) = 1$$

and the proof of the corollary is complete.

Generalizations of the above results are given in [419].

PROPOSITION 3.1.17. *Let $(S, \mathscr{F}, T)$ be a complete Menger space with continuous t-norm, let $g, h : S \to S$ be injective continuous mappings, let $f : S \to g(S) \cap h(S)$ be a continuous mapping* *and let g and h commute with f. Likewise, let the followiug three conditions be satisfied:*

(1) *For every $p \in S$, there exists an $n(p) \in \mathbb{N}$ such that, for every $q \in S$,*

$$F_{f^{n(p)}(p), f^{n(p)}(q)}(t) \geqslant \max\left\{F_{g(p), h(q)}\left(\frac{t}{k}\right), F_{h(p), g(q)}\left(\frac{t}{k}\right)\right\}$$

for every $t > 0$ and $k \in (0,1)$.

(2) *There exists a p_0 such that the sequence $\{f^{n(p_i)}(p_i)\}_{i \in \mathbb{N}}$ is probabilistic bounded, where the sequence $\{p_i\}_{i \in \mathbb{N}}$ is formed by the following two equalities:*

$$h(p_{2i-1}) = f^{n(p_{2i-2})}(p_{2i-2}), \quad g(p_{2i}) = f^{n(p_{2i-1})}(p_{2i-1}).$$

(3) *One of the following two conditions is satisfied:*

(i) *T is of $\mathscr{H}$-type.*

(ii) *The mapping f is u-semicompact, where $u \in \{g, h\}$.*

Then the mappings g, h and f have a unique common fixed point.

PROPOSITION 3.1.18. *Let $(S, \mathscr{F}, T)$ be a complete Menger space with continuous t-norm T, let $g, h : S \to S$ be injective continuous mappings and let $f : S \to g(S) \cap h(S)$ be a continuous mapping which commutes with g and h. Further, let the following two conditions be satisfied:*

(1) *For every $p \in S$, there exists an $n(p) \in \mathbb{N}$ such that, for every $q \in S$,*

$$F_{f^{n(p)}(p), f^{n(p)}(q)}(t) \geqslant \max\left\{F_{h(p), g(q)}\left(\frac{t}{k}\right), F_{g(p), h(q)}\left(\frac{t}{k}\right)\right\}$$

for every $t > 0$ and $k \in (0,1)$.

(2) *$g(S) \cap h(S)$ is a probabilistic bounded set.*

Then there exists a unique common fixed point for the mappings g, h and f.

O. Hadžić [190] has recently introduced the notion of Menger space with convex structure (see Definition 1.1.10.). She also introduced the notion of (W, p_0)-*convex mapping* as that mapping $g : S \to S$ such that

$$W(g(q), p_0, \lambda) = g(W(q, p_0, \lambda))$$

for every given $q \in S$, $\lambda \in [0,1]$ and $p_0 \in S$.

THEOREM 3.1.47. *Let $(S, \mathscr{F}, T)$ be a complete Menger space with a continuous t-norm T. Suppose that $(S, \mathscr{F}, T)$ has convex structure W which satisfies condition (P.II), let f, g and h be continuous mappings from S into S such that, for some $m \in \mathbb{N}$,*

$\overline{f^m(S)}$ is compact, let $f(S)$ be probabilistic bounded, $f(S) \subset g(S) \cap \cap h(S)$ and suppose that for every $p, q \in S$ and every $t \in \mathbb{R}^+$,

$$F_{f(p),f(q)}(t) \geqslant F_{g(p),h(q)}(t).$$

If there exists a $p_0 \in S$ such that g and h are (W, p_0)-convex, then there exists an $r \in S$ such that $r = f(r) = g(r) = h(r)$.

An interesting application of Kuratowski's function [56] is given bellow for the existence of common fixed points for pair mappings.

Let $(S, \mathscr{F})$ be a PM-space, and let Φ be an upper semicontinuous mapping of $S \times S$ into Δ with respect to the (ε, λ)-topology on S [42]. We say that two mappings $f, g : S \to S$ are *jointly Φ-contractions* if

$$\Phi(f(p), g(f(q))) > \Phi(p, f(q)) \text{ for all } p, q \in S \text{ with } p \neq f(q)$$

and

$$\Phi(g(p), f(g(q)) > \Phi(p, g(q)) \text{ for all } p, q \in S \text{ with } p \neq g(q).$$

THEOREM 3.1.48 [180]. *Let f and g be two continuous probabilistic densifying and jointly Φ-contractive mappings of a complete Menger space $(S, \mathscr{F}, T)$ onto itself such that* $\sup_{a<1} T(a, a) = 1$. *If, for some $p_0 \in S$, the sequence $\{p_n\}_{n\in\mathbb{N}}$, defined by*

$$p_{2n+1} = f(p_{2n}), \quad p_{2n+2} = g(p_{2n+1}),$$

is probabilistic bounded, then either f or g has a fixed point $r \in S$. Further, if r is a common fixed point of f and g, then it is the unique fixed point of f and g.

Proof. Let $A = \{p_{2n}\}_{n\in\mathbb{N}\cup\{0\}}$. Since $A = \{p_0\} \cup g(f(A))$, if $\alpha_A < H_0$, one gets $\alpha_A = \alpha(\{p_0\} \cup g(f(A)) = \min\{\alpha\{p_0\}, \alpha_{g(f(A))}\} = = \alpha_{g(f(A))} > \alpha_A$, i.e., a contradiction. Then $\alpha_A = H_0$ and A is relatively compact. Now, define $\varphi : \bar{A} \to \Delta$ by $\varphi(p) = \Phi(p, f(p))$ for each $p \in \bar{A}$. Let a net $\{p_d\} \subset \bar{A}$ be convergent to $p \in \bar{A}$. Since $\{(p_d, f(p_d))\}$ converges to $(p, f(p))$, it follows from the upper semicontinuity of Φ that, if $\varphi(p_d) \geqslant G \in \Delta$ for all d, then $\varphi(p) \geqslant G$. Thus, φ is upper semicontinuous. According to the dual of Birkhoff's Theorem 16 [42], it follows that φ assumes its maximal value for some $r \in \bar{A}$. If $f(r) \neq r$ and $g(f(r)) \neq f(r)$, one gets

$$\varphi(g(f(r))) = \Phi(g(f(r)), f(g(f(r))) > \Phi(f(r), g(f(r))) >$$
$$> \Phi(r, f(r)) = \varphi(r),$$

which is a contradiction. Therefore, $f(r) = r$ or $g(f(r)) = f(r)$. It can be proved that if r is a common fixed point of f and g, then it is the unique fixed point of f and g.

COROLLARY. *Let f and g be two continuous densifying and jointly φ-contractive mappings of a complete metric space (S, d) into itself. If, for some $p_0 \in S$, the sequence $\{p_n\}_{n \in \mathbb{N}}$, defined as in the above theorem, is bounded, then either f or g has a fixed point r. If r is a common fixed point of f and g, then it is unique with this property.*

Several interesting results in the study of pair mappings with Φ-contractive iterates were obtained by Shih-sen Chang [98]. Let us examine some of these.

THEOREM 3.1.49. *Let $(S, \mathscr{F}, T)$ be a complete Menger space with continuous t-norm T and let $\Phi : \mathbb{R}^+ \to \mathbb{R}^+$ be strictly increasing, $\Phi(0) = 0$ and $\Phi^n(t) \to \infty$ as $n \to \infty$, $\forall t > 0$. Suppose that f, g are self-mappings on $(S, \mathscr{F}, T)$, commutable and continuous and that, for each $p \in S$, the set*

$$O_{f,g}(p, 0, \infty) = \{f^i(g^j(p))\}_{i,\, j \in \mathbb{N} \cup \{0\}}$$

is probabilistic bounded. If there exist $m, m' \in \mathbb{Z}^+$, $m' + m \geqslant 1$ such that, for every $p \in S$ and any $t \geqslant 0$,

$$\inf_{r, s \in O_{f,g}(f^m(g^{m'}(p)), 0, \infty)} F_{r,s}(t) \geqslant \inf_{r, s \in O_{f,g}(p, 0, \infty)} F_{r,s}(\Phi(t))$$

holds, then, for every $p_0 \in S$, $f^n(g^n(p_0)) \to p_*$, $n \to \infty$, where p_* is a common fixed point of f and g.

Proof. Let $j = \max\{m, m'\}$ for any given $p_0 \in S$; then we have

$$\inf_{r, s \in O_{f,g}(f^j(g^j(p_0)), 0, \infty)} F_{r,s}(t) \geqslant \inf_{r, s \in O_{f,g}(f^m(g^{m'}(p_0)), 0, \infty)} F_{r,s}(t) \geqslant$$

$$\geqslant \inf_{r, s \in O_{f,g}(p_0, 0, \infty)} F_{r,s}(\Phi(t)) \geqslant \sup_{u < \Phi(t)} \inf_{r, s \in O_{f,g}(p_0, 0, \infty)} F_{r,s}(u)\,;\ \forall t \geqslant 0.$$

By induction, we can prove that

$$\inf_{r, s \in O_{f,g}(f^{nj}(g^{nj}(p_0)), 0, \infty)} F_{r,s}(t) \geqslant \inf_{r, s \in O_{f,g}(f^{(n-1)j}(g^{(n-1)j}(p_0)), 0, \infty)} F_{r,s}(\Phi(t)) \geqslant$$

$$\geqslant \ldots \geqslant \inf_{r, s \in O_{f,g}(p_0, 0, \infty)} F_{r,s}(\Phi^n(t)) \geqslant \sup_{u < \Phi^n(t)} \inf_{r, s \in O_{f,g}(p_0, 0, \infty)} F_{r,s}(u) =$$

$$= \begin{cases} \sup\limits_{u > 0} \inf\limits_{r, s \in O_{f,g}(p_0, 0, \infty)} F_{r,s}(u), & t > 0 \\ 0, & t = 0 \end{cases} = H_0(t),\ \forall t \geqslant 0.$$

Therefore, for any given $\varepsilon > 0$ and $\lambda > 0$, there exists an $n(\ \ \lambda) \in \mathbb{N}$ such that

$$\inf_{r, s \in O_{f,g}(f^{nj}(g^{nj}(p_0)), 0, \infty)} F_{r,s}(\varepsilon) > 1 - \lambda, \quad \forall n \geqslant n(\varepsilon, \lambda).$$

Hence, one concludes that all subsequences of $\{f^i(g^j(p_0))\}_{i,j\in\mathbb{N}\cup\{0\}}$ in which the subscripts i and j are both convergent to ∞, are Cauchy sequences. Since $(S, \mathscr{F}, T)$ is complete, the above subsequences have the same limit $p_* \in S$. In particular,

$$f^n(g^n(p_0)) \to p_*,\ f(f^n(g^n(p_0))) \to p_*,\ g(f^n(g^n(p_0))) \to p_*$$

as $n \to \infty$ and from the continuity of f and g, one gets

$$p_* = f(p_*) = g(p_*).$$

THEOREM 3.1.50. *Let* $(S, \mathscr{F}, T)$, f, g *and* Φ *be as in Theorem* 3.1.49. *Suppose that, for each* $p \in S$, *the set* $O_{f,g}(p, 0, \infty)$ *is probabilistic bounded. Suppose further that there exist* m, m', n, $n' \in \mathbb{Z}^+$ *with* $m + m' \geqslant 1$, $n + n' \geqslant 1$ *such that, for any* $p, q \in S$ *and any* $t \geqslant 0$,

$$(*)\qquad \inf_{r,s\in O_{f,g}(f^m(g^{m'}(p_0)),0,\infty)\cup O_{f,g}(f^n(g^{n'}(q)),0,\infty)} F_{r,s}(t) \geqslant$$

$$\geqslant \inf_{r,s\in O_{f,g}(p,0,\infty)\cup O_{f,g}(q,0,\infty)} F_{r,s}(\Phi(t)).$$

Then there exist a unique common fixed point p_* *of* f *and* g *in* S *and, for any* $p_0 \in S$, $f^n(g^n(p_0)) \to p_*$, $n \to \infty$.

Proof. Let $j = \max\{m, m', n, n'\}$ and put $p = q$ in $(*)$; then we have

$$\inf_{r,s\in O_{f,g}(f^j(g^j(p)),0,\infty)} F_{r,s}(t) \geqslant \inf_{r,s\in O_{f,g}(p,0,\infty)} F_{r,s}(\Phi(t)).$$

From Theorem 3.1.49, it follows that, for every $p_0 \in S$, $f^n(g^n(p_0)) \to p_*$, where p_* is a common fixed point for f and g.

If there exists another common fixed point q_* of f and g, then

$$F_{p_*,q_*}(t) = \inf_{r,s\in O_{f,g}(f^m(g^{m'}(p_*)),0,\infty)\cup O_{f,g}(f^n(g^{n'}(q_*)),0,\infty)} F_{r,s}(t) \geqslant$$

$$\geqslant \inf_{r,s\in O_{f,g}(p_*,0,\infty)\cup O_{f,g}(q_*,0,\infty)} F_{r,s}(\Phi(t)) = F_{p_*,q_*}(\Phi(t)),\ \forall t \geqslant 0.$$

It follows that

$$F_{p_*,q_*}(t) \geqslant F_{p_*,q_*}(\Phi^n(t)),\ \forall t \geqslant 0,\ n = 1, 2, \ldots,$$

and, for $n \to \infty$, one concludes that

$$F_{p_*,q_*}(t) = H_0(t),\quad \forall t \geqslant 0,\ \text{i.e.,}\ p_* = q_*.$$

COROLLARY. *Let* $(S, \mathscr{F}, T)$ *and* Φ *be as in Theorem* 3.1.50. *Suppose that* $f : S \to S$ *is continuous and that, for each* $p \in S$, *the set* $O_f(p, 0, \infty)$ *is probabilistic bounded. Suppose further that*

(i) *there exists an* $m \in \mathbb{Z}^+$ *such that, for all* $t \geqslant 0$,

$$\inf_{r, s \in O_f(f^m(p), 0, \infty)} F_{r,s}(t) \geqslant \inf_{r, s \in O_f(p, 0, \infty)} F_{r,s}(\Phi(t)), \quad \forall p \in S.$$

Then there exists a fixed point p_* *of* f *and* $f^n(p_0) \to p_*$, $n \to \infty$, *for every* $p_0 \in S$.

(ii) *there exist* $m, n \in \mathbb{Z}^+$ *such that, for any* $p, q \in S$ *and any* $t \geqslant 0$,

$$\inf_{r, s \in O_f(f^m(p), 0, \infty) \cup O_f(f^n(q), 0, \infty)} F_{r,s}(t) \geqslant$$

$$\geqslant \inf_{r, s \in O_f(p, 0, \infty) \cup O_f(q, 0, \infty)} F_{r,s}(\Phi(t)).$$

Then there exists a fixed point p_* *of* f, *and, for every* $p_0 \in S$, $f^n(p_0) \to p_*$, $n \to \infty$.

3.1.5. *Sequences of Probabilistic Contractions and Fixed Points*

Let $(S, \mathscr{F}, T)$ be a Menger space and let $f : S \to S$ be a mapping on S. The space $(S, \mathscr{F}, T)$ is said to be *f-orbitally complete* if every Cauchy sequence of the form $\{f^{n_i}(p)\}_{i \in \mathbb{N}}$ converges in S.

A sequence of mappings $f_n : (S, \mathscr{F}) \to (S, \mathscr{F})$ *converges uniformly* to a mapping $f : (S, \mathscr{F}) \to (S, \mathscr{F})$ iff, for every $\varepsilon > 0$ and $\lambda > 0$, there exists a positive integer $n(\varepsilon, \lambda)$ such that

$$F_{f(p), f_n(p)}(\varepsilon) > 1 - \lambda$$

for every $p \in S$ and $n > n(\varepsilon, \lambda)$.

THEOREM 3.1.51 [105]. *Let* $\{f_n\}_{n \in \mathbb{N}}$ *be a sequence of mappings on an f-orbitally complete Menger space* $(S, \mathscr{F}, \mathrm{Min})$ *and let* $f : S \to S$ *be a generalized contraction in the sense of Definition* 3.1.19. *If each* f_n, $n \in \mathbb{N}$, *has at least one fixed point* p_n *and if the sequence* $\{f_n\}_{n \in \mathbb{N}}$ *on the subset* $I = \{p \in S :$ *there exist some* f_i *such that* $p = f_i(p)\}$ *converges uniformly to* f, *then the sequence* $\{p_n\}_{n \in \mathbb{N}}$ *converges to a unique fixed point* p_* *of* f.

Proof. It is known [151] that f has a unique fixed point $p \in S$. Since $f_n \xrightarrow{\text{unif.}} f$, there exists an $n_0 \in \mathbb{N}$ such that

$$(*) \quad F_{f_n(p), f(p)}\left(\frac{1-k}{2}\, t\right) > 1 - \lambda \ \text{ for } n \geqslant n_0$$

for every $p \in S$. Let $p_i \in S$ with $f_i(p_i) = p_i$; then

$$F_{p_i, p}(t) \geqslant \operatorname{Min}\left(F_{f_i(p_i), f(p_i)}\left(\frac{1-k}{2} t\right), F_{f(p_i), p}\left(\frac{1+k}{2} t\right)\right)$$

and since f is a generalized contraction, one gets

$$F_{f(p_i), p}\left(k \frac{1+k}{2k} t\right) \geqslant$$

$$\geqslant \min\left\{F_{p_i, p}\left(\frac{1+k}{2k} t\right), F_{p_i, f(p_i)}\left(\frac{1+k}{2k} t\right),\right.$$

$$\left. F_{p, f(p)}\left(\frac{1+k}{2k} t\right), F_{p_i, f(p)}\left(\frac{1+k}{k} t\right), F_{p, f(p_i)}\left(\frac{1+k}{k} t\right)\right\} \geqslant$$

$$\geqslant \min\left\{F_{p_i, p}\left(\frac{1+k}{2k} t\right), F_{f_i(p_i), f(p_i)}\left(\frac{1+k}{2k} t\right),\right.$$

$$\left. F_{p, f(p_i)}\left(\frac{1+k}{k} t\right)\right\}.$$

But

$$F_{p, f(p_i)}\left(\frac{1+k}{k} t\right) \geqslant \operatorname{Min}\left\{F_{p, p_i}\left(\frac{1+k}{2k} t\right), F_{f_i(p_i), f(p_i)}\left(\frac{1+k}{2k} t\right)\right\} \geqslant$$

$$\geqslant \operatorname{Min}\left\{F_{p, p_i}\left(\frac{1+k}{2k} t\right), F_{f_i(p_i), f(p_i)}\left(\frac{1-k}{2} t\right)\right\}$$

and, hence,

$$F_{p_i, p}(t) \geqslant \min\left\{F_{f_i(p_i), f(p_i)}\left(\frac{1-k}{2} t\right), F_{p_i, p}\left(\frac{1+k}{2k} t\right)\right\} > 1-\lambda$$

for all $i \geqslant n_0$, $t > 0$. Indeed, since t is arbitrary, in the case

$$F_{p_i, p}(t) = F_{p_i, p}\left(\frac{1+k}{2k} t\right)$$

for all $i \geqslant n_0$ we conclude that

$$F_{p_i, p}(t) = F_{p_i, p}\left(\left(\frac{1+k}{2k}\right)^m t\right) \to 1 \text{ as } m \to \infty,$$

and, hence, it is not possible that $F_{p_i, p}(t) \leqslant 1 - \lambda$. The case $F_{p_i, p}(t) \geqslant F_{f_i(p_i), f(p_i)}\left(\frac{1-kt}{2}\right) > 1 - \lambda$ follows from (*) and the proof of the theorem is complete.

COROLLARY 1. *Let* $\{f_n\}_{n\in\mathbb{N}}$ *be a sequence of mappings on a Menger space* $(S, \mathscr{F}, \mathrm{Min})$ *such that* $f_n(p_n) = p_n$ *for some* $p_n \in S$ *and let* f *be a probabilistic contraction mapping on* S *with a fixed point* $p \in S$. *If* $f_n \xrightarrow{\text{unif.}} f$, *then* $p_n \xrightarrow{\mathscr{F}} p$, *as* $n \to \infty$.

Remark. The Corollary also holds if T is such that $\sup\limits_{a<1} T(a, a) = 1$.

COROLLARY 2. *Let* $\{f_n, g_n\}$ *be a sequence of pair mappings on* $(S, \mathscr{F}, T)$ *with* $\sup\limits_{a<1} T(a, a) = 1$. *Suppose that all pairs have common fixed points, i.e., there exists* $\{p_n\}_{n\in\mathbb{N}}$ *such that* $f_n(p_n) = g_n(p_n)$ *for all* $n \in \mathbb{N}$. *If* (f, g) *is a probabilistic contraction pair with a common fixed point* p *and* $f_n \xrightarrow{\text{unif.}} f$, $g_n \xrightarrow{\text{unif.}} g$, *as* $n \to \infty$, *then* $p_n \to p$ *as* $n \to \infty$.

PROPOSITION 3.1.19. *Let* $(S, \mathscr{F}, T)$ *be a Menger space for which* $\sup\limits_{a<1} T(a, a) = 1$. *Let* $\{f_n\}_{n\in\mathbb{N}}$ *and let* f *be contraction mappings such that there exist fixed points* p_n *for* f_n *and* p *for* f. *If the sequence* $\{f_n\}_{n\in\mathbb{N}}$ *converges pointwise to* f *and if there exists a subsequence* $\{p_{n_i}\}_{i\in\mathbb{N}}$ *which converges to* p_0, *then* $p_0 = p$.

Proof. For every $t > 0$ and $\delta > 0$, we have

$$F_{p_{n_i}, f(p_0)}(t) = F_{f_{n_i}(p_{n_i}), f(p_0)}(t) \geqslant$$

$$\geqslant T\left(F_{f_{n_i}(p_{n_i}), f_{n_i}(p_0)}\left(\frac{t}{2}\right), F_{f_{n_i}(p_0)}, \left(\frac{t}{2}\right)\right) \geqslant$$

$$\geqslant T\left(F_{p_{n_i}, p_0}\left(\frac{t}{2}\right), F_{f_{n_i}(p_0), f(p_0)}\left(\frac{t}{2}\right)\right) \geqslant$$

$$\geqslant T(1-\lambda, 1-\lambda) > 1-\delta$$

for $i > i(t, \lambda)$, where λ is conveniently chosen.

THEOREM 3.1.52 [98]. *Let* $(S, \mathscr{F}, T)$ *be a complete Menger space with a continuous t-norm* T *and let* $T \geqslant T_m$. *Let* $\Phi : \mathbb{R}^+ \to \mathbb{R}^+$ *be a strictly increasing function with* $\Phi(0) = 0$ *and* $\lim\limits_{n\to\infty} \Phi^n(t) = \infty$, $\forall t > 0$. *Let* $f_n : S \to S$, $n \in \mathbb{N}$ *and suppose that there exists a functional sequence* $\{m_n(p)\}_{n\in\mathbb{N}}$, $m_n : (S, \mathscr{F}, T) \to \mathbb{Z}^+$ *such that, for every* $n \in \mathbb{Z}^+$ *and every* $p \in S$, $m_n(p)/m_n(f_n(p))$, *and for any* $i, j \in \mathbb{Z}^+$, $i \neq j$ *and* $p, q \in S$, *the following holds:*

$$F_{f_i^{m_i(p)}(p), f_j^{m_j(q)}(q)}(t) \geqslant \min_{r, s \in \{p, q, f_i^{m_i(p)}(p), f_j^{m_j(q)}(q)\}} F_{r,s}(\Phi(t)), \ \forall t > 0.$$

Further, suppose that there exists some $p_0 \in S$ such that the set $\{p_n\}_{n\in\mathbb{Z}^+} \subset S$,

$$(*)\ p_n = f_n^{m_n(p_{n-1})}(p_{n-1}), \quad n = 1, 2, \ldots,$$

is probabilistic bounded. Then there exists a unique common fixed point $p_* \in S$, and $p_n \to p_*$ as $n \to \infty$.

Proof. If $p_0 \in S$, then the above-defined p_n is a Cauchy sequence of S. Indeed, for any $i, j \in \mathbb{R}^+$, $i \neq j$, and $t \geqslant 0$,

$$F_{p_i,p_j}(t) = F_{f_i^{m_i(p_{i-1})}(p_{i-1}),\, f^{m_j(p_{j-1})}(p_{j-1})}(t) \geqslant$$

$$\geqslant \min_{r,s\in\{p_{i-1},p_{j-1},p_i,p_j\}} F_{r,s}(\Phi(t)).$$

Therefore, for any $m, n \in \mathbb{Z}^+$, $m < n$,

$$\inf_{m\leqslant i,\, j\leqslant n} F_{p_i,p_j}(t) \geqslant \inf_{m-1\leqslant i,\, j\leqslant n} F_{p_i,p_j}(\Phi(t)).$$

Since n is arbitrary in $\mathbb{Z}^+$, $n > m$, it follows that

$$\inf_{i,j\geqslant m} F_{p_i,p_j}(t) \geqslant \inf_{i,j\geqslant m-1} F_{p_i,p_j}(\Phi(t)) \geqslant \sup_{u<\Phi(t)} \inf_{i,j\geqslant m-1} \Phi(u)$$

and, by induction,

$$\inf_{i,j\geqslant m} F_{p_i,p_j}(t) \geqslant \inf_{i,j\geqslant 0} F_{p_i,p_j}(\Phi^m(t)) \geqslant \sup_{u<\Phi^m(t)} \inf_{i,j\geqslant 0} F_{p_i,p_j}(u),$$

$$m = 1, 2, \ldots \ \forall t \geqslant 0.$$

Hence,

$$\lim_{n\to\infty} \inf_{i,j\geqslant m} F_{p_i,p_j}(t) \geqslant \lim_{n\to\infty} (\sup_{u<\Phi^n(t)} \inf_{i,j\geqslant 0} F_{p_i,p_j}(\Phi(u))) =$$

$$= \begin{cases} 0, & t = 0 \\ \sup_{u>0} \inf_{i,j\geqslant 0} F_{p_i,p_j}(u), & t > 0 \end{cases} = H_0(t), \ \forall t \geqslant 0.$$

Then, for every $\varepsilon > 0$ and $\lambda > 0$, there exists an $n(\varepsilon, \lambda) \in \mathbb{Z}^+$ such that

$$\inf_{i,j\geqslant m} F_{p_i,p_j}(\varepsilon) > 1 - \lambda, \quad \forall m \geqslant n(\varepsilon, \lambda),$$

and

$$F_{p_i,p_j}(\varepsilon) > 1 - \lambda, \quad \forall i, j \geqslant n(\varepsilon, \lambda),$$

i.e., $\{p_n\}_{n\in\mathbb{Z}^+}$ is a Cauchy sequence.

Let $p_* = \lim_{n\to\infty} p_n$; then, for any $i \in \mathbb{Z}^+$ and $n > i$, one gets

$$F_{p_n, f_i^{m_i(p_*)}(p_*)}(t) =$$

$$= F_{f_n^{m_n}(p_{n-1}), f_i^{m_i(p_*)}(p_*)}(t) \geqslant \min_{r,s \in \{p_{n-1}, p_n, p_*, f_i^{m_i(p_*)}(p_*)\}} F_{r,s}(\Phi(t))\}, \ \forall\, t \geqslant 0.$$

Let C_0 be the set of all discontinuity points of $F_{p_*, f_i^{m_i(p_*)}(p_*)}$. Then $C_n = \Phi^{-n}(C_0)$ is the set of all discontinuity points of $F_{p_*, f_i^{m_i(p_*)}(p_*)}(\Phi^n(t))$, $n = 1, 2, \ldots$, and

$$C = \bigcup_{n=0}^{\infty} C_n$$

is a countable set. If $t \in \mathbb{R}^+ - C$, it follows [385] that

$$F_{p_*, f_i^{m_i(p_*)}(p_*)}(t) = \lim_{n\to\infty} F_{p_n, f_i^{m_i(p_*)}(p_*)} \geqslant F_{p_*, f_i^{m_i(p_*)}(p_*)}(\Phi(t))$$

and, by induction,

$$F_{p_*, f_i^{m_i(p_*)}(p_*)}(t) \geqslant F_{p_*, f_i^{m_i(p_*)}(p_*)}(\Phi(t)) \geqslant \ldots \geqslant F_{p_*, f_i^{m_i(p_*)}(p_*)}(\Phi^n(t)).$$

Letting $n \to \infty$, one gets

$$F_{p_*, f_i^{m_i(p_*)}(p_*)}(t) = H_0(t), \quad \forall\, t \in \mathbb{R}^+ - C$$

If $t \in C$ and $t > 0$, there exist $t_1, t_2 \in \mathbb{R}^+ - C$ such that $0 < t_1 < t < t_2$, and then

$$1 = H_0(t) = F_{p_*, f_i^{m_i(p_*)}(p_*)}(t_1) \leqslant F_{p_*, f_i^{m_i(p_*)}(p_*)}(t) \leqslant$$

$$\leqslant F_{p_*, f_i^{m_i(p_*)}(p_*)}(t_2) = 1.$$

Hence,

$$F_{p_*, f_i^{m_i(p_*)}(p_*)}(t) = H_0(t), \quad \forall\, t \geqslant 0,$$

i.e.,

$$p_* = T_i^{m_i(p_*)}(p_*), \ i = 1, 2, \ldots .$$

To prove that p_* is the unique periodic point of $\{f_n\}_{n\in\mathbb{N}}$, let us suppose that there exists another periodic point q_* of some

T_j. Then, for any $i \in \mathbb{Z}^+$, $i \neq j$,

$$F_{p_*,q_*}(t) = F_{f_i^{m_i(p_*)}(p_*),f_j^{m_j(q_*)}(q_*)}(t) \geq \min_{r,s \in \{p_*,q_*\}} F_{r,s}(\Phi(t)) =$$

$$= F_{p_*,q_*}(\Phi(t)), \quad \forall\, t \geq 0,$$

and repeating the procedure, one gets

$$F_{p_*,q_*}(t) \geq F_{p_*,q_*}(\Phi^n(t)), \; n = 1, 2, \ldots, \quad \forall\, t \geq 0.$$

For $n \to \infty$,

$$F_{p_*,q_*}(t)\; p_* = H_0(t), \quad \forall\, t \geq 0, \text{ i.e., } p_* = q_*.$$

To prove that p_* is the unique common fixed point of $\{f_n\}_{n \in \mathbb{N}}$ let us observe that for $m_n(p)/m_n(f_n(p))$, $\forall\, p \in S$, for any $n \in \mathbb{Z}^+$, and for every $i = 1, 2, \ldots$ there exist some $k_i \in \mathbb{Z}^+$ such that

$$m_i(f_i(p_*)) = k_i m_i(p_*). \text{ Then}$$

$$f_i(p_*) = f_i(f_i^{m_i(p_*)}(p_*)) = f_i(f_i^{2m_i(p_*)}(p_*)) = \ldots = f_i(f_i^{k_i m_i(p_*)}(p_*)) =$$

$$= f_i(f_i^{m_i(f_i(p_*))}(p_*)) = f_i^{m_i(f_i(p_*))}(f_i(p_*)),$$

i.e., $f_i(p_*)$ is a periodic point of f_i. Since p_* is the unique periodic point of f_i, one gets $p_* = f_i(p_*)$, $i = 1, 2, \ldots$, and, hence, the proof is complete.

It is easy to see that the function $\Phi(t) = \dfrac{t}{k}$, $k \in (0, 1)$, $t \geq 0$ satisfies the conditions of the above theorem.

COROLLARY 1. *Let $\{f_n\}_{n \in \mathbb{N}}$, $(S, \mathcal{F}, T)$ and $\{m_n(p)\}_{n \in \mathbb{N}}$ be as in Theorem 3.1.52. If, for any $i, j \in \mathbb{Z}^+$, $i \neq j$, and for any $p, q \in S$, the relation*

$$F_{f_i^{m_i(p)}(p),\, f_j^{m_j(q)}(q)}(t) \geq \min_{r,s \in \{p,q,T_i^{m_i(p)}(p),T_j^{m_j(q)}(q)\}} F_{r,s}\left(\frac{t}{k}\right), \quad \forall\, t \geq 0$$

holds and if there exists a $p_0 \in S$ such that the set $\{p_n\}_{n \in \mathbb{Z}_+}$ defined by $(*)$ is probabilistic bounded, then the conclusion of Theorem 3.1.52 still holds.

COROLLARY 2. *Let $\{f_n\}_{n \in \mathbb{N}}$ be a sequence of set-mappings on $(S, \mathcal{F}, T)$. Suppose that there exists a sequence $\{m_n\}_{n \in \mathbb{N}} \subset \mathbb{Z}^+$ such that, for any $p, q \in S$ and for any $i, j \in \mathbb{Z}^+$, $i \neq j$, the relation*

$$F_{f_i^{m_i}(p),\, f_j^{m_j}(q)}(t) \geq \min_{r,s \in \{p,q,f_i^{m_i}(p),f_j^{m_j}(q)\}} F_{r,s}\left(\frac{t}{k}\right), \quad \forall\, t \geq 0$$

holds. If also there exists a $p_0 \in S$ *such that the set* $\{p_n\}_{n\in \mathbb{Z}^+}$, *defined by* $(*)$, *is probabilistic bounded, then the conclusion of Theorem* 3.1.52 *still holds.*

Remarks. (1) Theorem 3.1.24 is a particular case of Corollary 1 with $f_n = f$, $n \in \mathbb{N}$, $m(p) = m_i(p)$, $i \in \mathbb{N}$.

(2) Sehgal and Bharucha-Reid's main result [396] is a particular case of Corollary 2.

We shall now present several results regarding the sequences of generalized Menger-space mappings and the corresponding fixed points as obtained by I. Istrăţescu [219].

THEOREM 3.1.53. *Let* $(S, \mathscr{F}, \mathrm{Min} = T)$ *be a complete generalized Menger space and* $\{f_i\}_{i\in \mathbb{N}}$ *be a sequence of self-mappings of* S *such that*

(i) *There exist* $\varepsilon > 0$, $\lambda > 0$ *and* $k \in (0, 1)$ *such that*

$$F_{f_i(p), f_i(q)}(kt) \geqslant F_{p,q}(t) \text{ for all } t > 0, \ i \in \mathbb{N},$$

whenever $F_{p,q}(\varepsilon) > 1 - \lambda$ *(i.e., the distance between* p *and* q *is perhaps not infinite).*

(ii) $f_i(f_j) = f_j(f_i)$, $i, j \in \mathbb{N}$.

(iii) *If* $p_0 \in S$ *and* $p_n = f_n(p_{n-1})$, $n \in \mathbb{N}$, *then a positive integer* $n(p_0) = m$ *exists such that* $n \geqslant m$ *implies that*

$$F_{f_{n+r}(p_n), p_n}(\varepsilon) > 1 - \lambda, \ r \in \mathbb{N}.$$

Then there exists a $p \in S$ *such that* $p_n \to p$ *and* $f_n(p) \to p$ *as* $n \to \infty$. *Moreover,* $f_i(p) = p$, $i \in \mathbb{N}$, *and if* $f_n(p) \to p$ *and* $f_n(p') \to p'$ *implies that* $\lim_{t\to\infty} F_{p,p'}(t) = 1$, *then* p *is unique.*

Proof. Let ε, λ be positive reals; then

$$F_{p_{m+2}, p_{m+1}}(\varepsilon) = F_{f_{m+2}(p_{m+1}), f_{m+1}(p_m)}(\varepsilon) =$$

$$= F_{f_{m+2}(f_{m+1}(p_m)), f_{m+1}(p_m)}(\varepsilon) = F_{f_{m+1}(f_{m+2}(p_m)), f_{m+1}(p_m)}(\varepsilon) \geqslant$$

$$\geqslant F_{f_{m+2}(p_m), p_m}\left(\frac{\varepsilon}{k}\right).$$

Using the same argument repeatedly, one obtains, for $n \geqslant m$,

$$F_{p_{n+1}, p_n}(\varepsilon) > F_{f_{n+1}(p_m), p_m}\left(\frac{\varepsilon}{k^{n-m}}\right)$$

and

$$F_{p_n,p_{n+r}}(\varepsilon) \geqslant T(F_{p_n,p_{n+1}}(\varepsilon - k\varepsilon), F_{p_{n+1},p_{n+r}}(k\varepsilon)) \geqslant$$

$$\geqslant T\left(F_{f_{n+1}(p_m),p_m}\left(\frac{\varepsilon - k\varepsilon}{k^{n-m}}\right), T\left(F_{f_{n+2}(p_m),p_m}\left(\frac{k\varepsilon - k^2\varepsilon}{k^{n+1-m}}\right),\right.\right.$$

$$\left.\left.F_{p_{n+2},p_{n+r}}(k^2\varepsilon)\right)\right) \geqslant T\left(T\left(F_{f_{n+1}(p_m),p_m}\left(\frac{\varepsilon - k\varepsilon}{k^{n-m}}\right),\right.\right.$$

$$\left.\left.F_{f_{n+2}(p_m),p_m}\left(\frac{\varepsilon - k\varepsilon}{k^{n-m}}\right)\right)\right), T(F_{p_{n+2},p_{n+3}}(k^2\varepsilon - k^3\varepsilon),$$

$$\left.F_{p_{n+3},p_{n+r}}(k^3\varepsilon))\right) \geqslant T\left(\ldots\left(T\left(T\left(F_{f_{n+1}(p_m),p_m}\left(\frac{\varepsilon - k\varepsilon}{k^{n-m}}\right),\right.\right.\right.\right.$$

$$\left.F_{f_{n+2}(p_m),p_m}\left(\frac{\varepsilon - k\varepsilon}{k^{n-m}}\right), F_{f_{n+3}(p_m),p_m}\left(\frac{\varepsilon - k\varepsilon}{k^{n-m}}\right)\right)\right), \ldots$$

$$\left.\left.\ldots, F_{f_{n+r}(p_m),p_n}\left(\frac{k\varepsilon}{k^{n-m}}\right)\right)\right) > 1 - \lambda, \; r \in \mathbb{N}.$$

Hence, $\{p_n\}_{n\in\mathbb{N}}$ is a fundamental sequence. Since $(S, \mathscr{F}, T)$ is a complete generalized space, there exists a $p \in S$ such that $\lim_{n\to\infty} p_n = p$.

To prove that $f_n(p) \to p$, let n_0 be an integer such that $F_{p_n,p}\left(\frac{\varepsilon}{2}\right) > 1 - \lambda$ for all $n \geqslant n_0$; then

$$F_{f_n(p),p}(\varepsilon) \geqslant T\left(F_{f_n(p),p_n}\left(\frac{\varepsilon}{2}\right), F_{p_n,p}\left(\frac{\varepsilon}{2}\right)\right) =$$

$$= T\left(F_{f_n(p),f_n(p_{n-1})}\left(\frac{\varepsilon}{2}\right), F_{p_n,p}\left(\frac{\varepsilon}{2}\right)\right) \geqslant$$

$$\geqslant T\left(F_{p,p_{n-1}}\left(\frac{\varepsilon}{2k}\right), F_{p_n,p}\left(\frac{\varepsilon}{2}\right)\right) > 1 - \lambda \text{ for all } n \geqslant n_0 + 1,$$

and the assertion is proven.

To prove the last part, consider an arbitrary given point $p_0 \in S$ and let $n > n(p_0)$ be fixed too. Let A be the set of all $q \in S$ with the property that there exists a finite sequence $\{q = q_0, \; q_1, \ldots, q_j = p_n\} \subset S$ with $F_{q_i,q_{i-1}}(\varepsilon) > 1 - \lambda$, $i = 1, 2, \ldots, j =$ finite. Since

$$F_{f_{n+r}(q_i),\; f_{n+r}(q_{i-1})}(t) \geqslant F_{q_i,q_{i-1}}\left(\frac{\varepsilon}{k}\right) \geqslant F_{q_i,q_{i-1}}(\varepsilon) > 1 - \lambda, \; r \in \mathbb{N},$$

then, for every $q = \lim_{s\to\infty} q_s$, $q_s \in A$, we have $q \in A$ and $f_{n+2}(A) \subset A$.

Thus, A and f_{n+r}, $r \in \mathbb{N}$ satisfy the assumptions of Theorem 3.1.6.

It follows that, for every r, there exists a unique $p_r^* \in A$ such that $f_{n+r}(p_r^*) = p_r^*$.

To prove that $p_1^* = p_2^* = \ldots = p_r^* = p^*$, we observe that

$$f_{n+r}(f_{n+s}(p_r^*)) = f_{n+s}(f_{n+r}(p_r^*)) = f_{n+s}(p_r^*), \quad s \in \mathbb{N},$$

and since p_r^* is unique, it follows that $f_{n+s}(p_r^*) = p_r^*$ and since p_s^* is also unique, one gets $p_s^* = p_r^*$.

From the fact that, for every $r \in \mathbb{N}$,

$$f_r(p^*) = f_r(f_{n+s}(p^*)) = F_{n+s}(f_r(p^*)),$$

it follows that $f_r(p^*) = p^*$, i.e., no other point of A besides p^* can be a fixed point for all f_r, $r \in \mathbb{N}$.

Let us now prove that the common fixed points p^* and $p = \lim_{n\to\infty} f_n(p)$ coincide. To do this, suppose that there exists a $t > 0$ such that $a \geqslant \lim_{n\to\infty} F_{p,f_r^n(p)}(t)$, where $a \in [0, 1)$. If we choose r_0 such that

$$F_{r,f_r(p)}(t - kt) > a = 1 - \lambda \text{ for } r > r_0,$$

then

$$a < T(F_{p,f_r(p)}(t - kt), F_{p,f_r(p)}(t)) \leqslant T(F_{p,f_r(p)}(t-kt), F_{f_r(p),f_r^2(p)}(kt)) \leqslant$$

$$\leqslant F_{p,f_r^2(p)}(t); \quad T(F_{p,f_r(p)}(t - kt), F_{p,f_r^2}(t)) \leqslant$$

$$\leqslant T(F_{p,f_r(p)}(t - kt), F_{f_r(p),f_r^3(p)}(kt)) \leqslant F_{p,f_r^3(p)}(t);$$

. . . .

$$T(F_{p,f_r(p)}(t - kt), F_{p,f_r^{n-1}(p)}(t)) \leqslant T(F_{p,f_r(p)}(t - kt), F_{p,f_r^n(p)}(kt)) \leqslant$$

$$\leqslant F_{p,f_r^n(p)}(t) \to a,$$

a contradiction which implies that $f_r^n(p) \to p$ as $n \to \infty$. On the other hand, $\lim_{n\to\infty} f_r^n(p) \in A$ is a fixed point and from the uniqueness of $p_r^* = p^*$ it follows that $p^* = p$.

If we suppose that $f_n(p) \to p$ and $f_n(p') \to p'$ imply that the distance between p and p' is not perhaps infinite, i.e., $\lim_{t\to\infty} F_{p,p'}(t) = 1$, then $p = p'$. Indeed, if there exist $t > 0$ and

$a \in [0,1)$ such that $a = F_{p,p'}(t)$, then

$$a = F_{p,p'}(t) = F_{f_r^n(p), f_r^n(p')}(t) \geqslant F_{p,p'}\left(\frac{t}{k^n}\right)$$

and for $n \to \infty$, $F_{p,p'}\left(\frac{t}{k^n}\right) \to 1$, thus contradicting the selection of a and, therefore, $p = p'$.

In a similar way, one can prove

PROPOSITION 3.1.20. *Let $(S, \mathscr{F}, \mathrm{Min})$ be a complete generalized Menger space and let $\{f_i\}_{i\in\mathbb{N}}$ be a family of self-mappings on S, closed under composition, such that*

(i) *There exist $\varepsilon > 0$, $\lambda > 0$, $k \in (0,1)$ and $m \in \mathbb{N}$ such that, if $p, q \in S$ and $F_{p,q}(\varepsilon) > 1 - \lambda$, then*

$$F_{f_{m+i}(p), f_{m+i}(q)}(kt) \geqslant F_{p,q}(t) \textit{ for all } t > 0,\ i \in \mathbb{N}.$$

(ii) $f_i(f_j) = f_j(f_i)$, $i, j \in \mathbb{N}$.

(iii) *For $p_0 \in S$ arbitrary and $p_n = f_n(p_{n-1})$, $n \in \mathbb{N}$, there exists $n(p_0)$ such that $n > n(p_0)$ implies that*

$$F_{f_{n+i}(p_n), p_n}(\varepsilon) > 1 - \lambda \text{ for } i \in \mathbb{N},$$

and there exists a $p \in S$ such that $p_n \to p$ and $f_n(p) \to p$ as $n \to \infty$. Even further, $f_n(p') \to p'$, $f_n(p'') \to p''$ as $n \to \infty$ implies that $F_{p,q}(t) \to 1$ as $t \to \infty$; then p is unique and $f_{j+1}(p) = p$, $j = \max\{m, n(p_0)\}$, $j \in \mathbb{N}$.

Remark. For $m = 1$, one obtains the previous theorem, and one can show that hypothesis (i) is strictly more general than in that theorem.

If $f_i = g^i$, $i \in \mathbb{N}$, the conclusion of the proposition means that p is periodic under g with period not greater than $\max\{m, n(p_0)\}$.

An important property of mapping families is the continuity of the fixed point with respect to parameter variation. This property is used in solving systems of equations of the form $x = f(x, y)$, $y = g(x, y)$. In this respect, we give below some results from [186].

First, let us recall that for two given topological spaces X and Y, and for given multifunction $f: \Lambda \subset X \to 2^Y$, one says that the multifunction is upper semicontinuous at $\lambda_0 \in \Lambda$ if, for each open set $G \subset Y$ containing $f(\lambda_0)$, there exists an open neighbourhood $U(\lambda_0)$ of λ_0 in X such that $f(U(\lambda_0) \cap \Lambda) \subset G$.

THEOREM 3.1.54. *Let $(S, \mathscr{F}, T)$ be a complete Menger space with continuous t-norm T, let X be a complete metrizable topological space, let $\Lambda \subset X$, Γ be a closed, probabilistic, bounded sub-*

set of S and let $g : \Gamma \times \Lambda \to S$ be a given mapping such that the following conditions are satisfied:

(i) *$g(.,\lambda)$ is continuous for every $\lambda \in \Lambda$ and there exists a subset $\Lambda_0 \subset \Lambda$ such that, for every $\lambda_0 \in \Lambda_0, f(.,\lambda)$ is continuous at (x, λ_0) for every $x \in \Gamma$.*

(ii) *For every $\lambda \in \Lambda$, the equation $x = g(x, \lambda)$ has a solution in Γ.*

(iii) *For every $\Gamma' \subset \Gamma$ such that $\alpha_{\Gamma'} < H_0$, there exists an open neighbourhood $B(\Gamma)$ of Λ_0 such that, for any precompact set $\Lambda' \subset \Lambda \cap B$,*

$$\alpha_{g(\Gamma',\Lambda')} > \alpha_{\Gamma'}.$$

Then f is upper semicontinuous at each $\lambda \in \Lambda_0$, where, for each $\lambda \in \Lambda$, $f(\lambda) = \{x : x \in \Gamma,\ x = g(x, \lambda)\}$.

3.1.6. *Fixed Point Theorems for Mappings on Nonarchimedean PM-Spaces*

The importance and multitude of applications occurring in functional analysis over nonarchimedean spaces justifies a closer look into the probabilistic functional analysis. Among the outstanding problems, mention should be made of those connected with the existence and unicity of the fixed points of nonarchimedean probabilistic mappings. Such topics were dealt with by I. Istrăţescu [221], [218] [220], [224], [225] and O. Hadžić [172], [182]. Let us examine some of the main results obtained therein.

THEOREM 3.1.55. *Let $(S, \mathscr{F}, T)$ be a complete nonarchimedean Menger space with the archimedean strict t-norm T with additive generator g. Let f be a contraction mapping on S with respect to g and let $k \in (0, 1)$ be given by Definition* 3.1.1. *Then there exists a unique fixed point of f in S.*

Proof. Let q be an arbitrary element of S. Define a sequence $\{p_n\}_{n\in\mathbb{N}}$ inductively *via* $p_1 = f(q)$ and $p_{n+1} = f(p_n)$ for every $n \in \mathbb{N}$. It follows by induction that, for every positive integer n,

$$g(F_{p_n,p_{n+1}}(x)) \leqslant k^n g\left(F_{q,p_1}\left(\frac{x}{k^n}\right)\right).$$

Then, for $m > n$ and $x \in \mathbb{R}^+$, one writes

$$F_{p_n,p_m}(x) \geqslant T(F_{p_n,p_{n+1}}(x), F_{p_{n+1},p_n}(kx)) \geqslant T(T(F_{p_n,p_{n+1}}(x),$$
$$F_{p_{n+1},p_{n+2}}(kx)), F_{p_{n+2},p_m}(k^2x)) = g^{(-1)}[g(T(F_{p_n,p_{n+1}}(x),$$
$$F_{p_{n+1},p_{n+2}}(kx)) + g(F_{p_{n+2},p_m}(k^2x))] = g^{(-1)}\{g[g^{(-1)}(g(F_{p_n,p_{n+1}}(x)) +$$

$$+ g(F_{p_{n+1},p_{n+2}}(kx))] + g(F_{p_{n+2},p_m}(k^2x))\} \geqslant$$

$$\geqslant g^{(-1)}\left\{g\left[g^{(-1)}\left(k^n g\left(F_{q,p_1}\left(\frac{x}{k^n}\right)\right) + k^{n+1} g\left(F_{q,p_1}\left(\frac{x}{k^n}\right)\right)\right)\right] +\right.$$

$$\left. + k^{n+2} g\left(F_{q,p_{m-2}}\left(\frac{x}{k^n}\right)\right)\right\}.$$

Hence, we conclude that $\{p_n\}_{n\in\mathbb{N}}$ is a Cauchy sequence since $g^{(-1)}$ and g are continuous, $k^n \to 0$ as $n \to \infty$, $F_{p,q}(x) \to 1$ as $x \to \infty$ and $g^{(-1)}(0) = 1$. Since $(S, \mathscr{F}, T)$ is complete, there exists a point $p \in S$ such that $p_n \to p$ as $n \to \infty$. For every $x \in \mathbb{R}^+$ and for every $n \in \mathbb{N}$,

$$F_{p,f(p)}(x) \geqslant T(F_{p,p_n}(x), F_{p_n,f(p)}(x)) = g^{(-1)}(g(F_{p,p_n}(x)) +$$

$$+ g(F_{p_m,f(p)}x)) \geqslant g^{(-1)}\left(g(F_{p,p_n}(x)) + kg\left(F_{p_{n-1},p}\left(\frac{x}{k}\right)\right)\right) \geqslant$$

$$\geqslant \lim_{n\to\infty} g^{(-1)}\left(g(F_{p,p_n}(x) + kg\left(F_{p_{n-1},p}\left(\frac{x}{k}\right)\right)\right) = 1$$

i.e., $p = f(p)$ and the proof of existence is complete.

Next, suppose that there exists another $q \neq p$ in S such that $f(q) = q$, and assume that there exists an $x > 0$ such that $0 < F_{p,q}(x) < 1$. For this x,

$$g(F_{p,q}(x)) = f(F_{f(p)\ f(q)}(x)) \leqslant kg\left(F_{p,q}\left(\frac{x}{k}\right)\right) < g(F_{p,q}(x))$$

which is a contradiction. Thus, for every $x > 0$, either $F_{p,q}(x) = 0$ or $F_{p,q}(x) = 1$. Suppose that $F_{p,q}(x) = 0$ for some $x > 0$ and let

$$y = \sup\{x : F_{p,q}(x) = 0\}.$$

Then, since $\dfrac{y}{k} > y$,

$$g(F_{p,q}(y)) = g(F_{f(p),f(q)}(y)) \leqslant kg\left(F_{p,q}\left(\frac{y}{k}\right)\right) = kg(1) = 0,$$

whence $F_{p,q}(x) = 1$. But this contradicts the left-continuity of $F_{p,q}$ at y. Hence, $F_{p,q}(x) = 1$ for all $x > 0$, i.e., p is the unique fixed point. The proof is complete.

For contraction mappings in the sense of V. M. Sehgal (see Definition 3.1.2) developed on nonarchimedean PM-spaces, one states

THEOREM 3.1.56. *Let* $(S, \mathscr{F}, \mathrm{Min})$ *be a complete nonarchimedean Menger space. If* f *is any contraction mapping of* S *into itself, then there exists a unique* $p \in S$ *such that* $f(p) = p$. *Moreover,* $f^n(q) \to p$ *as* $n \to \infty$, *for each* $q \in S$.

Proof. To prove existence, consider an arbitrary $q \in S$ and define $p_n = f^n(q)$ for every $n \in \mathbb{N}$. Let us show that the sequence $\{p_n\}_{n \in \mathbb{N}}$ is fundamental in $(S, \mathscr{F}, \mathrm{Min} = T)$. Let ε, λ be positive reals. Since $F_{p,q}(\max\{x, y\}) \geqslant T(F_{p,r}(x), F_{r,q}(y))$ for every $p, q, r \in S$ and T is associative, it follows that, for $m > n$,

$$F_{p_n,p_m}(\varepsilon) \geqslant T(F_{p_n,p_{n+1}}(\varepsilon), F_{p_{n+1},p_m}(\varepsilon)) \geqslant T\left(F_{q,p_1}\left(\frac{\varepsilon}{k^n}\right),\right.$$

$$\left.F_{p_{n+1},p_m}(k\varepsilon)\right) \geqslant T\left(F_{q,p_1}\left(\frac{\varepsilon}{k^n}\right), T\left(F_{q,p_1}\left(\frac{\varepsilon}{k^n}\right), F_{p_{n+2},p_m}(k^2\varepsilon)\right)\right) \geqslant$$

$$\geqslant T\left(F_{q,p_1}\left(\frac{\varepsilon}{k^n}\right), F_{p_{n+2},p_m}(k^2\varepsilon)\right).$$

Using the same argument repeatedly, one obtains

$$F_{p_n,p_m}(\varepsilon) \geqslant T\left(F_{q,p_1}\left(\frac{\varepsilon}{k^n}\right), F_{p_{m-1},p_m}(k^{m-n-1}\varepsilon) \geqslant\right.$$

$$\geqslant T\left(F_{q,p_1}\left(\frac{\varepsilon}{k^n}\right), F_{q,p_1}\left(\frac{\varepsilon}{k^n}\right)\right) \geqslant F_{q,p_1}\left(\frac{\varepsilon}{k^n}\right).$$

If one chooses n_0 such that $F_{q,p_1}\left(\frac{\varepsilon}{n_0}\right) > 1 - \lambda$, it follows that $F_{p_n,p_m}(\varepsilon) > 1 - \lambda$ for all $n \geqslant n_0$. Hence, $\{p_n\}_{n \in \mathbb{N}}$ is a fundamental sequence. Since $(S, \mathscr{F}, T)$ is a complete nonarchimedean Menger space, there exists a $p \in S$ such that $p_n \to p$ as $n \to \infty$, i.e., $f^n(q) \to p$ as $n \to \infty$.

Let $U_{f(p)}(\varepsilon, \lambda)$ be any neighbourhood of $f(p)$. Then $p_n \to p$ as $n \to \infty$ implies the existence of n_0 such that $p_n \in U_p(\varepsilon, \lambda)$ for all $n \geqslant n_0$. Moreover,

$$F_{f(p_n),f(p)}(\varepsilon) \geqslant F_{p_n,p}\left(\frac{\varepsilon}{k}\right) \geqslant F_{p_n,p}(\varepsilon) > 1 - \lambda$$

for all $n \geqslant n_0$. Therefore, $f(p_n) \in U_{f(p)}(\varepsilon, \lambda)$ for all $n \geqslant n_0$ i.e., $f^n(q) \to f(p)$ as $n \to \infty$. Since $(S, \mathscr{F}, T)$ is a Hausdorff space [228], one concludes that $f(p) = p$. This proves the existence of the fixed point for f.

The proof of uniqueness is as in Theorem 3.1.2.

COROLLARY. *Let* (S, d) *be a complete nonarchimedean metric space and let* $f : S \to S$ *such that there exists a* $k \in (0, 1)$ *with*

$d(f(p), f(q)) \leqslant kd(p, q)$ for all $p, q \in S$. Then f has a unique fixed point $p_0 \in S$ and $f^n(q) \to p_0$ as $n \to \infty$, for each $q \in S$.

Proof. For $F_{p,q}(x) = H_0(x - d(p, q))$, $p, q \in S$, $x \in \mathbb{R}$, the space $(S, \mathscr{F}, \mathrm{Min})$ is a nonarchimedean Menger space, which is complete if (S, d) is complete. Since for each $x > 0$,

$$F_{f(p),f(q)}(kx) = H_0(kx - d(f(p), f(q)) \geqslant H_0(kx - kd(p, q)) =$$
$$= H_0(x - d(p, q)) = F_{p,q}(x),$$

it follows that f is a contraction on $(S, \mathscr{F}, \mathrm{Min})$. The conclusion of the corollary follows from the previous theorem.

THEOREM 3.1.57 [182]. *Let $(S, \mathscr{F}, T)$ be a complete nonarchimedean Menger space with the $\mathscr{H}$-type t-norm T. If f is any contraction mapping of S into S, then f has one and only one fixed point $p \in S$ and $p = \lim_{n\to\infty} f^n(q)$ for any arbitrary element $q \in S$.*

Proof. As in the previous theorem, consider $q \in S$ arbitrary and define $p_n = f^n(q)$ for every $n \in \mathbb{N}$. For $m > n$, $\varepsilon > 0$,

$$F_{p_n,p_m}(\varepsilon) \geqslant T(F_{p_n,p_{n+1}}(\varepsilon), F_{p_{n+1},p_n}(\varepsilon)) \geqslant T\left(F_{q,p_1}\left(\frac{\varepsilon}{k^n}\right),\right.$$

$$\left.F_{p_{n+1},p_m}(k\varepsilon)\right) \geqslant T\left(F_{q,p_1}\left(\frac{\varepsilon}{k^n}\right), T\left(F_{q,p_1}\left(\frac{\varepsilon}{k^n}\right), F_{p_{n+2},p_m}\left(k^2\varepsilon\right)\right)\right) =$$

$$= T\left(T\left(F_{q,p_1}\left(\frac{\varepsilon}{k^n}\right), F_{q,p_1}\left(\frac{\varepsilon}{k^n}\right)\right), F_{p_{n+2},p_m}(k^2\varepsilon)\right) \geqslant \ldots \geqslant$$

$$\geqslant T\left(T\left(\ldots T\left(T\left(F_{q,p_1}\left(\frac{\varepsilon}{k^n}\right), F_{q,p_1}\left(\frac{\varepsilon}{k^n}\right)\right), \ldots\right),\right.\right.$$

$$\left.\left.F_{p_{m-1},p_m}(k^{m-n-1}\varepsilon)\right)\right) \geqslant \left(T^{m-n-1}F_{q,p_1}\left(\frac{\varepsilon}{k^n}\right)\right).$$

Since the family $\{T^n(x)\}_{n\in\mathbb{N}}$ is equicontinuous at point $x = 1$, it follows that for every $\lambda \in (0, 1)$, there exists a $\delta(\lambda) \in (0, 1)$ such that

$$x > \delta(\lambda) \Rightarrow T_s(x) > 1 - \lambda, \text{ for every } s \in \mathbb{N};$$

if $n(\varepsilon, \lambda)$ is such that

$$F_{q,p_1}\left(\frac{\varepsilon}{k^n}\right) > \delta(\lambda) \quad \text{for every} \quad n \geqslant n(\varepsilon, \lambda),$$

then

$$T^{m-n-1}\left(F_{q,p_1}\left(\frac{\varepsilon}{k^n}\right)\right) > 1 - \lambda$$

for every $n \geqslant n(\varepsilon, \lambda)$ and every $m > n$. Hence, the sequence $\{p_n\}_{n\in\mathbb{N}}$ is a Cauchy sequence and if $p = \lim_{n\to\infty} p_n$ (as in the previous theorem), then $p = f(p)$ and $p' = f(p')$ imply that $p' = p$.

Results similar with those of sub-Sec. 1.1.2 can also be found in nonarchimedean probabilistic analysis. An example is in the corollary of Theorem 3.1.56. In order to derive a more subtle result, let us consider an abstract set S and a family of nonarchimedean pseudometrics D, i.e., for every $d \in D$:

1) $d(p, p) \geqslant 0$ for every $p \in S$,

2) $d(p, q) = d(q, p)$ for all $p, q \in S$,

3) $d(p, q) \leqslant \max \{d(p, r, d(r, q)\}$ for all $p, q, r \in S$.

The nonarchimedean uniformity generated by D is obtained by taking as a sub-base all sets of $S \times S$ of the form $U_{d,\varepsilon} = \{(p, q) : d(p, q) < \varepsilon\}$, $d \in D$ and $\varepsilon > 0$. Suppose that, for $p \neq q$ in S, there exists at least one $d' \in D$ for which $d'(p, q) > 0$, such that the uniformity generated by D is Hausdorff.

It is known [228] that if $(S, \mathscr{F}, \mathrm{Min})$ is a nonarchimedean Menger space, the collection of subsets of $S \times S$ defined by $U(\varepsilon, \lambda) = \{(p, q) : F_{p,q}(\varepsilon) > 1 - \lambda,\ \varepsilon > 0,\ \lambda > 0\}$ forms a basis for a Hausdorff nonarchimedean uniformity on S.

Let $d_\alpha : S \times S \to R$, $\alpha \in (0, 1)$ be defined by

$$d_\alpha(p, q) = \sup \{t : F_{p,q}(t) \leqslant 1 - \alpha\}.$$

Then a result similar to that in Proposition 3.1.6 is given by

PROPOSITION 3.1.21. *The function d_α is a nonarchimedean pseudometric for each $\alpha \in (0, 1)$.*

Proof. Since $F_{p,q}$ is nondecreasing, left-continuous and since $F_{p,q}(0) = 0$, $\sup_{t\in\mathbb{R}} F_{p,q}(t) = 1$, $F_{p,q} = F_{q,p}$, it follows that d_α is non-negative, finite and symmetric. Also, we have $d_\alpha(p, q) \leqslant \max \{d_\alpha(p, r), d_\alpha(r, q)\}$ for all $p, q, r \in S$, since if there exist $p, q, r \in S$ such that $d_\alpha(p, q) > \max \{d_\alpha(p, r), d_\alpha(r, q)\}$, one can choose $t_1 > d_\alpha(p, r)$ and $t_2 > d_\alpha(r, q)$ such that $\max\{t_1, t_2\} < d_\alpha(p, q)$. Hence,

$$F_{p,r}(t_1) = 1 - \alpha_1 > 1 - \alpha \text{ and } F_{r,q}(t_2) = 1 - \alpha_2 > 1 - \alpha.$$

Then

$$1 - \alpha \geqslant F_{p,q}(\max\{t_1, t_2\}) \geqslant T(F_{p,r}(t_1), F_{r,q}(t_2)) > 1 - \alpha,$$

which is a contradiction and the proof of the proposition is complete.

Remark. $F_{p,q}(\varepsilon) > 1 - \alpha$ if and only if $d_\alpha(p, q) < \varepsilon$ and hence the topology induced by the family $\{d_\alpha\}$ is like the (ε, λ)-topology.

Since, for fixed p and q, $d_\alpha(p, q)$ is a nonincreasing continuous function of α, it follows that, for any $\delta > 0$, the family $\{d_\alpha\}_{\alpha \in (0,\delta)}$ generates the same topology as that generated by the entire family $\{d_\alpha\}_{\alpha \in (0,1)}$. Moreover, we have

PROPOSITION 3.1.22 [172]. *Let $(S, \mathscr{F}, T)$ be a nonarchimedean Menger space with continuous t-norm of $\mathscr{H}$-type and every t_m, $m \in K$, from the decomposition of T (see Th. 1.1.1. herein) is strict. Then, there exists a sequence $\{a_n\}_{n \in \mathbb{N}} \subset (0,1)$ such that $\lim_{n\to\infty} a_n = 1$ and that the family of nonarchimedean pseudometrics $\{d_n\}_{n \in \mathbb{N}}$ generates the (ε, λ)-topology, where*

$$(*)\quad d_n(p, q) = \sup \{t : F_{p,q}(t) \leqslant a_n\}$$

for every $p, q \in S$ and $n \in \mathbb{N}$.

Proof. From Proposition 3.1.21, if $T = \mathrm{Min}$, d_n is a nonarchimedean pseudometric for arbitrary $a_n \in (0,1)$ and so in this case the sequence $\{a_n\}_{n \in \mathbb{N}} \subset (0,1)$ is an arbitrary sequence such that $\lim_{n\to\infty} a_n = 1$. Since $\lim_{n\to\infty} a_n = 1$, from $(*)$ it follows that the family $\{d_n\}_{n \in \mathbb{N}}$ generates the (ε, λ)-topology. If there exists a $b \in (0,1)$ such that the restriction $T|_{(b,1)} = \mathrm{Min}$, then $\{a_n\}_{n \in \mathbb{N}} \subset (b, 1)$. In case there exists no $b \in (0,1)$ such that $T|_{(b,1)} = \mathrm{Min}$ as in Proposition 3.1.7, there exists a subsequence $\{b_{m_k}\}_{k \in \mathbb{N}} \subset \{a_m\}_{m \in \mathbb{N}}$, where $J_m = (a_m, b_m)$, $m \in K$, such that $\lim_{k\to\infty} b_{m_k} = 1$. Then, by definition, we take $a_k \stackrel{\text{def}}{=} b_{m_k}$ for every $k \in \mathbb{N}$. If we suppose that, for some $n \in \mathbb{N}$ and $p, q, r \in S$, we have

$$d_n(p, q) > \max \{d_n(p, r), d_n(r, q)\},$$

then there exists a $t > 0$ such that

$$d_n(p, q) > t > \max \{d_n(p, r),\ d_n(r, q)\}.$$

Using the definition of d_n, it follows that

$$F_{p,q}(t) \leqslant a_n,\ F_{p,r}(t) > a_n,\ F_{r,q}(t) > a_n.$$

Since $t(a_n, a_n) = a_n$, for every $n \in \mathbb{N}$ and t_m is strict, for every $m \in K$, it follows that

$$a_n = t(a_n, a_n) < t(F_{p,r}(t), F_{r,q}(t)) \leqslant F_{p,q}(\max \{t, t\}) = \\ = F_{p,q}(t) \leqslant a_n,$$

which is a contradiction. So, for every $n \in \mathbb{N}$ and every $p, q, r \in S$, it follows that

$$d_n(p, q) \leqslant \max \{d_n(p, r), d_n(r, q)\},$$

and so for every $n \in \mathbb{N}$, d_n is a nonarchimedean pseudometric.

PROPOSITION 3.1.23. *Let $(S, \mathscr{P})$ be a sequentially complete nonarchimedean uniform space and let f be a mapping of S into S having the property that, for every $d_\alpha \in \mathscr{P}$, there exists a constant $k_\alpha \in (0,1)$ such that $d_\alpha(f(p), f(q)) \leqslant k_\alpha d_\alpha(p, q)$ for all $p, q \in S$. Then there exists a unique point $r \in S$ such that $f(r) = r$ and $f^n(p) \to r$, for $n \to \infty$ and $p \in S$.*

Proof. To prove existence, consider an arbitrary $p \in S$ and define $p_n = f^n(p)$, $n \in \mathbb{N}$. Let us show that the sequence $\{p_n\}_{n\in\mathbb{N}}$ is fundamental in S.

Let U be an arbitrary member of the basis, that is $U = \bigcup_{i=1}^{s} U_{\alpha_i, \varepsilon_i}$.

For each $i \in \{1,2,\ldots, s\}$ and $m > n$,

$$d_{\alpha_i}(p_n, p_m) \leqslant \max\{\max\{\ldots\{\max\{\max\{d_{\alpha_i}(p_n, p_{n+1}),$$

$$d_{\alpha_i}(p_{n+1}, p_{n+2})\}, \ldots\}, d_{\alpha_i}(p_{m-1}, p_m)\} \leqslant k^n_{\alpha_i}(p, p_1).$$

Choosing n sufficiently large, one gets

$$d_{\alpha_i}(p_n, p_m) < \varepsilon_i \text{ for } i \in \{1, 2, \ldots, s\}$$

and

$$(p_n, p_m) \in U.$$

Hence, $\{p_n\}_{n\in\mathbb{N}}$ is a fundamental sequence; since S is sequentially complete, the limit $r \in S$ exists, i.e., $f^n(p) \to r$, as $n \to \infty$. On the other hand, we have

$$d_\alpha(f^n(r), f(r)) \leqslant k_\alpha(f^{n-1}(r), r)$$

for each $d_\alpha \in \mathscr{P}$. Thus, $f^n(r) \to f(r)$ and $f(r) = r$ since S is a Hausdorff uniform space.

To prove uniqueness, let us observe that, for $r' \neq r$, such that $f(r) = r$ and $f(r') = r'$,

$$0 \leqslant d_\alpha(r, r') = d_\alpha(f(r), f(r')) \leqslant k_\alpha d_\alpha(r, r'),$$

and hence $d_\alpha(r, r') = 0$ since $k_\alpha \in (0, 1)$. As S is Hausdorff, it follows that $r = r'$.

Remarks. (1) Ref. [218, Th. 23] shows that if S is sequentially compact and if for $f: S \to S$ there exist a $d_\lambda \in \mathscr{P}$ and an $\varepsilon > 0$ such that $d_\alpha(f(p), f(q)) \leqslant d_\alpha(p, q)$ for all $d_\alpha \in \mathscr{P}$ whenever $d_\lambda(p, q) < \varepsilon$, then $d_\alpha(f(p), f(q)) = d_\alpha(p, q)$ whenever $d_\lambda(p, q) < \varepsilon$.

(2) If for $f: S \to f(S) \subset S$, there exist a $d_\lambda \in \mathscr{P}$ and an $\varepsilon > 0$ such that $d_\alpha(f(p), f(q)) < d_\alpha(p,q)$ for all $d_\alpha \in \mathscr{P}$ whenever $d_\lambda(p, q) < \varepsilon$ and $d_\alpha(p, q) > 0$, then the set $M = \{p \in S, \exists\, s \in \mathbb{N}, f^s(p) = p\}$ is finite and nonempty and $\lim_{n\to\infty} f^{ns}(q) \to p$ for some $p \in M$ and for every $q \in S$.

THEOREM 3.1.58. *Let* $(S, \mathscr{F}, \text{Min})$ *be a complete nonarchimedean Menger space, let* $f: S \to S$, *and suppose that there exists a* $\delta \in (0, 1)$ *such that for every* $\alpha \in (0, \delta)$ *there exists a* $k_\alpha \in (0, 1)$ *such that, for* $p, q \in S$,

$$F_{f(p), f(q)}(k_\alpha t) \geqslant F_{p,q}(t)$$

whenever $F_{p,q}(t) > 1 - \alpha$. *Then* f *has a unique fixed point* r *and* $f^n(p) \to r$ *for every* $p \in S$.

Proof. Let $\{d_\alpha\}$ be the family of nonarchimedean pseudometrics associated with $\mathscr{F}$. If $\alpha \in (0, \delta)$, $p, q \in S$ and $\varepsilon > 0$, one has

$$F_{p,q}(t) > 1 - \alpha$$

for every $t = d_\alpha(p, q) + \varepsilon$. Then from the remark of Proposition 3.1.21, we obtain

$$d_\alpha(f(p), f(q)) < k_\alpha t \leqslant k_\alpha d_\alpha(p, q)$$

and the theorem follows from Proposition 3.1.23.

Remarks. 1) Theorem 3.1.56 is a corollary of Theorem 3.1.58.

2) The result of Theorem 3.1.55 remains valid for the complete nonarchimedean Menger space of the type considered in Proposition 3.1.22 [172, Th. 2] or in generalized nonarchimedean spaces.

Likewise, every generalized complete nonarchimedean metric space of Luxemburg [286] may be regarded as a generalized complete nonarchimedean Menger space if we define

$$F_{p,q}(x) = \begin{cases} H_0(x - d(p, q)) & \text{if } d(p, q) < \infty \\ a \in [0, 1) \text{ for } x > 0 & \text{if } d(p, q) = \infty \end{cases}$$

and $T = \text{Min}$.

THEOREM 3.1.59. *Let* $(S, \mathscr{F}, \text{Min})$ *be a generalized complete nonarchimedean Menger space. For* f *any contraction mapping on* S *into itself and for* $p \in S$, *let us consider the "sequence of successive approximations with initial element* p": $p, f(p), f^2(p), \ldots, f^n(p), \ldots$. *Then the following alternative holds: either*

(A) *for every integer* $n = 0, 1, 2, \ldots$,

$$\lim_{x \to \infty} F_{f^n(p), f^{n+1}(p)}(x) = a \in [0, 1)$$

or

(B) *the sequence of successive approximations is* $\mathscr{F}$-*convergent to a fixed point of* f.

Proof. If (A) does not hold, then there exists an $m \in N$ such that

$$\lim_{x \to \infty} F_{f^m(p), f^{m+1}}(x) = 1.$$

Setting $q = f^m(p)$, one gets $\lim_{x\to\infty} F_{q\,f(q)}(x) = 1$. Then, for some $r \in S$, the restriction

$$f_r = f|_{\tilde{r}} : \tilde{r} \to \tilde{r}$$

is a contraction mapping, where $\tilde{r} = \{s \in S;\ \lim_{x\to\infty} F_{s,r}(x) = 1\}$ i.e., $\tilde{r}$ is a class of equivalence determined by the relation β ($r\beta s \Leftrightarrow \lim_{x\to\infty} F_{r,s}(x) = 1$) (see [220, Props. 2.2 and 2.3]). Indeed, q and $f(q)$ belong to the same $\beta(.)$ and, for every $r \in \beta(q)$, $f(r) \in \beta(q)$ since $F_{f(r),f(q)}(kx) \geqslant F_{r,q}(x)$ for all $x > 0$ implies that $f(r) \in \beta(f(q))$. Hence, $f(\beta(q)) \leqslant \beta(q)$ and the restriction $f|_{\beta(q)}$ is a contraction mapping from $\beta(q)$ into $\beta(q)$. Since $\beta(r) = \tilde{r}$ is a closed set ([220, Prop. 2.2]) in a complete PM-space, it is also a nonarchimedean complete Menger space and from Theorem 3.1.56 it follows that there exists a unique point $s \in \beta(r)$ such that $f|_{\beta(r)}(s) = s$. Moreover, $f^n|_{\beta(r)}(p) \to s$ as $n \to \infty$ and $p \in \beta(r)$. Since $f: S \to S$ has a fixed point if and only if $f|_{\beta(r)} : \beta(r) \to \beta(r)$ has a fixed point for some $r \in S$, one concludes that s is a fixed point and $s = \lim_{n\to\infty} f^n(p)$. The proof of the theorem is complete.

Now we shall define the generalized probabilistic contractions and shall give several results unifying and generalizing some of the theorems in refs. [101], [182], [186], [221], [396], [225].

THEOREM 3.1.60. *Let $(S, \mathcal{F}, T)$ be a complete nonarchimedean Menger space with a continuous $\mathcal{H}$-type t-norm T. If f is a self-mapping on $(S, \mathcal{F}, T)$ which satisfies:*

(i) $F_{f(p),f(q)}(t) \geqslant F_{p,q}(\Phi(t))$ *for every* $p, q \in S$ *and every* $t \geqslant 0$, *where $\Phi(t)$ is a function satisfying the following condition*

(ii) $\Phi : \mathbb{R}^+ \to \mathbb{R}^+$ *is strictly increasing,* $\Phi(0) = 0$ *and, for every* $t > 0$, $\lim_{n\to\infty} \Phi^n(t) = \infty$,

then f has a unique fixed point $p \in S$ and, for every $q \in S$, $p = \lim_{n\to\infty} f^n(q)$.

Proof. Let q be an arbitrary point of S and define $f^n(q) = p_n$ for every $n \in \mathbb{N}$. Then, for $m > n$, $t > 0$,

$$F_{p_n,p_m}(t) \geqslant T(F_{p_n,p_{n+1}}(t), F_{p_{n+1},p_m}(t)) \geqslant$$

$$\geqslant T(F_{p_{n-1},p_n}(\Phi(t)), (F_{p_{n+1},p_m}(\Phi^{-1}(t)) \geqslant$$

$$\geqslant T(F_{q,p_1}(\Phi^n(t)), F_{p_{n+1},p_m}(\Phi^{-1}(t))) \geqslant$$

$$\geqslant T(F_{q,p_1}(\Phi^n(t)), T(F_{p_{n+1},p_{n+2}}(\Phi^{-1}(t)), F_{p_{n+2},p_m}(\Phi^{-1}(t)))) \geqslant$$

$$\geqslant T(F_{q,p_1}(\Phi^n(t)),\ T(F_{q,p_1}(\Phi^{n+1}(\Phi^{-1})(t))),\ F_{p_{n+n},p_m}(\Phi^{-2}(t)))) =$$

$$= T(T(F_{q,p_1}(\Phi^n(t)),\ F_{q,p_1}(\Phi^n(t))),\ F_{p_{n+2},p_m}(\Phi^{-2}(t))) \geqslant$$

$$\geqslant \ldots \geqslant \underbrace{T(T(\ldots(T}_{m-n-1}(F_{q,p_1}(\Phi^n(t), F_{q,p_1}(\Phi^n(t))), \ldots), F_{q,p_1}(\Phi^n(t))),$$

$$F_{p_{m-1},p_m}(\Phi^{-(m-n-1)}(t))) \geqslant T^{m-n}(F_{q,p_1}(\Phi^n(t))).$$

Since T is $\mathscr{H}$-type, for every $\lambda \in (0,1)$ there exists a $\delta(\lambda) \in \in (0,1)$ such that

$$t > \delta(\lambda) \Rightarrow T^k(t) > 1 - \lambda, \text{ for every } k \in \mathbb{N},$$

and since $\lim_{n\to\infty} \Phi^n(t) = \infty$, there exists an $n(t, \lambda) \in \mathbb{N}$ such that

$$F_{q,p_1}(\Phi^n(t)) > \delta(\lambda) \text{ for every } n \geqslant n(t, \lambda).$$

It follows that

$$T^{m-n}(F_{q,p_1}(\Phi^n(t))) > 1 - \lambda \text{ for every } m \geqslant n(t, \lambda)$$

and every $m > n$. This means that the sequence $\{p_n\}_{n\in\mathbb{N}}$ is a Cauchy sequence in S. If $p = \lim_{n\to\infty} p_n$, then $p = f(p)$ (see also Theorem 3.1.57).

The following result generalizes Theorem 1 [101].

THEOREM 3.1.61. *Let $(S, \mathscr{F}, T)$ be a complete nonarchimedean Menger space with a continuous $\mathscr{H}$-type t-norm T. Let f be a mapping on S satisfying the following condition: for every $p \in S$, there exists a positive integer $n(p)$ such that, for all $q \in S$ and all $t \geqslant 0$,*

$$(1)\quad F_{f^{n(p)}(p),\, f^{n(p)}(q)}(t) \geqslant F_{q,p}(\Phi(t)),$$

where Φ satisfies condition (ii) *of Theorem* 3.1.60.

Then f has a unique fixed point r in S and, for every $q \in S$, the sequence $\{f^n(q)\}_{n\in\mathbb{N}}$ converges to r.

Proof. Let us first prove that, under the assumptions of this theorem,

$$(2)\quad \sup_{t>0}\ \inf_{O_f(p,0,\infty)} F_{q,p}(t) = 1,\ \forall p \in S.$$

Indeed, from the properties of the distribution functions and the properties of function Φ, it follows that, for any $p \in S$ and $\lambda \in (0,1)$, there exists a $t_\lambda > 0$ and $\delta_\lambda \in (0,1)$ such that

$$\min_{0\leqslant k\leqslant n(p)} F_{p,f^k(p)}(\Phi(t)) > \delta_\lambda,\ \forall t > t_\lambda.$$

For any $n \in \mathbb{N}$, $n \geqslant m(p)$, one writes $n = mn(p) + r$, where $m \in \mathbb{N}$ and $0 \leqslant r \leqslant n(p)$. Then in view of condition (1),

$$F_{p,f^n(p)}(\Phi(t)) = F_{p,f^{mn(p)+r}(p)}(\Phi(t)) \geqslant$$

$$\geqslant T(F_{p,f^{n(p)}(p)}(\Phi(t)), F_{f^{n(p)}(p), f^{mn(p)+r}(p)}(\Phi(t)) \geqslant$$

$$\geqslant T(F_{p,f^{n(p)}(p)}(\Phi(t)), F_{p,f^{(m-1)n(p)+r}(p)}(\Phi^2(t))) \geqslant \dots \geqslant$$

$$\geqslant T(F_{p,f^{n(p)}(p)}(\Phi(t)), T(F_{p,f^{n(p)}(p)}(\Phi^2(t)), F_{p,f^{(m-2)n(p)+r}(p)}(\Phi^3(t)))) =$$

$$= T(T(F_{p,f^{n(p)}(p)}(\Phi(t)), F_{p,f^{n(p)}(p)}(\Phi^2(t))), F_{p,f^{(m-2)n(p)+r}(p)}(\Phi^3(t))) \geqslant$$

$$\geqslant \dots \geqslant T(T(\dots(T(F_{p,f^{n(p)}(p)}(\Phi(t)), F_{p,f^{n(p)}(p)}(\Phi(t))), \dots),$$

$$, F_{p,f^{n(p)}(p)}(\Phi(t))), F_{p,f^r(p)}(\Phi^{m+1}(t))) \geqslant$$

$$\geqslant T(T(\dots(T(F_{p,f^{n(p)}(p)}(\Phi(t)), F_{p,f^{n(p)}(p)}(\Phi(t))),\dots), F_{p,f^{n(p)}(p)}(\Phi(t))),$$

$$, F_{p,f^r(p)}(\Phi(t))) \geqslant T^m(\min_{0 \leqslant k \leqslant n(p)} F_{p,f^k(p)}(\Phi(t))).$$

Since the family $\{T^n(x)\}_{n \in \mathbb{N}}$ is equicontinuous, one concludes that

$$F_{p,f^n(p)}(\Phi(t)) > 1 - \lambda$$

for all $n = 0, 1, 2, \dots$ and $t > t_\lambda$. Consequently,

$$\inf_{q \in O_f(p,0,\infty)} F_{p,q}(\Phi(t)) \geqslant 1 - \lambda, \ \forall t \geqslant t_\lambda,$$

i.e.,

$$\sup_{t>0} \inf_{q \in O_f(p,0,\infty)} F_{p,q}(t) \geqslant 1 - \lambda.$$

Since λ is arbitrary, relation (2) is implicit.

In a manner similar to that used in [101], let us now prove that, for every $p_0 \in S$, the sequence

$$\{p_m\}_{m=0}^{\infty} = \{p_0, \ p_1 = f^{n(p_0)}(p_0), \dots, p_{m+1} = f^{n(p_m)}(p_m), \dots\}$$

is Cauchy in S.

Let $n_i = n(p_i)$, $i = 0, 1, 2, \dots$ and $m \in \mathbb{N}$. Then

$$F_{p_m, p_{m+i}}(t) = F_{f^{n_{m-1}}(p_{m-1}), f^{n_{m+i-1}+\dots+n_m+n_{m-1}}(p_{m-1})}(t) \geqslant$$

$$\geqslant F_{p_{m-1}, f^{n_{m+i-1}+\dots+n_m}(p_{m-1})}(\Phi(t)) \geqslant \dots \geqslant$$

$$\geqslant F_{p_n, f^{n_{m+i-1}+\ldots+n_{m}}(p_0)}(\Phi(t)) \geqslant \inf_{q\in O_f(p_0,0,\infty)} F_{p_0,q}(\Phi^m(t)) \geqslant$$

$$\geqslant \sup_{u<\Phi^m(t)} \inf_{q\in O_f(p_0,0,\infty)} F_{p_0,q}(u).$$

In view of the conditions set in the definition of Φ and considering (2), it follows that

$$\lim_{m\to\infty} F_{p_m, p_{m+i}}(t) \geqslant \lim_{m\to\infty} \sup_{u<\Phi^m(t)} \inf_{q\in O_f(p_0,0,\infty)} F_{p_0,q}(u) =$$

$$= \begin{cases} 0 \text{ as } t=0 \\ \sup\limits_{u>0} \inf\limits_{q\in O_f(p_0,0,\infty)} F_{p_0,q}(u) \text{ as } t>0 \end{cases} = \begin{cases} 0 \text{ as } t=0 \\ 1 \text{ as } t>0 \end{cases},$$

which means that $\{p_n\}_{n\in\mathbb{N}}$ is a Cauchy sequence in S. Since $(S, \mathscr{F}, T)$ is a complete nonarchimedean Menger space, it follows that there exists a point $p\in S$ such that $\lim\limits_{n\to\infty} p_n = p$.

Let us now prove, as in [101], that p is the fixed point of $T^{n(p)}$. Indeed, for any positive integer i and $\forall t \geqslant 0$, condition (1) implies that

$$F_{p_i, f^{n(p)}(p_i)}(t) \geqslant F_{p_{i-1}, f^{n(p)}(p_{i-1})}(\Phi(t)) \geqslant \ldots \geqslant F_{p_0, f^{n(p)}(p_0)}(\Phi^i(t)) \geqslant$$

$$\geqslant \sup_{u<\Phi^i(t)} F_{p_0, f^{n(p)}(p_0)}(u).$$

Letting $i\to\infty$, one gets

$$(3)\quad \lim_{i\to\infty} F_{p_i, f^{n(p)}(p_i)}(t) \geqslant \lim_{i\to\infty} \sup_{u<\Phi^i(t)} F_{p_0, f^{n(p)}(p_0)}(u) =$$

$$= \begin{cases} 0, \ t=0 \\ \sup\limits_{u>0} F_{p_0, f^{n(p)}(p_0)}(u), \ t>0 \end{cases} = \begin{cases} 0, \ t=0 \\ 1, \ t>0 \end{cases}$$

so that, $\forall t\geqslant 0$

$$(4)\quad \lim_{i\to\infty} F_{p, f^{n(p)}(p_i)}(t) \geqslant \lim_{i\to\infty} T(F_{p,p_i}(t), F_{p_i, f^{n(p)}(p_i)}(t)) = 1.$$

It follows from (1) that

$$F_{p,f^{n(p)}(p)}(t) \geqslant T(F_{p,f^{n(p)}(p_i)}(t), F_{f^{n(p)}(p), f^{n(p)}(p_i)}(t)) \geqslant$$

$$\geqslant T(F_{p,f^{n(p)}(p_i)}(t), F_{p,p_i}(\Phi(t))), \quad \forall t>0.$$

Letting $i\to\infty$ on both sides and using (4),

$F_{p,f^{n(p)}(p)}(t) = 1$, $\forall t>0$, i.e., $p = f^{n(p)}(p)$.

If another point $q \in S$ such that $q = f^{n(p)}(q)$ exists, then

$$F_{p,q}(t) = F_{f^{n(p)}(p),\, f^{n(p)}(q)}(t) \geqslant F_{p,q}(\Phi(t)) \geqslant F_{p,q}(\Phi^2(t)) \geqslant \ldots \geqslant$$
$$\geqslant F_{p,q}(\Phi^n(t)), \ \forall t > 0,$$

and, hence, for $n \to \infty$, $F_{p,q}(t) = 1$, $\forall t > 0$, i.e., $p = q$, and $f^{n(p)}$ has a unique fixed point in S. But

$$f(p) = f(f^{n(p)}(p)) = f^{n(p)}(f(p)),$$

i.e., $f(p)$ is also a fixed point of $f^{n(p)}$ and, hence, $f(p) = p$.

To prove that the sequence $\{f^n(q)\}_{n \in \mathbb{N}}$ converges to p for any $q \in S$, let us observe that for any integer $n > n(p)$, $n = m\, n(p) + s$, $0 \leqslant s \leqslant n(p)$ and

$$F_{p, f^n(q)}(t) = F_{f^{n(p)}(p),\, f^{m n(p)+s}(q)}(t) \geqslant F_{p, f^{(m-1)n(p)+s}(q)}(\Phi(t)) \geqslant \ldots$$
$$\ldots \geqslant F_{p, f^s(q)}(\varphi^n(t)), \ \forall t > 0.$$

Let $m \to \infty$ such that $n \to \infty$. Then

$$\lim_{n \to \infty} F_{p, f^n(q)}(t) \geqslant \lim_{m \to \infty} F_{p, f^s(q)}(\Phi^m(t)) = 1, \ \forall t > 0,$$

which implies $f^n(q) \to p$ and the proof of the theorem is complete.

Regarding the problem of common fixed points, let us now consider a result which generalizes those in [187, Th. 2] and [419, Th. 1].

THEOREM 3.1.62. *Let $(S, \mathcal{F}, T)$ be a complete nonarchimedean Menger space with a continuous $\mathcal{H}$-type t-norm T. Let g and h be injective and continuous mappings on S. Let $f : S \to g(S) \cap h(S)$ be a continuous mapping which commutes with g and h.*

Suppose that the following conditions are satisfied:

(i) *For every $p \in S$, there exists an $n(p) \in \mathbb{N}$ such that, for every $q \in S$,*

$$F_{f^{n(p)},\, f^{n(p)}(q)}(t) \geqslant \max\{F_{(g)p,\, h(q)}(\Phi(t)),\ F_{h(p),\, g(q)}(\Phi(t))\},$$

where Φ satisfies condition (ii) *of Theorem* 3.1.60 *and* $t > 0$;

(ii) *For some $p_0 \in S$, the set*

$$f^k h^m(p_0),\ k \in \{0, 1, \ldots, 2n(p_0) - 1\},\ m \in \mathbb{N}$$

is probabilistic bounded.

Then one and only one element $r \in S$ exists such that $r = f(r) = g(r) = h(r)$.

Proof. Let $p_0 \in S$ and form the sequence $\{p_i\}_{i\in\mathbb{N}}$ such that

$$g(p_{2k-1}) = f^{n(p_{2k-2})}(p_{2k-2})\,;\ \ h(p_{2k}) = f^{n(p_{2k-1})}(p_{2k-1}),\ k\in\mathbb{N}.$$

For any $n\in\mathbb{N}$, $n \geqslant n(p_1)$, one has $n = m\,n(p_1) + s$, where $m\in N$, $0 \leqslant s \leqslant n(p_0)$. Hence,

$$\begin{aligned}
&F_{g(p_1),f^n(p_1)}(\Phi(t)) \geqslant T(F_{g(p_1),f^{n(p_1)}(p_1)}(\Phi(t)),\\
&\quad , F_{f^{n(p_1)}(p_1),f^{mn(p_1)+s}(p_1)}(\Phi(t))) \geqslant\\
&\quad \geqslant T(F_{g(p_1),f^{n(p_1)}(p_1)}(\Phi(t)),\ F_{g(p_1),hf^{(m-1)n(p_1)+s}(p_1)}(\Phi^2(t))) \geqslant\\
&\quad \geqslant T(F_{g(p_1),f^{n(p_1)}(p_1)}(\Phi(t)),\ T(F_{g(p_1),f^{n(p_1)}(p_1)}(\Phi^2(t)),\\
&\quad , F_{f^{n(p_1)}(p_1),hf^{(m-1)n(p_1)+s}(p_1)}(\Phi^2(t)))) \geqslant\\
&\quad \geqslant T(F_{g(p_1),f^{n(p_1)}(p_1)}(\Phi(t)),\ T(F_{g(p_1),f^{n(p_1)}(p_1)}(\Phi^2(t)),\\
&\quad , F_{g(p_1),h^2f^{(m-2)n(p_1)+s}}(\Phi^3(t)))) \geqslant \ldots \geqslant T(T(\ldots\\
&\quad \ldots(T(T(F_{g(p_1),f^{n(p_1)}(p_1)}(\Phi(t)),\ F_{g(p_1),f^{n(p_1)}(p_1)}(\Phi(t))),\\
&\quad , F_{g(p_1),f^{n(p_1)}(p_1)}(\Phi(p_1))),\ldots),\ F_{g(p_1),f^{n(p_1)}(p_1)}(\Phi(t))),\\
&\quad , F_{g(p_1),h^m f^s(p_1)}(\Phi(t))) \geqslant\\
&\quad \geqslant T^m(\min\{F_{g(p_1),f^{n(p_1)}(p_1)}(\Phi(t)),\ F_{g(p_1),f^sh^m(p_1)}(\varphi(t))\}).
\end{aligned}$$

Since

$$F_{g(p_1),f^sh^m(p_1)}(\Phi(t)) = F_{f^{n(p_0)}(p_0),f^{s+n(p_0)}h^m(p_0)}(\Phi(t))$$

and since $\Phi(t) > t$, it follows that, for every $\delta\in(0,1)$, $t_\delta > 0$ exists such that

$$\inf_{s\in\{0,1,\ldots,n(p_0)-1\}} F_{f^{n(p_0)}(p_0),f^{s+n(p_0)}h^m(p_0)}(t_\delta) > \delta.$$

On the other hand, the properties of distribution functions imply that there exists a t'_δ such that $F_{g(p_1),f^{n(p_1)}(p_1)}(\Phi(t'_\delta)) > \delta$. Thus, one can choose $\delta = \delta_\lambda \in (0,1)$ such that $x > \delta_\lambda \Rightarrow$ $\Rightarrow T^m(x) > 1 - \lambda$ for every $m\in\mathbb{N}$ and every $\lambda\in(0,1)$. Hence, one concludes that

$$\sup_{u>0}\inf_{n\in\mathbb{N}} F_{g(p_1),f^n(p_1)}(u) = 1.$$

Let

$$r_m = \begin{cases} g(p_{2k-1}), & m = 2k-1 \\ h(p_{2k}), & m = 2k, \end{cases} \qquad k \in \mathbb{N}.$$

Let us prove that $\{r_m\}_{m\in\mathbb{N}}$ is a Cauchy sequence. For every $k \in \mathbb{N}$,

$$F_{r_{2k-1}, r_{2k}}(t) = F_{g(p_{2k-1}),\, h(p_{2k})}(t) = F_{f^{n(p_{2k-2})}(p_{2k-2}),\, f^{n(p_{2k-1})}(p_{2k-1})}(t) =$$

$$= F_{f^{n(p_{2k-2})}(p_{2k-1}),\, f^{n(p_{2k-1})}g^{-1}g(p_{2k-1})}(t) =$$

$$= F_{f^{n(p_{2k-2})}(p_{2k-2}),\, f^{n(p_{2k-1})}g^{-1}f^{n(p_{2k-2})}(p_{2k-2})}(t) \geqslant$$

$$\geqslant F_{h(p_{2k-2}),\, f^{n(p_{2k-1})}(p_{2k-2})}(\Phi(t)) =$$

$$= F_{f^{n(p_{2k-3})}(p_{2k-3}),\, f^{n(p_{2k-1})}(p_{2k-2})}(\Phi(t)) =$$

$$= F_{f^{n(p_{2k-3})}(p_{2k-3}),\, f^{n(p_{2k-1})}h^{-1}h(p_{2k-2})}(\Phi(t)) =$$

$$= F_{f^{n(p_{2k-3})}(p_{2k-3}),\, f^{n(p_{2k-1})}h^{-1}f^{n(p_{2k-3})}(p_{2k-3})}(\Phi(t)) \geqslant$$

$$\geqslant F_{g(p_{2k-3}),\, f^{n(p_{2k-1})}(p_{2k-3})}(\Phi^2(t)) \geqslant \ldots \geqslant F_{h(p_2),\, f^{n(p_{2k-1})}(p_2)}(\Phi^{2k-3}(t))$$

$$\geqslant F_{g(p_1),\, f^{n(p_{2k-1})}(p_1)}(\Phi^{2k-2}(t)).$$

According to relation (2) above, for every $\lambda \in (0,1)$, there exists $k_1(\lambda, t) \in \mathbb{N}$ such that

$$F_{g(p_1),\, f^{n(p_{2k-1})}(p_1)}(\Phi^{2k-2}(t)) > 1 - \lambda \text{ for every } k > k_1(\lambda, t).$$

Similarly,

$$F_{r_{2k}, r_{2k+1}}(t) > 1 - \lambda \text{ for every } k > k_2(\lambda, t)$$

and

$$F_{r_{2k-1}, r_{2k+1}}(t) \geqslant T(F_{r_{2k-1}, r_{2k}}(t),\ F_{r_{2k}, r_{2k+1}}(t)) \geqslant \ldots \geqslant$$

$$\geqslant T(F_{g(p_1),\, f^{n(p_{2k-1})}(p_1)}(\Phi^{2k-2}(t)),\ F_{g(p_1),\, f^{n(p_{2k})}(p_1)}(\Phi^{2k-1}(t))) > 1-\lambda$$

for $k > \max\{k_1(1-\delta, t),\ k_2(1-\delta, t)\}$, since, for every $\lambda \in (0,1)$, there exists $\delta \in (0,1)$ such that $T(x, x) > 1 - \lambda$ for $x \geqslant \delta$.

These facts imply that $\{r_m\}_{m\in\mathbb{N}}$ is a Cauchy sequence. Indeed,

$$F_{r_m, r_{m+j}}(t) \geqslant T(T(\ldots T(T(F_{r_m, r_{m+1}}(t), F_{r_{m+1}, r_{m+2}}(t)), F_{r_{m+2}, r_{m+3}}(t)),$$

$$\ldots), F_{r_{m+j-1}, r_{m+j}}(t)) \geqslant T^j(F_{u,v}(\Phi^{2k-2}(t))) > 1 - \lambda$$

for every $\lambda \in (0,1)$, $t>0$, $j \in \mathbb{N}$ and $k > k_0$, where

$$F_{u,v}(\Phi^{2k-2}(t)) = \inf_{n\in\mathbb{N}} F_{g(p_1), f^n(p_1)}(\Phi^{2k-2}(t)).$$

Since $\{g(p_{2k-1})\}_{k\in\mathbb{N}}$ and $\{h(p_{2k})\}_{k\in\mathbb{N}}$ are subsequences of the sequence $\{r_m\}_{m\in\mathbb{N}}$, it follows that

$$\lim_{k\to\infty} g(p_{2k-1}) = \lim_{k\to\infty} h(p_{2k}) = \lim_{m\to\infty} r_m = y.$$

But also

$$F_{g(p_{2k-1}), f(p_{2k-1})}(t) = F_{f^{n(p_{2k-2})}(p_{2k-2}), f g^{-1} g(p_{2k-1})}(t) =$$

$$= F_{f^{n(p_{2k-2})}(p_{2k-2}), f g^{-1} f^{n(p_{2k-2})}(p_{2k-2})}(t) \geqslant F_{h(p_{2k-2}), f(p_{2k-2})}(\Phi(t)) =$$

$$= F_{f^{n(p_{2k-3})}(p_{2k-3}), f h^{-1} h(p_{2k-2})}(\Phi(t)) =$$

$$= F_{f^{n(p_{2k-3})}(p_{2k-3}), f h^{-1} f^{n(p_{2k-3})}(p_{2k-3})}(\Phi(t)) \geqslant$$

$$\geqslant F_{g(p_{2k-3}), f(p_{2k-3})}(\Phi^2(t)) \geqslant \ldots \geqslant F_{g(p_1), f(p_1)}(\Phi^{2k-2}(t)) \to 1$$

for every $t>0$ and $k\to\infty$.

Furthermore,

$$F_{g(p_{2k-1}), f^2(p_{2k-1})}(t) = F_{f^{n(p_{2k-2})}(p_{2k-2}), f^2 g^{-1} g(p_{2k-1})}(t) \geqslant$$

$$\geqslant F_{h(p_{2k-2}), f^2(p_{2k-2})}(\Phi(t)) \geqslant \ldots \geqslant F_{g(p_1), f^2(p_1)}(\Phi^{2k-2}(t)) \to 1$$

for every $t>0$, $k\to\infty$. Hence,

$$r = \lim_{k\to\infty} g(p_{2k-1}) = \lim_{k\to\infty} h(p_{2k}) = \lim_{n\to\infty} f(p_n) = \lim_{n\to\infty} f^2(p_n).$$

Since the mappings f, g and h are continuous and g and h commute with f,

$$f(r) = f(\lim_{k\to\infty} h(p_{2k})) = \lim_{k\to\infty} f(h(p_{2k})) = h(\lim_{k\to\infty} f(p_{2k})) = h(r).$$

Similarly, $f(r) = g(r)$; from the equality $r = f(r)$, one gets that r is a common fixed point for the mappings f, g and h.

If one supposes that there exists an $r' \in S$, $r' \neq r$, which is also a common fixed point for the mappings f, g and h, then $r = f^{n(r)}(r)$, $r' = f^{n(r)}(r')$ and

$$F_{r,r'}(t) = F_{f^{n(r)}(r),\, f^{n(r)}(r')}(t) \geqslant F_{g(r),\, h(r')}(\Phi(t)) = F_{r,r'}(\Phi(t)) \geqslant \ldots \geqslant$$

$$\geqslant F_{r,r'}(\Phi^n(t)) \to 1, \text{ as } n \to \infty,$$

for every $t > 0$, which means that $r = r'$. Hence, the common fixed point for f, g and h is unique and the proof of the theorem is complete.

COROLLARY. *Let $(S, \mathcal{F}, T)$, f, g, h and Φ be defined as before. If $g(S) \cap h(S)$ is a probabilistic bounded set, then the mappings f, g and h have an unique common fixed point.*

Regarding the convergence of fixed point sequences for convergent contraction sequences over nonarchimedean probabilistic metric spaces, one states :

PROPOSITION 3.1.24. *Let $(S, \mathcal{F}, T)$ be a nonarchimedean Menger space with a t-norm T such that $\sup_{x<1} T(x, x) = 1$. Let $f_n : (S, \mathcal{F}, T) \to (S, \mathcal{F}, T)$, $n \in \mathbb{N}$, be a sequence of probabilistic contractions, each of which has at least one fixed point $p_n \in S$. Then,*

i) if $\{f_n\}_{n\in\mathbb{N}}$ is pointwise convergent to f, for $n \to \infty$, where f is a probabilistic contraction in S, then the limit p of any subsequence $\{p_{n_k}\}_{k\in\mathbb{N}} \subset \{p_n\}_{n\in\mathbb{N}}$ is a fixed point of f,

ii) if $\{f_n\}_{n\in\mathbb{N}}$ converges uniformly towards a probabilistic contraction f having a fixed point $p \in S$, then $p_n \to p$ in the (ε, λ)-topology.

3.1.7. *Fixed Point Theorems for Multivalued Mappings in PM-Spaces*

Results regarding the existence of fixed points for multivalued mappings are well known for metric and vector spaces, where as in the uniform-space case, some important results have been obtained over the last two decades. A short but interesting and substantial review of such results for multivalued mappings in uniform and *PM*-spaces was given by O. Hadžić [191]. The authoress also developed the main results on multivalued mappings on *PM*-spaces [177], [189], [191]. Other results in this respect are due to Do Hong Tan [424], [423]. Both succeeded in utilizing the existing results for uniform spaces in order to derive new results of *PM*-spaces. The main dificulty is that the topology in *PM*-spaces depends on a parameter $\lambda \in (0,1)$, so that one cannot choose an iterative sequence independently of λ without imposing additional assumptions.

We present next some results from refs. [189], [424] but with a simplified proof.

THEOREM 3.1.63. *Let* $(S, \mathscr{F}, \mathrm{Min})$ *be a compact Menger space and let* $f: S \to CB(S)$ *be such that, for every* $p, q \in S$ *and* $t > 0$,

$$(*) \quad F_{f(p), f(q)}(kt) \geqslant F_{p,q}(t), \quad k \in (0,1),$$

where $CB(S)$ *is the family of all nonempty closed probabilistic bounded subsets of* S.

Then f *has a fixed point.*

Proof. First let us recall a result presented in ref. [177].

LEMMA 3.1.9. *Let* S *be a nonempty compact uniform space such that the family* $\{d_i\}_{i \in I}$ *of pseudometrics* d_i *generates the uniformity of* S. *Let* $f: S \to S$ *be a multivalued mapping. Suppose that, for every* $i \in I$, *there exists a constant* $k_i \in [0,1)$ *such that*

$$D_i(f(p), f(q)) \leqslant k_i d_i(p, q), \text{ for all } p, q \in S,$$

where

$$D_i(A, B) = \max \{\sup_{p \in A} \inf_{q \in B} d_i(p, q), \sup_{q \in B} \inf_{p \in A} d_i(p, q)\}$$

for every $A, B \subseteq S$.

Then f *has a fixed point.*

Proof. Let us prove that $(*)$ implies that

$$(**) \quad D_\alpha(f(p), f(q)) \leqslant k d_\alpha(p,q) \text{ for all } p, q \in S \text{ and all } \alpha \in (0,1),$$

where $d_\alpha(p, q) = \sup \{t : F_{p,q}(t) \leqslant 1 - \alpha\}$, $\alpha \in (0, 1)$.

Suppose that, for some $\alpha \in (0, 1)$ and some $p, q \in S$,

$$D_\alpha(f(p), f(q)) > k d_\alpha(p,q)$$

and let $t = D_\alpha(f(p), f(q))/k$. Then $t > d_\alpha(p, q)$. From the definition of the pseudometrics $d_\alpha(p, q)$, it follows that $F_{p,q}(t) > > 1 - \alpha$. This implies that $F_{f(p), f(q)}(kt) > 1 - \alpha$ and since $kt = = D_\alpha(f(p), f(q))$, it follows that

$$F_{f(p), f(q)}(D_\alpha(f(p), f(q))) > 1 - \alpha.$$

From the definition of $F_{A,B}$,

$$F_{f(p), f(q)}(v) = \sup_{u < v} \mathrm{Min} \{\inf_{p \in f(p)} \sup_{q \in f(q)} F_{p,q}(u), \inf_{q \in f(q)} \sup_{p \in f(p)} F_{p,q}(u)\},$$

where $F_{A,B}$ is a left-continuous function. Then there exists a $u_0 < D_\alpha(f(p), f(q))$ such that $F_{f(p),f(q)}(u_0) > 1 - \alpha$ and

$$\text{Min}\ \{\inf_{p\in f(p)} \sup_{q\in f(q)} F_{p,q}(u),\ \inf_{q'\in f(q)} \sup_{p'\in f(p)} F_{p',q'}(u)\} > 1 - \alpha$$

for a $u < u_0$. Hence,

$$\inf_{p\in f(p)} \sup_{q\in f(q)} F_{p,q}(u) > 1 - \alpha,\ \inf_{q'\in f(q)} \sup_{p'\in f(p)} F_{p',q'}(u) > 1 - \alpha.$$

Then, for any $p \in f(p)$, there exists a $q \in f(q)$ and, for any $q' \in f(q)$, there exists a $p' \in f(p)$ such that

$$F_{p,q}(u) > 1 - \alpha,\ F_{p',q'}(u) > 1 - \alpha.$$

Hence, for any $p \in f(p)$, there exists a $q \in f(q)$ and, for any $q' \in f(q)$, there exists a $p' \in f(p)$ such that

$$d_\alpha(p,q) \leqslant u,\ d_\alpha(p', q') \leqslant u.$$

Therefore, $\inf_{q\in f(q)} d_\alpha(p, q) \leqslant u$ for all $p \in f(p)$ and $\inf_{p'\in f(p)} d_\alpha(p', q') \leqslant u$ for all $q' \in f(q)$. Hence,

$$\sup_{p\in f(p)} \inf_{q\in f(q)} d_\alpha(p, q) \leqslant u \text{ and } \sup_{q'\in f(q)} \inf_{p'\in f(p)} d_\alpha(p', q') \leqslant u, \text{ i.e.,}$$

$$D_\alpha(f(p), f(q)) \leqslant u,$$

which is a contradiction. Therefore, relation (**) is satisfied. From this, it follows that there exists a fixed point of the mapping f since it is easy to see that, for every $\alpha \in (0,1)$ and $\varepsilon > 0$, the set $\{p \in S,\ d_\alpha(p, f(p)) < \varepsilon\}$ is nonempty.

Let $(S, d_\alpha)_{\alpha\in I}$ be a space with a family of pseudometrics. Let ρ_α denote *the largest distance* between two subsets $A, B \subset S$ defined as

$$\rho_\alpha(A, B) = \sup\ \{d_\alpha(p, q) : p \in A,\ q \in B\}$$

and let $\delta_\alpha(A) = \rho_\alpha(A, A)$ be the α-*diameter* of A. Let $B(S)$ be the class of all nonempty subsets of S with finite α-diameter for each $\alpha \in I$. In what follows, we assume that the family $\{d_\alpha\}_{\alpha\in I}$ has the following property :

$$d_\alpha(x, y) = 0,\ \forall\, \alpha \in I \Rightarrow x = y.$$

Then

THEOREM A [424]. *Let* $(S, d_\alpha)_{\alpha\in I}$ *be a complete space with a family of pseudometrics, and let* $f : S \to B(S)$ *be a mapping satisfying the following condition : for each* $\alpha \in I$, *there exists a*

$k_\alpha \in [0,1)$ *such that*

$$\rho_\alpha(f(p), f(q)) \leqslant$$
$$\leqslant k_\alpha \max\{d_\alpha(p, q), \rho_\alpha(p, f(p)), \rho_\alpha(q, f(q)), \rho_\alpha(p, f(q)), \rho_\alpha(q, f(p))\}$$

for every $p, q \in S$.

Then f *has a unique fixed point* p_*, *which, in fact, is a stationary point*, i.e., $f(p_*) = p_*$.

As in the definition of the probabilistic diameter given in [424], we introduce the function

$$\delta_{A,B}(t) = \sup_{s<t} \inf_{p\in A,\ q\in B} F_{p,q}(s),$$

where A and B are probabilistic bounded sets.

It is known [424] that $\delta_{A,B} \in \mathscr{D}^+$, $\delta_{A,B} = H_0$ implies that $A = B$. Also, if

(α) $d_\alpha(p, q) = \sup\{t : F_{p,q}(t) \leqslant 1 - \alpha\}$, $\alpha \in (0,1)$,

the inequality

(δ) $\delta_{A,B}(\rho_\alpha(A, B)) \leqslant 1 - \alpha$

holds for every $A, B \in K(S)$ (the class of all nonempty compact subsets of S in the (ε, λ)-topology).

Then an interesting result for multivalued quasi-contractions in PM-spaces can be stated :

THEOREM 3.1.64 [424]. *Let* $(S, \mathscr{F}, \mathrm{Min})$ *be a complete Menger space and let* $f : S \to K(S)$ *be a mapping satisfying the following condition* :

(q) $\delta_{f(p),f(q)}(kt) \geqslant \min\{F_{p,q}(t), \delta_{p,f(p)}(t), \delta_{q,f(q)}(t), \delta_{p,f(q)}(t), \delta_{q,f(p)}(t)\}$,

for every $p, q \in S$ *and* $t > 0$, $k \in [0,1)$.

Then there exists a unique fixed point of f, *which is also a stationary point.*

Proof. If d_α is defined by relation (α), then (S, d_α) is a complete space possessing a family of pseudometrics. Let us prove that f satisfies the conditions of Theorem A.

Indeed, if $p, q \in S$ and $\alpha \in (0,1)$ exist such that

$$\rho_\alpha(f(p), f(q)) > k \max\{d_\alpha(p, q), d_\alpha(p, f(p)), d_\alpha(p, f(q)), d_\alpha(q, f(q)),$$
$$d_\alpha(q, f(p))\},$$

then, putting $t = \rho_\alpha(f(p), f(q))/k$, one gets $t > \max\{d_\alpha(p, q), d_\alpha(p, f(p)), d_\alpha(q, f(q)), d_\alpha(p, f(q)), d_\alpha(q, f(p))\}$. Using ($\delta$), one obtains

$$\min\{F_{p,q}(t), \delta_{p,f(p)}(t), \delta_{q,f(q)}(t), \delta_{p,f(q)}(t), \delta_{q,f(p)}(t)\} > 1 - \alpha.$$

According to (q),

$$\delta_{f(p),f(q)}(\rho_\alpha(f(p),\ f(q)) = \delta_{f(p),f(q)}(kt) > 1 - \alpha,$$

which contradicts the relation (δ). The proof of the theorem is complete.

Remark. It can be proved [177] that if $f, g : S \to K(S)$ and if a, b, $c \in [0,1)$ exist such that $a + 2b + 2c < 1$ and if, for every $p, q \in S$,

$$\delta_{f(p),g(q)}(r_1 + r_2 + r_3 + r_4 + r_5) \geqslant \min\{F_{p,q}(r_1/a),\ \delta_{p,f(p)}(r_2/b),$$
$$,\ \delta_{q,g(q)}(r_3/b),\ \delta_{p,g(q)}(r_4/c),\ \delta_{q,f(p)}(r_5/c)\},$$

where $r_i \geqslant 0$, $i = \overline{1,5}$, then there exists one and only one $p \in S$ such that $p \in f(p)$ and $p \in g(p)$ and $\{p\} = f(p) = g(p)$.

Next, we introduce the notion of probabilistic proximate fixed-point property for multivalued mappings, which, like the δ-continuity of multivalued mappings introduced by Klee [261], is based on the probabilistic δ-continuity.

Let us consider a topological space X and a Menger space $(S, \mathscr{F}, T)$ with a continuous t-norm T. For $A \subset S$ and $p \in S$, let us define the distance between p and A as

$$F_{p,A}(t) = \sup_{s<t}\ \sup_{q\in A} F_{p,q}(s) \text{ for all } s \in \mathbb{R}.$$

It is easy to see that $F_{p,A} \in \mathscr{D}^+$ and $F_{r,A}(t) \geqslant T(F_{r,q}(t-\varepsilon),\ F_{q,A}(\varepsilon))$ for all $t > 0$, $\varepsilon > 0$ and $q \in S$. It is also true that the set

$$N_A(\delta,\ \lambda) = \{p : F_{p,A}(\delta) > 1 - \lambda\}$$

is an open set of $(S, \mathscr{F}, T)$ for all $\delta > 0$ and $\lambda \in (0,1)$. The set

$$\bar{N}_A(\delta,\ \lambda) = \{p : F_{p,A}(\delta) \geqslant 1 - \lambda\}$$

is a closed set of $(S, \mathscr{F}, T)$.

DEFINITION 3.1.21. Let f be a multivalued mapping from the topological space X into the Menger space $(S, \mathscr{F}, T)$ and let δ and λ be positive reals.

(1) The mapping f is called *probabilistic lower δ-continuous* (*plδ-c*) iff, for each open set $V \subset S$, the set $\{u \in X : f(u) \cap N_V(\delta, \varepsilon) \neq \emptyset\}$ is open.

(2) The mapping f is called *probabilistic upper δ-continuous* (*puδ-c*) iff, for each closed set C, the set $\{u \in X :\ f(u) \cap N_C(\delta, \varepsilon) \neq \emptyset\}$ is closed.

(3) The mapping f is called *probabilistic δ-continuous* (*pδ-c*) iff it is both *plδ-c* and *puδ-c*.

PROPOSITION 3.1.25. *Let $f: X \to \mathscr{P}((S, \mathscr{F}, T))$ be a plδ-c multivalued mapping. If $\varepsilon > 0$, $\lambda > 0$, $u \in X$ and $p \in f(u)$, then there exists an open set $N(u)$ containing u such that $F_{p,f(v)}(\delta + \varepsilon) > 1 - \lambda$ for all $v \in N(u)$. Consequently, if f is plδ-c for all $\delta > 0$, then f is lower semicontinuous (lsc).*

Proof. Let $p \in f(u)$ and $B = N_p(\varepsilon, \lambda)$; then B and $N_B(\delta, \lambda)$ are open sets. Then, the set

$$\{v : f(v) \cap N_B(\delta, \lambda) \neq \emptyset\} = D.$$

is the required open set.

Indeed, since $p \in D$ and since f is $pl\delta$-c, it follows that D is an open set. If $v \in D$, then, for every $r \in f(v) \cap N_B(\delta, \lambda)$,

$$F_{p,f(v)}(\delta + \varepsilon) \geqslant F_{p,r}(\delta + \varepsilon) \geqslant T(F_{p,B}(\varepsilon), F_{B,r}(\delta)) = T(1, F_{B,r}(\delta)) > 1 - \lambda.$$

To prove the last part, let V be an open set of S and let $p \in f(u) \cap V$. Let also $\lambda' > 0$ be such that $N_p(2\varepsilon, \lambda') \subset V$. Then there exists an open set $V(u)$ of $u \in X$ such that

$$F_{p,f(v)}(2\varepsilon) > 1 - \lambda$$

for all $v \in V(u)$. Then there exists a point $r \in f(v)$ such that $F_{p,r}(2\varepsilon) > 1 - \lambda$, which implies that $r \in N_p(2\varepsilon, \lambda)$. Choosing λ' such that $\lambda' > \lambda$, one gets $N_p(2\varepsilon, \lambda) \subset N_p(2\varepsilon, \lambda') \subset V$ and $r \in V$. Hence, $f(v) \cap V \neq \emptyset$ for all $v \in V(u)$ and hence f is *lsc*.

PROPOSITION 3.1.26. *Let f have the property that $f(u)$ is compact for each $u \in X$. Then $f \in pu\delta$-c for every $\delta > 0$ implies that $f \in uc$.*

Let δ and λ be positive reals and let $f \in pu\delta$-c and $u \in X$; then there exists an open set $V(u)$ containing u such that, for every $v \in V(u)$ and every $p \in f(v)$, $F_{p,f(u)}(\delta) > 1 - \lambda$.

Proof. Let us first observe that, for $A \subset S$, a point p belongs to $\bar{A}$ iff $F_{p,A} = H_0$. Indeed, if $F_{p,A} = H_0$ then, for all $\varepsilon > 0$ and $\lambda > 0$, $\sup_{r \in A} F_{p,r}(\varepsilon) > 1 - \lambda$ and, hence, for every positive integer n, $\sup_{r \in A} F_{p,r}\left(\frac{1}{n}\right) > 1 - \frac{1}{n}$ and, hence, there exists a $p_n \in A$ such that $F_{p,p_n}\left(\frac{1}{n}\right) > 1 - \frac{1}{n}$. Therefore, $p_n \to p$ and $p \in \bar{A}$. If $p \in \bar{A}$, it also follows that, for all $\varepsilon > 0$ and $\lambda > 0$, there exists a q_n such that $F_{p,q_n}(\varepsilon) > 1 - \lambda$ and hence $\sup_{r \in A} F_{p,r}(\varepsilon) \geqslant \sup_{q_n \in A} F_{p,q_n}(\varepsilon) \geqslant F_{p,q_n}(\varepsilon) > 1 - \lambda$. Since $\varepsilon > 0$ and $\lambda > 0$ are arbitrary, it follows that $F_{p,A} = H_0$.

Let B be a closed subset of S. Since $f(u)$ is compact and since B is closed, it follows that $\{u : (f(u) \cap B \neq \emptyset\} = \bigcap_{\delta>0} \{u : f(u) \cap \cap \bar{N}_B(\delta, \lambda) \neq \emptyset\}$ for every $\lambda > 0$. Indeed, $B \subset \bar{N}_B(\delta, \lambda)$ implies that $B \subset \bigcap_{\delta>0} \bar{N}_B(\delta, \lambda)$. Conversely, let $p \in \bar{N}_B(\delta, \lambda)$ for all $\delta > 0$ and $\lambda > 0$; then $F_{p,B}(\delta) \geqslant 1 - \lambda$ for all $\delta > 0$ and $\lambda > 0$, hence $p \in \bar{B}$.

To prove the last part, let $\eta = \delta + \varepsilon$, $u \in X$ and $A = S - N_{f(u)}(\eta, \lambda) = S - \{p : F_{p,f(u)}(\delta + \varepsilon) > 1 - \lambda\}$. Since $f \in pu\delta$-c, it follows that $\{v : f(v) \cap \bar{N}_A(\delta, \lambda') = \emptyset\}$, $\lambda' > \lambda$, is an open set containing u. Indeed, if there exists an $r \in f(u) \cap \bar{N}_A(\delta, \lambda')$, then $F_{r,A}(\delta) \geqslant 1 - \lambda'$. On the other hand, $F_{q,f(u)}(\delta + \varepsilon) \leqslant 1 - \lambda$ for every fixed q, $q \in A$. For every $q \in A$, there exists an $r \in f(u)$ such that $F_{q,r}(\delta + s) \leqslant 1 - \lambda$ which implies that $F_{r,A}(\delta + \varepsilon) \leqslant 1 - \lambda$ and hence $1 - \lambda' \leqslant F_{r,A}(\delta) \leqslant F_{r,A}(\delta + \varepsilon) \leqslant \leqslant 1 - \lambda$. This is a contradiction, since $\lambda' > \lambda$.

Let $v \in X$ and let $p \in f(v)$ be such that $F_{p,f(u)}(\delta + \varepsilon) \leqslant 1 - \lambda$; then $p \notin N_{f(u)}(\eta, \lambda)$ and $p \in A$. Since $p \in f(v)$ and since $p \in A$, it follows that $f(v) \cap A \neq \emptyset$ and then $f(v) \cap \bar{N}_A(\delta, \lambda') \neq \emptyset$ and $v \notin V(u)$. If $v \in V(u)$, it follows that $F_{p,f(v)}(\delta + \varepsilon) > 1 - \lambda$ since $p \in f(v)$. The proof of the proposition is complete.

DEFINITION 3.1.22. A PM-space $(S, \mathscr{F}, T)$ is said to have the *probabilistic proximate fixed point property for multivalued mappings* iff, for every $\varepsilon > 0$ and $\lambda > 0$, there exists a $\delta > 0$ such that, for every multivalued function $f : S \to S$ and $f \in p\delta$-c there exists a point $p \in S$ such that

$$F_{p,f(p)}(\varepsilon) > 1 - \lambda.$$

If $(S, \mathscr{F}, T)$ is compact and if it has the probabilistic proximate fixed point property for mappings f such that $f(p)$ is probabilistic compact for every $p \in S$, then S has the probabilistic fixed point property for continuous multivalued mappings f such that $f(p)$ is compact for every $p \in S$.

We define

$$F'_{A,B}(t) = \sup_{\substack{p \in A \\ q \in B}} F_{p,q}(t) \text{ for all } A, B \subset S \text{ and all } t > 0.$$

DEFINITION 3.1.23. A Menger space $(S, \mathscr{F}, T)$ is *probabilistic δ-chainable* iff, for given $\varepsilon > 0$ and $\lambda \in (0, 1)$, there exists a finite decomposition $\{A_1, \ldots, A_n\}$ of open sets of S such that

i) $\bigcup_{i=1}^{n} A_i = S$,

ii) $D_{A_i}(\varepsilon) > 1 - \lambda$ for $i = 1, \ldots, n$,

iii) $F'_{A_i, A_j}(2\delta) \geqslant 1 - \lambda$ iff $|i - j| \leqslant 1$, where $\delta =$

$$= \frac{1}{2} \max_{|i-j|>1} F'_{A_i, A_j}(\varepsilon).$$

THEOREM 3.1.65. *Each probabilistic δ-chainable continuous Menger space $(S, \mathscr{F}, T)$ has the probabilistic proximate fixed point property.*

Proof. It is easy to see that $\delta > 0$. Indeed, (i) and (ii) imply that S is probabilistic precompact. It is known [56] that if T is a continuous t-norm, then S is a probabilistic precompact space iff, for every $\varepsilon, \lambda > 0$, there exists a finite subset $A_{\varepsilon,\lambda} \subset S$ such that, for every $p \in S$, there exists a $q = q(p) \in A_{\varepsilon,\lambda}$ such that $F_{p,q}(\varepsilon) > 1 - \lambda$.

Let f be a $p\delta$-c multivalued mapping. Define the sets A and B by : $A = \{p \in S : \text{for some } i, p \in A_i \text{ and for all } j < i, f(p) \cap \cap N_{A_j}(\delta, \lambda) = \emptyset\}$ and $B = \{q \in S : \text{for some } i, q \in A_i \text{ and for some } j < i, f(q) \cap \bar{N}_{\bar{A}_j}(\delta, \lambda) \neq \emptyset\}$. Then it follows that $A_1 \subset A$ and that A and B are closed sets. Since for every $p \in S - A$ there exists an i such that $p \in A_i$ and $f(p) \cap N_{A_j}(\delta, \lambda) \neq \emptyset$ for some $j < i$, it follows that $p \in B$. Therefore, $S = A \cup B$.

If one supposes that $B \neq \emptyset$, then $f(p) \subset A_n$ for every $p \in A_n$. Hence, $F_{p,f(p)}(\varepsilon) > 1 - \lambda$.

Let $B \neq \emptyset$. Since S is connected and since A, B are closed, there exists a $p_0 \in A \cap B$. Let i be the least index such that $p_0 \in A_i$. Then $f(p_0) \cap N_{A_j}(\delta, \varepsilon) = \emptyset$ for all $j < i$, but there exists a $j < i$ such that $f(p_0) \cap \bar{N}_{\bar{A}_j}(\delta, \varepsilon) \neq \emptyset$. Let $f(p_0) \cap \bar{N}_{A_{i-1}}(\delta, \lambda) \neq \emptyset$; then $f(p_0) \cap A_i \neq \emptyset$ since if $f(p_0) \cap A_j = \emptyset$ for $j = 1, 2, \ldots, i-1$ and if $f(p_0) \cap A_i = \emptyset$, then $f(p_0) \subset \subset \bigcup_{j=i+1}^{n} A_j$. If there exists a q such that $q \in f(p_0) \cap \bar{N}_{\bar{A}_{i-1}}(\delta, \lambda) \cap \cap A_{i+1}$, then $q \in \bar{N}_{\bar{A}_{i+1}}\delta, \lambda)$. Since $(i+1) - (i-1) = 2 > 1$, then $F'_{\bar{A}_{i-1}, \bar{A}_{i+1}}(2\delta) < 1 - \lambda$. On the other hand,

$$F'_{\bar{A}_{i-1}, \bar{A}_{i+1}}(2\delta) \geqslant T(F_{\bar{A}_{i-1}, q}(\delta), F'_{q, \bar{A}_{i+1}}(\delta)) =$$

$$= T(F_{\bar{A}_{i-1}, q}(\delta), 1) = F'_{\bar{A}_{i-1}, q}(\delta) \geqslant 1 - \lambda,$$

which is a contradiction.

In a similar way, one obtains a contradiction if $q \in A_j$ with $j > i + 1$. Therefore, $f(p_0) \cap A_i \neq \emptyset$ and for $r \in A_i \cap f(p_0)$ one gets $F_{p_0,r}(\varepsilon) > 1 - \lambda$. Furthermore, if $f(p_0) \cap (A_i \cap A_{i+1}) =$

$= \emptyset$, then $f(p_0) \in \bigcup_{j=i+2}^{n} A_j$. Since $p_0 \in B \cap A_i$, there exist $j < i$ and $r \in f(p_0) \cap N_{\bar{A}_j}(\delta, \lambda)$. On the other hand, if $r \in A_m$ with $m \geqslant i + 2$, then $r \in \bar{A}_m$ and

$$F'_{\bar{A}_j, \bar{A}_m}(2\delta) \geqslant T(F'_{\bar{A}_j, r}(\delta), F'_{r, \bar{A}_m}(\delta)) \geqslant 1 - \lambda,$$

which contradicts the relation $F_{\bar{A}_j, \bar{A}_m}(2\delta) < 1 - \lambda$ for $|j - m| > 1$. Therefore, $f(p_0) \cap (A_i \cup A_{i+1}) \neq \emptyset$. If $f(p_0) \cap A_i = \emptyset$, then q exists such that $q \in f(p_0) \cap A_{i+1} \subset f(p_0) \cap \bar{A}_{i+1}$. Then $f(p_0) \cap \bar{N}_{\bar{A}_j}(\delta, \lambda) = \emptyset$ for $j = 1, 2, \ldots, i - 1$ and $f(p_0) \cap \bar{N}_{\bar{A}_i}(\delta, \lambda) \neq \emptyset$, which implies that $p_0 \in A_{i+1}$ and $f(p_0) \cap A_{i+1} \neq \emptyset$. Hence, there exists a $q_0 \in f(p_0)$ such that $F_{p_0, q_0}(\varepsilon) > 1 - \lambda$. The proof of the theorem is complete.

3.2. FIXED POINT THEOREMS FOR MAPPINGS ON RANDOM NORMED SPACES

There exist hosts of applications of the results regarding the existence and uniqueness of fixed points for mappings on random normed spaces. Many of these are connected with similar applications for the probabilistic metric spaces and particularly for probabilistic locally convex spaces. We shall next present some of the outstanding results obtained by Gh. Bocşan [56], [45], [49], O. Hadžić [169], [196], [195], [175], [180], [184] [188], [186], [185], [191], Gh. Constantin [109] and I. Istrăţescu [224], [115].

Let $(S, \mathscr{F}, \tau)$ be a random normed space (RN-space) with the (ε, λ)-topology defined by the random norm $\mathscr{F}$.

DEFINITION 3.2.1. A continuous mapping $f: S \to S$ is called a probabilistic α-contraction iff for every probabilistic bounded subset $A \subset S$,

i) $f(A)$ is *probabilistic bounded,*

ii) there exists a $k \in (0,1)$ such that

$$\alpha_{f(A)}(kx) \geqslant \alpha_A(x), \text{ for every } x > 0.$$

THEOREM 3.2.1 [49]. *Let $(S, \mathscr{F}, \tau)$ be a complete RN-space and let A be a nonempty probabilistic bounded, closed and convex subset of S. If $\tau = \tau_{\text{Min}}$, then every probabilistic α-contraction f with $f(A) \subset A$ has a fixed point in A.*

Proof. Let $A_1 = A$ and $A_{n+1} = \overline{\text{co}}\, f(A_n)$, $n = 1, 2, \ldots$. Clearly, every A_n is a nonempty probabilistic bounded, closed and convex set and $A_n \supseteq A_{n+1}$, $n = 1, 2, \ldots$. From the properties of Kuratowski's function (see Propositions 1.2.1. and 1.2.2) and from Definition 3.2.1, it follows that

$$\alpha_{A_n}(x) \geqslant \alpha_A\left(\frac{x}{k^n}\right), \quad x > 0,$$

and therefore,

$$\alpha_{A_n} \xrightarrow{L} H.$$

From Theorem 1.2.5, it follows that $A_0 = \bigcap_{n=1}^{\infty} A_n$ is a nonempty compact convex subset of A and $f(A_0) \subset A_0$.

Since the (ε, λ)-topology defined by random norm $\mathscr{F}$ is a locally convex topology, we can apply the Schauder-Tychonoff theorem for f and A_0 and, hence, f has a fixed point in A_0.

Remark. a) One can substitute α in Definition 3.2.1 by a random measure γ of noncompactness, invariant to closed convex hull, and thus one obtains a γ-contraction mapping.

If a convergence theorem in the space $(2^s, \mathscr{F}_1, \mathrm{Min})$ of the type of Theorem 1.2.5 holds for γ, the above result is still valid, as if, say γ is a Hausdorff function.

b) Any probabilistic contraction, as stated in Definition 3.1.2, in an RN-space, is a probabilistic α-contraction.

Indeed, it follows immediately that such a mapping is continuous. From

$$D_{f(A)}(kx) \geqslant D_A(x),$$

one obtains then condition (i) of Definition 3.2.1.

In order to prove (ii), consider an element $F \in \mathscr{D}^+ \cap K_A$ as defined in Proposition 1.2.2. Let $\{A_i\}_{i\in I}$, I = finite, a covering of A having the property that $D_{A_i}(x) \geqslant F(x)$ for any $x > 0$. Denoting $B_i = f(A_i)$, it follows that $\{B_i\}_{i\in I}$ is a finite covering of $f(A)$. Then

$$D_{B_i}(x) \geqslant D_{A_i}\left(\frac{x}{k}\right) \geqslant F\left(\frac{x}{k}\right),$$

i.e., $k \circ F \in K_{f(A)}$ and, hence $k \circ F \leqslant \alpha_{f(A)}$. From Proposition 1.2.2, follows ii).

Following this reasoning, one can conclude that Theorems 3.1.2 and 3.1.3 can be regarded as a corollary of Theorem 3.2.1.

An important application of random measures of noncompactness towards the theory of fixed points is the following theorem (a generalization of S. N. Sadovski's [370] result):

THEOREM 3.2.2. *Let* $(S, \mathscr{F}, \tau)$ *be a complete RN-space,* $\tau = \tau_{\mathrm{Min}}$, *and let* $A \subseteq S$ *be a closed, convex and probabilistic bounded subset. Let* $f : S \to S$ *be a* γ*-condensing mapping such that* $f(A) \subseteq A$. *If* γ *is a nonsingular invariant to the closed convex hull measure of noncompactness, then* f *has a fixed point.*

Proof. Since $\tau = \tau_{\mathrm{Min}}$, then from Remark 4, Chap. 1, Sec. 1, it follows that the (ε, λ)-topology defined by the random norm $\mathscr{F}$ is locally convex. It is known [370] that if one defines the transfinite net A_α by

$$A_0 = A, \quad A_\alpha = \begin{cases} \mathrm{co}\, f(A) \text{ if } \alpha \text{ is not a limit ordinal,} \\ \bigcap\limits_{\beta<\alpha} A_\beta \text{ if } \alpha \text{ is a limit ordinal,} \end{cases}$$

then

1) every A_α is a closed and convex set,

2) $f(A_\alpha) \subseteq A_\alpha$,

3) $\alpha < \beta$ implies that $A_\beta \subseteq A_\alpha$,

4) there exists an ordinal number δ such that $A_\alpha = A_\delta$ for all $\alpha \geqslant \delta$.

The set A_δ is called the codomain limit of f and it has the property that $A_\delta = A_{\delta+1} = \mathrm{co}\,(f(A_\delta))$ and $\gamma_{A_\delta}(x) = \gamma_{f(A_\delta)}(x)$. Since f is a γ-condensing mapping, it follows that A_δ is probabilistic precompact. Also, S is a complete metrizable locally convex topological vector space and, hence, A_δ is a compact set. Let us denote by $\mathscr{F}_1/A$ the restriction to A of the random metric $\mathscr{F}_1$ induced by $\mathscr{F}$. From Proposition 3.1.11, it follows that there exists a $K \subseteq A$, $K \neq \emptyset$, such that $f(K) = K$.

By transfinite induction one shows that $K \subseteq A_\delta$. Applying Schauder-Tychonoff's fixed point theorem to the set A_δ, which is nonempty, convex, compact and $f(A_\delta) \subseteq A_\delta$, the theorem is proven.

Remarks. 1) For $\gamma = \alpha =$ the Kuratowski function, the proof of the theorem is as given in ref. [45]. For $\gamma = \beta =$ the Hausdorff function, the above result is a generalization to the *RN*-space of Sadovski's result from [370].

2) Since every probabilistic γ-contraction is a probabilistic γ-condensing, Darbo's theorem [122] is obtained from the above theorem.

V. Klee introduced the notion of admissible topological vector space. This concept is very important in the fixed point theory :

DEFINITION 3.2.2 [191]. Let S be a Hausdorff topological vector space and let $\mathscr{V}$ be the fundamental system of neighbourhoods of zero in S. A subset $M \subseteq S$ is said to be *admis-*

sible if, for every compact subset K of M and every $U \in \mathscr{V}$, there exists a continuous mapping $h : K \to M$ so that :

i) $\dim(\text{span}(h(K))) < \infty$,

ii) for every $x \in K$, $x - h(x) \in U$.

In ref. [201], S. Hahn and K. F. Pötter proved the following fixed point theorem :

THEOREM 3.2.3 [201]. *If M is a closed, convex and admissible subset of a Hausdorff topological vector space S and $f : M \to M$ is a compact mapping, then f has a fixed point in M.*

This theorem is a generalization of Tichonoff's fixed point theorem since every convex subset of a locally convex space is admissible.

THEOREM 3.2.4 [196]. *Let $(S^1, \mathscr{F}^1, T^1)$ be a complete RN-space with continuous t-norm T^1, let $(S^2, \mathscr{F}^2, T^2)$ be an RN-space with continuous t-norm T^2 and let U be a closed and probabilistic bounded subset of S^1. Let V be a closed, convex, compact and admissible subset of S^2 and let $H : U \to V$ and $K : U \times V \to$ $\to V$ be continuous mappings such that the following condition is satisfied :*

$$F_{H(x_1, y) - H(x_2, y)}(kt) \geqslant F_{x_1 - x_2}(t), \quad 0 < k < 1, \text{ for every}$$

$$x_i \in U,\ i = 1, 2, \text{ every } y \in V \text{ and every } t > 0.$$

Then there exists at least one element $z_0 \in U \times V$ such that :

$$Gz_0 = z_0, \text{ where } G : U \times V \to U \times V \text{ and } Gz = (H(z), K(z))$$

for every $z = (x,y) \in U \times V$.

Proof. We shall prove that there exists one and only one continuous mapping $R : V \to U$ such that

$R_v = H(R_v, v)$ for every $v \in V$

$R_v = H(R_v, v)$.

Let us denote by $C(V, S^1)$ the set of all continuous mappings from the compact set V into the space S^1. The triplet $(C(V, S^1), \tilde{\mathscr{F}}, T^1)$ can be considered a complete RN-space if the mapping $\tilde{\mathscr{F}} : C(V, S^1) \to \mathscr{D}^+$ is defined as

$$\tilde{\mathscr{F}}_{\tilde{x}}(t) = \sup_{s < t} \inf_{v \in V} F^1_{x(v)}(s)$$

for every $\tilde{x} = \{x(v)\} \in C(V, S^1)$.

It is also easy to see that if U is a probabilistic bounded subset of S^1, then the condition

$$\sup_{t \in \mathbb{R}} \inf_{\tilde{x}, \tilde{y} \cdot C(V, S^1)} \tilde{F}_{\tilde{x} - \tilde{y}}(t) = 1$$

is satisfied.

Let $\widetilde{E}:C(V, U) \to C(V, S^1)$ be the mapping defined by

$(\widetilde{E}(\tilde{x}))(v) = H(x(v), v)$ for every $v \in V$ and every $\tilde{x} = \{x(v)\} \in \in C(V, U)$.

Then it can be proven that

$$\widetilde{F}_{\widetilde{E}(\tilde{x})-\widetilde{E}(\tilde{y})}(kt) \geqslant \widetilde{F}_{\tilde{x}-\tilde{y}}(t), \quad 0 < k < 1$$

for every $\tilde{x}, \tilde{y} \in C(V, U)$ and every $t > 0$. Applying Theorem 3.1.3 for $\widetilde{E}$, one concludes that there exists an $\widetilde{R} = \{R(v)\}$ such that $\widetilde{E}(\widetilde{R}) = \widetilde{R}$. Since $\overline{K(U, V)} \subseteq V$, the mapping $f: V \to V$, $f(v) = K(R_v, v)$, $v \in V$, satisfies all the conditions of Theorem 3.2.3 and, hence, there exists at least one fixed point for f and, thus, at least one fixed point z_0 for the mapping G. The proof is complete.

Remark. A similar result is given in [169, Th. 4], for $T_1 = T_2 = \text{Min}$, U a closed, convex subset of S^1, V a closed, convex and probabilistic bounded subset of S^2, $H: U \times V \to U$ a uniformly continuous and generalized probabilistic contraction with respect to the first variable and K a densifying mapping.

The above result yields as particular cases some well-known Banach space theorems :

THEOREM 3.2.5 [196]. *Let $(S, \mathscr{F}, T)$ be a complete RN-space with continuous t-norm T, let A be a probabilistic bounded, closed and convex subset of S and let $f: A \to A$ be a continuous mapping such that the following condition is satisfied :*

$$\forall C \subseteq A, \quad \alpha_C \geqslant \alpha_{\overline{\text{co}} f(C)} \Rightarrow \alpha_C = H_0.$$

If every closed and convex subset of A is admissible, then f has a fixed point.

Proof. Let $a \in A$ and let K be the set of limit points of $O_f(a, 1, \infty)$. Then $O_f(a, 1, \infty)$ is a compact set since, from the relation $\alpha_{O_f(a,1,\infty)} \neq H_0$, it follows that

$$\alpha_{O_f(a,1,\infty)} < \alpha_{\overline{\text{co}} f(O_f(a,1,\infty))},$$

which is a contradiction because

$$\alpha_{O_f(a,1,\infty)} = \alpha_{\{f(a)\} \cup f(O_f(a,1,\infty))} = \min\{\alpha_{f(a)}, \alpha_{f(O_f(a,1,\infty))}\} =$$
$$= \alpha_{f(O_f(a,1,\infty))} \geqslant \alpha_{\overline{\text{co}} f(O_f(a,1,\infty))}.$$

Therefore, K is a nonempty and compact subset of A such that $K \subseteq f(K)$.

Let

$$\mathscr{A} = \{C : C \subseteq A,\ C = \overline{\text{co}}\ C, K \subseteq C,\ f(C) \subseteq C\}.$$

It is easy to see that $\mathscr{A} \neq \emptyset$ and if $\mathscr{C}$ is a chain in the ordered set $(\mathscr{A}, \subseteq)$, then $\bigcap_{C \in \mathscr{A}} C$ is a lower bounded set of $\mathscr{C}$. Hence, according to Zorn's lemma, $\mathscr{C}$ has a minimal element C_0. Since $f(C_0) \subseteq C_0$ and since C_0 is closed and convex, it follows that $\overline{\text{co}}\ f(C_0) \subseteq C_0$ and

$$f(\overline{\text{co}}\ f(C_0)) \subseteq f(C_0) \subseteq \overline{\text{co}}\ f(C_0).$$

From $K \subseteq \overline{\text{co}}\ f(C_0)$, it follows that $\overline{\text{co}}\ f(C_0) \in \mathscr{C}$. Since C_0 is a minimal element of $\mathscr{C}$, it follows that $f(C_0) = C_0$ and then $\alpha_{\overline{\text{co}}f(C_0)} = \alpha_{C_0}$. Hence, C_0 is precompact. From the completeness of S and the closedness of C_0 we conclude that C_0 is compact and, hence, $f|_{C_0}$ is a compact mapping of C_0 into C_0. From Theorem 3.2.3, it follows that there exists an $x_0 \in C_0$ such that $f(x_0) = x_0$.

COROLLARY 1. *Let $(S, \mathscr{F}, T)$ be a complete RN-space with continuous t-norm T of $\mathscr{H}$-type, let A be as in Theorem 3.2.5 and let $f: A \to A$ be a continuous mapping such that the following condition is satisfied:*

For every $C \subseteq A$, $C \neq \emptyset$ such that $C = \overline{\text{co}}C$ and $\alpha_{\overline{\text{co}}f(C)} \leqslant \alpha_C$, it follows that C is compact.

Then f has a fixed point.

COROLLARY 2. *Let $(S, \mathscr{F}$, Min) be a complete RN-space, let A be as in Theorem 3.2.5 and let $f: A \to A$ be a probabilistic densifying mapping. Then f has a fixed point.*

Regarding the continuous dependence of the fixed point with respect to a parameter, the following result [180] can be given:

THEOREM 3.2.6. *Let A be a probabilistic bounded, closed and convex subset of a complete RN-space $(S, \mathscr{F}, T)$ with continuous t-norm T, let Λ be a Hausdorff topological space and let $\lambda_0 \in \Lambda$. Suppose that f is a mapping $(f: A \times \Lambda \to A)$ satisfying the following conditions:*

i) *f is continuous at (x, λ_0) for every $x \in A$;*

ii) *if $\alpha_C \subset H_0$, where $C \subseteq A$, then there exists a neighbourhood $V(C)$ of λ_0 such that $\alpha_{\overline{\text{co}}f(C \times \Lambda')} > \alpha_C$ for every relatively compact subset Λ' of $V(C)$;*

iii) *for each $\lambda \in \Lambda$, there exists an element $x(\lambda) \in A$ such that $x(\lambda) = f(x(\lambda), \lambda)$ and $x(\lambda_0)$ is the unique fixed point of the mapping $x \to f(x, \lambda_0)$.*

Then, for every sequence $\{\lambda_k\}_{k\in\mathbb{N}}$ *converging to* λ_0, *the sequence* $\{x(\lambda_k)\}_{k\in\mathbb{N}}$ *converges to* $x(\lambda_0)$.

Proof. Let $C = \{x(\lambda_k)\}\ k \in \mathbb{N}$ and suppose that $\alpha_C < H_0$. Then there exists a neighbourhood $V(C)$ of λ_0 such that $\alpha_{\overline{\mathrm{co}}f(C\times\Lambda')} > \alpha_C$ for every relatively compact subset Λ' of $V(C)$. Let $n \in \mathbb{N}$ be such that

$$\Lambda' = \{\lambda_k : k > n\} \subset V(C).$$

Then from

$$C = \{x(\lambda_1), \ldots, x(\lambda_n)\} \cup \{f(x(\lambda_k), \ \lambda_k) : \ k > n\} \subseteq$$
$$\subseteq \{x(\lambda_1), \ldots, x(\lambda_n)\} \cup f(C \times \Lambda'),$$

it follows that

$$\alpha_C \geqslant \min\{H_0, \alpha_{f(C\times\Lambda')}\} = \alpha_{f(C\times\Lambda')} \geqslant \alpha_{\overline{\mathrm{co}}f(C\times\Lambda')},$$

which is a contradiction. Hence, $\alpha_C = H_0$ and so C is relatively compact. Supposing that the sequence $\{x(\lambda_k)\}$ does not converge to $x(\lambda_0)$, it follows that there exists a neighbourhood U of $x(\lambda_0)$ and a subsequence $\{\lambda_{k(i)}\}_{i\in\mathbb{N}}$ of $\{\lambda_k\}_{k\in\mathbb{N}}$ such that $x(\lambda_{k(i)}) \notin U$, for all $i \in \mathbb{N}$. Since C is relatively compact, the sequence $\{\lambda_{k(i)}\}_{i\in\mathbb{N}}$ has a subsequence $\{\lambda_{k(ij)}\}_{j\in\mathbb{N}}$ such that the sequence $\{x(\lambda_{k(ij)})\}$ converges to some $x_0 \in \bar{C} \subseteq A$. Then, from the continuity of f at (x_0, λ_0), it follows that the sequence $\{f(x(\lambda_{k(ij)}), \ \lambda_{k(ij)})\}_{j\in\mathbb{N}}$ converges to $f(x_0, \ \lambda_0)$. Therefore, $x_0 = f(x_0, \lambda_0)$, which shows that x_0 is a fixed point of the mapping $x \to f(x, \ \lambda_0)$. Thus, we obtain that $x_0 = x(\lambda_0)$ which is a contradiction and the theorem is proven.

Remark. The results of the above two theorems, together with some adequate hypotheses regarding the admissibility of certain compact sets, the weakly Φ-contractive mapping for $f : A \times \Lambda \to \Lambda$, and the continuous dependence of the fixed point on some parameter were used by O. Hadžić [196, Th. 4], [195], [186] to obtain certain theorems regarding the existence of a solution for the system $x = H(x, \ y)$, $y = K(x, \ y)$ in RN-spaces.

In order to present a fixed point theorem of Krasnosel'skiĭ type in RN-spaces, we recall the following result for topological vector-spaces.

THEOREM [186, Th. B]. *Let* X *be a Hausdorff topological vector space, let* K *be a nonempty, closed and convex subset of* X, *let* $f : X \to X$ *be an affine, continuous mapping and let* $h : K \to X$

be a continuous mapping such that $h(K)$ is compact. Suppose that the following conditions are satisfied:

i) *For every $y \in \overline{\text{co}}\, h(K)$, there exists one and only one solution x, $y \in K$ of the equation $x = f(x) + y$ and the set $\{x(y)\}_{y \in \overline{h(K)}}$ is compact.*

ii) *For every V from the fundamental family $\mathscr{U}$ of neighbourhoods of zero in X and every $x \in h(K)$, there exists a $U \in \mathscr{U}$ such that*

$$\text{co}(x + U) \cap h(K) \subseteq x + V.$$

Then there exists an $x \in K$ such that $x = f(x) + h(x)$.

THEOREM 3.2.7 [186]. *Let $(S, \mathscr{F}, T)$ be a complete RN-space with a continuous t-norm T, let K be a closed, convex and probabilistic bounded subset of S, let $f : S \to S$ be a linear continuous mapping, let $h : K \to S$ be a compact mapping and let the following conditions be satisfied:*

i) *There exist $n \in \mathbb{N}$ and $k \in (0, 1)$ such that, for every $\varepsilon > 0$ and very x, $y \in K$:*

$$F_{f^n(x)-f^n(y)}(kx) \geqslant F_{x-y}(\varepsilon).$$

ii) $f(K) + \overline{\text{co}}\, h(K) \subseteq K.$

iii) *For every $\varepsilon > 0$, every $\delta \in (0, 1)$ and every $x \in \overline{h(K)}$, there exist $\varepsilon' > 0$ and $\delta' \in (0,1)$ such that:*

$$\text{co}(x + N_0(\varepsilon', \delta')) \cap \overline{h(K)} \subseteq x + N_0(\varepsilon, \delta).$$

Then there exists an $x \in K$ such that $x = f(x) + h(x)$.

Proof. For every $x \in \overline{\text{co}}\, h(K)$, let the mapping $H_y : K \to K$ be defined by

$$H_y(x) = f(x) + y, \quad x \in K.$$

Then

$$H_y^n(x) - H_y^n(z) = f^n(x) - f^n(z), \quad x, z \in K.$$

Since f is linear and K is a probabilistic bounded subset of S, it follows that there exists one and only one element $R(y)$ such that $H_y(R(y)) = R(y)$ for every $y \in \overline{\text{co}}\, h(K)$.

Now, let us prove that the mapping $R : \overline{h(K)} \to K$ is continuous. Denote by $\mathscr{C}(h(K), M)$, $M \subseteq S$, the set of all continuous mappings from $\overline{h(K)}$ into M. Then the triplet $(\mathscr{C}(\overline{h(K)}, S), \mathscr{F}, T)$ is a complete RN-space, where the mapping $\widetilde{\mathscr{F}} : \mathscr{C}(\overline{h(K)}, S) \to \Delta^+$ is defined by

$$\widetilde{\mathscr{F}}(\tilde{x}) = \widetilde{F}_{\tilde{x}} \text{ with } \widetilde{F}^{\sim}(t) = \sup_{s<t} \inf_{y \in \overline{h(K)}} F_{\tilde{x}(y)}(s)$$

for every $\tilde{x} \in \mathscr{C}(\overline{h(K)}, S)$. If one defines the mapping $\widetilde{H} : \mathscr{C}(\overline{h(K)}, K) \to \mathscr{C}(\overline{h(K)}, S)$ by

$$(\widetilde{H}\tilde{x})(y) = f(\tilde{x}(y)) + y$$

for every $y \in \overline{h(K)}$ and every $\tilde{x} \in \mathscr{C}(\overline{h(K)}, K)$, then

$$(\widetilde{H}^n\tilde{x})(y) = f^{(n)}(\tilde{x}(y)) + \sum_{i=0}^{n-1} f^i(y).$$

Since

$$\widetilde{F}_{\widetilde{H}^n(\tilde{x}) - \widetilde{H}^n(\tilde{z})}(ku) \geqslant \widetilde{F}_{\tilde{x}-\tilde{z}}(u)$$

for every $u > 0$ and every $\tilde{x}, \tilde{z} \in C(\overline{h(K)}, K)$ and since

$$\sup_t \inf_{\tilde{x},\tilde{z} \in \mathscr{C}(\overline{h(K)},K)} \widetilde{F}_{\tilde{x}-\tilde{z}}(t) = 1,$$

it follows (Th. 3.1.34) that there exists one and only one element $\tilde{x}^* \in \mathscr{C}(\overline{h(K)},K)$ such that $\widetilde{H}(\tilde{x}^*) = \tilde{x}^*$. Then $\tilde{x}(y) = R(y)$ for every $y \in \overline{h(K)}$, which implies that the mapping R is continuous. Since the set $\overline{h(K)}$ is compact, condition (i) from the above theorem is satisfied. The proof is complete.

COROLLARY. *Let $(S, \mathscr{F}, T)$ be a complete RN-space such that T is continuous and of $\mathscr{H}$-type, let $f : S \to S$ be a linear continuous mapping and let $h : K \to S$ be a compact mapping such that conditions* (i) *and* (ii) *of Theorem* 3.2.7 *are satisfied. Then there exists an $x \in K$ such that $x = f(x) + h(x)$.*

Remark. A fixed point theorem for maps on RN-spaces with a contractive iterate at a point is given in ref. [224].

Regarding the existence of common fixed points, the following results follow easily from Theorem 3.1.49 :

COROLLARY. *Let $(S, \mathscr{F}, T)$ be a complete RN-space with continuous t-norm T and let g and h be continuous mappings from S into S. If f is a continuous mapping from S into $g(S) \cap h(S)$ which commutes with g and h, if $f(S)$ is bounded in the (ε, λ)-topology and if*

$$F_{f(x)-f(y)}(kt) \geqslant F_{g(x)-h(y)}(t) \text{ for every } t > 0$$

and every $x, y \in S$, where $k \in (0, 1)$, then f, g and h have a unique common fixed point.

Using a reasoning similar to that made in the PM-space case, one can prove

THEOREM 3.2.8 [184]. *Let $(S, \mathscr{F}, T)$ be a complete RN-space with continuous t-norm T, let g and h be linear continuous* *mappings from S into S. Further, let f be a continuous mapping which commutes with g and h such that $f(S) \subseteq g(S) \cap h(S)$, $f(S)$ is bounded in the (ε, λ)-topology and there exists an $m \in \mathbb{N}$ such that $f^m(S)$ is relatively compact. If, for every $x, y \in S$ and every $t > 0$,*

$$F_{f(x)-f(y)}(t) \geqslant F_{g(x)-h(y)}(t),$$

then there exists an $x \in S$ such that $x = f(x) = g(x) = h(x)$.

Proof. Let $\{r_n\}_{n\in\mathbb{N}} \subseteq (0,1)$ such that $\lim\limits_{n\to\infty} r_n = 1$. For every $n \in \mathbb{N}$, let $f_n(x) = r_n f(x)$, $x \in S$. From $f(S) \subseteq g(S) \cap h(S)$ and from the fact that g and h are linear, it follows that $f_n(S) \subseteq g(S) \cap h(S)$. On the other hand, $f_n(g(x)) = r_n f(g(x)) = r_n g(f(x))$ and $g(f_n(x)) = g(r_n f(x)) = r_n g(f(x))$ for every $x \in S$. Hence, f_n and g are commutative and so are f_n and h. As for every $x, y \in S$,

$$F_{f_n(x)-f_n(y)}(t) = F_{r_n f((x)-f(y))}(t) =$$

$$= F_{f(x)-f(y)}\left(\frac{t}{r_n}\right) \geqslant F_{(g(x)-h(y))}\left(\frac{t}{r_n}\right).$$

From the above Corollary, one concludes that, for every $n \in \mathbb{N}$, there exists an $x_n \in S$ such that

$$x_n = f_n(x_n) = g(x_n) = h(x_n)$$

and $\lim\limits_{n\to\infty} (x_n - f(x_n)) = 0$. Let us now prove that $\lim\limits_{n\to\infty} (x_n - f^m(x_n)) = 0$. First, let us show that, for every k, $n \in \mathbb{N}$ and every $t > 0$,

$$F_{f^k(x_n)-f^{k+1}(x_n)}(t) \geqslant F_{x_n-f(x_n)}(t).$$

We use induction in k. For $k = 1$ and $n \in \mathbb{N}$, we have

$$F_{f(x_n)-f^2(x_n)}(t) \geqslant F_{g(x_n)-h(f(x_n))}(t) = F_{g(x_n)-f(h(x_n))}(t) = F_{x_n-f(x_n)}(t).$$

If one supposes that, for some k and for every $n \in \mathbb{N}$,

$$F_{f^k(x_n)-f^{k+1}(x_n)}(t) \geqslant F_{x_n-f(x_n)}(t),$$

then

$$F_{f^{k+1}(x_n)-f^{k+2}(x_n)}(t) \geqslant F_{g(f^k(x_n))-h(f^{k+1}(x_n))}(t) =$$

$$= F_{f^k(g(x_n))-f^{k+1}(h(x_n))}(t) = F_{f^k(x_n)-f^{k+1}(x_n)} \geqslant F_{x_n-f(x_n)}(t)$$

and the inequality is proven.

On the other hand, from the definition of an RN-space, it follows that

$$F_{x_n-f^m(x_n)}(t) \geqslant T\left(F_{x_n-f(x_n)}\left(\frac{t}{2}\right), F_{f(x_n)-f^m(x_n)}\left(\frac{t}{2}\right)\right) \geqslant$$

$$\geqslant T\left(F_{x_n-f(x_n)}\left(\frac{t}{2}\right), T\left(F_{f(x_n)-f^2(x_n)}\left(\frac{t}{4}\right), F_{f^2(x_n)-f^m(x_n)}\left(\frac{t}{4}\right)\right)\right) \geqslant$$

$$\geqslant T\left(F_{x_n-f(x_n)}\left(\frac{t}{2}\right), T\left(F_{x_n-f(x_n)}\left(\frac{t}{4}\right), T\left(F_{x_n-f(x_n)}\left(\frac{t}{2^3}\right), \ldots\right.\right.\right.$$

$$\left.\left.\left.\ldots, F_{x_n-f(x_n)}\left(\frac{t}{2^{m-1}}\right)\right)\right)\right).$$

Since $T(1,1) = 1$, T is continuous and

$$\lim_{n\to\infty} F_{x_n-f(x_n)}\left(\frac{t}{2^p}\right) = 1, \quad p = 1, 2, \ldots, m-1,$$

it follows that

$$\lim_{n\to\infty} F_{x_n-f^m(x_n)}(t) = 1 \text{ for every } t>0,$$

which means that

$$\lim_{n\to\infty} (x_n - f^m(x_n)) = 0.$$

On the other hand, it can be easily seen that since $f^m(S)$ is relatively compact, it follows that f^m is α-densifying and $\{x_n\}_{n\in\mathbb{N}}$ is compact. Suppose that $x^* = \lim_{k\to\infty} x_{n_k}$. Then $x^* = \lim_{k\to\infty} x_{n_k} = f(\lim_{k\to\infty} x_{n_k}) = g(\lim_{k\to\infty} x_{n_k}) = h(\lim_{k\to\infty} x_{n_k})$ and so x^* is a common fixed point for the mappings f, g and h.

Before concluding this brief exposition regarding the existence of fixed points for mappings on RN-spaces, we shall give some results [109] concerning the existence of the nearly invariant points for mappings on RN-spaces, which are similar with the results obtained by V. Klee [261].

DEFINITION 3.2.3. A mapping f on an RN-space $(S, \mathscr{F}, \mu)$ into an RN-space $(S', \mathscr{F}', \mu')$ will be called (ε, λ)-*continuous* $(\varepsilon>0,\ 0<\lambda<1)$ provided that each point $p\in S$ admits an (ε', λ')-neighbourhood $N_p(\varepsilon', \lambda')$ such that

$$D_{f(N_p(\varepsilon',\lambda'))}(\varepsilon) > 1 - \lambda.$$

DEFINITION 3.2.4. A point $p \in S$ is called a (δ, λ)-*invariant point* ($\delta > 0$, $0 < \lambda < 1$) for a mapping $g : S \to S$ if $F_{p,f(p)}(\delta) > 1 - \lambda$; f will be called a (δ, λ)-*mapping* provided each point of S is (δ, λ)-invariant for f.

PROPOSITION 3.2.1. *Let X and Y be two topological spaces, let $(S, \mathscr{F}, \mu)$ be an RN-space with $\mu \in (F, G)(x) = \mu_1(F, G)(x) = \sup_{t \in [0,1]} \mathrm{Min}\{F(tx), G((1-t)x)\}$, $x \in \mathbb{R}$, let f be a continuous mapping of X into Y, let g be an (ε, λ)-continuous mapping of Y into S, and let h be a (δ, λ)-mapping of S into S. Then $h \circ g \circ f$ is an $(\varepsilon + 2\delta, \lambda)$-continuous mapping of X into S.*

Proof. Let p be an arbitrary point of X. Since g is (ε, λ)-continuous, there exists a neighbourhood $V = N_{f(p)}(\varepsilon', \lambda')$ such that $D_{g(V)}(\varepsilon) > 1 - \lambda$ and since f is continuous, there exists a neighbourhood U_p of p such that $f(U_p) \subseteq V$. Then $D_{(g \circ f)(U_p)}(\varepsilon) > 1 - \lambda$.

On the other hand, h is a (δ, λ)-mapping; therefore, for all r, $s \in U_p$, one has

$$F_{(h \circ g \circ f)(r), (h \circ g \circ f)(s)}(\delta + \varepsilon) \geqslant \mathrm{Min}(F_{(h \circ g \circ f)(r), (g \circ f)(r)}(\delta),$$

$$F_{(g \circ f)(r), (h \circ g \circ f)(s)}(\delta + \varepsilon)) \geqslant \mathrm{Min}(F_{(h \circ g \circ f)(r), (g \circ f)(r)}(\delta), \mathrm{Min}(F_{(g \circ f)(r), (g \circ f)(s)}(\varepsilon),$$

$$F_{(g \circ f)(s), (h \circ g \circ f)(s)}(\delta))) \geqslant \mathrm{Min}(1 - \lambda, \mathrm{Min}(1 - \lambda, 1 - \lambda)) = 1 - \lambda$$

and the proof is complete.

PROPOSITION 3.2.2. *Let P be a compact convex polyhedron in a finite-dimensional RN-space and let g be an (ε, λ)-continuous mapping of P into P. Then there exists a continuous mapping f of P into P such that*

$$F_{f(p) - g(p)}(\varepsilon) > 1 - \lambda.$$

Proof. From the compactness of P and from the (ε, λ)-continuity of g, it follows that P must admit a finite open covering $\mathscr{U}$ such that

$$D_{g(U)}(\varepsilon) > 1 - \lambda \quad \text{for each} \quad U \in \mathscr{U}.$$

Let $\mathscr{S}$ be a simplicial subdivision of P such that every member of $\mathscr{S}$ lies in some member of $\mathscr{U}$. Let $f(s) = g(s)$ for each vertex s of. $\mathscr{S}$ and then extend f affinely over all the simplices in $\mathscr{S}$.

It is easy to see that if (L, σ) is a finite-dimensional topological linear T_1-space and if there exists an $\mathscr{F}$ such that $(L, \mathscr{F}, \mu)$ is an *RN-space*, then (L, σ) and $(L, \mathscr{F}, \mu)$ are linearly homeomorphic. Then, since the extension $\tilde{f}$ of f over all the simplices of $\mathscr{S}$ is a continuous map of P into P, it follows, from

Brouwer's fixed-point theorem and from the invariant property of the existence of fixed-point of continuous mappings on homeomorphic spaces, that there exists a $p_0 \in P$ such that $f(p_0) = p_0$.

Let $p \in P$ be an arbitrary point. Then there exists an $S \in \mathscr{S}$ such that $p \in S$. Let V be the set of all vertices of the carrier of p in S. It follows that

$$D_{g(p) \cup g(V)}(\varepsilon) > 1 - \lambda.$$

Then $g(V) \subseteq N_{g(p)}(\varepsilon, \lambda)$ and $N_{g(p)}(\varepsilon, \lambda)$ is convex. Indeed, if $r, s \in N_0(\varepsilon, \lambda) = \{p : F_p(\varepsilon) > 1 - \lambda\}$, it follows that

$$F_{tr+(1-t)s}(\varepsilon) \geqslant \sup_{t' \in [0,1]} [\mathrm{Min}(F_{tr}(t'\varepsilon), F_{(1-t)s}((1 - t')\varepsilon))] =$$

$$= \sup_{t' \in [0,1]} \left[\mathrm{Min}\left(F_r\left(\frac{t'}{t}\varepsilon\right), F_s\left(\frac{1-t'}{1-t}\varepsilon\right)\right)\right] \geqslant$$

$$\geqslant \mathrm{Min}(F_r(\varepsilon), F_s(\varepsilon)) > \mathrm{Min}(1 - \lambda, 1 - \lambda) = 1 - \lambda.$$

Therefore, $g(V) \subseteq N_{g(p)}(\varepsilon, \lambda)$. Since $f(p) \in \mathrm{co}\, f(V) = \mathrm{co}\, g(V)$, it follows that $F_{f(p)-g(p)}(\varepsilon) > 1 - \lambda$.

COROLLARY. *If P and g are as in Proposition* 3.2.2, *then there exists an* (ε, λ)*-invariant point of P under g.*

Indeed, in particular, we have

$$F_{f(p_0)-g(p_0)}(\varepsilon) > 1 - \lambda \text{ if and only if } F_{p_0 - g(p_0)}(\varepsilon) > 1 - \lambda.$$

THEOREM 3.2.9. *If C is a compact convex subset of an RN-space $(S, \mathscr{F}, \mu)$ and if g is an (ε, λ)-continuous mapping of C into C, then there exists an (ε', λ)-invariant point of C.*

Proof. Let $\delta \in \left(0, \dfrac{\varepsilon' - \varepsilon}{3}\right)$. Since C is compact, it follows that C is precompact and there exists a finite set $E \subset C$ and $f : C \to E$ such that, for each $p \in C$, there exists an $f(p) \in E$ for which $F_{p-f(p)}(\delta) > 1 - \lambda$. If $P = \mathrm{co}E$, then it follows from Proposition 3.2.1 and from the fact that f is a (δ, λ)-mapping of C into E that $f \circ g$ is an $(\varepsilon + 2\delta, \lambda)$-continuous mapping of C into P. According to Proposition 3.2.2, there exists a $q \in P$ such that

$$F_{(f\circ g)(q)-q}(\varepsilon + 2\delta) > 1 - \lambda$$

and

$$F_{g(q)-q}(\varepsilon') \geqslant \sup_{t \in [0,1]} [\mathrm{Min}(F_{g(q)-(f\circ g)(q)}(t\varepsilon'), F_{(f\circ g)(q)-q}(1 - t)\varepsilon')] \geqslant$$

$$\geqslant \mathrm{Min}(F_{g(q)-(f\circ g)(q)}(\delta), F_{(f\circ g)(q)-q}(\varepsilon + 2\delta)) >$$

$$> \mathrm{Min}(1 - \lambda, 1 - \lambda) = 1 - \lambda.$$

We intend now to present some results regarding the existence of fixed points for multivalued mappings in RN-spaces, as obtained by O. Hadžić [177], [175], [188], [181], [191] and G. L. Cain [83].

Let E be a topological vector space, let M be a subset of E and let $\mathscr{R}(M)$ denote the family of all nonempty, closed and convex subsets of M.

A result which generalizes a theorem of Browder's to the case of RN-spaces is given in

THEOREM 3.2.10 [181]. *Let $(S, \mathscr{F}, T)$ be a complete RN-space with continuous t-norm T, let K be a nonempty, convex and compact subset of S and let f be an affine mapping from K into K such that*

$$F_{f(p)-f(q)}(kt) \geqslant F_{p-q}(t) \quad \textit{for every } p, q \in M$$

and every $t > 0$, where $k \in (0,1)$. If $g : K \to \mathscr{R}(K)$ is an upper semicontinuous mapping such that $f(K) + g(K) \subseteq K$ and if the set $g^{-1}(y)$ is open for every $y \in S$, then there exists at least one fixed point of the mapping $f + g$.

The notion of $\mathscr{S}$-admissible subset of a topological vector space E is given in ref. [191, p. 126].

Let H be a topological space and let $M = \overline{M} \subseteq E$ and $\mathscr{S}(M)$ be a nonempty system of subsets of M. The set M is said to be *$\mathscr{S}$-admissible* if for each compact mapping $g : H \to \mathscr{S}(M)$ and for each neighbourhood V of zero in E, there exists a finite-dimensional vector subspace E_V of E and a compact mapping $g_V : H \to \mathscr{S}(M)$ such that $g_V(H) \subseteq E_V$ and $g_V(x) \subseteq g(x) + V$ for every $x \in H$.

It is known that every closed, convex subset of a locally convex space is $\mathscr{S}$-admissible.

Let E and H be two Hausdorff topological spaces and let $\mathscr{S}(H)$ be a system of nonempty subsets of H. A set-valued (multivalued) mapping $f : E \to \mathscr{S}(H)$ is said to be *upper semicontinuous* at a given point $p_0 \in E$ if and only if, for each open subset $G \subseteq H$ such that $f(p_0) \subseteq G$, there exists a neighbourhood V of p_0 such that $f(V) \subseteq G$. The mapping $f : E \to \mathscr{S}(H)$ is upper semicontinuous on E if f is upper semicontinuous at each point of E. If f is upper semicontinuous on E and $f(E)$ is compact in H, f is said to be *compact.*

Considering the above facts, we now give:

THEOREM 3.2.11 [175]. *Let $(S, \mathscr{F}, T)$ be a complete RN-space with continuous t-norm T, let M be a closed, convex and probabilistic bounded subset of S, let $f : M \to S$ be an affine probabilistic contraction maapping, let $g : M \to \mathscr{R}(S)$ be a compact map-*

ping and let $f(p) + g(q) \subseteq M$ for every $p, q \in M$. If M is $\mathscr{R}$-admissible, then the mapping $f + g$ has at least one fixed point.

A similar result can be derived for mappings that can be expressed as the sum of a nonexpansive and a demicompact mapping. We recall that a mapping on an RN-space $(S, \mathscr{F}, T)$ is said to be a *nonexpansive mapping* if

$$F_{f(p)-f(q)}(t) \geqslant F_{p-q}(t), \text{ for every } p, q \in S \text{ and } t > 0.$$

Let E be a topological vector space and let $M \subseteq E$. A mapping $f : M \to E$ is said to be *demicompact* if, for every bounded sequence $\{p_n\}_{n \in \mathbb{N}}$ for which the sequence $\{p_n - f(p_n)\}_{n \in \mathbb{N}}$ is convergent, there exists a convergent subsequence $\{p_{n_k}\}_{k \in \mathbb{N}}$.

THEOREM 3.2.12 [175]. *Let $(S, \mathscr{F}, T)$, M and g be as in Theorem* 3.2.11 and *let f be a nonexpansive, demicompact affine mapping from M into S such that $f(p) + g(q) \subseteq M$ for every $p, q \in M$. Then there exists at least one fixed point of the mapping $f + g$.*

The class of demicompact mappings can be introduced for multivalued mappings too, using

DEFINITION 3.2.5. The mapping $f : M \to 2^M$($M \subseteq E$, where E is a topological vector space) is *strongly demicompact* if, for every sequence $\{p_n\}_{n \in \mathbb{N}} \subseteq M$ for which the sequence $\{p_n - q_n\}_{n \in \mathbb{N}}$, $q_n \in f(p_n)$ is convergent, there exists a subsequence $\{p_{n_k}\}_{k \in \mathbb{N}}$ which is convergent.

The mapping $g : M \to 2^E$ is *h-strongly demicompact*, where $h : M \to E$, if, for every sequence $\{p_n\}_{n \in \mathbb{N}} \subseteq M$ for which the sequence $\{h(p_n) - q_n\}_{n \in \mathbb{N}}$, $q_n \in g(p_n)$ is convergent, there exists a subsequence $\{(p_{n_k})\}_{k \in \mathbb{N}}$ which is convergent.

Remark. If $f = Id_M$ and g is a singlevalued mapping, then if g is compact it follows that g is I-strongly demicompact.

THEOREM 3.2.13 [177]. *Let $(S, \mathscr{F}, T)$ be a complete RN-space with continuous t-norm T and let M be a closed subset of S, $f : M \to 2^M$ such that for every $p \in M$ the set $f(p)$ is closed in the (ε, λ)-topology. Further, for every $u, v \in M$, every $p \in f(u)$ and every $\delta > 0$, there exists a $q \in f(v)$ such that*

$$F_{p-q}(t) \geqslant F_{u-v}\left(\frac{t-\delta}{k}\right) \text{ for every } t > 0,$$

where $k \in (0, 1)$.

If the mapping f is strongly demicompact or if the t-norm T is of $\mathscr{H}$-type, then there exists at least one fixed point of f.

COROLLARY. *Let (S, d) be a complete metric space, let M be a closed subset of S and let $f: M \to CB(M)$, where $CB(M)$ is the family of all closed and bounded subsets of M, such that, for every $p, q \in M$, we have*

$$D(f(p), f(q)) \leqslant k\, d(p, q), \quad k \in (0, 1),$$

where D is the Hausdorff-Pompei metric. Then there exists a $p \in M$ such that $p \in f(p)$.

A generalization of the above result is given by

THEOREM 3.2.14 [200]. *Let $(S, \mathscr{F}, T)$ be a complete RN-space with continuous t-norm T, let M be a closed subset of S, let f and g be continuous mappings from M into M, and let $h: M \to 2^{(f(M) \cap g(M))}$ be a closed mapping such that, for every $u, v \in M$, every $p \in h(u)$ and every $\delta > 0$, there exists a $q \in h(v)$ such that*

$$F_{p-q}(t) \geqslant \max\left\{F_{f(u)-g(v)}\left(\frac{t-\delta}{k}\right), F_{g(u)-f(v)}\left(\frac{t-\delta}{k}\right)\right\}$$

for every $t > 0$, where $k \in (0, 1)$. Further, let f and g be commutative with h. If the mapping h is φ-strongly demicompact, where $\varphi \in \{f, g\}$, or if the t-norm T is of $\mathscr{H}$-type, then there exists an $r \in M$ such that $\{f(r), g(r)\} \subseteq h(r)$.

THEOREM 3.2.15 [200]. *Let $(S, \mathscr{F}, T)$ be a complete RN-space with continuous t-norm T, let f and g be continuous linear mappings, let $h: S \to 2^{f(S) \cap g(S)}$ be a closed and compact mapping which is commutative with f and g and, for every $u, v \in S$, every $p \in h(u)$ and every $\delta > 0$, there exists a $q \in h(v)$ such that, for every $t > 0$:*

$$F_{p-q}(t) \geqslant \max\{F_{f(u)-g(v)}(t-\delta), F_{g(u)-f(v)}(t-\delta)\}.$$

Then there exist $r, s \in S$ such that $f(r) \in h(r)$ and $g(s) \in h(s)$.

Let A be a probabilistic bounded subset of an RN-space $(S, \mathscr{F}, T)$. We recall that a mapping $f: A \to 2^A$ is *strongly condensing* iff the following implication holds:

$$M \subseteq A \text{ and } \alpha_{\overline{\operatorname{co}} f(M)} \leqslant \alpha_M \Rightarrow \bar{M} \text{ is compact.}$$

It is easy to see that if the t-norm $T = \text{Min}$, then the mapping f is strongly condensing iff f is condensing.

The notions of α-contraction and of strongly condensing mapping can be applied to the theory of the existence of fixed points for multivalued mappings in RN-spaces as illustrated in

THEOREM 3.2.16 [185]. *Let $(S, \mathscr{F}, T)$ be an RN-space with continuous t-norm T, let A be a probabilistic bounded, closed* *and convex subset of S and let $f : A \to \mathcal{R}(A)$ be a strongly condensing, closed mapping on every convex and compact subset of A. If, for every $\varepsilon > 0$ and $\delta \in (0, 1)$, there exist $\varepsilon' > 0$* and *$\delta' \in (0, 1)$ such that*

$$\mathrm{co}(U(\varepsilon', \delta') \cap (f(A) - f(A)) \subseteq U(\varepsilon, \delta),$$

then there exists a $p \in A$ such that $p \in f(p)$.

In what follows, we intend to extend some results of [83] by introducing a measure of noncompactness and a corresponding class of probabilistic condensing mutivalued mappings.

Let us denote $S(t, \lambda) = \{p \in S : F_p(t) > 1 - \lambda]$ and $\bar{S}(t, \lambda) = \{p \in S : F_p(t) \geqslant 1 - \lambda\}$.

PROPOSITION 3.2.3. *Let $(S, \mathcal{F}, T)$ be an RN-space and let T be a continuous of $\mathcal{H}$-type t-norm. Then, for $\lambda_n \in \mathbb{R}^+$, $\{\lambda_n\}$ a monotone, decreasing, convergent-to-zero sequence and $T(1 - \lambda_n, 1 - \lambda_n) = 1 - \lambda_n$, the family $\{\bar{S}(t, \lambda_n)\}_{n \in \mathbb{N}}$ is a generalized basis of the neighbourhood system of the origin $\mathcal{N}_0$. This neighbourhood system determines a Hausdorff locally convex topological vector space.*

Proof. It is known [173] that any RN-space with a continuous $\mathcal{H}$-type t-norm T is in the (ε, λ)-topology a locally convex topological space.

The family $\{\bar{S}(t, \lambda_n)\}_{n \in \mathbb{N}}$ having the properties mentioned in the hypothesis is a generalized basis for $\mathcal{N}_0$ since, for every $V \in \mathcal{N}_0$, there exist an $S(\varepsilon, \lambda') \subset V$ and an $\lambda'_n < \lambda'$ such that

$$p \in \bar{S}(\varepsilon, \lambda'_n) \Leftrightarrow F_p(\varepsilon) \geqslant 1 - \lambda'_n > 1 - \lambda' \Rightarrow p \in S(\varepsilon, \lambda') \Rightarrow$$
$$\Rightarrow p \in \bar{V}.$$

On the other hand, $\bar{S}(\varepsilon, \lambda) \supset S(\varepsilon, \lambda)$ implies that $\bar{S}(\varepsilon, \lambda) \in \mathcal{N}_0$.

The existence for every $\lambda' > 0$ of a λ'_n such that $T(\lambda'_n, \lambda'_n) = \lambda'_n$ follows from the characterization of t-norm $\mathcal{H}$-type (see Lemma 1.1.1) i.e., since T is continuous and of $\mathcal{H}$-type, it follows that for every $a > 0$ there exists a $b > a$ such that $T(b, b) = b < 1$.

To prove convexity, take $p, q \in \bar{S}(t, \lambda_n)$ and let $r = up + (1 - u)q$, $0 \leqslant u \leqslant 1$, and then consider

$$F_r(t) = F_{up+(1-u)q}(ut + (1-u)t) \geqslant T(F_{up}(ut), F_{(1-u)q}((1-u)t)) =$$
$$= T(F_p(t), F_q(t)) \geqslant T(1 - \lambda_n, 1 - \lambda_n) = 1 - \lambda_n.$$

Thus, $r \in \bar{S}(t, \lambda_n)$.

Remark. One writes

$$\bar{S}(t_1, \lambda_n) + \bar{S}(t_2, \lambda_n) \subset \bar{S}(t_1 + t_2, \lambda_n)$$

since for every $p \in \bar{S}(t_1, \lambda_n)$ and $q \in \bar{S}(t_2, \lambda_n)$

$$F_{p+q}(t_1 + t_2) \geqslant T(F_p(t_1), F_q(t_2)) \geqslant 1 - \lambda_n.$$

LEMMA 3.2.1. *A subset $A \subset S$ is bounded if and only if $A \subseteq \bar{S}(t, \lambda_n)$ for some t and λ_n.*

Let $\mathscr{S} = \{\bar{S}(t, \lambda_n)\}_{n \in N}$, where $\{\lambda_n\}_{n \in N} = \Lambda$ has the form stated in the above Proposition. For a subset $A \subset S$ and for each $\lambda_k \in \Lambda$, define

$$\gamma_{\lambda_k}(A) = \inf\{t : A \subset F + \bar{S}(t, \lambda_k), \text{ for a finite set } F \subset S\}.$$

LEMMA 3.2.2. *A set $A \subset S$ is precompact if and only if $\gamma_{\lambda_k}(A) = 0$ for every $\lambda_k \in \Lambda$.*

DEFINITION 3.2.6. Let $(S, \mathscr{F}, T)$ be an RN-space with a continuous $\mathscr{H}$-type t-norm T. A multivalued mapping $f : C \to 2^S$ defined on a subset C of S is said to be *probabilistic condensing* if, for every bounded non-precompact set $A \subset C$,

$$\gamma_{\lambda_k}(f(A)) \leqslant \gamma_{\lambda_k}(A)$$

for every $\lambda_k \in \Lambda$ and $\gamma_{\lambda'_k}(f(A)) < \gamma_{\lambda'_k}(A)$ for at least one $\lambda_{k'} \in \Lambda$.

THEOREM 3.2.17. *Let $(S, \mathscr{F}, T)$ be an RN-space with a continuous $\mathscr{H}$-type t-norm T and let C be a complete convex subset of S. Suppose that $f : C \to 2^C$ is a probabilistic condensing upper semicontinuous multivalued mapping such that $f(p)$ is closed and convex for each $p \in C$ and $f(C)$ is bounded. Then f has a fixed point in C.*

Proof. Let us recall the fixed point result of Himmelberg, Porter, and Van Vleck [211]. The multivalued mapping f is closed-valued and upper semicontinuous, so it has a closed graph. Then it is sufficient to show that f is condensing in the sense of [211] with respect to the basis $\mathscr{S}$.

Indeed, let $A \subset C$ be a bounded and non-precompact subset of C.

Then

$Q(A) = \{\bar{S}(t, \lambda_k) \in \mathscr{S} : A \subset K + \bar{S}(t, \lambda_k)$ for some precompact $K\}$.

To prove that $Q(A)$ is properly contained in $Q(f(A))$, choose $\bar{S}(\varepsilon, \lambda) \in Q(A)$. There exists a precompact K for which

$$A \subset K + \bar{S}(\varepsilon, \lambda).$$

and $\gamma_\lambda(A) \leqslant \varepsilon$.

Indeed, if $\gamma_\lambda(A) > \varepsilon$, we choose t such that $\varepsilon + t < \gamma_\lambda(A)$. Since the set K is precompact, there exists a finite set F for which

$$K \subset F + \bar{S}(t, \lambda).$$

Hence,

$$A \subset K + \bar{S}(\varepsilon, \lambda) \subset F + \bar{S}(t, \lambda) + \bar{S}(\varepsilon, \lambda) \subset F + \bar{S}(t + \varepsilon, \lambda),$$

which contradicts that $t + \varepsilon < \gamma_\lambda(A)$. Thus,

$$\gamma_\lambda(A) \leqslant \varepsilon,$$

and from the condensing hypothesis, it follows that

$$\gamma_\lambda(f(A)) \leqslant \gamma_\lambda(A) \leqslant \varepsilon.$$

This, of course, implies that

$$\bar{S}(\varepsilon, \lambda) \in Q(f(A)).$$

Hence, $Q(A) \in Q(f(A))$. This inclusion is proper since if λ is chosen such that $\gamma_\lambda(f(A)) < \gamma_\lambda(A)$ and $\bar{t} \in (\gamma_\lambda(f(A)), \gamma_\lambda(A))$, then $\bar{S}(\bar{t}, \lambda) \in Q(f(A))$ but $\bar{S}(\bar{t}, \lambda) \notin Q(A)$.

The existence of the fixed point follows from

THEOREM B [211]. *Let C be a nonempty complete convex subset of a separated locally convex space E and let $f : C \to 2^C$ be a condensing multivalued mapping with convex values, closed graph and bounded range. Then f has a fixed point.*

This result inclines one to study the connection between condensing mappings as defined by Himmelberg, Porter and Van Vleck and γ-condensing mappings in general, where γ is a random measure of noncompactness. To this purpose, let us note that the result given in [49, Th. 2.2] is valid for more general situations too, i.e.,

THEOREM 3.2.18. *Let $(S, \mathscr{F}, T)$ be an RN-space with a continuous $\mathscr{H}$-type t-norm T. Then there exists a random measure of noncompactness such that every mapping is condensing in the sense of Himmelberg, Porter and Van Vleck with respect to $\mathscr{S}$ if and only if it is probabilistic γ-condensing.*

Proof. Let $f : (S, \mathscr{F}, T) \to 2^S$ be a condensing mapping with respect to $\mathscr{S}$. Then there exist (t_0, λ_0) and a non-precompact set $A \subset S$ such that $f(A) \subseteq \bar{S}(t_0, \lambda_0) + K_0$ for a precompact set $K_0 \subset S$ but $A \not\subset \bar{S}(t_0, \lambda_0) + K$, whatever the precompact set $K \subset S$. Hence, there exists a $t \in \mathbb{R}$ such that

$$\sup_{t'<t} \sup\{\lambda \in \Lambda : \exists K = \text{precompact subset}, A \subseteq \bar{S}(t', \lambda) + K\} <$$

$$< \sup_{t'<t} \sup\{\lambda \in \Lambda : \exists K = \text{precompact subset},$$

$$f(A) \subseteq \bar{S}(t', \lambda) + K\}.$$

Define

$$\gamma_A(t) = \sup_{t'<t} \sup\{\lambda \in \Lambda, \exists K = \text{precompact subset}, A \subseteq \bar{S}(t', \lambda) + $$
$$+ K\}$$ and prove that γ_A is a random measure of noncompactness on $\mathscr{P}(S)$ whereas f is a probabilistic γ-condensing mapping.

First let us show that $\gamma_A = H$ if and only if A is a precompact set. Indeed, let $t \in \mathbb{R}^+$ and $\lambda \in \Lambda$. Choose $t' \in \mathbb{R}^+$ and $\lambda' \in \Lambda$ such that

$$\bar{S}(t', \lambda') + \bar{S}(t', \lambda') \subseteq \bar{S}(t, \lambda).$$

From $\gamma_A = H_0$, it follows that there exists a precompact subset K' such that $A \subseteq \bar{S}(t', \lambda') + K'$. Since K' is a precompact subset, there exists also a finite set F' such that $K' \subseteq \bar{S}(t', \lambda') + F'$ and hence $A \subseteq \bar{S}(t, \lambda) + F'$, i.e., A is a precompact subset.

Conversely, if A is a precompact subset, then for every (t, λ) there exists a finite set F such that $A \subseteq \bar{S}(t, \lambda) + F$; hence, $\gamma_A = H_0$.

It is also true that $\gamma_A \in \Delta^+$. The fact that $\gamma_A \in \mathscr{D}^+$ when A is a probabilistic bounded set follows from the property that $A \subset S(t, \lambda)$ for some t, λ.

Moreover, γ is even monotone, subadditive and invariant with respect to closure and to a convex hull.

Proof. The first three properties follow immediately. For invariance to a convex hull, it is sufficient to prove that $\gamma_{\text{co}A} \geqslant \gamma_A$. So, let $\bar{S}(t, \lambda) \in \mathscr{S}$ and let K be a precompact subset such that $A \subseteq \bar{S}(t, \lambda) + K$. Then co$K$ is also precompact. Since $\bar{S}(t, \lambda)$ is convex, then $\bar{S}(t, \lambda) + \text{co}K$ is also convex and $\bar{S}(t, \lambda) + \text{co}K \supseteq \text{co}A$. Hence,

$$\gamma_{\text{co}A}(t) = \sup_{t'<t} \sup\{\lambda \in \Lambda, \exists K = \text{precompact}, \text{co}A \subseteq \bar{S}(t', \lambda) +$$

$$+ K\} \geqslant \gamma_A(t)$$

and thus we have proven that γ_A is a random measure of noncompactness associated to the condensing mapping f.

The fact that f is γ-condensing follows from the existence of the pair (t_0, λ_0) such that there exists a precompact set K_0 for which $f(A) \subseteq \bar{S}(t_0, \lambda_0) + K_0$ but $A \not\subset \bar{S}(t_0, \lambda_0) + K$, whatever the precompact set $K \subset S$. Indeed, in this case $\gamma_{f(A)}(t_0) > > \gamma_A(t_0)$ and generally $\gamma_{f(A)}(t) \geqslant \gamma_A(t)$.

Conversely, let f be a probabilistic γ-condensing mapping. If $A \in \mathscr{P}(S)$ is such that $Q(A) \neq \mathscr{S}$, then A is not a precompact set, i.e., $\gamma_A < H_0$. Since f is γ-condensing, then $\gamma_{f(A)}(t) > \gamma_A(t)$, and, hence, there exists a $t' < t$ such that $f(A) \subseteq \bar{S}(t', \lambda) + K_0$ for a precompact subset K_0, but $A \not\subset \bar{S}(t', \lambda) + K$, whatever the precompact set K. Therefore, $\bar{S}(t', \lambda) \in Q(f(A))$ but $\bar{S}(t', \lambda) \notin \notin Q(A)$, i.e., f is a condensing mapping with respect to $\mathscr{S}$, and the proof of the theorem is complete.

Remark. The above result allows one to state fixed point theorems for a condensing mapping as defined by Himmelberg, Porter and Van Vleck with respect to $\mathscr{S}$, which are also true for *RN*-spaces with respect to γ-condensing. Conversely, the fixed point theorem of Himmelberg, Porter and Van Vleck in *RN*-spaces can be derived from the previous theorem for a probabilistic γ-condensing mapping, where γ is as stated above.

In order to give an example of how the probabilistic condensing property can be used to derive new fixed point theorems, let us recall

DEFINITION 3.2.7. A multivalued mapping (multifunction) $f: C \subset S \to 2^S$ is a *probabilistic contraction* if there exists a constant k, $0 < k < 1$ such that, for $p, q \in C$ and $r \in f(p)$, there exists a point $s \in f(q)$ with

$$F_{r-s}(kt) \geqslant F_{p-q}(t) \text{ for all } t > 0,$$

Let us also recall that a multivalued mapping $h: C \to 2^S$ is called *completely continuous* if $h(B)$ is precompact whenever B is a bounded subset of C.

A probabilistic and multifunction analogy of Krasnosel'skiĭ's theorem for *RN*-spaces with $\mathscr{H}$-type t-norm T is

THEOREM 3.2.19. *Suppose C is a complete convex subset of the RN-space $(S, \mathscr{F}, T)$. If $f: C \to 2^C$ is an upper semicontinuous compact convex multivalued mapping of C into itself, $f = = g + h$, where g is a compact-valued contraction and h is completely continuous, then f has a fixed point.*

Proof. Let us first consider a result which is of interest also by itself.

LEMMA 3.2.3. *Let $f: C \to 2^S$ be a multivalued mapping such that $f = g + h$, where g is a compact-valued contraction and h is completely continuous. Then f is probabilistic condensing.*

Proof. Let B be a bounded non-precompact subset of the domain of f. We shall show that, for each $\lambda \in \Lambda$, there exists a $t' \leqslant \gamma_\lambda(B)$ such that

$$f(B) \subseteq F + \bar{S}(t', \lambda)$$

for some finite set F. This will prove that f is probabilistic condensing.

Let $\varepsilon > 0$ be an arbitrary positive number and let k be the contraction constant of g. We choose t and t' such that

$$kt < t' < \gamma_\lambda(B) + \varepsilon = t.$$

Since $t > \gamma_\lambda(B)$, there exists a finite set G such that

$$B \subseteq G + \bar{S}(t, \lambda).$$

Also, since $h(B)$ is contained in a precompact set, there exists a finite set H such that

$$h(B) \subseteq H + \bar{S}((t' - 2k)/2, \lambda).$$

Let I_r be a finite set for each $r \in G$ such that

$$g(r) \subset I_r + \bar{S}((t' - kt)/2, \lambda).$$

Now, define $J = H + \cup \{I_r : r \in G\}$, which is clearly finite. We shall prove next that

$$f(B) \subseteq J + \bar{S}(t', \lambda).$$

Let $p \in B$ and let q be an arbitrary element of $g(p)$. Let $r \in G$ be such that

$$F_{p-r}(t) > 1 - \lambda,$$

and choose $s \in g(r)$ such that

$$F_{s-q}(kt) \geqslant F_{p-r}(t) > 1 - \lambda.$$

There exists a $u \in I_r$ for which that

$$F_{u-s}((t' - kt)/2) \geqslant 1 - \lambda$$

is true. Thus,

$$F_{u-q}((t' + kt)/2) \geqslant T(F_{u-s}((t' - kt)/2), F_{s-q}(kt)) \geqslant$$
$$\geqslant T(1 - \lambda, 1 - \lambda) = 1 - \lambda$$

i.e., in other words,

$$g(B) \subseteq \cup \{I_r : r \in G\} + \bar{S}((t' + kt)/2, \lambda).$$

We can now conclude that

$$f(B) = g(B) + h(B) \subseteq \cup \{I_r, r \in G\} + \bar{S}((t' + kt)/2, \lambda) + \\ + H + \bar{S}((t' - kt)/2, \lambda) \subseteq J + \bar{S}(t', \lambda).$$

Thus, $\gamma_\lambda(f(B)) \leqslant t' < \gamma_\lambda(B) + \varepsilon$ which, of course, means that $\gamma_\lambda(f(B)) \leqslant \gamma_\lambda(B)$.

To see that $\gamma_\lambda(f(B)) < \gamma_\lambda(B)$ for at least one $\lambda \in \Lambda$, note that if $\gamma_\lambda(B) > 0$, then ε can be chosen to be zero. Hence, f is condensing.

This lemma and Theorem 3.2.16 show that f has a fixed point. Hence, the proof of Theorem 3.2.19 is complete.

Some other applications of condensing multivalued mappings on RN-spaces to the problem of solution stability of some classes of multivalued mappings will be given in Chap. 5.

3.3. SOME RESULTS REGARDING THE PROBABILISTIC INNER PRODUCT SPACES

Difficulties arising from the theory of probabilistic inner product spaces (PIP-spaces) were shown in Chap. 1, together with examples illustrating various attempts at introducing such spaces. Now, we shall deal with PIP-spaces as given by Definition 1.1.12 and we shall present the fixed point theorems stated by S. Chang in ref. [99].

Examples applying these theorems to the study of the existence and uniqueness of the solution of Urysohn's integral equation will be given in Chap. 5.

THEOREM 3.3.1. *Let f and g be a pair of continuous self-mappings on the complete PIP-space $(S, \mathscr{F}, T)$ satisfying the following condition:*

$$\sum_{n=0}^{\infty} (\sup_{t \in \mathbb{R}} \{t : F_{f^n(p)-g^n(q), f^n(p)-g^n(q)}(t)\})^{1/2} < \infty, \ \forall p, q \in S.$$

Then there exists a point $p_0 \in S$ such that

(i) *p_0 is the unique common fixed point of f and g;*

(ii) *p_0 is the unique fixed point of each f and g;*

(iii) *for any given $q_0 \in S$, the iterative sequences $\{f^n(q_0)\}_{n \in \mathbb{N} \cup \{0\}}$, $\{g^n(q_0)\}_{n \in \mathbb{N} \cup \{0\}}$ converge to the same point p_0.*

Proof. For any $q_0 \in S$, let us define the sequence $p_n = f^n(q_0)$, $n \in \mathbb{N}$. According to Proposition 1.1.4,

$$(\sup_{t \in \mathbb{R}} \{t : F_{p_{i+j}-p_i, p_{i+j}-p_i}(t) < t\})^{1/2} \leqslant$$

$$\leqslant \sum_{m=i}^{i+j-1} (\sup \{t : F_{p_{m+1}-p_m, p_{m+1}-p_m}(t) < 1\})^{1/2} \leqslant$$

$$\geqslant \sum_{m=i}^{i+j-1} [(\sup_{t \in \mathbb{R}}\{t : F_{f^m(p_1)-g^m(p_1), f^m(p_1)-g^m(p_1)}(t) < 1\})^{1/2} +$$

$$+ \sup_{t \in \mathbb{R}} \{t : F_{g^m(p_1)-f^m(p_0), g^m(p_1)-f^m(p_0)}(t) < 1\})^{1/2}].$$

Letting $i \to \infty$, one gets

$$\lim_{i \to \infty} (\sup_{t \in \mathbb{R}} \{t : F_{p_{i+j}-p_i, p_{i+j}-p_i}(t) < 1\})^{1/2} = 0.$$

Hence, $\{p_n\}_{n \in \mathbb{N}}$ is a Cauchy sequence in S. Let $p_0 = \lim_{n \to \infty} p_n$, and from the continuity of f it follows that $\lim_{n \to \infty} f(p_n) = f(\lim_{n \to \infty} p_n) = f(p_0) = p_0$, i.e., p_0 is a fixed point of f.

Similarly, one can prove that for any given $q_0 \in S$, the sequence $\{g^n(q_0)\}_{n \in \mathbb{N}}$ converges to some point $\bar{p}_0 \in S$ and that $\bar{p}_0$ is a fixed point of g.

To prove that $p_0 = \bar{p}_0$, consider $\mathrm{Fix}(f)$ and $\mathrm{Fix}(g)$ to be the sets of all fixed points of f and g respectively. Since $p_0 \in \mathrm{Fix}(f)$, $\bar{p}_0 \in \mathrm{Fix}(g)$, it follows that $\mathrm{Fix}(f) \neq \emptyset$ and $\mathrm{Fix}(g) \neq \emptyset$. On the other hand, for any $p \in \mathrm{Fix}(f)$ and $q \in \mathrm{Fix}(g)$, one writes

$$\sum_{n=0}^{\infty} (\sup_{t \in \mathbb{R}} \{t : F_{f^n(p)-g^n(q), f^n(p)-g^n(q)}(t) < 1\})^{1/2} =$$

$$= \sum_{n=0}^{\infty} (\sup_{t \in \mathbb{R}}\{t : F_{p-q, p-q}(t) < 1\})^{1/2} < \infty.$$

This implies that

$$\sup_{t \in \mathbb{R}}\{t : F_{p-q, p-q}(t) < 1\} = 0, \quad \forall\, p \in \mathrm{Fix}(f),\ q \in \mathrm{Fix}(g).$$

Hence, $p = q$ for every $p \in \mathrm{Fix}(f)$ and every $q \in \mathrm{Fix}(g)$, i.e., $\mathrm{Fix}(f) = \mathrm{Fix}(g) = \{p\} = \{q\}$.

COROLLARY. *Let $\{f_n\}_{n \in \mathbb{N}}$ be a sequence of continuous mappings on a complete PIP-space $(S, \mathcal{F}, T)$. Suppose that, for any positive integer $m, n, m \neq n$, one has*

$$\sum_{i=0}^{\infty} (\sup_{t \in \mathbb{R}}\{t : F_{f_m^i(p)-f_n^i(q), f_m^i(p)-f_n^i(q)}(t) < 1\})^{1/2} < \infty$$

$\forall\, p, q \in S$. Then there exists a point $p_0 \in S$ such that:

(i) *p_0 is the unique common fixed point of $\{f_n\}_{n\in\mathbb{N}}$;*

(ii) *p_0 is the unique fixed point of each $f \in \{f_n\}_{n\in\mathbb{N}}$;*

(iii) *for any given $q_0 \in S$ and for any positive integer m, the iterative sequence $\{f_m^n(q_0)\}_{n\in\mathbb{N}\cup\{0\}}$ converges to the same point p_0.*

The proof of this corollary is immediate from the above theorem.

THEOREM 3.3.2. *Let f and g be a pair of self-mappings on a complete PIP-space $(S, \mathscr{F}, T)$. If there exists a constant $k \in (0, 1)$ such that*

$$F_{f(p)-g(q),f(p)-g(q)}(kt) \geqslant F_{p-q,p-q}(t), \quad \forall\, p, q \in S,\ t \in \mathbb{R},$$

then the conclusion of the above theorem still holds.

Proof. Following the assumptions made with respect to f and g, one states that

$$\sup_{t\in\mathbb{R}}\{t : F_{f(p)-g(q),f(p)-g(q)}(t) < 1\} \leqslant \sup_{t\in\mathbb{R}}\{t : F_{p-q,p-q}(t) < 1\} =$$
$$= k \sup_{t\in\mathbb{R}}\{t : F_{p-q,p-q}(t) < 1\}, \quad \forall\, p, q \in S.$$

Then for any $n \in \mathbb{N}$ one can write

$$\sup_{t\in\mathbb{R}}\{t : F_{f^n(p)-g^n(q),f^n(p)-g^n(q)}(t) < 1\} \leqslant k^n \sup_{t\in\mathbb{R}}\{t : F_{p-q,p-q}(t) < 1\},$$
$$\forall\, p, q \in S.$$

Therefore,

$$\sum_{i=0}^{\infty} (\sup_{t\in\mathbb{R}}\{t : F_{f^i(p)-g^i(q),f^i(p)-g^i(q)}(t) < 1\})^{1/2} \leqslant$$
$$\leqslant \sum_{i=0}^{\infty} c^{i/2}(\sup_{t\in\mathbb{R}}\{t : F_{p-q,p-q}(t) < 1\})^{1/2} < \infty, \quad \forall\, p, q \in S.$$

On the other hand,

$$(\sup_{t\in\mathbb{R}}\{t : F_{f(p)-f(q),f(p)-f(q)}(t) < 1\})^{1/2} \leqslant$$
$$\leqslant (\sup_{t\in\mathbb{R}}\{t : F_{f(p)-g(p),f(p)-g(p)}(t) < 1\}) +$$
$$+ (\sup_{t\in\mathbb{R}}\{t : F_{g(p)-f(q),g(p)-f(q)}(t) < 1\})^{1/2} \leqslant$$
$$\leqslant \left(\sup_{t\in\mathbb{R}}\left\{t : F_{p-p,p-p}\left(\frac{t}{k}\right) < 1\right\}\right)^{1/2} +$$
$$+ \left(\sup_{t\in\mathbb{R}}\left\{t : F_{p-q,p-q}\left(\frac{t}{k}\right) < 1\right\}\right)^{1/2} =$$
$$= 0 + (k \sup_{t\in\mathbb{R}}\{t : F_{p-q,p-q}(t) < 1\})^{1/2}, \quad \forall\, p, q \in S.$$

This implies that f and g are continuous self-mappings on $(S, \mathscr{F}, T)$. Thus, all the conditions of Theorem 3.2.1 are satisfied and the proof of Theorem 3.2.2 follows immediately therefrom.

PROPOSITION 3.3.1. *Let f and g be a pair of self-mappings on a complete PIP-space $(S, \mathscr{F}, T)$ and let $\{c_n\}_{n\in\mathbb{N}}$ be a sequence of positive constants satisfying $\sum_{n=1}^{\infty} c_n^{1/2} < \infty$. If for every $n \in \mathbb{N}$,*

$$F_{f^n(p)-g^n(q),f^n(p)-g^n(q)}(c_n t) \geqslant F_{p-q,p-q}(t), \quad \forall\, p, q \in S,\ t \in \mathbb{R},$$

then the conclusions of Theorem 3.3.1 still hold.

Proof. Reasoning as in Theorem 3.3.2 and setting $n = 1$, one can prove that f and g are continuous self-mappings on $(S, \mathscr{F}, T)$ and that the following inequality holds :

$$\sum_{n=1}^{\infty} (\sup_{t\in\mathbb{R}}\{t : F_{f^n(p)-g^n(q),f^n(p)-g^n(q)}(t) < 1\})^{1/2} \leqslant$$

$$\leqslant \sum_{n=1}^{\infty}\left((\sup_{t\in\mathbb{R}}\left\{t : F_{p-q,p-q}\left(\frac{t}{c_n}\right) < 1\right\}\right)^{1/2} =$$

$$= \sum_{n=1}^{\infty} c_n^{1/2}(\sup_{t\in\mathbb{R}}\{t : F_{p-q,p-q}(t) < 1\})^{1/2} < \infty, \quad \forall\, p, q \in S.$$

Hence, the proof is complete.

3.4. SOME FIXED POINT THEOREMS FOR MAPPINGS ON PROBABILISTIC LOCALLY CONVEX SPACES

The study of the existence and uniqueness of the fixed points of various classes of mappings on probabilistic locally convex spaces (*PLC*-spaces) is much more cumbersome but nevertheless important. Next, we shall present some of the results due to O. Hadžić [170], [167], [168], [171], [174], [191], [192], and Do Hong Tan [423].

Let $(S, \mathscr{F}^i, T)$ be a probabilistic locally convex space in which the family of neighbourhoods of zero is of the form $\{U^i(\varepsilon, \lambda)\}$, $i \in I$, $\varepsilon > 0$, $\lambda \in (0, 1)$, where $U^i(\varepsilon, \lambda) = \{p \in S : F_p^i(\varepsilon) > 1 - \lambda\}$.

THEOREM 3.4.1 [174]. *Let $(S, \mathscr{F}^i, T)$, $i \in I$, be a Hausdorff sequentially complete PLC-space, where T is a continuous t-norm and f is a mapping from S into S such that the following conditions are satisfied :*

(i) *For every $i \in I$, there exist $v(i) > 0$ and $u(i) \in I$ such that $F^i_{f(p)-f(q)}(v(i)t) \geqslant F^{u(i)}_{p-q}(t)$ for every $t > 0$, $p, q \in S$,*

(ii) $\overline{\lim_{n\to\infty}}\, v(u^n(i))<1$ *for every* $i\in I$,

(iii) *There exists a* $p_0\in S$ *such that, for every* $\lambda\in(0,1)$ *and* $i\in I$, *there exist* $t_{\lambda,i}$ *and* $n(i,\lambda)$ *such that*: $G^j_{p_0}(t_{\lambda,i})>1-\lambda$ *for every* $j\in\{u^m(i):\ m\geqslant n(i,\lambda)\}$, *where*

$$G^i_{p_0}(t)=\inf_{n\in\mathbb{N}} F^i_{p_n-p_0},\quad p_n=f(p_{n-1}),\ n\in\mathbb{N}.$$

Then there exists a $p\in S$ *such that* $f(p)=p$. *From* $f(q)=q$ *and from* $\lim_{t\to\infty} F^{u^{m(i)}}_{q-p_0}(t)=1$ *uniformly with respect to* $m\in\mathbb{N}$, *it follows that* $q=\lim_{n\to\infty} p_n$.

Proof. Condition (i) implies that, for every $m,n\in\mathbb{N}$,

$$F^i_{p_{n+m}-p_n}(t)\geqslant F^{u(i)}_{p_{n+m-1}-p_{n-1}}\left(\frac{t}{v(i)}\right)\geqslant\ \dots\ \geqslant F^{u^n(i)}_{p_m-p_0}\left(\frac{t}{\prod_{j=0}^{n-1} v(u^j(i))}\right).$$

Condition (ii) shows that there exist $n_i\in\mathbb{N}$ and $w(i)\in(0,1)$ such that

$$v(u^n(i))\leqslant w(i)\ \text{for every}\ n\geqslant n_i.$$

Then, for $n\geqslant n_i$,

$$F^i_{p_{n+m}-p_n}(t)\geqslant F^{u^n(i)}_{p_m-p_0}\left(\frac{t}{\prod_{j=0}^{n_i-1} v(u^j(i))[w(i)]^{n-n_i}}\right)=$$

$$=F^{u^n(i)}_{p_m-p_0}\left(\frac{a(i)}{w(i)^n}\right)\geqslant G^{u^n(i)}_{p_0}\left(\frac{a(i)}{w(i)^n}\right),$$

where

$$a(i)=\frac{t[w(i)]^{n_i}}{\prod_{j=0}^{n_i-1} v(u^j(i))}.$$

Let $\lambda\in(0,1)$, $i\in I$ and $t>0$. According to condition (iii), it follows that there exist $t_{\lambda,i}>0$ and $n(i,\lambda)$ such that

$$G^j_{p_0}(t_{\lambda,i})>1-\lambda\ \text{for every}\ j\in\{u^m(i):m\geqslant n(i,\lambda)\}.$$

Let $n_1(i,t,\lambda)$ be such that

$$\frac{a(i)}{[w(i)]^n}\geqslant t_{\lambda,i}\ \text{for every}\ n\geqslant n_1(i,t,\lambda);$$

then

$$G^j_{p_0}\left(\frac{a(i)}{[w(i)]^n}\right) \geqslant G^j_{p_0}(t_{\lambda,i}) > 1 - \lambda \text{ for every } n \geqslant n_1(i, t, \lambda)$$

and, thus,

$$F^i_{p_{n+m}-p_n}(t) > 1 - \lambda$$

for every $n \geqslant n^*(i, t, \lambda) = \max\{n_i, n_1(i, t, \lambda), n(i, \lambda)\}$. This means that $\{p_n\}_{n\in\mathbb{N}}$ is a Cauchy sequence. Then there exists a $p^* = \lim_{n\to\infty} p_n$.

On the other hand,

$$F^i_{p_{n+1}-f(p^*)}(t) \geqslant F_{p_n-p^*}\left(\frac{t}{v(i)}\right)$$

and then $\lim_{n\to\infty} p_{n+1} = p^*$. According to condition (iii),

$$F^{u^m(i)}_{p_n-p_0}(t_{\lambda,i}) > 1 - \lambda \text{ for every } n \in \mathbb{N}$$

and $m > n(i, \lambda)$. Since $\lim_{n\to\infty} p_n = p^*$, there exists an $n(i, \lambda, m)$ such that

$$F_{p_{n(i,\lambda,m)}-p^*}(t_{\lambda,i}) > 1 - \lambda$$

and, hence,

$$F^{u^m(i)}_{p^*-p_0}(2t_{\lambda,i}) \geqslant T(F^{u^m(i)}_{p^*-p_{n(i,\lambda,m)}}(t_{\lambda,i}),\ F_{p_{n(i,\lambda,m)}-p_0}(t_{\lambda,i})).$$

According to the continuity of T, for every $\delta \in (0, 1)$, there exists a $\lambda_\delta \in (0, 1)$ such that

$$T(a, b) > 1 - \delta \text{ for every } a, b > 1 - \lambda_\delta$$

and, thus,

$$F^{u^m(i)}_{p^*-p_0}(2t_{\lambda,i}) > 1 - \delta \text{ for every } m > n(i, \lambda_\delta).$$

Let now $q = f(q)$ and $\lim_{t\to\infty} F^{u^m(i)}_{q-p_0}(t) = 1$ uniformly in respect to $m \in \mathbb{N}$.

Then, one can write

$$F^i_{p^*-q}(t) \geqslant F^{u^m(i)}_{p^*-q}\left(\frac{t}{v(i)\ \ldots\ v(u^{m-1}(i))}\right) \geqslant$$

$$\geqslant T\left(F^{u^m(i)}_{p^*-p_0}\left(\frac{t}{2\prod_{j=0}^{m-1} v(u^j(i))}\right),\ F^{u^m(i)}_{q-p_0}\left(\frac{t}{2\prod_{j=0}^{m-1} v(u^j(i))}\right)\right).$$

On the other hand, one can choose m such that

$$F^{u^m(i)}_{p^*-p_0}\left(\frac{t}{2\prod_{j=0}^{m-1} v(u^j(i))}\right) > 1 - \lambda,$$

$$F^{u^m(i)}_{q-p_0}\left(\frac{t}{2\prod_{j=0}^{m-1} v(u^j(i))}\right) > 1 - \lambda,$$

where $T(1-\lambda, 1-\lambda) > 1-\delta$. Hence, for every $\delta \in (0,1)$,

$$F^i_{p^*-q}(t) > 1 - \delta$$

and, thus, $F^i_{p^*-q}(t) = 1$ for every $i \in I$ and $t > 0$. Since $(S, \mathscr{F}, T)$ is a Hausdorff topological space, one concludes that $p^* = q$.

This theorem yields also the main result of [170] using the

COROLLARY. *Let* $(S, \mathscr{F}^i, \text{Min})$, $i \in I$, *be a Hausdorff sequentially complete PLC-space and let* $f: S \to S$ *be such that conditions* (i) *and* (ii) *of the above theorem are satisfied. Suppose that there exists a* $p_0 \in S$ *such that, for every* $i \in I$, $\lim_{t\to\infty} F^i_{j(p_0)-p_0}(t) = 1$ *uniformly with respect to* $j \in \{u^m(i), m \in \mathbb{N} \cup \{0\}\}$. *Then there exists one and only one element* $p \in S$ *such that*:

(1) $p = f(p)$,

(2) $\lim_{t\to\infty} F^j_{p-p_0}(t) = 1$ *uniformly with respect to* $j \in \{u^m(i), m \in \mathbb{N} \cup \{0\}\}$.

Proof. One can write

$$F^j_{p_n-p_0}\left(\frac{t}{w(i)^k}\right) \geqslant$$

$$\geqslant \text{Min}\left(F^j_{p_n-p_{n-1}}\left(\frac{t}{y(i)^{n-1}w(i)^k}\right),\ F^j_{p_{n-1}-p_{n-2}}\left(\frac{(y(i)-1)t}{y(i)^{n-1}w(i)^k}\right), \ldots,\right.$$

$$\left.\ldots, F^j_{p_1-p_0}\left(\frac{(y(i)-1)t}{y(i)w(i)^k}\right)\right), \text{ where } y(i) = \frac{1}{w(i)}.$$

From condition (i), it follows that

$$F^{j}_{p_n-p_0}\left(\frac{t}{w(i)^k}\right) \geqslant$$

$$\geqslant \operatorname{Min}\left(F^{u^{n-1}(j)}_{p_1-p_0}\left(\frac{t}{y(i)^{n-1}\prod_{m=0}^{n-1} v(u^m(j))w(i)^k}\right),\right.$$

$$\left., F^{u^{n-2}}_{p_1-p_0}\left(\frac{y(i)-1}{y(i)^{n-1}}\frac{t}{\prod_{m=0}^{n-2} v(u^m(j))w(i)^k}\right), \ldots, F^{j}_{p_1-p_0}\left(\frac{(y(i)-1)t}{y(i)w(i)^k}\right)\right).$$

If $j \in \{u^m(i),\ m \geqslant n_i\}$, then $v(u^m(j)) \leqslant w(i) < 1$ for every $m = 0,1,2,\ldots$. Since $y(i)w(i) = 1$, one obtains

$$F^{j}_{p_n-p_0}\left(\frac{t}{w(i)^k}\right) \geqslant \operatorname{Min}\left(F^{u^{n-1}(j)}_{p_1-p_0}\left(\frac{t}{w(i)^k}\right),\right.$$

$$\left. F^{u^{n-2}(j)}_{p_1-p_0}\left(\frac{(y(i)-1)}{y(i)}\frac{t}{w(i)^k}\right), \ldots, F^{j}_{p_1-p_0}\left(\frac{y(i)-1}{y(i)}\frac{t}{w(i)^k}\right)\right).$$

As $\lim_{t\to\infty} F^{j}_{p_1-p_0}(t) = 1$ uniformly with respect to $j \in \{u^m(i),\ m \in N \cup \{0\}\}$, it follows that

$$F^{j}_{p_n-p_0}\left(\frac{t}{w(i)^k}\right) > 1 - \frac{\lambda}{2} \text{ for every } n \in \mathbb{N}, \text{ and } j \in \{u^m(i):$$

$n \geqslant ;\ n_i\}$, $k \geqslant (i, t, \lambda)$. Hence,

$$G^{j}_{p_0}\left(\frac{t}{w(i)^k}\right) > 1 - \lambda \text{ for every } k \geqslant k(i, t, \lambda)$$

and $j \in \{u^m(i) : m \geqslant n_i\}$.

Remark. The above theorem yields as a corollary the existence and uniqueness of the fixed point of a probabilistic contraction on the RN-space with a continuous t-norm T.

Another important result is given by

THEOREM 3.4.2 [167]. *Let $(S, \mathscr{F}^i, T)$, $i \in I$, be a sequentially complete Hausdorff PLC-space, where T is a continuous t-norm, let A be a closed probabilistic bounded and star-convex subset of S and let f be a mapping from A into A such that both (i) from the above theorem and $\overline{\lim_{n\to\infty}}\, v(u^n(i)) = 1$ hold and the following con-*

ditions are satisfied:

(i) *For every* $i \in I$, *there exist* $y(i) \in I$ *and* $\varphi_i : \mathbb{R}^+ \to \mathbb{R}^+$ *such that*

(a) $\lim_{t\to\infty} \varphi_i(t) = \infty$,

(b) $F_p^{u^n(i)}(t) \geqslant F_p^{y(i)}(\varphi_i(t))$ *for every* $t > 0$, $p \in S$, $n \in \mathbb{N}$.

(ii) $\overline{f^{n_0}(A)}$ *is sequentially compact in the* (ε, λ)-*topology*.

Then $Fix(f) = \{p \in A : f(p) = p\} \neq \emptyset$.

If $u(i) = i$, for every $i \in I$, the relation $\overline{\lim_{n\to\infty}} v(u^n(i)) = 1$ means that the mapping f is a nonexpansive mapping and there exist many fixed point theorems for such classes of mappings if S is a locally convex space.

THEOREM 3.4.3 [171]. *Let* $(S, \mathscr{F}^i, T)$, $i \in I$, *be a Hausdorff sequentially complete PLC-space, let* T *be a continuous t-norm and let* f *be a continuous mapping from* S *into* S *such that the following conditions are satisfied*:

(i) *For every* $i \in I$, *there exist* $v(i) > 0$ *and* $u(i) \in I$ *such that*

$$F_{f^{n(p)}(p)-f^{n(p)}(q)}(v(i)t) \geqslant F_{p-q}^{u(i)}(t)$$

for every $(t, p, q) \in \mathbb{R}^+ \times S \times S$, *where* $n(p)$ *is a positive integer which depends on* p.

(ii) *For every* $i \in I$,

$$\overline{\lim_{n\to\infty}}\ v(u^n(i)) < 1.$$

(iii) *There exists a* $p_0 \in S$ *such that, for every* $(\lambda, i) \in (0, 1) \times I$, *there exist* $t_{\lambda,i}$ *and* $n(\lambda, i)$ *such that*

$$G_{p_0}^j(t_{\lambda,i}) > 1 - \lambda \text{ for every } j \in \{u^m(i),\ m \geqslant n(\lambda, i)\}.$$

where

$$G_{p_0}^j(t) = \inf_{m\in\mathbb{N}} F_{f^m(p_0)-p_0}(t).$$

Then there exists one and only one element $p_* \in S$ *such that* $p_* = f(p_*)$ *and* $\lim_{t\to\infty} F_{p_*-p_0}^{u^m(i)}(t) = 1$, *for every* $i \in I$, *uniformly with respect to* $m \in \mathbb{N}$.

Proof. Let us define the sequence $\{p_n\}_{n\in\mathbb{N}}$ as $p_n = f^{n(p_{n-1})}(p_{n-1})$, $n \in \mathbb{N}$, and let p_0 be given by condition (iii). From (ii) it follows that there exist $w(i) \in (0, 1)$ and $n(i) \in \mathbb{N}$ such that

$$v(u^m(i)) \leqslant w(i) \text{ for every } m \geqslant n(i).$$

Now, for $n > n(i)$,

$$F^i_{p_{n+k}-p_n}(t) = F^i_{f^{n(p_{n+k-1})}(f^{n(p_{n+k-2})}(\ldots(f^{n(p_{n-1})}(p_{n-1}))\ldots)-f^{n(p_{n-1})}(p_{n-1})}(t) \geqslant$$

$$\geqslant F^{u(i)}_{f^{\sum_{j=n}^{n+k-1} n(p_j)}(p_{n-1})-p_{n-1}}\left(\frac{t}{v(i)}\right) \geqslant \ldots$$

$$\ldots \geqslant F^{u^n(i)}_{f^{\sum_{j=n}^{n+k-1} n(j)}(p_0)-p_0}\left(\frac{a(i)}{w(i)^n}\right), \text{ where } a(i) = \frac{w(i)^{n(i)}t}{\prod_{j=0}^{n(i)-1} v(u^j(i))}.$$

Let $(\lambda, i, t) \in (0,1) \times I \times \mathbb{R}^+$. From condition (iii) it follows that there exist $t_{\lambda,i} \in \mathbb{R}^+$ and $n(\lambda, i) \in \mathbb{N}$ such that

$$G^j_{p_0}(t_{\lambda,i}) > 1 - \lambda \text{ for every } j \in \{u^m(i), m \geqslant n(\lambda, i)\}.$$

Let $n_1(\lambda, i, t)$ be a positive integer such that

$$\frac{a(i)}{w(i)^n} \geqslant t_{\lambda,i} \text{ for every } n \geqslant n_1(\lambda, i, t).$$

Then

$$G^j_{p_0}\left(\frac{a(i)}{w(i)^n}\right) \geqslant G^j_{p_0}(t_{\lambda,i}) \text{ for every } n \geqslant n_1,$$

and, thus, $F^i_{p_{n+k}-p_n}(t) > 1-\lambda$ for every $n \geqslant \max\{n(i), n_1(\lambda, i, t), n(\lambda, i)\}$. This means that $\{p_n\}_{n\in\mathbb{N}}$ is a Cauchy sequence and hence there exists a $p_* = \lim_{n\to\infty} p_n$ which is the fixed point of the mapping f. More precisely, one has

$$F^i_{f(p_n)-p_n}(t) \geqslant G^{u(i)}_{p_0}\left(\frac{t}{\prod_{j=0}^{n-1} v(u^j(i))}\right)$$

and, according to conditions (iii), it follows that $\lim_{n\to\infty} f(p_n) = \lim_{n\to\infty} p_n$. The mapping f is continuous and, hence, $p_* = f(\lim_{n\to\infty} p_n) = f(p_*)$.

It follows also from condition (iii) that

$$F^{u^m i}_{p_*-p_0}(t_{\lambda,i}) > 1 - \lambda \text{ for every } m \geqslant n(\lambda, i) \text{ and every } n \in \mathbb{N}.$$

On the other hand, since $\lim_{n\to\infty} p_n = p_*$, there exists an $n(\lambda, i, t) \in \mathbb{N}$ such that

$$F^{u^m(i)}_{p_{n(\lambda,i,t)}-p_*}(t_{\lambda,i}) > 1 - \lambda.$$

Let $\delta \in (0,1)$ and $\lambda_\delta \in (0,1)$ be such that

$$a, b > 1 - \lambda_\delta \Rightarrow T(a, b) > 1 - \delta$$

Then one can write

$$F^{u^m(i)}_{p_*-p_0}(2t_{\lambda,i}) \geqslant T(F^{u^m(i)}_{p_*-p_{n(\lambda,i,t)}}(t_{\lambda,i}), F^{u^m(i)}_{p_{n(\lambda,i,t)}-p_0}(t_{\lambda,i}))$$

and

$$F^{u^m(i)}_{p_*-p_0}(2t_{\lambda_\delta,i}) \geqslant 1 - \delta \text{ for every } m \geqslant n(\lambda_\delta, i),$$

i.e.,

$$\lim_{t\to\infty} F^{u^m(i)}_{p_*-p}(t) = 1 \text{ uniformly with respect to } m \in \mathbb{N}.$$

Let us suppose that $q = f(q)$ and that $\lim_{t\to\infty} F^{u^m(i)}_{q-p_0}(t) = 1$ uniformly with respect to $m \in \mathbb{N}$. Then

$$F^i_{p_*-q}(t) = F^i_{f^{n(p_*)}(p_*)-f^{n(p_*)}(q)}(t) \geqslant F^{u(i)}_{p_*-q}\left(\frac{t}{v(i)}\right) \geqslant \ldots \geqslant$$

$$\geqslant F^{u^n(i)}_{p_*-q}\left(\frac{t}{\prod_{j=0}^{n-1} v(u^j(i))}\right) \geqslant T\left(F^{u^n(i)}_{p_*-p_0}\left(\frac{t}{\prod_{j=0}^{n-1} v(u^j(i))}\right),\right.$$

$$\left. F^{u^n(i)}_{q-p_0}\left(\frac{t}{2\prod_{j=0}^{n-1} v(u^j(i))}\right)\right).$$

According to the continuity of T, it follows that $F^i_{p_*-q}(t) = 1$ for every $t > 0$, i.e., $p_* = q$.

To prove that $p_* = \lim_{n\to\infty} f^n(p_0)$, suppose that $n \geqslant n(p_*)$, $n = m\ n(p_*) + r$, $0 \leqslant r < n(p_*)$. If $m \geqslant n(i) + 1$, it follows that

$$F^i_{f^n(p_0)-p_*}(t) = F^i_{f^{mn(p_*)+r}(p_0)-f^{n(p_*)}(p_*)}(t) \geqslant$$

$$\geqslant F^{u(i)}_{f^{(m-1)n(p_*)+r}(p_0)-p_*}\left(\frac{t}{v(i)}\right) \geqslant \ldots \geqslant F^{u^m(i)}_{f^r(p_0)-p_*}\left(\frac{t}{\prod_{j=0}^{m-1} v(u^j(i))}\right) \geqslant$$

$$\geqslant T\left(F^{u^m(i)}_{f^r(p_0)-p_*}\left(\frac{t}{2\prod_{j=0}^{m-1} v(u^j(i))}\right), F^{u^m(i)}_{p_*-p_0}\left(\frac{t}{2\prod_{j=0}^{m-1} v(u^j(i))}\right)\right) \geqslant$$

$$\geqslant T\left(F^{u^m(i)}_{f^r(p_0)-p_*}\left(\frac{a(i)}{2w(i)^m}\right), F^{u^m(i)}_{p_*-p_0}\left(\frac{a(i)}{2w(i)^m}\right)\right).$$

Then, from conditions (ii) and (iii) and from the property $\lim_{t\to\infty} F^{u^m(i)}_{p_*-p_0}(t) = 1$ for every $i \in I$, uniformly with respect to $m \in \mathbb{N}$, we conclude that

$$F^i_{f^n(p_0)-p_*}(\varepsilon) > 1 - \lambda \text{ for every } n \geqslant n(\lambda, i, t)$$

and, hence, $\lim_{n\to\infty} f^n(p_0) = p_*$.

COROLLARY. *Let* $(S, \mathscr{F}^i, \text{Min})$, $i \in I$, *be a Hausdorff sequentially complete PLC-space and let* f *be a mapping from* S *into* S *such that conditions* (i) *and* (ii) *of the above theorem hold. If there exists a* $p_0 \in S$ *such that, for every* $i \in I$,

$$\lim_{t\to\infty} F^{u^m(i)}_{f^k(p_0)-p_0}(t) = 1, \quad k = 1, 2, \ldots, n(p_0)$$

uniformly with respect to $m \in \mathbb{N}$, *then there exists one and only one element* $p_* \in S$ *such that* $p_* = f(p_*)$ *and*

$$\lim_{t\to\infty} F^{u^m(i)}_{p_*-p_0}(t) = 1$$

uniformly with respect to $m \in \mathbb{N}$.

Proof. If $sn(p_0) \leqslant m \leqslant (s+1)n(p_0)$, one concludes [170] that

$$F^{u^n(i)}_{f^m(p_0)-p_0}(t) \geqslant$$

$$\geqslant \text{Min}\{F^{u^{n+s}(i)}_{f^{m-sn(p_0)}(p_0)-p_0}(t), F^{u^{n+s-1}(i)}_{f^{n(p_0)}(p_0)-p_0}(t(1 - v(u^{n+s-1}(i)))), \ldots$$

$$\ldots, F^{u^n(i)}_{f^{n(p_0)}(p_0)-p_0}(t(1 - v(u^n(i)))\},$$

For $n > n(i)$, it follows that

$$F^{u^n(i)}_{f^m(p_0)-p_0}(t) \geqslant$$

$$\geqslant \text{Min}\{F^{u^{n+s}(i)}_{f^{m-sn(p_0)}(p_0)-p_0}(t), F^{u^{n+s-1}(i)}_{f^{n(p_0)}(p_0)-p_0}(t(1 - w(i))), \ldots$$

$$\ldots, F^{u^n(i)}_{f^{n(p_0)}(p_0)-p_0}(t(1 - w(i)))\}.$$

This inequality shows that, for every $i \in I$ and $\lambda \in (0,1)$, there exists $t'(\lambda, i)$ such that

$$F^{u^n(i)}_{f^m(p_0)-p_0}(t) > 1 - \frac{\lambda}{2} \text{ for every } t \geqslant t'(\lambda, i)\,;\ m > n(p_0),\ n \geqslant n(i)$$

and, thus,

$$G^{u^n(i)}_{p_0}(t) > 1 - \lambda \text{ for every } n \geqslant n(\lambda, i), \quad t \geqslant t'(\lambda, i).$$

Hence all the conditions of the Theorem are satisfied and the proof is complete.

Remark. Similar results are obtained for the nonarchimedean *PLC*-spaces [167].

Let us now consider some fixed point theorem of Krasnosel'skiĭ type over *PLC*-spaces. First consider

THEOREM 3.4.4 [166]. *Let* $(S, \mathscr{F}^i, \text{Min})$, $i \in I$, *be a sequentially complete Hausdorff PLC-space, let* Λ *be a topological space and let* f *be a continuous mapping of* $S \times \Lambda$ *into* S *such that the following conditions are satisfied*:

(i) *For every* $i \in I$, *there exist* $v(i) > 0$ *and* $u(i) \in I$ *such that*

$$F'_{f(p,\lambda)-f(q,\lambda)}(v(i)t) \geqslant F^{u(i)}_{p-q}(t) \text{ for every } t > 0, \text{ every } p, q \in S \text{ and}$$

every $\lambda \in \Lambda$.

(ii) *For every* $i \in I$, $\overline{\lim\limits_{n\to\infty}}\, v(u^n(i)) < 1$.

(iii) *For every* $i \in I$, *there exist* $y(i) \in I$ *and* $M(i) > 0$ *such that*

$$F^{u^n(i)}_p(t) \geqslant F^{y(i)}_p(M(i)t) \text{ for every } p \in S \text{ and every } t > 0,$$

$n = 0, 1, 2, 3, \ldots$

Then there exists one and only one continuous mapping $p(\lambda)$ *of* Λ *into* S *such that*

$$p(\lambda) = f(p(\lambda), \lambda) \text{ for every } \lambda \in \Lambda.$$

Proof. The existence and uniqueness of the fixed point $p(\lambda)$ of the mapping $p \xrightarrow{f_\lambda} f(p, \lambda)$ follows from the above theorem since

$$F^{u^n(i)}_{f_\lambda(p)-p}(t) \geqslant F^{y(i)}_{f_\lambda(p)-p}(M(i)t)$$

and, thus,

$$1 \geqslant \lim_{m\to\infty} F^{u^n(i)}_{f_\lambda(p)-p}\left(\frac{t}{w(i)^m}\right) \geqslant \lim_{m\to\infty} F^{y(i)}_{f_\lambda(p)-p}\left(\frac{M(i)t}{w(i)^m}\right) = 1.$$

Consider an element $p_{0,\lambda}$ and let $\{p_{n,\lambda}\}_{n\in\mathbb{N}\cup\{0\}}$ be defined as $p_{n,\lambda} = f_\lambda(p_{n-1,\lambda})$ and $p(\lambda) = \lim_{n\to\infty} p_{n,\lambda}$. It remains to prove that the mapping $\tilde{p}$, defined by $\lambda \xrightarrow{\tilde{p}} p(\lambda)$, is continuous at every point $\lambda_0 \to \Lambda$, which means that for every $i \in I$, every $t > 0$ and every $\varepsilon \in (0,1)$, there exists a neighbourhood V of λ_0 such that $\lambda \in V$ implies that

$$F^i_{p(\lambda)-p(\lambda_0)}(t) > 1 - \varepsilon.$$

Let $p_{0,\lambda}$ be $p(\lambda_0)$ for every $\lambda_0 \in \Lambda$. From condition (ii), it follows that there exists an $n(i) \in \mathbb{N}$ such that

$$v(u^n(i)) \leqslant w(i) < 1 \text{ for every } n \geqslant n(i) \text{ and every } i \in I.$$

For $n \geqslant n(i)$,

$$F^{u^n(i)}_{p_{m,\lambda}-p(\lambda_0)}(t) \geqslant$$

$$\geqslant \operatorname{Min}\{F^{u^n(i)}_{p_{m,\lambda}-p_{1,\lambda}}(v(u^n(i))t), F^{u^n(i)}_{p_{1,\lambda}-p(\lambda_0)}((1-v(u^n(i)))t)\} =$$

$$= \operatorname{Min}\{F^{u^n(i)}_{f(p_{m-1,\lambda},\lambda)-f(p_{0,\lambda},\lambda)}(v(u^n(i))t), F^{u^n(i)}_{p_{1,\lambda}-p(\lambda_0)}((1-v(u^n(i)))t)\} \geqslant$$

$$\geqslant \operatorname{Min}\{F^{u^{m+1}(i)}_{p_{m-1,\lambda}-p(\lambda_0)}(t), F^{u^n(i)}_{p_{1,\lambda}-p(\lambda_0)}((1-v(u^n(i)))t)\} \geqslant$$

$$\geqslant \operatorname{Min}\{F^{u^{m+2}(i)}_{p_{m-2,\lambda}-p(\lambda_0)}(t), F^{u^{m+1}(i)}_{p_{1,\lambda}-p(\lambda_0)}((1-v(u^{m+1}(i)))t),$$

$$F^{u^n(i)}_{p_{1,\lambda}-p(\lambda_0)}((1-v(u^n(i)))t)\} \geqslant \ldots$$

$$\ldots \geqslant \operatorname{Min}\{F^{u^{n+m-1}(i)}_{p_{1,\lambda}-p(\lambda_0)}(t), F^{u^{n+m-2}(i)}_{p_{1,\lambda}-p(\lambda_0)}((1-v(u^{n+m-2}(i)))t), \ldots$$

$$\ldots, F^{u^n(i)}_{p_{1,\lambda}-p(\lambda_0)}((1-v(u^n(i)))t)\} \geqslant \ldots$$

$$\ldots \geqslant \operatorname{Min}\{F^{u^{n+m-1}}_{p_{1,\lambda}-p(\lambda_0)}((1-w(i))t), F^{u^{n+m-2}(i)}_{p_{1,\lambda}-p(\lambda_0)}((1-w(i))t), \ldots$$

$$\ldots, F^{u^n(i)}_{p_{1,\lambda}-p(\lambda_0)}((1-w(i))t)\} \geqslant F^{\gamma(i)}_{p_{1,\lambda}-p(\lambda_0)}(M(i)(1-w(i))t).$$

Still further,

$$\liminf_{m\to\infty} F^i_{p_{m,\lambda}-p(\lambda_0)}(t) = F^i_{p(\lambda)-p(\lambda_0)}(t)$$

and, hence,

$$(*)\quad F^{u^n(i)}_{p(\lambda)-p(\lambda_0)}(t) \geqslant F^{\gamma(i)}_{p_{1,\lambda}-p(\lambda_0)}(M(i)(1-w(i))t)$$

for every $n \geqslant n(i)$ and $\lambda \in \Lambda$.

On the other hand,

$$F^i_{p(\lambda)-p(\lambda_0)}(t) = F^i_{f(p(\lambda),\lambda)-f(p(\lambda_0),\lambda_0)}(t) \geqslant$$

$$\geqslant \operatorname{Min}\left\{F^i_{f(p(\lambda),\lambda)-f(p(\lambda_0),\lambda)}\left(\frac{t}{2}\right), F^i_{f(p(\lambda_0)-f(p(\lambda_0),\lambda_0)}\left(\frac{t}{2}\right)\right\} \geqslant$$

$$\geqslant \operatorname{Min}\left\{F^{u(i)}_{p(\lambda)-p(\lambda_0)}\left(\frac{t}{2v(i)}\right), F^i_{f(p(\lambda_0),\lambda)-f(p(\lambda_0),\lambda_0)}\left(\frac{t}{2}\right)\right\} \geqslant$$

$$\geqslant \left\{F^{u^2(i)}_{p(\lambda)-p(\lambda_0)}\left(\frac{t}{2^2 v(i)v(u(i))}\right), F^{u(i)}_{f(p(\lambda_0),\lambda)-f(p(\lambda_0),\lambda_0)}\left(\frac{t}{2^2 v(i)}\right),\right.$$

$$\left., F^i_{f(p(\lambda_0),\lambda)-f(p(\lambda_0),\lambda_0)}\left(\frac{t}{2}\right)\right\} \geqslant \ldots$$

$$\ldots \geqslant \operatorname{Min}\left\{F^{u^{n(i)}(i)}_{p(\lambda)-p(\lambda_0)}\left(\frac{t}{2^{n(i)}\prod_{k=0}^{n(i)-1} v(u^k(i))}\right),\right.$$

$$F^{u^{n(i)-1}(i)}_{f(p(\lambda_0),\lambda)-f(p(\lambda_0),\lambda_0)}\left(\frac{t}{2^{n(i)}\prod_{k=0}^{n(i)-2} v(u^k(i))}\right)\ldots$$

$$\left.\ldots, F^i_{f(p(\lambda_0),\lambda)-f(p(\lambda_0),\lambda_0)}\left(\frac{t}{2}\right)\right\}.$$

Using relation (*), one writes

$$F^i_{p(\lambda)-p(\lambda_0)}(t) \geqslant$$

$$\geqslant \operatorname{Min}\left\{F^{y(i)}_{f(p(\lambda_0),\lambda)-f(p(\lambda_0),\lambda_0)}\left(M(i)(1 - w(i))\frac{t}{2^{n(i)}\prod_{k=0}^{n(i)-1} v(u^k(i))}\right),\right.$$

$$F^{u^{n(i)-1}(i)}_{f(p(\lambda_0),\lambda)-f(p(\lambda_0),\lambda_0)}\left(\frac{t}{2^{n(i)}\prod_{k=0}^{n(i)-2} v(u^k(i))}\right), \ldots$$

$$\left.\ldots, F^i_{f(p(\lambda_0),\lambda)-f(p(\lambda_0),\lambda_0)}\left(\frac{t}{2}\right)\right\}.$$

The mapping f is continuous and, thus, for every $j \in I$, every $t > 0$, every $\lambda_0 \in \Lambda$ and every $\varepsilon \in (0, 1)$, there exist $V(\lambda_0, j, t, \varepsilon)$ such that

$$F^i_{f(p(\lambda_0),\lambda)-f(p(\lambda_0),\lambda_0)}(t) > 1 - \varepsilon$$

for every $\lambda \in V(\lambda_0, j, t, \varepsilon)$.

Let $V_0, V_1, V_2, \ldots, V_{n(i)}$ be defined as

$$V_0 = V\left(\lambda_0, y(i), M(i)(1 - w(i)) \frac{t}{2^{n(i)} \prod_{k=0}^{n(i)-1} v(u^k(i))}, \varepsilon\right)$$

$$V_j = V\left(\lambda_0, u^{n(i)-j}(i), \frac{t}{2^{n(i)+1-j} \prod_{k=0}^{n(i)-j-1} v(u^k(i))}, \varepsilon\right)$$

$j = 1, 2, \ldots, n(i)$ and $W = \bigcap_{k=0}^{n(i)} V_k$.

Then from $\lambda \in W$, it follows that

$$F^i_{p(\lambda)-p(\lambda_0)}(t) > 1 - \varepsilon$$

and then the mapping $\lambda \xrightarrow{\tilde{p}} p(\lambda)$ is continuous.

PROPOSITION 3.4.1 [166]. *Let A be a closed and convex subset of the complete Hausdorff PLC-space $(S, \mathscr{F}, \mathrm{Min})$, let f be a continuous mapping from A into the topological space Λ and let G be a continuous mapping from $A \times \overline{f(A)}$ into A such that the following conditions are satisfied:*

(i) *For every $i \in I$, there exist $v(i) > 0$ and $u(i) \in I$ so that*

$$F^i_{G(p_1,q)-G(p_2,q)}(v(i)t) \geqslant F^{u(i)}_{p_1-p_2}(t)$$

for every $p_1, p_2 \in A$, every $q \in \overline{f(A)}$ and every $t > 0$;

(ii) *For every $i \in I$, $\lim_{n\to\infty} v(u^n(i)) < 1$;*

(iii) *The set $\overline{f(A)}$ is compact.*

Then there exists at least one element $p \in A$ such that $p = G(p, f(p))$.

Proof. Following a path similar with that used in the proof of Theorem 3.2.4, one constructs the mapping $R : \overline{f(A)} \to A$, $R_y = G(R_y, y)$ for every $y \in \overline{f(A)}$. The mapping $p \xrightarrow{R^*} R(f(p))$ defined on A satisfies all the conditions of Tychonoff's fixed point theorem and, thus, there exists a $p \in A$ such that $R^*(p) = p$. Then

$$p = R^*(p) = R(f(p)) = G(R(f(p)), f(p)) = G(p, f(p)),$$

and the proof of the proposition is complete.

Let $(S_1, \mathscr{F}^i_1, T_1)$ be a Hausdorff sequentially complete *PLC*-space with continuous t-norm T_1, let $(S_2, \mathscr{F}^i_2, T_2)$ be a *PLC*-space with continuous t-norm T_2 and let $U \subseteq S_1$, $V \subseteq S_2$, $H : U \times V \to U$ and $K : U \times V \to V$ be continuous mappings.

Remark. It is said that a set $A \subset S$ is star-shaped if there exists a $p_0 \in A$ such that

$$tp + (1 - t)p_0 \in A, \text{ for every } p \in A \text{ and every } t \in [0, 1].$$

The above proposition holds also if $(S, \mathscr{F}^i, T)$ is a Hausdorff sequentially complete PLC-space, where T is a continuous t-norm, A is a closed, probabilistic bounded subset of S, f is a compact mapping $f: A \to \Lambda$, $G: A \times f(A) \to A$ and G is continuous. Suppose that condition (i) is satisfied,

$$\lim_{n\to\infty} \prod_{k=0}^{n} v(u^k(i)) = 0 \text{ for every } i,$$

and that there exists a $y(i) \in I$ such that

$$(*) \quad F_p^{u^n(i)}(t) \geqslant F_p^{y(i)}(t)$$

for every $p \in S$, every $t > 0$ and every $n \in \mathbb{N} \cup \{0\}$.

If either condition (a) or (b) is satisfied, then there exists a $p \in A$ such that $p = G(p, f(p))$, where:

(a) The set A is admissible and convex,

(b) S is admissible and A is star-shaped closed neighbourhood retract.

A fixed point theorem of Krasnosel'skiĭ type follows also from this result [198]:

Let $(S, \mathscr{F}^i, T)$, $i \in I$, and let A be as in the above remark, let $h: A \to S$ be a compact and let $g: A \to S$ be a continuous mapping such that the following conditions are satisfied:

(1) either condition (a) or (b) above are satisfied;

(2) for every $p, q \in A$, $g(p) + h(q) \in A$;

(3) for every $i \in I$, there exist $v(i) > 0$ and $u(i) \in I$ such that

$$F^i_{g(p)-g(q)}(v(i)t) \geqslant F^{u(i)}_{p-q}(t) \text{ for every } p, q \in A \text{ and } t > 0,$$

where

$$\lim_{n\to\infty} \prod_{k=0}^{n} v(u^k(i)) = 0,$$

and for every $i \in I$ there exists a $y(i) \in I$ such that $(*)$ is satisfied. Then, there exists at least one element $p \in A$ such that $p = g(p) + h(p)$.

Indeed, we can choose $f = h$ and $G(p, q) = g(p) + q$. Then the equation $p = G(p, f(p))$ becomes $p = g(p) + h(p)$.

One also notices that $\operatorname{Fix}(g + \varphi) \neq \emptyset$ if φ is a demi-compact mapping, that condition $(*)$ is satisfied and that

$$\prod_{k=0}^{n} v(u^k(i)) \leqslant w(i) < \infty \text{ for every } i \in I \text{ and } n \in \mathbb{N}.$$

THEOREM 3.4.5 [192]. *Suppose that* U *is closed and that* $\sup_{t>0}\sup_{u<t}\inf_{p,q\in U} F^i_{p-q}(u)=1$ *for every* $i\in I$, V *is closed, convex and compact and also that the following two conditions are satisfied:*

(i) *For every* $i\in I$, *there exist* $v(i)>0$ *and* $u(i)\in I$ *such that*

$$F^i_{H(p_1,q)-H(p_2,q)}(v(i)t) \geqslant F^{u(i)}_{p_1-p_2}(t)$$

for every $p_1, p_2\in U$, *every* $q\in V$ *and* $t>0$, *where for every* $i\in I$

$$\lim_{n\to\infty}\prod_{j=1}^{n} v(u^j(i)) = 0\,;$$

(ii) *For every* $i\in I$, *there exists a* $y(i)\in I$ *such that*

$$F^{u^n(i)}_p(t) \geqslant F^{y(i)}_p(t) \text{ for every } p\in S, \text{ every } t>0 \text{ and } n=0,1,2,\ldots$$

If T_2 *is of* $\mathscr{H}$*-type, then there exists at least one fixed point of the mapping* $G: U\times V\to U\times V$, *where* $G(z)=(H(z), K(z))$ *for every* $z\in U\times V$.

Proof. As in the proof of Theorem 3.2.4, it is shown that there exists one and only one continuous mapping $\tilde{R}: V\to U$ such that $\tilde{R}(q)=H(\tilde{R}(q), q)$ for every $q\in V$. If $R^*: V\to V$ is given by $R^*(q)=K(\tilde{R}(q), q)$ for every $q\in V$, then R^* is a compact mapping and since $(S^2, \mathscr{F}^i_2, T_2)$ is a locally convex vector space, it follows (using Tychonoff's fixed point theorem) that there exists at least one fixed point q_0 of the mapping R^*. Then $(\tilde{R}(q_0), q_0)$ is a fixed point of the mapping G.

COROLLARY. *Let* $(S_1, \mathscr{F}_1, T_1)$ *be a complete RN-space with continuous t-norm* T_1, *let* $(S_2, \mathscr{F}_2, T_2)$ *be an RN-space with continuous and* $\mathscr{H}$*-type t-norm* T_2. *Let* $U\subseteq S_1$, $V\subseteq S_2$, *and* $H: U\times V\to U$, $K: U\times V\to V$ *be such that*·

(i) $F_{H(p_1,q)-H(p_2,q)}(kt)\geqslant F_{p_1-p_2}(t)$ *for every* $(p_1, p_2, q)\in U\times U\times V$ *and every* $t>0$, *where* $k\in(0,1)$.

(ii) H *and* K *are continuous mappings,* $U=\overline{U}$, $V=\overline{\mathrm{co}}\,V$ *and* V *is compact.*

Then there exists at least one fixed point of the mapping G,

$G(z)=(H(z), K(z))$ *for every* $z\in U\times V$.

Another way of obtaining the Krasnosel'skiĭ-type theorem is to use measures of noncompactness on *PLC*-spaces and γ-condensing mappings. The following theorems are representative in this respect :

THEOREM 3.4.6 [423]. *Let A be a complete convex set in a PLC-space $(S, \mathscr{F}^i, \mathrm{Min})$, $i \in I$, and let $f: A \to 2^A$ be an upper-semicontinuous probabilistic γ-condensing mapping with non-empty closed convex values and bounded range. Then f has a fixed point in A (γ is an α- or β-measure of noncompactness).*

THEOREM 3.4.7 [423]. *Let A be a complete convex set in a PLC-space $(S, \mathscr{F}^i, \mathrm{Min})$, $i \in I$, and let $f: A \to 2^A$ be a probabilistic contraction with nonempty convex, compact values and bounded range. Then f has a fixed point in A.*

Remark. Since the sum of a condensing mapping and a compact mapping is again a condensing mapping, one can state the following result:

Let D be a complete convex set in a PLC-space S and let $f = g + h$ be a mapping of A into 2^A, where g is defined on S and satisfies all conditions in Theorem 3.4.7, and where h is a compact mapping with closed convex values. Then f has a fixed point in A.

Regarding the existence of common fixed points of two mappings on PLC-spaces, a result similar with that of RN-spaces can be given [417]:

THEOREM 3.4.8. *Let $(S, \mathscr{F}^i, T)$, $i \in I$, be a sequentially complete PLC-space with continuous t-norm T and let f and g be continuous mappings of S into S.*

The mappings f and g have a unique common fixed point in S if and only if there exists a continuous mapping $h: S \to f(S) \cap h(S)$ which commutes with f and g such that $h(S)$ is a probabilistic bounded subset of S and satisfies the following conditions:

(i) *For every $i \in I$, there exist $v(i) > 0$ and $u(i) \in I$ such that*

$$F^i_{h(p)-h(q)}(v(i)t) \geqslant F^{u(i)}_{f(p)-g(q)}(t),$$

for every $t > 0$ and every $p, q \in S$;

(ii) *For every $i \in I$, there exist numbers $n_i \in \mathbb{N}$ and $w(i) \in (0,1)$ such that*

$$v(u^n(i)) \leqslant w(i) < 1 \text{ for every } n > n_i;$$

(iii) *For every $i \in I$, there exists a $y(i) \in I$ such that*

$$F^{u^n(i)}_p(t) \geqslant F^{y(i)}_p(t) \text{ for every } t > 0, \text{ every } p \in S \text{ and every } n \in \mathbb{N}.$$

Then there exists one and only one element $p_* \in S$ such that $h(p_*)$ is the unique common fixed point for the mappings h, f and g.

A proof of this theorem is given in ref. [191, p. 273], considering the space S to be a Hausdorff quasi-uniformizable space.

Chapter 4

Random fixed point theorems

The study of fixed point theorems for random operators was initiated by A. Špaček [413], [208] and O. Hanš [203], [205], [207], [208]. Important contributions to the random fixed point theory were made by : A. T. Bharucha-Reid [25], [27], [31], [32], [24], A. Mukherjea [300], [301], [302], [303], [299], M. Z. Nashed [309], [310], R. Kannan [255], [249], [254], [252], H. Salehi [252], [253], [254], [255], S. Itoh [237], [241], [242], H. W. Engl [132], [133], [134], [135], [136], [137], [138], [139], A. C. Lee and Padgett [276], [277], A. Nowak [314], [315], [316], W. Römish [364], [365], [366], [367], Gh. Bocşan [50], [51], [52], [53], [49], [54], [55], Gh. Constantin [110], [111], [112], [116], V. Radu [346], [348], [350], [351], O. Hadžić [187], [191], [193], [199], Chang Sih-sen [95], [96], [94], [98].

The importance of random fixed points for the study of the theory of random operator equations is well known and need not be overstressed. Thus, for example, it is sufficient to mention the random differential and integral equations with hosts of applications to real-world problems (e.g. applications in hereditary mechanics [31], [431], telephone traffic theory, diffusion theory, mathematical statistics and physics, turbulence theory, population dynamic, stochastic control, biology, chemical kinetics, etc.).

The first part of this chapter will introduce several techniques using probabilistic structures to analyse the problem of random fixed points. The following parts deal with several random fixed point theorems for abstract random continuous operators on separable metric or Banach spaces. The last part of the chapter presents random fixed points theorems for continuous random operators with stochastic domain or for set-valued random operators with stochastic domain.

4.1. SOME APPLICATIONS OF PROBABILISTIC STRUCTURES TO THE RANDOM FIXED POINT THEORY

This paragraph introduces some classical results of the random fixed point theory as obtained during the pioneering stage of this theory by A. Špaček [413], [208], O. Hanš [203], [205], [207], [208], A. T. Bharucha-Reid [25], [27], [24], [31], [32], A. Mukherjea [299], [300], [301], [302], R. Kannan and H. Salehi [255] and B. L. S. Prokasa Rao [31]. Several generalizations of these results are given then as applications of some theorems from the theory of the probabilistic structures presented in Chap. 3.

The working techniques based on probabilistic structures rely on the concepts of E-spaces introduced by H. Sherwood [405] and RN-spaces of applications introduced by D. H. Muštari [304]. An important step towards reducing some important problems for random operators on separable Banach spaces to similar problems for operators on RN-spaces was achieved by Gh. Bocşan [50] by association of Nemytskii operators.

The following theorems were presented with proofs and explanations in ref. [31] [32]. We shall only state them here for a direct comparison with the generalizations deriving thereof.

Given the wide applicability of Banach's contraction mapping theorem in the study of deterministic and probabilistic operator equations, Špaček and Hanš studied the probabilistic versions and used their results to the problem of the existence, uniqueness, and measurability of solution of some classes of integral equations. Let us mention some of their results:

THEOREM 4.1.1 [207]. *Let X be a separable Banach space and let $T(\omega)$ be a continuous random operator from $\Omega\times X$ to X (for $T(\omega, x)$, we also write $T(\omega)x$) such that*

$$P\left(\bigcup_{m=1}^{\infty}\bigcup_{n=1}^{\infty}\bigcap_{x_1\in X}\bigcap_{x_2\in X}\left\{\omega : \|T^n(\omega)x_1 - T^n(\omega)x_2\| \leqslant \right.\right.$$
$$\left.\left.\leqslant\left(1-\frac{1}{m}\right)\|x_1 - x_2\|\right\}\right) = 1,$$

where, for every $\omega\in\Omega$, $x\in X$, and $n = 1, 2, \ldots$, we put $T^1(\omega)x = T(\omega)x$, and $T^{n+1}(\omega)x = T(\omega)(T^n(\omega)x)$. Then, there exists an X-valued random variable $x(\omega)$ which is the unique fixed point of $T(\omega)$; since $T(\omega)x(\omega) = x(\omega)$ a.s. and if $y(\omega)$ is another fixed point, then $x(\omega) = y(\omega)$ a.s.

DEFINITION 4.1.1. A random operator $T(\omega)$ on a Banach space X with domain $\mathscr{D}(T(\omega))$ is said to be a *random contraction operator* if there exists a non-negative real-valued random variable $k(\omega)$ such that $k(\omega) < 1$, and such that

$$\text{(c)} \quad P(\{\omega : \| T(\omega)x_1 - T(\omega)x_2 \| \leqslant k(\omega) \| x_1 - x_2 \|\}) = 1$$

for all $x_1, x_2 \in \mathscr{D}(T(\omega))$. If $k(\omega) = k$ (a constant) for all $\omega \in \Omega$, then $T(\omega)$ is called a *uniform random contraction operator.*

A veritable random contraction mapping theorem is

THEOREM 4.1.2 [207]. *Let $T(\omega)$ be a continuous random operator on $\Omega \times X$ to X, where X is a separable Banach space, and let $k(\omega)$ be a non-negative real-valued random variable such that $k(\omega) < 1$ a.s. and*

$$\| T(\omega)x_1 - T(\omega)x_2 \| \leqslant k(\omega) \| x_1 - x_2 \|$$

for every two elements $x_1, x_2 \in X$. Then there exists an X-valued random variable $x(\omega)$ which is the unique fixed point of $T(\omega)$.

Such problems as the existence, uniqueness and measurability of the solutions of the random operator equation of the form $T(\omega)x(\omega) = y(\omega)$ for a given X-valued random variable $y(\omega)$, are frequently posed for an arbitrary X-valued random variable $y(\omega)$. This is equivalent to the question of existence and measurability of the inverse of a random operator $T(\omega)$, and then the above theorem can be used to establish solutions of random eigenproblems of the form $(T(\omega) - \lambda I)x = y(\omega)$, I denoting the identity operator.

THEOREM 4.1.3. *Let $k(\omega)$ be a non-negative real-valued random variable which is bounded a.s. and let $T(\omega)$ be a continuous random operator on a separable Banach space X satisfying condition (c). Then, for every real $\lambda \neq 0$ such that $k(\omega) < |\lambda|$ a.s there exists a random operator $S(\omega)$ which is the inverse of $T(\omega) - \lambda I$.*

An important topological fixed point theorem in analysis was given by Schauder [378]. Several probabilistic versions of this theorem were given by A. Mukherjea [299], [300], [301], A. T. Bharucha-Reid [32], R. Kannan and H. Salehi [252] and others, as is, for example,

THEOREM 4.1.4 [300]. *Let $(\Omega, \mathscr{K}, P)$ be an atomic probabilistic measure space and let E be a compact (or closed and bounded) convex subset of a separable Banach space X. Let $T(\omega)$ be a compact random operator mapping E into itself. Then there exists an E-valued random variable $x(\omega)$ such that $T(\omega)x(\omega) = x(\omega)$ a.s.*

A more interesting theorem is due to A. T. Bharucha-Reid and A. Mukherjea:

THEOREM 4.1.5 [32]. *Let E be a compact convex subset of a Banach space and let $T(\omega)$ be a continuous random operator mapping E into itself. Then there exists an E-valued random variable $x(\omega)$ such that $T(\omega)x(\omega) = x(\omega)$ a.s.*

Another probabilistic version of Schauder's theorem having the above conditions relaxed is:

THEOREM 4.1.6 [252]. *Let $T(\omega)$ be a stochastically continuous random operator on a separable Banach space X into itself. Suppose that for each $\omega \in \Omega$, $\{x : T(\omega)x = x\} \neq \emptyset$. Then there exists a measurable multivalued map $x(\omega) : \Omega \to 2^X$ such that $x(\omega) = \{x : T(\omega)x = x\}$ a.s.*

Several applications of fixed point theorems in probabilistic analysis are based on the assumption that a random operator $T(\omega)$ satisfies the hypotheses of Schauder's theorem for each $\omega \in \Omega$. Then, if $T(\omega)$ is a continuous random operator, it is also stochastically continuous and separable. Hence, the Kannan-Salehi theorem corresponds to a probabilistic version of Schauder's theorem. This theorem has important applications to random nonlinear differential and integral equations with monotonic nonlinearities (cf. [254], [255]).

It is of interest to prove the existence of the Krasnosel'skiĭ-type theorem for perturbed operator equations and the probabilistic versions of this type of theorems.

THEOREM 4.1.7 [31], [32]. *Let $(\Omega, \mathscr{K}, P)$ be a probability measure space and let E be a compact and convex subset of a separable Banach space X. Let $T(\omega)$ and $S(\omega)$ be continuous random operators mapping X into itself such that: (i) $S(\omega)x_1 + T(\omega)x_2 \in E$ for all $x_1, x_2 \in E$ and $\omega \in \Omega$; (ii) there exists a non-negative real-valued random variable $k(\omega)$ such that $\|S(\omega)x_1 - S(\omega)x_2\| \leqslant \leqslant k(\omega)\|x_1 - x_2\|$ for all $x_1, x_2 \in E$ and $k(\omega) < 1$ a.s. Then there exists an X-valued random variable $x(\omega)$ such that $S(\omega)x(\omega) + T(\omega)x(\omega) = x(\omega)$ for all $\omega \in \Omega$.*

In what follows, we shall introduce several results regarding the fixed points of random operators on complete metric spaces. The E-space technique will be used over a separable complete metric space (M, d). Several results (due to Shih-sen Chang and generalizing Theorems 4.1.1 and 4.1.2) will be given in the sequel.

PROPOSITION 4.1.1. *Let $(S, \mathscr{F}, T_m)$ be an E-space over the separable complete metric space (M, d). Suppose that $T(\omega) : M \to M$ is a d-continuous random operator. Then $T(\omega)$ maps S into S and*

$$T(\omega)x_n(\omega) \xrightarrow{P} T(\omega)x(\omega)$$

for every $\{x_n(\omega)\}_{n \in \mathbb{N}} \subset S$, $x_n(\omega) \xrightarrow{P} x(\omega) \in S$.

Proof. Since (M, d) is a separable metric space, it is easy to see that, by virtue of a well-known result, $T(\omega): M \to M$ is a d-continuous random operator (see, for example, [207, Th. 3] or [31, Th. 2.14]). Then $T(\omega)$ maps every random variable of S into a random variable of S.

To prove the second part, let $\{x_n(\omega)\}_{n \in \mathbb{N}} \subset S$, $x_n(\omega) \xrightarrow{P} x(\omega) \in S$ and let $\varepsilon > 0$, and for every $n \in \mathbb{N}$

$$G_n = \{\omega \in \Omega : \text{ for every } u \in M,\ d(x(\omega), u) < \frac{1}{n} \text{ implies that } d(T(\omega)x(\omega), T(\omega)u) < \varepsilon\}.$$

Then, every G_n is P-measurable and

$$G_1 \subset G_2 \subset \ldots \subset G_n \subset \ldots \subset \Omega.$$

Let us prove that $\Omega = \bigcup_{n \in \mathbb{N}} G_n$. Suppose the contrary and then there exists an $\omega_0 \in \Omega$ such that $\omega_0 \notin G_n$ for every $n \in \mathbb{N}$. This means that for every $n \in \mathbb{N}$ there exists a $u_n \in M$ such that

$$d(x(\omega_0), u_n) < \frac{1}{n} \quad \text{and} \quad d(T(\omega_0)x(\omega_0), T(\omega_0)u_n) \geqslant \varepsilon,$$

which shows that $x(\omega_0)$ is a discontinuous point of $T(\omega_0)$, i.e., a contradiction. The fact that $\lim_{n \leftarrow \infty} P(G_n) = P(\lim_{n \to \infty} G_n) = P(\Omega) = 1$ implies that, for every $\lambda \in (0, 1)$, there exists an $n_0 \in \mathbb{N}$ such that $P(G_{n_0}) > 1 - \frac{\lambda}{2}$. Since $x_n(\omega) \xrightarrow{P} x(\omega)$, it follows that $\lim_{n \to \infty} P(B_n) = 1$, where

$$B_n = \left\{\omega \in \Omega : d(x_n(\omega), x(\omega)) < \frac{1}{n_0}\right\}, \quad n \in \mathbb{N}.$$

Let $m_0 \in \mathbb{N}$ be such that

$$P(B_n) > 1 - \frac{\lambda}{2}, \text{ for every } n \geqslant m_0.$$

If we now consider the sequence of random variables $\{T(\omega)x_n(\omega)\}_{n \in \mathbb{N}} \subset S$, and if we denote

$$D_n = \{\omega \in \Omega : d(T(\omega)x_n(\omega), T(\omega)x(\omega)) < \varepsilon\}, \quad n \in \mathbb{N},$$

it follows that $G_{n_0} \cap B_n \subset D_n$, $n \in \mathbb{N}$ and

$$P(D_n) \geqslant P(G_{n_0} \cap D_n) \geqslant P(G_{n_0}) - P(CB_n) > 1 - \frac{\lambda}{2} - \frac{\lambda}{2} = 1 - \lambda \quad \forall n \geqslant m_0.$$

This means that $\{T(\omega)x_n(\omega)\}_{n\in\mathbb{N}}$ converges in probability to $T(\omega)x(\omega)$.

We shall now give some common random fixed point theorems due to Shih-sen Chang [95] with applications to a class of random operator equations.

In what follows, we shall assume that $T(\omega)$, $U(\omega): M \to M$ are d-continuous random operators and Φ is a real function satisfying conditions (Φ_1) from Theorem 3.1.36, Chap. 3. Also, we denote

$$O_T(x(\omega), 0, \infty) = \{x(\omega), T(\omega)x(\omega), \ldots, T^n(\omega)x(\omega), \ldots\}$$

for every $T(\omega): M \to M$ (i.e., $T: \Omega \times M \to M$) and $\{x(\omega)\} \in S$.

THEOREM 4.1.8. *Let (M, d) be a separable complete metric space, let $(\Omega, \mathscr{K}, P)$ be a complete probability measure space and let $(S, \mathscr{F}, T_m)$ be an E-space over (M, d). Suppose that $T(\omega)$, $U(\omega): M \to M$ are two d-continuous random operators, and assume that there exist some positive integers i and j such that, for any $x(\omega) \in S$ and any $t \geqslant 0$,*

$$\inf_{p,q\in O_T(T^i(\omega,x(\omega)),0,\infty)\cup O_U(U^j(\omega,x(\omega)),0,\infty)} F_{p,q}(t) \geqslant$$

$$\geqslant \inf_{p,q\in O_T(x(\omega),0,\infty)\cup O_U(x(\omega),0,\infty)} F_{p,q}(\Phi(t)).$$

Then, for any $x_0(\omega) \in S$, the sequences $\{T^n(\omega, x_0(\omega))\}_{n\in\mathbb{N}\cup\{0\}}$ and $\{U^n(\omega, x_0(\omega))\}_{n\in\mathbb{N}\cup\{0\}}$ of M-valued random variables converge in the (ε, λ)-topology to a common random fixed point $x_*(\omega) \in S$ of $T(\omega,.)$ and $U(\omega,.)$.

Proof. For any $x_0(\omega) \in S$, Proposition 4.1.1 implies that $O_T(x_0(\omega), 0, \infty) \subset S$ and $O_U(x_0(\omega), 0, \infty) \subset S$. Let $m = \max\{i, j\}$. Then, for every $n \in \mathbb{N}$,

$$\inf_{p,q\in O_T(T^{nm}(\omega,x_0(\omega)),0,\infty)\cup O_U(U^{nm}(\omega,x_0(\omega)),0,\infty)} F_{p,q}(t) \geqslant$$

$$\geqslant \inf_{p,q\in O_T(T^{(n-1)m}(\omega,x_0(\omega),0,\infty)\cup O_U(U^{(n-1)m}(\omega,x_0(\omega,0)),\infty)} F_{p,q}(\Phi(t)) \geqslant$$

$$\geqslant \ldots \geqslant \inf_{p,q\in O_T(x_0(\omega),0,\infty)\cup O_U(x_0(\omega),0,\infty)} F_{p,q}(\Phi^n(t)).$$

On the other hand, since $\Phi^n(t) > \Phi^{n-1}(t) > \ldots > \Phi(t) > t$, $\forall\, t > 0$, then, for any $t \geqslant 0$, the sequence

$$\left\{\inf_{p,q\in O_T(x_0(\omega),0,\infty)\cup O_U(x_0(\omega),0,\infty)} F_{p,q}(\Phi^n(t))\right\}_{n\in\mathbb{N}\cup\{0\}}$$

is nondecreasing and therefore it has a limit. We also have

$$\lim_{n\to\infty} \inf_{p,q\in O_T(T^{nm}(\omega,x_0(\omega)),0,\infty)\cup O_U(U^{nm}(\omega,x_0(\omega)),0,\infty)} F_{p,q}(t) \geqslant$$

$$\geqslant \lim_{n\to\infty} \inf_{p,q\in O_T(x_0(\omega),0,\infty)\cup O_U(x_0(\omega),0,\infty)} F_{p,q}(\Phi^n(t)) =$$

$$= \inf_{p,q\in O_T(x_0(\omega),0,\infty)\cup O_U(x_0(\omega),0,\infty)} F_{p,q}(\lim_{n\to\infty} \Phi^n(t)) = H_0(t), \quad \forall\ t \geqslant 0.$$

From this, one can easily see that for any $\varepsilon > 0$ and $\lambda > 0$, there exists $n(\varepsilon, \lambda)$ such that

$$\inf_{p,q\in O_T(T^{nm}(\omega,x_0(\omega)),0,\infty)\cup O_U(U^{nm}(\omega,x_0(\omega)),0,\infty)} F_{p,q}(\varepsilon) >$$

$$> 1 - \lambda, \quad \forall\, n \geqslant n(\varepsilon, \lambda),$$

i.e., for any $p, q \in O_T(T^{nm}(\omega, x_0(\omega)), 0, \infty)$ and $p', q' \in O_U(U^{nm}(\omega, x_0(\omega)), 0, \infty)$, one has

$$F_{p,q}(\varepsilon) > 1 - \lambda,\ F_{p',q'}(\varepsilon) > 1 - \lambda,$$

which implies that $\{T^n(\omega, x_0(\omega))\}_{n\in\mathbb{N}}$ and $\{U^n(\omega, x_0(\omega))\}_{n\in\mathbb{N}}$ are Cauchy sequences. Since $(S, \mathscr{F}, T_m)$ is a complete Hausdorff space, it follows that these sequences converge to the same limit $x_*(\omega) \in S$.

Also,

$$F_{x_*(\omega),T(\omega,x_*(\omega))}(t) \geqslant \max\left\{F_{x_*(\omega),T^n(\omega,x_0(\omega))}\left(\frac{t}{2}\right) + \right.$$

$$\left. + F_{T(\omega,T^{n-1}(\omega,x_0(\omega)),T(\omega,x_*(\omega))}\left(\frac{t}{2}\right) - 1, 0\right\},$$

and from the d-continuity of T, one gets

$$F_{x_*(\omega),T(\omega,x_*(\omega))}(t) \geqslant \lim_{n\to\infty} \max\left\{F_{x_*(\omega),T^n(\omega,x_0(\omega))}\left(\frac{t}{2}\right) + \right.$$

$$\left. + F_{T(\omega,T^{n-1}(\omega,x_0(\omega)),T(\omega,x_*(\omega))}\left(\frac{t}{2}\right) - 1, 0\right\} = \max\left\{F_{x_*(\omega),x_*(\omega)}\left(\frac{t}{2}\right) + \right.$$

$$\left. + F_{T(\omega,x_*(\omega)),T(\omega,x_*(\omega))}\left(\frac{t}{2}\right) - 1, 0\right\} = \max\left\{2H_0\left(\frac{t}{2}\right) - 1, 0\right\},$$

$$\forall\, t \geqslant 0$$

i.e.,

$$x_*(\omega) = T(\omega, x_*(\omega)),$$

which means that $x_*(\omega)$ is a random fixed point of T.

One can similarly obtain the result that $x_*(\omega)$ is a random fixed point of U, i.e., $x_*(\omega) \in S$ is a common random fixed point of $T(\omega,.)$ and $U(\omega,.)$.

COROLLARY. *Let (M, d), $(\Omega, \mathscr{K}, P)$, $(S, \mathscr{F}, T_m)$ and $T(\omega)$ be as in Theorem 4.1.8 and suppose that there exists a $j \in \mathbb{N}$ such that one of the following conditions holds:*

(a) *for any $x(\omega) \in S$ and any $t \geqslant 0$,*

$$\inf_{p,q \in O_T(T^j(\omega,x(\omega)),0,\infty)} F_{p,q}(t) \geqslant \inf_{p,q \in O_T(x(\omega),0,\infty)} F_{p,q}(\Phi(t));$$

(b) *for any $x(\omega) \in S$, any $i \in \mathbb{N}$ and any $t \geqslant 0$,*

$$F_{T^j(\omega,x(\omega)),T^{j+i}(\omega,x(\omega))}(t) \geqslant \inf_{p,q \in O_T(x(\omega),0,\infty)} F_{p,q}(\Phi(t)).$$

Then, for any $x_0(\omega) \in S$, the sequence $\{T^n(\omega, x_0(\omega))\}_{n \in \mathbb{N} \cup \{0\}}$ of M-valued random variables converges in the (ε, λ)-topology to a random fixed point of $T(\omega,.)$.

Proof. If condition (a) holds and assuming that $T = U$ and $i = j$ in Theorem 4.1.8, then the conclusion of the corollary follows easily.

Suppose that condition (b) holds and that, for every $m, n \in \mathbb{N}$, $m \geqslant n$,

$$F_{T^{j+n}(\omega,x(\omega)),T^{j+m}(\omega,x(\omega))}(t) =$$

$$= F_{T^j(\omega,T^n(\omega,x(\omega))),T^{j+m-n}(\omega,T^n(\omega,x(\omega)))}(t) \geqslant$$

$$\geqslant \inf_{p,q \in O_T(T^n(\omega,x(\omega)),0,\infty)} F_{p,q}(\Phi(t)) \geqslant \inf_{p,q \in O_T(\omega,x(\omega),0,\infty)} F_{p,q}(\Phi(t))$$

for every $t \geqslant 0$.

Hence,

$$\inf_{m,n \geqslant 0} F_{T^{j+n}(\omega,x(\omega)),T^{j+m}(\omega,x,(\omega))}(t) =$$

$$= \inf_{p,q \in O_T(T^j(\omega,x(\omega)),0,\infty)} F_{p,q}(t) \geqslant \inf_{p,q \in O_T(x(\omega),0,\infty)} F_{p\,q}(\Phi(t))$$

for every $t \geqslant 0$, i.e., condition (a) holds and the corollary follows from Theorem 4.1.8.

PROPOSITION 4.1.2. *Let (M, d), $(\Omega, \mathscr{K}, P)$, $(S, \mathscr{F}, T_m)$ and $T(\omega,.)$, $U(\omega,.)$ be as in Theorem 4.1.8. Suppose that there exist $i,j \in \mathbb{N}$ such that for any $x(\omega)$, $y(\omega) \in S$ and any $t \geqslant 0$, we* *have*

$$\inf_{p,q\in O_T(T^i(\omega,x(\omega)),0,\infty)\cup O_U(U^j(\omega,y(\omega)),0,\infty)} F_{p,q}(t) \geqslant$$

$$\geqslant \inf_{p,q\in O_T(x(\omega),0,\infty)\cup O_U(y(\omega),0,\infty)} F_{p,q}(\Phi(t)).$$

Then there exists a unique common random fixed point $x_*(\omega)\in S$ of $T(\omega,.)$ and $U(\omega,.)$ and for any $x_0(\omega)\in S$, the sequences $\{T^n(\omega, x_0(\omega)\}_{n\in\mathbb{N}\cup\{0\}}$ and $\{U^n(\omega,x_0(\omega)\}_{n\in\mathbb{N}\cup\{0\}}$ of M-valued random variables converge in the (ε, λ)*-topology to the fixed point.*

Proof. Theorem 4.1.8 implies that for any $x_0(\omega)\in S$ the sequences $\{T^n(\omega, x_0(\omega))\}_{n\in\mathbb{N}\cup\{0\}}$ and $\{U^n(\omega, x_0(\omega))\}$ of random variables converge in the (ε, λ)-topology to some common fixed point $x_*(\omega)\in S$. In order to prove that $x_*(\omega)$ is the unique common random fixed point of $T(\omega,.)$ and $U(\omega,.)$ in S, we consider another common random fixed point $y_*(\omega)\in S$. Then, we have

$$F_{x_*(\omega),y_*(\omega)}(t) = \inf_{p,q\in O_T(T^i(\omega,x_*(\omega)),0,\infty)\cup O_U(U^j(\omega,y_*(\omega)),0,\infty)} F_{p,q}(t) \geqslant$$

$$\geqslant \inf_{p,q\in O_T(x_*(\omega),0,\infty)\cup O_U(y_*(\omega),0,\infty)} F_{p,q}(\Phi(t)) =$$

$$= F_{x_*(\omega),y_*(\omega)}(\Phi(t)), \quad \forall\, t \geqslant 0.$$

Since $\Phi(t)$ satisfies condition (Φ_1), one repeats the above procedure and obtains

$$F_{x_*(\omega),y_*(\omega)}(t) \geqslant \lim_{n\to\infty} F_{x_*(\omega),y_*(\omega)}(\Phi^n(t)) = H_0(t), \ \forall\, t\geqslant 0,$$

which implies that $x_*(\omega) = y_*(\omega)$.

COROLLARY 1. *Let (M, d), $(\Omega, \mathscr{K}, P)$, $(S, \mathscr{F}, T_m)$ be as in Theorem 4.1.8 and let $\mathscr{M}$ be a family of d-continuous random operators from $\Omega\times M$ into M. Suppose that there exist functions $i:\mathscr{M}\to\mathbb{N}$ and $j:\mathscr{M}\to\mathbb{N}$ such that, for any different random operators $T(\omega,.)$ and $U(\omega,.)\in\mathscr{M}$, and any $x(\omega)$, $y(\omega)\in S$, the following holds:*

$$\inf_{p,q\in O_T(T^{i(T)}(\omega,x(\omega)),0,\infty)\cup O_U(U^{j(U)}(\omega,y(\omega)),0,\infty)} F_{p,q}(t) \geqslant$$

$$\geqslant \inf_{p,q\in O_T(x(\omega),0,\infty)\cup O_U(y(\omega),0,\infty)} F_{p,q}(\Phi(t)), \quad \forall\, t\geqslant 0.$$

Then there exists a unique common random fixed point of $\mathscr{M}$ in S, and, for any $x_0(\omega)\in S$, and any $T(\omega,.)\in\mathscr{M}$, the sequence $\{T^n(\omega, x_0(\omega))\}_{n\in\mathbb{N}}\subset S$ converges in the (ε, λ)*-topology to the fixed point.*

COROLLARY 2. *Let (M, d), $(\Omega, \mathcal{K}, P)$, $(S, \mathcal{F}, T_m)$ and $T(\omega,.)$ be as in Theorem 4.1.8. Suppose that there exist $i, j \in \mathbb{N}$ such that, for any $x(\omega)$, $y(\omega) \in S$ and any $t \geqslant 0$, the following holds:*

$$\inf_{p,q \in O_T(T^i(\omega, x(\omega)), 0, \infty) \cup O_T(T^j(\omega, y(\omega)) 0, \infty)} F_{p,q}(t) \geqslant$$

$$\geqslant \inf_{p,q \in O_T(x(\omega), 0, \infty) \cup O_T(y(\omega), 0, \infty)} F_{p,q}(\Phi(t)).$$

Then $T(\omega,.)$ has a unique random fixed point $x_*(\omega) \in S$ and, for any $x_0(\omega) \in S$, the sequence $\{T^n(\omega, x_0(\omega))\}_{n \in \mathbb{N}}$ of random variables converges in the (ε, λ)-topology to $x_*(\omega)$.

PROPOSITION 4.1.3. *Let (M, d), $(\Omega, \mathcal{K}, P)$, $(S, \mathcal{F}, T_m)$ and $T(\omega,.)$ be as in Theorem 4.1.8. Suppose that there exist $i, j \in \mathbb{N}$ such that, for any $x(\omega)$, $y(\omega) \in S$ and any $t \geqslant 0$,*

$$F_{T^i(\omega, x(\omega)), T^j(\omega, y(\omega))}(t) \geqslant \inf_{p,q \in O_T(x(\omega), 0, \infty) \cup O_T(y(\omega), 0, \infty)} F_{p,q}(\Phi(t)).$$

Then there exists a unique random fixed point $x_*(\omega)$ of $T(\omega,.)$ in S and, for any $x_0(\omega) \in S$, the sequence $\{T^n(\omega, x_0(\omega))\}_{n \in \mathbb{N}}$ of random variables converges in the (ε, λ)-topology to $x_*(\omega)$.

Proof. Suppose that $i \geqslant j$. For any $z(\omega) \in S$, let $x(\omega) = T^n(\omega, z(\omega))$, $y(\omega) = T^{i-j+m}(\omega, z(\omega))$, where m, n are arbitrary in $\mathbb{N}$. Then, we have

$$F_{T^{n+i}(\omega, z(\omega)), T^{i+m}(\omega, z(\omega))}(t) \geqslant$$

$$\geqslant \inf_{p,q \in O_T(T^n(\omega, z(\omega)), 0, \infty) \cup O_T(T^{i-j+m}(\omega, z(\omega)), 0, \infty)} F_{p,q}(\Phi(t)) \geqslant$$

$$\geqslant \inf_{p,q \in O_T(z(\omega), 0, \infty)} F_{p,q}(\Phi(t)), \quad \forall\, t \geqslant 0, \quad \forall\, z(\omega) \in S.$$

Hence,

$$\inf_{m,n \geqslant 0} F_{T^{i+n}(\omega, z(\omega)), T^{i+m}(\omega, z(\omega))}(t) =$$

$$= \inf_{p,q \in O_T(T^i(\omega, z(\omega)), 0, \infty)} F_{p,q}(t) \geqslant \inf_{p,q \in O_T(z(\omega), 0, \infty)} F_{p,q}(\Phi(t))$$

for every $z(\omega) \in S$ and $t \geqslant 0$.

The uniqueness of the fixed point of $T(\omega,.)$ follows as in Proposition 4.1.2.

Before concluding this paragraph on the use of the E-spaces in the study of the existence and uniqueness of random fixed points, we shall give bellow an application of Proposition 3.1.18 as:

PROPOSITION 4.1.4 [419]. *Let $(S, \mathcal{F}, T_m)$ be an E-space over (M, d), let T and U be injective continuous random operators on*

$\Omega \times M$ and let V be a continuous random operator $V(\Omega \times M) \subset T(\Omega \times M) \cap U(\Omega \times M)$, where T and U commute with V. If for every $x(\omega) \in S$ there exists an $n(x) \in \mathbb{N}$ such that, for every $y(\omega) \in S$,

$$F_{V^{n(x)}(\omega,x),V^{n(x)}(\omega,y)}(kt) \geqslant$$

$$\geqslant \max{}'\{F_{T(\omega,x),U(\omega,y)}(t),\ F_{U(\omega,x),T(\omega,y)}(t)\},$$

where $k \in (0, 1)$, then there exists a unique common random fixed point for the random operators T, U and V.

The proof of this proposition is identical with the proof of Proposition 3.1.18, with the only difference that now the assertion that the set $T(\Omega \times M) \cap U(\Omega \times M)$ is probabilistic bounded is no longer needed.

Applications of the above and other results to the study of the existence and uniqueness of random solutions of some classes of random operator equations are given in Chapter 5.

We shall now use the two fundamental ideas adopted in the organization of the space $\tilde{\mathscr{V}}_B$ (see Sec. 2.1) i.e., that of an RN-space and that of associating to every continuous random operator T on X the Nemytskiĭ operator $\tilde{T}$ on the space of all classes of random variables $\tilde{\mathscr{V}}_B$. This was done by G. Bocşan [50] to obtain some general results regarding the fixed points of a large class of random operators or of multivalued mappings. These results also stimulated the development (by O. Hadžić and M. Stojaković) of some fixed point theorems for various operator classes.

Let $(\Omega, \mathscr{K}, P)$ be a complete probability measure space, let $(X, \|\ .\ \|)$ be a separable Banach space and let $(\tilde{\mathscr{V}}_B, \mathscr{F}, \tilde{T}_m)$ be the associated RN-space.

PROPOSITION 4.1.5 [193]. *Let $T : \tilde{\mathscr{V}}_B \to \tilde{\mathscr{V}}_B$ be such that the following conditions are satisfied:*

(i) *There exists a $c > 0$ such that, for every $\tilde{x} \in \tilde{\mathscr{V}}_B$,*

$$P(\{\omega : \| T(\tilde{x})(\omega) \| \leqslant c\}) = 1;$$

(ii) *There exists a $k \in (0, 1)$ such that, for every $\tilde{x}, \tilde{y} \in \tilde{\mathscr{V}}_B$,*

$$P(\{\omega : \| T(\tilde{x})(\omega) - T(\tilde{y})(\omega) \| \leqslant k \| \tilde{x}(\omega) - \tilde{y}(\omega) \|\}) = 1.$$

Then there exists one and only one element $\tilde{x}_* \in \tilde{\mathscr{V}}_B$ such that $T(\tilde{x}_*) = \tilde{x}_*$, i.e., $T(\tilde{x}_*)(\omega) = \tilde{x}_*(\omega)$, for every $\omega \in \Omega_0$, $P(\Omega_0) = 1$. Further, $\tilde{x}_n \xrightarrow{P} \tilde{x}_*$, where $\tilde{x}_n = T(\tilde{x}_{n-1})$, $n \in \mathbb{N}$, and x_0 is an arbitrary element from $\tilde{\mathscr{V}}_B$.

Proof. We shall prove that all the conditions of Theorem 3.1.3 are satisfied.

Let us denote

$$G_1(\tilde{x}, \tilde{y}, t) = \{\omega : \| \tilde{x}(\omega) - \tilde{y}(\omega) \| < t\},$$

$$G_2(\tilde{x}, \tilde{y}) = \{\omega : \| T(\tilde{x})(\omega) - T(\tilde{y})(\omega) \| \leqslant k \| \tilde{x}(\omega) - \tilde{y}(\omega) \|\}.$$

Let $\omega \in G_1(\tilde{x}, \tilde{y}, t/k) \cap G_2(\tilde{x}, \tilde{y})$; then $\omega \in G_1(T(\tilde{x}), T(\tilde{y}), t)$ and so

$$G_1(\tilde{x}, \tilde{y}, t/k) \cap G_2(\tilde{x}, \tilde{y}) \subset G_1(T(\tilde{x}), T(\tilde{y}), t).$$

Hence,

$$P(G_1(\tilde{x}, \tilde{y}, t/k) \cap G_2(\tilde{x}, \tilde{y})) = P(G_1(\tilde{x}, \tilde{y}, t/k)) \leqslant$$
$$\leqslant P(G_1(T(\tilde{x}), T(\tilde{y}), t)),$$

i.e., from the definition of $F_{\tilde{x}}$, it follows that

$$F_{T(\tilde{x})-T(\tilde{y})}(t) \geqslant F_{\tilde{x}-\tilde{y}}(t/k).$$

The proof of

$$\sup_t \inf_{n \in \mathbb{N}} F_{T^n(\tilde{x}_0)-\tilde{x}_0}(t) = 1,$$

where $\tilde{x}_0 \in \tilde{\mathscr{V}}_B$, is equivalent to the proof of

$$\sup_t \{\inf_{n \in \mathbb{N}} P(\{\omega : \| T^n(\tilde{x}_0)(\omega) - \tilde{x}_0(\omega) \| < t\})\} = 1.$$

Let us prove that for every $\lambda \in (0, 1)$ there exists an $\varepsilon(\lambda) > 0$ such that

$$\inf_{n \in \mathbb{N}} P(\{\omega : \| T^n(\tilde{x})(\omega) - \tilde{x}(\omega) \| < \varepsilon(\lambda)\} > 1 - \lambda.$$

Indeed, since $\lim_{t\to\infty} F_{\tilde{x}}(t) = 1$ for every $\lambda' < \lambda$, there exists a $\delta(\lambda')$ such that

$$F_{\tilde{x}}(\delta(\lambda')) = P(\{\omega : \| \tilde{x}(\omega) \| < \delta(\lambda')\} > 1 - \lambda'.$$

For every $n \in \mathbb{N}$, let

$$A_n = \{\omega : \| T^n(\tilde{x})(\omega) - \tilde{x}(\omega) \| < c + \delta(\lambda')\}$$

$$B_n = \{\omega : \| T^n(\tilde{x})(\omega) \| \leqslant c\}$$

and

$$D = \{\omega : \| \tilde{x}(\omega) \| < \delta(\lambda')\}.$$

Then, $B_n \cap D = A_n$ and since $P(B_n) = 1$, then $P(D) = P(B_n \cap D) \leqslant P(A_n)$. From $P(D) > 1 - \lambda$, it follows that

$P(A_n) > 1 - \lambda'$, and so

$$\inf_{n \in \mathbb{N}} P(A_n) \geqslant 1 - \lambda' > 1 - \lambda,$$

i.e., $\varepsilon(\lambda) = c + \delta(\lambda')$. Hence, the conditions of Theorem 3.1.3 are satisfied and $\operatorname{Fix}(T) \neq \emptyset$.

Using Theorem 3.4.2 and reasoning as in the above proposition, one can prove

PROPOSITION 4.1.6. *Let $\tilde{\mathscr{V}}_B$ be as in Proposition 4.1.5, and let M be a star-convex and closed subset of $\tilde{\mathscr{V}}_B$ in the (ε, λ)-topology such that the following conditions are satisfied:*

(i) $\sup_t \inf_{n \in \mathbb{N}} P(\{\omega : \| \tilde{x}(\omega) - \tilde{y}(\omega) \| < t\}) = 1$;

(ii) *For every* $\tilde{x}, \tilde{y} \in \tilde{\mathscr{V}}_B$, $P(\{\omega : \| T(\tilde{x})(\omega) - T(\tilde{y})(\omega) \| \leqslant$
$\leqslant \| \tilde{x}(\omega) - \tilde{y}(\omega) \|\}) = 1$;

(iii) *There exists an $m \in \mathbb{N}$ such that $T^m(M)$ is compact in the (ε, λ)-topology.*

Then there exists a $z \in M$ such that $T(z) = z$.

From among the results regarding the existence of the common fixed points of PM-spaces with applications to the case of mappings on $\tilde{\mathscr{V}}_B$, one should mention:

PROPOSITION 4.1.7 [187]. *Let T, U and V be continuous mappings from $\tilde{\mathscr{V}}_B$ into $\tilde{\mathscr{V}}_B$, let V commute with T and U and let $V(\tilde{\mathscr{V}}_B) \subseteq T(\tilde{\mathscr{V}}_B) \cap U(\tilde{\mathscr{V}}_B)$. Suppose that the following two conditions are satisfied:*

(1) *For any* $\tilde{x}, \tilde{y} \in \tilde{\mathscr{V}}_B$,

$$P(\{\omega : \| V(\tilde{x})(\omega) - V(\tilde{y})(\omega) \| \leqslant k \| T(\tilde{x})(\omega) - U(\tilde{y})(\omega) \|\}) = 1,$$

$k \in (0, 1)$;

(2) *There exists a $c > 0$ so that, for every $\tilde{x} \in \tilde{\mathscr{V}}_B$, $P(\{\omega : \| V(\tilde{x})(\omega) \| \leqslant c\}) = 1$.*

Then there exists one and only one common fixed point for mappings T, U and V.

Proof. We shall show that (1) implies (i) and (2) implies (ii) of Theorem 3.1.46. Indeed, if we denote

$$G_{\tilde{x},t} = \{\omega \in \Omega : \| \tilde{x}(\omega) \| \leqslant t\}$$

and

$$H_{\tilde{x},\tilde{y}} = \{\omega \in \Omega : V\| (\tilde{x})(\omega) - V(\tilde{y})(\omega) \| \leqslant$$
$$\leqslant k \| T(\tilde{x})(\omega) - U(\tilde{y})(\omega) \|\},$$

then it is obvious that

$$G_{T\tilde{x}-U\tilde{y}t/k} \cap H_{\tilde{x},\tilde{y}} \subseteq G_{V\tilde{x}-V\tilde{y},t}.$$

Hence,

$$P(G_{T\tilde{x}-U\tilde{y},t/k} \cap H_{\tilde{x},\tilde{y}}) = P(G_{T\tilde{x}-U\tilde{y},t/k}) \leqslant P(G_{V\tilde{x}-V\tilde{y},t})$$

and so (1) $\Rightarrow$ (i) since $P(G_{\tilde{x},t}) = F_{\tilde{x}}(t)$. Also, since (2) implies that $F_{V\tilde{x}}(c) = 1$, for every $\tilde{x} \in \tilde{\mathscr{V}}_B$ and

$$F_{V\tilde{x}-V\tilde{y}}(2c) \geqslant T_m(F_{V\tilde{x}}(c), F_{V\tilde{y}}(c)) = T_m(1, 1) = 1$$

for every $\tilde{x}, \tilde{y} \in \tilde{\mathscr{V}}_B$, and hence the proof is complete.

PROPOSITION 4.1.8 [419]. *Let T and U be injective continuous mappings of $\tilde{\mathscr{V}}_B$ into $\tilde{\mathscr{V}}_B$, let V be a continuous mapping of $\tilde{\mathscr{V}}_B$, into $T(\tilde{\mathscr{V}}_B) \cap U(\tilde{\mathscr{V}}_B)$ and let T and U commute with V. Let also the following condition be satisfied:*

For every $\tilde{x} \in \tilde{\mathscr{V}}_B$ there exists $n(\tilde{x})$ such that, for every $\tilde{y} \in \tilde{\mathscr{V}}_B$,

$$P(\{\omega \in \Omega : \| V^{n(\tilde{x})}(\tilde{x})(\omega) - V^{n(\tilde{x})}(\tilde{y})(\omega) \| < k \| T(\tilde{x})(\omega) - U(\tilde{y})(\omega) \|\}) = 1$$

$$P(\{\omega \in \Omega : \| V^{n(\tilde{x})}(\tilde{x})(\omega) - V^{n(\tilde{x})}(\tilde{y})(\omega) \| < k \| U(\tilde{x})(\omega) - T(\tilde{y})(\omega) \|\}) = 1,$$

where $k \in (0, 1)$.

Then, there exists a unique common fixed point for the mappings T, U and V.

To prove this proposition, it is sufficient to show that the above conditions imply conditions (1) and (2) in Proposition 3.1.18.

Before concluding the present outline of the results regarding the fixed points of mappings on $\tilde{\mathscr{V}}_B$, we shall recall a theorem on probabilistic γ-condensing mappings, where γ is a random measure of noncompactness, such that $\gamma_{A \cup \{\tilde{x}\}} = \gamma_A$ for every probabilistic bounded subset of $\tilde{\mathscr{V}}_B$.

THEOREM 4.1.9 [199]. *Let $T : \tilde{\mathscr{V}}_B \to \tilde{\mathscr{V}}_B$ be a probabilistic γ-condensing mapping such that the following conditions are satisfied:*

(i) *There exists a $c > 0$ such that, for every $\tilde{x} \in \tilde{\mathscr{V}}_B$,*

$$P(\{\omega \in \Omega : \| T(\tilde{x})(\omega) \| \leqslant c\}) = 1;$$

(ii) *For every* $\tilde{x}, \tilde{y} \in \tilde{\mathscr{V}}_B$,

$$P(\{\omega \in \Omega : \| T(\tilde{x})(\omega) - T(\tilde{y})(\omega) \| \leqslant \| \tilde{x}(\omega) - \tilde{y}(\omega) \|\}) = 1$$

and there exists an $\varepsilon_{\tilde{x},\tilde{y}} > 0$ *such that*

$$P(\{\omega \in \Omega : \| T(\tilde{x})(\omega) - T(\tilde{y})(\omega) \| < \varepsilon_{\tilde{x},\tilde{y}} \leqslant \| \tilde{x}(\omega) - \tilde{y}(\omega) \|\}) > 0.$$

Then there exists one and only one fixed point of the mapping T.

Proof. We shall prove that all the conditions of Theorem 3.1.32 are satisfied (see also [199], Th. 1), where $\Phi : \tilde{\mathscr{V}}_B \times \tilde{\mathscr{V}}_B \to \mathscr{D}$ is defined by $\Phi(\tilde{x}, \tilde{y}) = F_{\tilde{x}-\tilde{y}}$ for every $\tilde{x}, \tilde{y} \in \tilde{\mathscr{V}}_B$. It is easy to prove that Φ is upper semicontinuous in the sense of Birkhoff and $\Phi(T\tilde{x}, T\tilde{y}) > \Phi(\tilde{x}, \tilde{y})$ for every $\tilde{x} \neq \tilde{y}, \tilde{x}, \tilde{y} \in \tilde{\mathscr{V}}_B$. Indeed, if $G_1(\tilde{x}, \tilde{y}, t)$ is as in Proposition 4.1.5 and $G_3(\tilde{x}, \tilde{y}) = \{\omega \in \Omega : \| T(\tilde{x})(\omega) - T(\tilde{y})(\omega) \| \leqslant \| \tilde{x}(\omega) - \tilde{y}(\omega) \|\}$, it is obvious that $G_1(\tilde{x}, \tilde{y}, t) \cap G_3(\tilde{x}, \tilde{y}) \subseteq G_1(T\tilde{x}, T\tilde{y}, t)$ for every $\tilde{x}, \tilde{y} \in \tilde{\mathscr{V}}_B$, and so (ii) implies that

$$F_{\tilde{x},\tilde{y}}(t) = P(G_1(\tilde{x}, \tilde{y}, t)) \leqslant P(G_1(T\tilde{x}, T\tilde{y}, t)) = F_{T\tilde{x},T\tilde{y}}(t).$$

We shall prove that

$$F_{\tilde{x}-\tilde{y}}(\varepsilon_{\tilde{x},\tilde{y}}) < F_{T\tilde{x}-T\tilde{y}}(\varepsilon_{\tilde{x},\tilde{y}})$$

and for this we shall denote

$$G_4(\tilde{x}, \tilde{y}) = \{\omega \in \Omega : \| T(\tilde{x})(\omega) - T(\tilde{y})(\omega) \| < \varepsilon_{\tilde{x},\tilde{y}} \leqslant \\ \leqslant \| \tilde{x}(\omega) - \tilde{y}(\omega) \|\}.$$

We observe that

$$G_4(\tilde{x}, \tilde{y}) = \bar{G}_1(\tilde{x}, \tilde{y}, \varepsilon_{\tilde{x},\tilde{y}}) \cap G_1(T\tilde{x}, T\tilde{y}, \varepsilon_{\tilde{x},\tilde{y}})$$

and

$$G_1(T\tilde{x}, T\tilde{y}, \varepsilon_{\tilde{x},\tilde{y}}) = (G_1(\tilde{x}, \tilde{y}, _{\tilde{x},\tilde{y}}) \cap G_1(T\tilde{x}, T\tilde{y}, \varepsilon_{\tilde{x},\tilde{y}})) \cup \\ \cup (\bar{G}_1(\tilde{x}, \tilde{y}, \varepsilon_{\tilde{x},\tilde{y}}) \cap G_1(T\tilde{x}, T\tilde{y}, \varepsilon_{\tilde{x},\tilde{y}})),$$

and so

$$P(G_1(T\tilde{x}, T\tilde{y}, \varepsilon_{\tilde{x},\tilde{y}})) = P(G_1(\tilde{x}, \tilde{y}, \varepsilon_{\tilde{x},\tilde{y}}) \cap \\ \cap G_1(T\tilde{x}, T\tilde{y}, \varepsilon_{\tilde{x},\tilde{y}})) + P(G_4(\tilde{x}, \tilde{y})).$$

Since $P(G_4(\tilde{x}, \tilde{y}) > 0$,

$$P(G_1(T\tilde{x}, T\tilde{y}, \varepsilon_{\tilde{x},\tilde{y}})) > P(G_1(\tilde{x}, \tilde{y}, \varepsilon_{\tilde{x},\tilde{y}}) \cap G_1(T\tilde{x}, T\tilde{y}, \varepsilon_{\tilde{x},\tilde{y}})).$$

On the other hand,

$$G_1(\tilde{x}, \tilde{y}, \varepsilon_{\tilde{x},\tilde{y}}) = (G_1(\tilde{x}, \tilde{y}, \varepsilon_{\tilde{x},\tilde{y}}) \cap G_3(\tilde{x}, \tilde{y})) \cup (G_1(\tilde{x}, \tilde{y}, \varepsilon_{\tilde{x},\tilde{y}}) \cap \cap \bar{G}_3(\tilde{x}, \tilde{y}),$$

and so

$$G_1(\tilde{x}, \tilde{y}, \varepsilon_{\tilde{x},\tilde{y}}) \cap G_1(T\tilde{x}, T\tilde{y}, \varepsilon_{\tilde{x},\tilde{y}}) = (G_1(\tilde{x}, \tilde{y}, \varepsilon_{\tilde{x},\tilde{y}}) \cap \cap G_1(T\tilde{x}, T\tilde{y}, \varepsilon_{\tilde{x},\tilde{y}}) \cap G_3(\tilde{x}, \tilde{y})) \cup G_1(\tilde{x},\tilde{y}, \varepsilon_{\tilde{x},\tilde{y}}) \cap \cap G_1(T\tilde{x}, T\tilde{y}, \varepsilon_{\tilde{x},\tilde{y}}) \cap \bar{G}_3(\tilde{x}, \tilde{y})).$$

Because $P(G_3(\tilde{x}, \tilde{y})) = 1$ and $G_1(\tilde{x}, \tilde{y}, \varepsilon_{\tilde{y},\tilde{x}}) \cap G_3(\tilde{x}, \tilde{y}) \subseteq \subseteq G_1(T\tilde{x}, T\tilde{y}, \varepsilon_{\tilde{x},\tilde{y}})$, it follows that

$$P(G_1(\tilde{x}, \tilde{y}, \varepsilon_{\tilde{x},\tilde{x}}) \cap G_1(T\tilde{x}, T\tilde{y}, \varepsilon_{\tilde{x},\tilde{y}})) = P(G_1(\tilde{x}, \tilde{y}, \varepsilon_{\tilde{x},\tilde{y}})$$

and

$$\Phi(T\tilde{x}, T\tilde{y}) = P(G_1(T\tilde{x}, T\tilde{y}, \varepsilon_{\tilde{x},\tilde{y}})) > P(G_1(\tilde{x}, \tilde{y}, \varepsilon_{\tilde{x},\tilde{y}}) = = \Phi(\tilde{x}, \tilde{y}).$$

Also, as in the second part of the proof of Proposition 4.1.5, it can be proved that

$$\sup_{t>0} \inf_{n\in\mathbb{N}} P(\{\omega \in \Omega : \| T^n(\tilde{x})(\omega) - \tilde{x}(\omega)\| < t\}) = 1$$

for every $\tilde{x} \in \tilde{\mathscr{V}}_B$.

We note the special importance of the following applications of the probabilistic structures to the random fixed point theory. The important result is that the mapping $x \to \tilde{x}$ of $\mathscr{V}_B$ onto $\tilde{\mathscr{V}}_B$ is an isomorphism which preserves the random norms. Moreover, if $T(\omega)$ is a continuous random operator on X, then [50] *the Nemytskiĭ operator* $\tilde{T}$ on $\tilde{\mathscr{V}}_B$, defined by

$$(\tilde{T}\tilde{x})\ (\omega) = T(\omega, \tilde{x}(\omega)), \tilde{x} \in \tilde{\mathscr{V}}_B,$$

is continuous on $\tilde{\mathscr{V}}_B$ with the topology of the convergence in probability. Hence, $\tilde{T}$ is (ε, λ)-continuous on the random normed space $(\tilde{\mathscr{V}}_B, \mathscr{F}, \tau_m)$.

Moreover, the fact that x is the fixed point of $T(\omega)$, i.e., $T(\omega, x(\omega)) = x(\omega)$ a.s. is equivalent with $\tilde{T}\tilde{x} = \tilde{x}$. Thus, we obtain

PROPOSITION 4.1.9. *A continuous random operator $T(\omega)$ on the separable Banach space X has a fixed point $\tilde{x} \in \tilde{\mathscr{V}}_B$ if and only if $\tilde{x}$ is the fixed point of the Nemytskiĭ operator $\tilde{T}$ in the random normed space $(\tilde{\mathscr{V}}_B, \mathscr{F}, \tau_m)$.*

Clearly, this proposition is true for an arbitrary complete probability measure space. In what follows, $(\Omega, \mathscr{K}, P)$ is an atomic probability measure space, i.e., $\Omega = \bigcup_{n=1}^{\infty} \Omega_n$, $\{\Omega_n\}_{n \in \mathbb{N}}$ being the distinct atoms.

The operator $\tilde{T}$ on $\tilde{\mathscr{V}}_B$ is said to be *condensing* if for all probabilistic bounded subsets $A \in \tilde{\mathscr{P}}(\tilde{\mathscr{V}}_B)$ such that $\tilde{\alpha}_A < H_0$, we have $\tilde{\alpha}_{\tilde{T}(A)} > \tilde{\alpha}_A$, where $\tilde{\alpha}$ is the random measure of noncompactness introduced in Proposition 2.1.3. The random operator $T(\omega)$ on the separable Banach space X is called a *random condensing* operator if

$$P(\{\omega \in \Omega : T(\omega) \text{ is condensing}\}) = 1.$$

THEOREM 4.1.10 [50]. *If $T : \Omega \times X \to X$ is a random condensing operator on the separable Banach space X, then the Nemytskiĭ operator $\tilde{T} : \tilde{\mathscr{V}}_B \to \tilde{\mathscr{V}}_B$ is a condensing operator on the RN-space $(\tilde{\mathscr{V}}_B, \mathscr{F}, \tau_m)$.*

Moreover, if for a measurable multivalued application $\omega \to E(\omega)$ whose values are nonempty, bounded, convex and closed subsets of X we have

$$T(\omega, E(\omega)) \subseteq E(\omega) \ P\text{-a.s.},$$

then

$$T\mathscr{S} \subseteq \mathscr{S},$$

where $\mathscr{S}$ is the set of all measurable selectors of the mapping E.

Proof. We can suppose that all the operators $T(\omega, .) = T(\omega)$ are condensing on the separable Banach space X, i.e., for each $\omega \in \Omega$, we have

$$(*) \quad \alpha(T(\omega, B)) \leqslant \alpha(B)$$

for any bounded subset B of X, the inequality being strict if $\alpha(B) > 0$.

Now, let A be a probabilistic bounded subset of $\tilde{\mathscr{V}}_B$ such that $\tilde{\alpha}_A < H$. From Proposition 2.1.1, it follows that $A(\omega) = \{\tilde{x}(\omega) : \tilde{x} \in A\}$ is a bounded subset of X for all $\omega \in \Omega$. Hence, relation $(*)$ implies that

$$\alpha(T(\omega, A(\omega))) \leqslant \alpha(A(\omega)) \text{ for every } \omega \in \Omega,$$

i.e.,

$$\alpha((\tilde{T}A)(\omega)) \leqslant \alpha(A(\omega))$$

and

$$(**)\ \tilde{\alpha}_{\tilde{T}(A)}(t) \geqslant \tilde{\alpha}_A(t),\ t \in \mathbb{R}.$$

From $\tilde{\alpha}_A < H$ and Proposition 2.1.3, we conclude that A is not probabilistic precompact. From Theorem 1.2.3 and the equivalence of the (ε, λ)- and Γ-topology (Proposition 2.1.1), it follows that there exists an n_0 such that $A(\omega)$ is not precompact in X, for all $\omega \in \Omega_{n_0}$. This implies that $P(\Omega_0) > 0$, where $\Omega_0 = \{\omega \in \Omega : A(\omega)$ is not precompact in $X\}$.

We shall prove that $\tilde{\alpha}_{\tilde{T}(A)}(t) = \tilde{\alpha}_A(t)$, $t \in \mathbb{R}$, leads to a contradiction, which means that in $(**)$ we have the strict inequality. Indeed, if $\alpha_{\tilde{T}(A)}(t) = \tilde{\alpha}_A(t)$, $t \in \mathbb{R}$, then $\alpha(T(\omega, A(\omega))) = \alpha(A(\omega))$ a.s. and since $T(\omega,.)$ is condensing, $A(\omega)$ is precompact. This contradicts $P(\Omega_0) > 0$ and, hence, $\tilde{T}$ is a condensing operator on the RN-space $(\tilde{\mathscr{V}}_B, \tilde{\mathscr{F}}, \tau_m)$.

That $\tilde{T}\mathscr{S} \subseteq \mathscr{S}$ is obvious.

From the theorem there follows

THEOREM 4.1.11 [50]. *If $T : \Omega \times X \to X$ is a random condensing operator on a separable Banach space X such that*

$$T(\omega, E(\omega)) \subseteq E(\omega) \ a.s.,$$

where $\omega \to E(\omega)$ is a measurable multivalued application whose values are nonempty, bounded, convex and closed subsets of X, then T has a fixed point.

Proof. From the above theorem, it follows that the Nemytskiĭ operator $\tilde{T}$ is condensing with respect to the random measure of noncompactness $\tilde{\alpha}$ and $\tilde{T}\mathscr{S} \subseteq \mathscr{S}$. Using Proposition 2.1.3 and Remark 2.2.2, and the fact that the RN-space $(\tilde{\mathscr{V}}_B, \tilde{\mathscr{F}}, \tau_m)$ is a complete locally convex space, we conclude that $\tilde{T}$ has a fixed point in $\tilde{\mathscr{V}}_B$ (see Theorem 3.2.2). From the above theorem, it follows that the random operator T has a fixed point.

Remark. Since the random compact operators and the random operators of Krasnosel'skiĭ type are random condensing operators, the above theorem implies as a special case the results of Mukherjea (Theorem 4.1.4) and Prakasa Rao (Theorem 4.1.7).

Moreover, since a random contraction operator is a random condensing operator, Hanš and Špaček's fixed point theorem for the case when the probability measure space is atomic is immediate.

Let (M, d) be a separable metric space. The mapping $T: \Omega \times M \to M$ is a random contraction if $\omega \to T(\omega, x)$, $\omega \in \Omega$, is an element of $\mathscr{V}_M$ and, for all x and y in M,

$$P(\{\omega \in \Omega : d(T(\omega, x), T(\omega, y)) \leqslant k(\omega)d(x,y)\}) = 1,$$

where $k(.)$ is a non-negative random variable such that $k(\omega) < < 1$ a.s. If $k(\omega)$ is constant a.s., then T is called a uniform random contraction.

As an answer to a Bharucha-Reid problem (see [32, § 5, Problem d]), we give

THEOREM 4.1.12 [50] *Let $\{T_n(\omega)\}_{n \in \mathbb{N}}$ be a sequence of uniform random contraction mappings on the complete separable metric space (M, d) and let x_n be a fixed point of $T_n(\omega)$.*

$$\textit{If } T_n(\omega) \xrightarrow{u.P} T_0(\omega), \textit{ then } x_n \xrightarrow{P} x_0.$$

Proof. We recall that the sequence of random mappings $\{T_n(\omega)\}_{n \in \mathbb{N}}$ converges uniformly in probability to the random mapping T_0 (i.e., $T_n \xrightarrow{u.P} T_0$) if, for every $\varepsilon > 0$ and $\lambda \in (0,1)$, there exists an $n_0(\varepsilon, \lambda)$ such that

$$P(\{\omega \in \Omega : d(T_n(\omega, x), T_0(\omega, x)) < \varepsilon\}) > 1 - \lambda$$

for all $n \geqslant n_0(\varepsilon, \lambda)$ and all $x \in M$.

We consider the E-space $(\tilde{\mathscr{V}}_M, \mathscr{F}, \tau_m)$ over (M, d), where

$$F_{\tilde{x}, \tilde{y}}(t) = P(\{\omega \in \Omega : d(\tilde{x}(\omega), \tilde{y}(\omega)) < t\}), \quad t \in \mathbb{R}.$$

According to Slutsky's theorem, $\tilde{\mathscr{V}}_M$ is complete in probability, hence the Menger space $(\tilde{\mathscr{V}}_M, \mathscr{F}, \tau_m)$ is complete.

Since every $\tilde{T}_n(\omega)$ is a strict contraction (see Definition 3.1.11) on the E-space $(\tilde{\mathscr{V}}_M, \mathscr{F})$, from Theorem 3.1.38 it follows that each $\tilde{T}_n(\omega)$ has a fixed point, i.e., there exists an $\tilde{x}_n \in \tilde{\mathscr{V}}_M$ with $\tilde{T}_n\tilde{x}_n = \tilde{x}_n$. Condition $T_n(\omega) \xrightarrow{u.P} T_0(\omega)$ implies that $\tilde{T}_n$ converges to $\tilde{T}_0$ uniformly in the Menger space $(\tilde{\mathscr{V}}_M, \mathscr{F}, \tau_m)$ (see Sub-sec. 1.1.5). From Theorem 3.1, we conclude that $\{\tilde{x}_n\}_{n \in \mathbb{N}}$ converges to $\tilde{x}_0$ in the (ε, λ)-topology on $\tilde{\mathscr{V}}_M$, hence it converges in probability.

Using the above techniques, one can also obtain some interesting corollaries of other theorems, as is, for instance, the following corollary of Theorem 4.1.9 :

COROLLARY [199]. *Let $(X, \|\cdot\|)$ be a separable Banach space, let $(\Omega, \mathscr{K}, P)$ be an atomic probability measure space and let*

$T : \Omega \times X \to X$ be a random nonexpansive α-condensing operator such that there exists a bounded subset $B \subseteq X$ with

$$P(\{\omega \in \Omega : T(\omega, X) \subseteq B\}) = 1.$$

If, for each two elements $\tilde{x}, \tilde{y} \in \tilde{\mathscr{V}}_B$, there exists an $\varepsilon(\tilde{x}, \tilde{y}) > 0$ such that

$$P(\{\omega \in \Omega : \| T(\omega, \tilde{x}(\omega)) - T(\omega, \tilde{y}(\omega)) \| < \varepsilon(\tilde{x}, \tilde{y}) \leqslant \| \tilde{x}(\omega) - \tilde{y}(\omega) \|\}) > 0,$$

then there exists one and only one $\tilde{z} \in \tilde{\mathscr{V}}_B$ such that

$$T(\omega, \tilde{z}(\omega)) = \tilde{z}(\omega) \text{ a.s.}$$

As a corollary of Theorem 3.2.6, we give the following

COROLLARY. *Let A be a probabilistic bounded and closed subset of the RN-space $(\tilde{\mathscr{V}}_B, \mathscr{F}, \tau_m)$, let Λ be a metrizable topological space, let γ be a random monotone measure of noncompactness and let $T : A \times \Lambda \to A$ be a continuous mapping such that the following conditions are satisfied:*

(i) *For every $\tilde{x}, \tilde{y} \in A$ and every $\lambda \in \Lambda$,*

$$P(\{\omega \in \Omega : \| T(\tilde{x}, \lambda)(\omega) - T(\tilde{y}, \lambda)(\omega) \| \leqslant \| \tilde{x}(\omega) - \tilde{y}(\omega) \|\}) = 1;$$

(ii) *For every $\tilde{x}, \tilde{y} \in A$, there exists an $\varepsilon_{\tilde{x},\tilde{y}} > 0$ such that*

$$P(\{\omega \in \Omega : \| T(\tilde{x}, \lambda)(\omega) - T(\tilde{y}, \lambda)(\omega) \| < \varepsilon_{\tilde{x},\tilde{y}} \leqslant \| \tilde{x}(\omega) - \tilde{y}(\omega) \|\}) > 0;$$

(iii) *For $X = A$, condition* (ii) *of Theorem* 3.2.6 *is satisfied.*

Then there exists one and only one continuous mapping $f : \Lambda \to A$ such that

$$f(\lambda)(\omega) = T(f(\lambda), \lambda)(\omega) \text{ for every } \lambda \in \Lambda \text{ a.s.}$$

Now, let $E : \omega \to E(\omega)$ be a mapping from Ω into $CB(X)$ (the set of all nonempty, closed and bounded subsets of X). The mapping E is measurable if and only if for every closed subset $C \subset X$, $\{\omega \in \Omega : E(\omega) \cap C \neq \emptyset\} \in \mathscr{K}$. If $(\Omega, \mathscr{K}, P)$ is an atomic probability measure space, then E is constant a.s. on every atom.

Using the above corollary, we can prove the following

COROLLARY. *For every $\lambda \in \Lambda$, let $T(\lambda, \cdot, \cdot)$ be a random operator on X such that for every $\lambda \in \Lambda$ and every $\omega \in \Omega$, $T(\lambda, \omega, E(\omega)) \subseteq E(\omega)$ and $P(\{\omega \in \Omega : (\lambda, x) \to T(\lambda, \omega, x)$ is continuous on $\Lambda \times E(\omega)\}) = 1$. Suppose that the following conditions are satisfied:*

(i) *For every* $\lambda \in \Lambda$,

$P(\{\omega \in \Omega : \| T(\lambda, \omega, x_1) - T(\lambda, \omega, x_2) \| \leqslant \| x_1 - x_2 \|$, *for every* $x_1, x_2 \in E(\omega)\}) = 1$;

(ii) *For each two measurable selectors* x, y *of* E, *there exists an* $\varepsilon_{x,y} > 0$ *such that for every* $\lambda \in \Lambda$

$$P(\{\omega \in \Omega : \| T(\lambda, \omega, x(\omega)) - T(\lambda, \omega, y(\omega)) \| < \varepsilon_{x,y} \leqslant \leqslant \| x(\omega) - y(\omega) \| \}) > 0;$$

(iii) *For every* $\omega \in \Omega$ *and for every* $B \subseteq E(\omega)$ *and* $\lambda \in \Lambda$, *there exists a neighbourhood* $V(\lambda)$ *of* $\lambda \in \Lambda$ *so that for every compact* $C \subseteq V$, $\alpha(T(C, \omega, B)) \leqslant \alpha(B)$ *with strict inequality if* $\alpha(B) > 0$.

Then there exists one and only one continuous mapping $f : \Lambda \to \mathscr{S}$ ($\mathscr{S}$ = *the set of all measurable selectors with convergence in probability*) *such that*

$$f(\lambda)(\omega) = T(\lambda, \omega, f(\lambda)(\omega)) \quad a.s.$$

4.2. SOME RANDOM FIXED POINT THEOREMS FOR CONTINUOUS RANDOM OPERATORS

The aim of this section is to present several random fixed point theorems for abstract continuous random operators. After considering some results which generalize and unify the results already presented in the foregoing paragraph, we shall give two theorems which ensure the existence of a random fixed point under rather general conditions provided that the corresponding deterministic fixed point problem is solvable [314], [53].

We shall first recall a result given be G. Bocşan, based on the characterization of separable random operators in Banach spaces [303], [31] or in metric spaces [55].

THEOREM 4.2.1. *Let* $(\Omega, \mathscr{K}, P)$ *be a complete probability measure and let* E *be a nonempty convex, closed and bounded subset of a separable Banach space* X.

If T *is a random condensing operator on* X *and* $T(\omega, E) \subseteq \subseteq E$ *a.s., then there exists an* E*-valued variable* x *such that*

$$T(\omega, x(\omega)) = T(\omega)x(\omega) = x(\omega).$$

Proof. Let Ω_0 be the set of all $\omega \in \Omega$ for which $T(\omega,.)$ is condensing and $T(\omega,E) \subseteq E$.

We now state two well-known fixed point theorems.

THEOREM A [153, Th. 1]. *Let* (X, d) *be a metric space and let* $T : D \to X$ *be a densifying (i. e.,* α*-condensing) mapping defin-*

ed on a complete subset $D \subset X$. Then, any bounded sequence $\{x_n\}_{n \in \mathbb{N}}$, such that $d(x_n, Tx_n) \to 0$, is compact and all the limit points of $\{x_n\}_{n \in \mathbb{N}}$ are fixed for T.

THEOREM B [155, Th. 3]. *Let Q be a bounded, closed and convex subset of a Banach space X and let $T : Q \to Q$ be a densifying (i.e., α-condensing) mapping. Then T has at least one fixed point.*

If $\omega \in \Omega_0$, then, according to Theorem B, the set

$$F(\omega) = \{x \in E : T(\omega, x) = x\}$$

is nonempty. Since E is a bounded and closed subset of the Banach space X, then Theorem A implies that $F(\omega)$ is a compact subset of E. Thus, $F :(\Omega_0, \mathscr{K} \cap \Omega_0, P) \to K(E)$, where $(\Omega_0, \mathscr{K} \cap \Omega_0, P)$ is a complete probability measure space and E is the metric space induced by the norm of X.

Let $\mathscr{F}(E)$ be the countable class of all sets of the form $\{x : \rho_E(x, x_i) \leqslant r\}$, $\{x : \rho_E(x, x_i) \geqslant r\}$ and their finite intersection, where $\{x_i\}_{i \in \mathbb{N}}$ is dense in E and r are rational numbers. Then, for all $A \in \mathscr{F}(E)$, we have

$$(1)\quad \{\omega \in \Omega_0 : F(\omega) \cap A \neq \emptyset\} = \{\omega \in \Omega_0 : \inf_{x \in A} \| x - T(\omega)x \| = 0\}.$$

Indeed, if we denote

$$y(\omega) = \inf_{x \in A} \| x - T(\omega)x \|, \ \omega \in \Omega_0,$$

then, for every $\omega \in \Omega_0$ such that $y(\omega) = 0$, it follows that $F(\omega) \cap A = 0$. This assertion follows from the fact that $T(\omega) : A \to E$ is a condensing operator and A is a complete subset of E. According to Theorem B, there exists an $x \in A$ such that $T(\omega, x) = x$, i.e., $F(\omega) \cap A \neq \emptyset$. The converse inclusion is obvious and hence (1) holds. Thus,

$$(2)\quad \{\omega \in \Omega, F(\omega) \cap A \neq \emptyset\} \in \mathscr{K}$$

if y is a random variable.

To prove that $y(.) : \Omega_0 \to R$ is a random variable, we define

$$y(\omega, x) = \| x - T(\omega)x \|, \ \omega \in \Omega_0, \ x \in E.$$

Since T is a random condensing operator, it is a continuous random operator, and hence $y(\omega,.)$ is a continuous random operator on E with values in $[0, \infty)$. Then, by Theorem 2.1.3, $y(\omega,.)$ is separable and since E is a bounded set, it follows that $y(\omega, E) \subset [0, M]$ for a certain $M > 0$ and for all $\omega \in \Omega_0$. Thus, according to Lemma 2.2.1, there exists a negligible set $N \subset \Omega_0$ such that if $\omega \notin N$

$$(3)\quad y(\omega, x) \in \bigcap_{A:\, x \in A} \overline{y(\omega, A \cap S)}$$

for every $x \in E$, where S is a separant for $y(.,.)$ and $A \in \mathscr{F}(E)$.

If $\varepsilon > 0$ is given, then, since $y(\omega) = \inf_{x \in A} y(\omega, x)$, there exists an $x_0 \in A$ such that

$$y(\omega) \leqslant y(\omega, x_0) < y(\omega) + \frac{\varepsilon}{2},$$

and, using (3), one gets $y(\omega, x_0) \in \overline{y(\omega, A \cap S)}$. Hence, there exists an $x_1 \in A \cap S$ such that

$$|y(\omega, x_0) - y(\omega, x_1)| < \frac{\varepsilon}{2},$$

and hence

$$y(\omega, x_1) \leqslant |y(\omega, x_1) - y(\omega, x_0)| + y(\omega, x_0) < y(\omega) + \varepsilon.$$

Since ε is an arbitrary positive number, it follows that

$$y(\omega) = \inf_{x \in A \cap S} y(\omega, x), \quad \omega \notin N,$$

and since $P(N) = 0$, it follows that y is a random variable.

As E is a separable metric space, then, for all fixed closed sets $C \subseteq E$, there exists a net $\{A_n\}_{n \in \mathbb{N}} \subset \mathscr{F}(E)$ such that $C = \bigcap_{n \in \mathbb{N}} A_n$. Considering the properties of $\mathscr{F}(E)$, it follows that the net $\{A_n\}_{n \in \mathbb{N}}$ can be considered nonincreasing. Now we state that

$$(4) \quad \{\omega \in \Omega_0,\ F(\omega) \cap C \neq \emptyset\} = \bigcap_{n \in \mathbb{N}} \{\omega \in \Omega_0, F(\omega) \cap A_n \neq \emptyset\}.$$

Indeed, if ω is such that $B_n = F(\omega) \cap A_n = \emptyset$ for all $n \in \mathbb{N}$, then $\bigcap_{n \in \mathbb{N}} B_n \neq \emptyset$ and thus $F(\omega) \cap C \neq \emptyset$. Since the inclusion $\subseteq$ is obvious, equality (4) is proved. Using (2) and (4), we conclude that

$$\{\omega \in \Omega_0,\ F(\omega) \cap C \neq \emptyset\} \in \mathscr{K}$$

for every closed $C \subseteq E$.

From the selection theorem of Kuratowski-Ryll-Nardzewski (Th. 2.2.4), it follows that there exists a measurable function $x : \Omega_0 \to E$ with $x(\omega) \in F(\omega)$ for all $\omega \in \Omega_0$ and $T(\omega, x(\omega)) = x(\omega)$, $\omega \in \Omega_0$.

Remark. The above theorem gives a direct proof of Theorem 4.1.11 and the remark thereof remains still valid.

Since a random α-contraction is a random condensing mapping, the above result can be considered as a generalization of a well-known theorem due to Darbo [122].

A more general result applying to the case of general measurable spaces $(\Omega, \mathscr{K})$ is given by

THEOREM 4.2.2 [242]. *Let E be a separable closed convex subset of a Banach space E and let $T : \Omega \times E \to E$ be a random con-*

densing operator. Suppose that, for any $\omega \in \Omega$, $T(\omega, E)$ is bounded. Then there exists a random fixed point $x : \Omega \to E$ of $T(\omega)$.

Proof. Let $\{x_n\}_{n \in \mathbb{N}}$ be a dense subset of E. Define the mapping $F : \Omega \to 2^X$ by $F(\omega) = \{x \in E : T(\omega, x) = x\}$. Then, according to Theorem B above, $F(\omega) \in \mathscr{K}(E)$ for all $\omega \in \Omega$. Let us show that F is measurable. For any nonempty subset C of E, denote

$$L(C) = \bigcap_{n \in \mathbb{N}} \bigcup_{x_i \in C_n} \left\{ \omega \in \Omega : \| T(\omega, x_i) - x_i \| < \frac{2}{n} \right\},$$

where $C_n = \left\{ x \in E : d(x, C) < \dfrac{1}{n} \right\}$ and $d(x, C_n) = \inf \{\| x - y \|, y \in C\}$. Then $F^{-1}(C) = L(C)$. Indeed, it is easy to see that $F^{-1}(C) \subset L(C)$. Conversely, if $\omega \in L(C)$, then for each n, there exists an $x_{i(n)}$ such that $d(x_{i(n)}, C) < \dfrac{1}{n}$ and $\| T(\omega, x_{i(n)}) - x_{i(n)} \| < \dfrac{2}{n}$. The subset $B = \{x_{i(n)}, n \in \mathbb{N}\} \subset E$ (see Theorem A above). Thus, there exists a subsequence of B converging to some $y_0 \in E$. We have $y_0 \in C$ and $T(\omega, y_0) = y_0$, showing that $\omega \in F^{-1}(C)$. Hence, $F^{-1}(C) = L(C)$ and F is measurable. According to Kuratowski and Ryll-Nardzewski's theorem, there exists a measurable selector $x : \Omega \to E$ of F which is the desired random fixed point of $T(\omega)$.

As a corollary, one can obtain a random fixed point of Schauder type which generalizes Theorem 4.1.5.

COROLLARY. *Let E be a compact (or separable and closed) convex subset of a Banach space X and let $T : \Omega \times E \to E$ be a compact random operator. Then $T(\omega)$ has a random fixed point.*

A result similar to that in Theorem 4.2.2 is valid for a separable, bounded and complete metric space E if $T : \Omega \times E \to E$ is a condensing random operator with $F(\omega) \neq \emptyset$ for all $\omega \in \Omega$.

Two random fixed point theorems of Krasnosel'skiĭ type which generalize Theorem 4.1.7 are given below.

THEOREM 4.2.3 [242]. *Let E be a closed convex subset of a separable Banach space X, let $T : \Omega \times E \to X$ be a random operator such that for any $\omega \in \Omega$, $T(\omega)$ is a $k(\omega)$-random contraction and let $S : \Omega \times E \to X$ be a compact random operator (i. e., S is continuous and $S(\omega)(X)$ is precompact a.s.). Suppose that for any $\omega \in \Omega$, $T(\omega, x) + S(\omega, y) \in E$. Then $T + S$ has a random fixed point.*

Proof. For each fixed $y \in E$, define a mapping $R_y : \Omega \times E \to E$ as $R_y(\omega, x) = T(\omega, x) + S(\omega, y)$; then, since E is separable, R_y is a random operator such that for each $\omega \in \Omega$, $R_y(\omega,.)$ is a $k(\omega)$-contraction. Thus, according to Theorem 4.1.2, there exists a unique random fixed point ξ_y of R_y. If we define the random operator $U : \Omega \times E \to E$ by $U(\omega, x) = \xi_x(\omega)$, then U is continuous and, for every $x,y \in E$, one has

$$\|S(\omega, x) - S(\omega, y)\| = \|\xi_x(\omega) - T(\omega, \xi_x(\omega)) - \xi_y(\omega) +$$

$$+ T(\omega, \xi_y(\omega))\| \geqslant \|\xi_x(\omega) - \xi_y(\omega)\| - \|T(\omega, \xi_x(\omega) -$$

$$- T(\omega, \xi_y(\omega))\| \geqslant \|\xi_x(\omega) - \xi_y(\omega))\| - k(\omega)\|\xi_x(\omega) - \xi_y(\omega)\| =$$

$$= (1 - k(\omega))\|U(\omega, x) - U(\omega, y)\|.$$

Since S is compact, it follows from the above inequality that U is compact. From the above corollary it follows that there exists a random fixed point ξ of U which, evidently, is a random fixed point of $T + S$.

In order to give another random fixed point theorem of Krasnosel'skiĭ type, we recall that a Banach space X is said to satisfy *Opial's condition* if the following holds : if $\{x_n\}_{n \in \mathbb{N}}$ converges weakly to x_0, and $x \neq x_0$, then $\lim \inf\|x_n - x\| > \lim \inf \|x_n - x_0\|$. Banach spaces satisfying Opial's condition include Hilbert spaces and $l^p(1 \leqslant p < \infty)$ spaces. A mapping $f : E \subset X \to X$ is called *completely continuous* if $\{x_n\}_{n \in \mathbb{N}}$ converges weakly to x_0 : then $\{f(x_n)\}_{n \in \mathbb{N}}$ converges to $f(x_0)$; f is called *demiclosed* if $\{x_n\}_{n \in \mathbb{N}} \subset E$ is such that $\{x_n\}_{n \in \mathbb{N}}$ converges weakly to x_0 and $\{f(x_n)\}_{n \in \mathbb{N}}$ converges to $y_0 \in X$; then $f(x_0) = y_0$.

THEOREM 4.2.4 [242]. *Let E be a weakly compact convex subset of a separable Banach space X, let $T : \Omega \times E \to X$ be a nonexpansive random operator (i.e., for all $x, y \in E$ and $\omega \in \Omega$, $\|T(\omega, x) - T(\omega, y)\| \leqslant \|x - y\|$), and let $S : \Omega \times E \to X$ be a completely continuous random operator. Suppose that X is either uniformly convex or satisfying Opial's condition and that, for any $\omega \in \Omega$, $T(\omega, x) + S(\omega, x) \in E$ if $x \in E$. Then, there exists a random fixed point of $T + S$.*

Proof. Let $y \in E$ be a fixed element of E and let $\{k_n\}_{n \in \mathbb{N}}$ be a sequence of real numbers such that $k_n \in (0, 1)$ and $k_n \to 0$ as $n \to \infty$. For each $n \in \mathbb{N}$, define a mapping $R_n : \Omega \times E \to E$ by

$$R_n(\omega, x) = k_n y + (1 - k_n)(T(\omega, x) + S(\omega, x));$$

then R_n is a $(1 - k_n)$-α-contraction random operator. According to Theorem 4.2.2, there exists a random fixed point ξ_n of R_n. For each $n \in N$, define $F_n : \Omega \to WK(E)$ as

$$F_n(\omega) = w\text{-cls } \{\xi_i(\omega), i \geqslant n\},$$

where w-cls(C) is the weak closure of C. Let $F : \Omega \to WK(E)$ be a mapping defined by

$$F(\omega) = \bigcap_{n \in \mathbb{N}} F_n(\omega);$$

then, since the weak topology on E is a metric topology (see [127, p. 14–16]), F is w-measurable (Theorem 2.2.2). Thus, according to Kuratowski and Ryll-Nardzewski's theorem (Theorem 2.2.4), there exists a w-measurable selector ξ of F. For any x^* of the dual X^*, $x^*(\xi(.))$ is measurable as a random variable on Ω. Since X is separable, ξ is measurable (cf. [31, p. 16]). Moreover, ξ is a random fixed point of $T + S$. Indeed, let $\omega \in \Omega$ be fixed, and then there exists $\{\xi_{n_k}\}_{k \in \mathbb{N}} \subset \{\xi_n\}_{n \in \mathbb{N}}$ such that $\xi_{n_k}(\omega)$ converges weakly to $\xi(\omega)$. Hence, $\{S(\omega, \xi_{n_k}(\omega)\}_{n \in \mathbb{N}}$ converges to $S(\omega, \xi(\omega))$. On the other hand, we have

$$\xi_{n_k}(\omega) - T(\omega, \xi_{n_k}(\omega)) = k_{n_k}(y - T(\omega, \xi_{n_k}(\omega)) + \\ + S(\omega, \xi_{n_k}(\omega)) + S(\omega, \xi_{n_k}(\omega))$$

and

$$\xi_{n_k}(\omega) - T(\omega, \xi_{n_k}(\omega)) \to S(\omega, \xi(\omega)) \text{ as } k \to \infty.$$

Since $T(\omega.,)$ is nonexpansive , $I - T(\omega,.)$ is demiclosed by Browder [72] if X is uniformly convex or by Opial [324] if X satisfies Opial's condition, where $I(x) = x$ for every $x \in X$. Therefore, it follows that $\xi(\omega)$ is a random fixed point of $T + S$.

In the Hilbert-space case, S. Itoh gives more precise results for nonexpansive mappings. Thus,

THEOREM 4.2.5 [242]. *Let E be a bounded closed convex subset of a separable Hilbert space H and let $T : \Omega \times E \to E$ be a nonexpansive random operator. Then, for each $\omega \in \Omega$, $M(\omega) = \{x \in E : T(\omega, x) = x\}$ is nonempty, closed and convex and the mapping $P : \Omega \times E \to E$, defined by $P(\omega, x) = P_{M(\omega)}(x)$, is a nonexpansive random operator. Further, for any $x \in E$, $P(.,x)$ is a random fixed point of T.*

A large class of generalized probabilistic contractions was introduced and studied by S. Chang [96]. Thus, one gets results that contain Theorem 4.1.1 and the main results due to Hanš [205], [207] and Špaček [413] as particular cases.

Let $(\Omega, \mathscr{K}, P)$ be a complete probability measure space and let (X, d) be a separable complete metric space. Let $O_T(x(\omega), 0, \infty)$ denote the random orbit of $x(\omega) \in X$ under T and for any $i, j \in \mathbb{N} \cup \{0\}$, $j \geqslant i$, we put

$$O_T(x(\omega), i, j) = \{x_i(\omega), x_{i+1}(\omega), \ldots, x_j(\omega)\} = \\ = \{T(\omega, x_{i-1}(\omega)), \ldots, T(\omega, x_{j-1}(\omega))\},$$

and

$$O_T(x(\omega), y(\omega), i, j) = \{x_i(\omega), y_i(\omega), x_{i+1}(\omega), y_{i+1}(\omega), \ldots,$$
$$\ldots, x_j(\omega), y_j(\omega)\},$$

where $x_0(\omega) = x(\omega)$, $y_0(\omega) = y(\omega)$.

For any subset $A \subset X$, let $\delta(A)$ be the diameter of A, i.e.

$$\delta(A) = \sup_{x,y \in A} d(x, y).$$

Let $\Phi : \Omega \times [0, \infty) \to [0, \infty)$ be a random function satisfying the following conditions:

(Φ_1) for any fixed $\omega \in \Omega$, $\Phi(\omega, t)$ is nondecreasing and right-continuous with respect to t;

(Φ_2) $P(\{\omega \in \Omega : \lim_{t\to\infty} (t - \Phi(\omega, t)) = \infty\}) = P(F) = 1$;

(Φ_3) for any $t > 0$,

$$P(\{\omega \in \Omega : \lim_{n\to\infty} \Phi^n(\omega, t) = 0\}) = P(G_t) = 1,$$

where Φ^n (ω, t) is the n-th iterate of $\Phi(\omega, t)$.

LEMMA 4.2.1. *Let the random functions $\Phi(\omega, t)$ satisfy conditions (Φ_1) and (Φ_3). Then,*

(1) *for any sequence $\{t_n(\omega)\}_{n\in\mathbb{N}}$ satisfying the conditions*

(a) $t_n(\omega) : \Omega \to [0, \infty)$, $n \in \mathbb{N}$,

(b) $t_n(\omega) \leqslant t_{n-1}(\omega)$, $n \geqslant 2$,

(c) $P(\{\omega \in \Omega : t_{n+1}(\omega) \leqslant \Phi(\omega, t_n(\omega))\}) = P(W_n) = 1, \quad n \in \mathbb{N}$,

we have

$$\lim_{n\to\infty} t_n(\omega) = 0 \ a.s.;$$

(2) *if $t(\omega) : \Omega \to [0, \infty)$ satisfies*

$$t(\omega) \leqslant \Phi(\omega, t(\omega)) \ a.s.,$$

then we have $t(\omega) = 0$ a.s., and

(3) *for any $t > 0$, we have $\Phi(\omega, t) < t$ a.s.*

Proof. (1) From the right continuity of $\Phi(\omega, t)$ with respect to t and from condition (c), one gets

$$W = \left(\bigcap_{n=1}^{\infty} W_n\right) \cap \left(\bigcap_{t>0}^{\infty} G_t\right) \in \mathscr{K} \text{ and } P(W) = 1,$$

where the sets W_n, $n \in \mathbb{N}$ and G_t, respectively, are defined by conditions (c) and (Φ_3).

Let $\bar{\omega} \in W$ be arbitrary but fixed; then, from (c), we have

$$0 \leqslant t_{n+1}(\bar{\omega}) \leqslant \Phi(\bar{\omega},\ t_n(\bar{\omega})) \leqslant \Phi^2(\bar{\omega},\ t_{n-1}(\bar{\omega})) \leqslant \ldots$$

$$\ldots \leqslant \Phi^n(\bar{\omega},\ t_1(\bar{\omega})).$$

From (Φ_3), it follows that

$$0 \leqslant \lim_{n\to\infty} t_{n+1}(\bar{\omega}) \leqslant \lim_{n\to\infty} \Phi^n(\bar{\omega},\ t_1(\bar{\omega})) = 0,$$

and hence

$$\lim_{n\to\infty} t_n(\omega) = 0 \text{ a.s.}$$

(2) If we put $t_n(\omega) = T(\omega)$, $n \in \mathbb{N}$ in (1), it follows that $t(\omega) = 0$ a.s.

(3) If there exists a $t_0 \in (0, \infty)$ such that $t_0 \leqslant \Phi(\omega, t_0)$ a.s., it follows from (2) that $t_0 = 0$. This contradiction proves (3).

THEOREM 4.2.6 [96]. *Let $T : \Omega \times X \to X$ be a continuous random operator. Suppose that there exists a positive integer m such that, for any $x \in X$, the following holds:*

$$P(\{\omega \in \Omega :\ \delta(O_T(x,\ m,\ \infty)) \leqslant \Phi(\omega,\ \delta(O_T(x, 0, \infty))\}) =$$

$$= P(E_x) = 1,$$

where $\Phi : \Omega \times [0, \infty) \to [0, \infty)$ is a random function satisfying conditions (Φ_1) — (Φ_3) above.

Then, for any random variable $x_0 : \Omega \to X$, the sequence of X-valued random variables $\{x_n(\omega)\}_{n\in\mathbb{N}\cup\{0\}}$, defined by

$$x_n(\omega) = T(\omega,\ x_{n-1}(\omega)),\ n \in \mathbb{N},$$

converges almost surely to a random fixed point of T.

Proof. For any X-valued random variable $x_0(\omega)$, it follows from the continuity of T that the random orbit of $x_0(\omega)$ under T, i.e. $O_T(x_0(\omega),\ 0,\ \infty) = \{x_n\}_{n\in\mathbb{N}\cup\{0\}}$, is a sequence of X-valued random variables. From the separability of X, the continuity of $T(\omega,.)$ and the right continuity of $\Phi(\omega,.)$, it follows that:

$$H = \{(\bigcap_{x\in X} E_x) \cap F \cap (\bigcap_{t>0} G_t)\} \in \mathscr{K} \text{ and } P(H) = 1.$$

For the random variable $x_0(\omega) \in X$, the assumption in this theorem implies that

$$\delta(O_T(x_0(\omega),\ m,\ \infty)) \leqslant \Phi(\omega,\ \delta(O_T(x_0(\omega),\ 0,\ \infty))),\ \forall\omega \in H.$$

It also follows that

$$(*)\ \delta(O_T(x_0(\omega), nm, \infty)) = \delta(O_T(x_{(n-1)m}(\omega), m, \infty)) \leqslant$$
$$\leqslant \Phi(\omega, \delta(O_T(x_{(n-1)m}(\omega), 0, \infty))) =$$
$$= \Phi(\omega, \delta(O_T(x_0(\omega), (n-1)m, \infty))), \omega \in H, n \in \mathbb{N}.$$

However,

$$\delta(O_T(x_0(\omega), (n-1), m, \infty)) =$$
$$= \max\{\delta(O_T(x_0(\omega), (n-1), m, nm)),$$
$$\delta(O_T(x_0(\omega), nm+1, \infty)), \sup_{\substack{(n-1)m \leqslant r \leqslant nm \\ s \geqslant nm+1}} d(x_r(\omega), x_s(\omega))\} \leqslant$$
$$\leqslant \max\{\delta(O_T(x_0(\omega), (n-1)m, nm)), \delta(O_T(x_0(\omega), nm, \infty)),$$
$$\sup_{(n-1)n \leqslant r \leqslant nm} d(x_r(\omega), x_{nm}(\omega)) + \sup_{s \geqslant nm+1} d(x_{nm}(\omega), x_s)\} \leqslant$$
$$\leqslant \delta(O_T(x_0(\omega), (n-1)m, nm)) +$$
$$+ \delta(O_T(x_0(\omega), nm, \infty)).$$

Hence, we obtain

$$\delta(O_T(x_0(\omega), (n-1)m, \infty)) \leqslant \delta(O_T(x_0(\omega), (n-1)m, nm)) +$$
$$+ \Phi(\omega, \delta(O_T(x_0(\omega), (n-1)m, \infty))), \omega \in H$$

and, for $n = 1$, one has

$$(**)\ \delta(O_T(x_0(\omega), 0, \infty)) \leqslant \delta(O_T(x_0(\omega), 0, m)) +$$
$$+ \Phi(\omega, \delta(O_T(x_0(\omega), 0, \infty))), \omega \in H.$$

On the other hand, for every $\omega \in H$, we have

$$\delta(O_T(x_0(\omega), 0, \infty)) < \infty.$$

Indeed, suppose that there exists an $\overline{\omega} \in H$ such that

$$\delta(O_T(x_0(\overline{\omega}), 0, \infty)) = \infty$$

and denote

$$A_n(\overline{\omega}) = \delta(O_T(x_0(\overline{\omega}), 0, n)), n \in \mathbb{N}.$$

Then $\{A_n(\overline{\omega})\}_{n \in \mathbb{N}}$ is an increasing sequence of positive numbers and

$$\lim_{n \to \infty} A_n(\overline{\omega}) = \delta(O_T(x_0(\overline{\omega}), 0, \infty)) = \infty.$$

On the other hand, from conditions (Φ_2) and $(**)$, it follows that

$$\infty = \lim_{n\to\infty} (A_n(\bar{\omega}) - \Phi(\bar{\omega}, A_n(\bar{\omega}))) \leqslant \delta(O_T(x_0(\bar{\omega}), 0, m)) < \infty.$$

Putting

$$t_n(\omega) = \begin{cases} \delta(O_T(x_0(\omega), \ nm, \ \infty)) & \text{for } \omega \in H \\ 0 & \text{for } \omega \in \Omega - H, \ n \in \mathbb{N}, \end{cases}$$

from $(*)$ it follows that $\{t_n(\omega)\}_{n\in\mathbb{N}}$ satisfies all conditions of Lemma (1) above. Therefore,

$$\lim_{n\to\infty} \delta(O_T(x_0(\omega), \ nm, \ \infty)) = 0 \ \text{ a.s.}$$

This implies that $\{x_n(\omega)\}_{n\in\mathbb{N}}$ is a Cauchy sequence of X-valued random variables. Then there exists an X-valued random variable $x_*(\omega) = \lim_{n\to\infty} x_n(\omega)$ a.s., and from the continuity of T it follows that $x_*(\omega) = T(\omega, \ x_*(\omega))$ a.s.

PROPOSITION 4.2.1. *Let $\Phi(\omega, t)$ be as in Theorem 4.2.6 and let $T : \Omega \times X \to X$ be a continuous random operator. Suppose that there exists an $m \in \mathbb{N}$ such that, for any $s \in \mathbb{N}$ and any $x \in X$, the following holds*:

$$P(\{\omega \in \Omega : d(T^m(\omega, x), \ T^{m+s}(\omega, x)) \leqslant$$
$$\leqslant \Phi(\omega, \delta(O_T(x_0(\omega), \ 0, \ \infty)))\}) = P(E_{xs}) = 1.$$

Then the conclusion of Theorem 4.2.6 still holds.

Proof. From the assumptions on X, T and Φ, it is easy to see that

$$E = \bigcap_{x\in X} \bigcap_{s\in\mathbb{N}} E_{xs} \in \mathscr{K} \text{ and } P(E) = 1.$$

Also, for any pair r, $s \in \mathbb{N}$, $r \leqslant s$, and any $x \in X$, it follows that for any $\omega \in E$

$$d(T^{m+r}(\omega, x), \ T^{m+s}(\omega, \ x)) = d(T^m(\omega, \ T^r(\omega, x)),$$
$$T^{m+s-r}(\omega, \ T^r(\omega, \ x))) \leqslant \Phi(\omega, \delta(O_T(T^r(\omega, x), 0, \infty))) =$$
$$= \Phi(\omega, \delta(O_T(x, r, \infty))) \leqslant \Phi(\omega, \delta(O_T(x, 0, \infty))).$$

Consequently,

$$\delta(O_T(x, m, \infty)) = \sup_{r,s\geqslant 0} d(T^{m+r}(\omega, x), \ T^{m+s}(\omega, \ x)) \leqslant$$
$$\leqslant \Phi(\omega, \delta(O_T(x, 0, \infty))),$$

i.e. the condition of Theorem 4.2.6 holds.

COROLLARY. *Let* $T : \Omega \times X \to X$ *be a continuous random operator. Suppose that there exists an* $m \in \mathbb{N}$ *such that, for any* $x \in X$, *one of the following conditions holds:*

(I) $P(\{\omega \in \Omega : \delta(O_T(x, m, \infty)) \leqslant k(\omega)\delta(O_T(x, 0, \infty))\}) = 1$;

(II) *for any* $s \in \mathbb{N}$, $P(\{\omega \in \Omega : d(T^m(\omega, x), T^{m+s}(\omega, x)) \leqslant$

$$\leqslant k(\omega)\delta(O_T(x, 0, \infty))\}) = 1,$$

where $k : \Omega \to [0, \infty)$ *is a random variable satisfying* $P(\{\omega \in \Omega : 0 < k(\omega) < 1\}) = 1$.

Then the conclusion of Theorem 4.2.6 *holds.*

Proof. Taking $\Phi(\omega, t) = k(\omega)\, t$, $t \geqslant 0$, it as easy to see that conditions $(\Phi_1) - (\Phi_3)$ are satisfied.

THEOREM 4.2.7 [96]. *Let* $T : \Omega \times X \to X$ *be a continuous random operator. Let* Φ *be as in Theorem* 4.2.6. *Suppose that there exist* $r, s \in \mathbb{N}$ *such that, for any* $x, y \in X$,

$$P(\{\omega \in \Omega : d(T^r(\omega, x), T^s(\omega, y)) \leqslant$$

$$\leqslant \Phi(\omega, \delta(O_T(x, y, 0, \infty)))\}) = P(E_{xy}) = 1.$$

Then T *has a unique random fixed point* $x_*(\omega)$, *and, for any* X*-valued random variable* $x_0(\omega)$, *the sequence*

$$x_n(\omega) = T(\omega, x_{n-1}(\omega)), \quad n \in \mathbb{N}$$

converges almost surely to $x_*(\omega)$.

Proof. Let $r \geqslant s$ and $G = \bigcap_{x \in X} \bigcap_{y \in X} E_{xy}$; then $G \in \mathscr{K}$ and $P(G) = 1$. For any $x \in X$, and $m \in \mathbb{N}$ and any $\omega \in G$, let us denote $y = T^{r-s+m}(\omega, x)$. Then

$$d(T^r(\omega, x), T^s(\omega, T^{r-s+m}(\omega, x))) = d(T^r(\omega, x), T^{r+m}(\omega, x)) \leqslant$$

$$\leqslant \Phi(\omega, \delta(O_T(x, T^{r-s+m}(\omega, x), 0, \infty))) =$$

$$= \Phi(\omega, \delta(O_T(x, 0, \infty))).$$

According to Theorem 4.2.6, for any random variable $x_0(\omega)$, the sequence

$$x_n(\omega) = T(\omega, x_{n-1}(\omega)), \quad n \in \mathbb{N}$$

converges a.s. to a random fixed point of T.

To prove the uniqueness of the random fixed point of T, let us consider two such points $x_*(\omega)$, $\bar{x}_*(\omega)$. Then

$$d(x_*(\omega), \bar{x}_*(\omega)) = d(T^r(\omega, x_*(\omega)), T^s(\omega, \bar{x}_*(\omega)) \leqslant$$

$$\leqslant \Phi(\omega, \delta(O_T(x_*(\omega), \bar{x}_*(\omega), 0, \infty))) =$$

$$= \Phi(\omega, d(x_*(\omega), \bar{x}_*(\omega))) \quad \text{a.s.}$$

Using Lemma 4.2.1, (2), we get $d(x_*(\omega), \bar{x}_*(\omega)) = 0$ a.s., i.e., $x_*(\omega) = \bar{x}_*(\omega)$ a.s.

COROLLARY 1. *Let $T: \Omega \times X \to X$ be a continuous random operator. Suppose that there exist $r, s \in \mathbb{N}$ such that for any pair $x, y \in X$*

$$P(\{\omega \in \Omega : d(T^r(\omega, x), T^s(\omega, y)) \leqslant$$

$$\leqslant k(\omega)\, \delta(O_T(x, y, 0, \infty))\}) = 1,$$

where $k(\omega): \Omega \to [0, \infty)$ is a random variable such that $P(\{\omega \in \Omega : 0 < k(\omega) < 1\}) = 1$.

Then the conclusion of Theorem 4.2.7 holds.

COROLLARY 2. *Let $T: \Omega \times X \to X$ be a continuous random operator. If there exist $r, s \in \mathbb{N}$ such that, for any pair $x, y \in X$,*

$$P(\{\omega \in \Omega : d(T^r(\omega, x), T^s(\omega, y)) \leqslant$$

$$\leqslant \Phi(\omega, \max\{d(x, y), d(x, T^r(\omega, x)),$$

$$d(y, T^s(\omega, y)), d(x, T^s(\omega, y)), d(y, T^s(\omega, x))\}\})\}) = 1,$$

then the conclusion of Theorem 4.2.7 holds.

PROPOSITION 4.2.2. *Let T be a continuous random operator satisfying the following conditions:*

$$P\left(\bigcup_{r=1}^{\infty} \bigcup_{s=1}^{\infty} \bigcap_{x \in X} \bigcap_{y \in X} \{\omega \in \Omega : d(T^r(\omega, x), T^s(\omega, y)) \leqslant \right.$$

$$\left. \leqslant \Phi(\omega, \delta(O_T(x, y, 0, \infty)))\}\right) = P(Q) = 1,$$

where $\Phi(\omega, t)$ is as in Theorem 4.2.6.

Then the conclusion of Theorem 4.2.7 holds.

THEOREM 4.2.8 [94]. *Let $T: \Omega \times X \to X$ be a continuous random operator, let $S_n: \Omega \times X \to X$ be a sequence of continuous operators which commutes with T and satisfies the following conditions:*

(I) *for every $S_n(\omega, x)$, $n \in \mathbb{N}$ and any X-valued random variable $y(\omega)$, there exists an X-valued random variable $z_n(\omega)$ such that*

$$P(\{\omega \in \Omega : S_n(\omega, y(\omega)) = T(\omega, z_n(\omega))\}) = 1;$$

(II) *there exists a sequence* $\{m_i\}_{i\in\mathbb{N}} \subset \mathbb{N}$ *such that, for any* $r, s \in \mathbb{N}$ *and* $x, y \in X$, *the following holds:*

$$P(\{\omega \in \Omega : d(S_i^{m_i}(\omega, x), S_j^{m_j}(\omega, y)) \leqslant$$

$$\leqslant \Phi(\omega, \max\{d(T(\omega, x), T(\omega, y)),$$

$$d(T(\omega, x), S_i^{m_i}(\omega, x)), d(T(\omega, y), S_j^{m_j}(\omega, y)),$$

$$d(T(\omega, y), S_i^{m_i}(\omega, x)), d(T(\omega, x), S_j^{m_j}(\omega, y))\})\}) = 1.$$

Then T *and* S_i, $i \in \mathbb{N}$, *have a unique common random fixed point.*

We shall conclude this paragraph by giving two general results due to A. Nowak [314] and G. Bocşan [53]. These results furnish a uniform approach to probabilistic analogues of fixed point theorems. They ensure the existence of a random fixed point of a continuous random operator under rather general conditions provided that the corresponding deterministic fixed-point problem is solvable.

THEOREM 4.2.9 [315, Th. 1]. *Let* $(\Omega, \mathscr{K}, P)$ *be a complete probability space and let* X *be an analytic subset of a Polish space. Let* $T : \Omega \times X \to X$ *be a continuous random operator with*

$$F(\omega) = \{x \in X : T(\omega, x) = x\} \neq \emptyset.$$

Then there exists random fixed point for T, *i.e., an* X-*valued random variable* $x(a)$ *such that* $T(\omega, x(\omega)) = x(\omega)$ *a.s.*

Proof. Let $A = \{\omega \in \Omega : F(\omega) \neq \emptyset\}$ and let $T(\omega.,)$ be continuous. From the hypotheses, $A \in \mathscr{K}$ and $P(A) = 1$.

In order to prove that the multivalued mapping F restricted to A has a measurable selector, we examine the graph of F

$$\mathrm{Gr}(F) = \{(\omega, x) \in A \times X : x \in F(\omega)\} =$$

$$= \{(\omega, x) \in A \times X : d(x, T(\omega, x)) = 0\},$$

where d is the metric of X. The function $x : A \times X \to R$, defined by $u(\omega, x) = d(x, T(\omega, x))$, is measurable in ω and continuous in x. Hence, u is measurable with respect to the product σ-field $(A \cap \mathscr{K}) \times \mathscr{B}_X$ which implies that $\mathrm{Gr}(F) \in (A \cap \mathscr{K}) \times \mathscr{B}_X$.

From the selection theorem of Sainte-Beuve ([373, Th. 3]), it follows that there exists a measurable function $x : A \to X$ such that $x(\omega) \in F(\omega)$ for all $\omega \in A$. For $\omega \in \Omega - A$, we put $x(\omega) = x_0$, where x_0 is a point of X. Then, $x(\omega)$ is the desired random fixed point of T.

Remark. Using Theorem 4.2.9, one can obtain the probabilitsic analogue to any determinstic fixed point theorem. As a first example, we recall that A. Nowak gave a random version to Schauder's Theorem (i.e., Theorem 4.1.5). Also, Theorem 4.2.9 is very close to the result of Kannan and Salehi (Theorem 4.1.6).

We present now a result given in ref. [57] regarding a certain class of contractions. Let us introduce

DEFINITION 4.2.1. A mapping $T:(X, d) \to (X, d)$ is called a *φ-generalized contraction* if

$$d(Tx, Ty) \leqslant \varphi\left(\max\left\{d(x, y),\ d(x, Tx),\ d(y, Ty),\ \frac{1}{2}[d(x, Ty) + d(y, Tx)]\right\}\right)$$

for all $x, y \in X$ and for φ a nondecreasing upper semicontinuous function from the right on $[0, \infty)$ and such that $\varphi(t) < t$ for $t > 0$.

Next, we present a generalized probabilistic version of the Boyd and Wong fixed point theorem.

THEOREM 4.2.10 [57]. *A φ-generalized contraction T on the complete metric space (X, d) has a unique fixed point x_0 and $T^n x \to x_0$ for each $x \in X$.*

A probabilistic analogue is as follows:

DEFINITION 4.2.2. A random operator $T:\Omega \times X \to X$ on a Banach space $(X, \|\cdot\|)$ is said to be a random *φ-generalized contraction operator* if there exists a random continuous function $\varphi(\omega, t):\Omega \times [0, \infty) \to [0, \infty)$ satisfying the conditions: $\varphi(\omega, t) < t$ a.s. for $t > 0$, $\varphi(\omega, 0) = 0$ a.s., where $\varphi(\omega, t)$ is nondecreasing a.s. and such that

$$\|T(\omega)x - T(\omega)y\| \leqslant \varphi(\omega, \max\{\|x - y\|, \|x - T(\omega)y\|, \|y - T(\omega)y\|, \frac{1}{2}[\|x - T(\omega)y\| + \|y - T(\omega)x\|]\}) \text{ a.s.}$$

THEOREM 4.2.11. *Let $T:\Omega \times X \to X$ be a continuous random φ-generalized contraction from a separable Banach space X into itself. Then there exists a unique random fixed point $x_0(\omega)$ for $T(\omega)$. Moreover, if for any $\omega \in \Omega$, $x \in X$ and $n \in \mathbb{N}$, we put $T^1(\omega)x = T(\omega)x$ and $T^{n+1}(\omega)x = T(\omega)(T^n(\omega)x)$, then all successive approximations $\{T^n(\omega)\}_{n \in \mathbb{N}}$ converge to $x_0(\omega)$ a.s. as $n \to \infty$.*

Proof. Let $A = \{\omega \in \Omega : T(\omega)$ is a continuous random φ-generalized contraction$\}$. Then $A \in \mathscr{K}$ and $P(A) = 1$. Let the mapping $x_0 : \Omega \to X$ be defined as follows: for any $\omega \in A$, $x_0(\omega)$ is equal to the unique fixed point of $T(\omega)$ from Theorem 4.2.10, and for every $\omega \in A$ put $x_0(\omega) = 0 \in X$. Then, $T(\omega) x_0(\omega) = x_0(\omega)$ a.s.

In order to establish the measurability of the fixed point $x_0(\omega)$, we show first that all successive approximations $\{T^n(\omega)x\}$, where $x \in X$, converge to $x_0(\omega)$ a.s.

Let us consider the sequence

$$c_n(\omega) = \| T^n(\omega)x - T^{n-1}(\omega)x \|,$$

and from

$$c_{n+1}(\omega) \leqslant \varphi(\omega, \ M_{n-1}(\omega)) \quad \text{a.s.} \leqslant \varphi(\omega, c_n(\omega)) \quad \text{a.s.} \leqslant c_n(\omega) \quad \text{a.s.}$$

where

$$M_{n-1}(\omega) = \max\{\| T^{n-1}(\omega)x - T^n(\omega)x \|, \| T^n(\omega)x - T^{n+1}(\omega)x,$$
$$\frac{1}{2} \| T^{n-1}(\omega)x - T^{n+1}x \|\},$$

it follows that $c_n(\omega)$ is a decreasing sequence converging to zero.

Now, let us show that, for every $x \in X$, $\{T^n(\omega)x\}$ is a Cauchy sequence. If it were not, then, for all positive integer r, there exist $\varepsilon > 0$ and $m(r)$ and $n(r)$ with $m(r) > n(r) \geqslant r$ such that

$$(*) \ d_r(\omega) = \| T^{m(r)}(\omega) - T^{n(r)}(\omega)x \| \geqslant \varepsilon \text{ a.s.} \quad \text{for } r \in \mathbb{N}.$$

Indeed, as in the deterministic case, we first assume that $\| T^{m(r)-1}(\omega)x - T^{n(r)}(\omega)x \| < \varepsilon$ a.s.

By choosing $m(r)$ to be the smallest number exceeding $n(r)$ for which $(*)$ holds, we conclude that $d_r(\omega) \to \varepsilon_+$ a.s. for $r \to \infty$.

On the other hand, we have

$$\varepsilon_+ \leqslant \lim_{r \to \infty} d_r(\omega) \quad \text{a.s.} \leqslant \varphi(\omega, \ \varepsilon_+) < \varepsilon_+ \quad \text{a.s.}$$

which is a contradiction for $\varepsilon > 0$. This implies that $\{T^n(\omega)x\}_{n \in \mathbb{N}}$, $x \in X$, is a Cauchy sequence. Let $x_0(\omega)$ be the limit of this sequence. Then $x_0(\omega)$ is an X-valued random variable (see Chap. 2).

In order to show that $x_0(\omega)$ is a fixed point of $T(\omega)$, we observe that, as in the deterministic case (see [57]), one has

$$\| T(\omega)x_0(\omega) - x_0(\omega) \| \leqslant \| T(\omega)x_0(\omega) - T^n(\omega)x \| +$$
$$+ \| T^n(\omega)x - x_0(\omega) \| \text{ a.s.},$$

where

$$\| T(\omega)x_0(\omega) - T^n(\omega)x \| \leqslant \varphi(\omega, M_{n-1}(\omega)) \quad \text{a.s.},$$

which implies that

$$\| T(\omega)x_0(\omega) - T^n(\omega)x \| \to 0 \quad \text{a.s.} \quad \text{for } n \to \infty.$$

Hence, $T(\omega)x_0(\omega) = x_0(\omega)$ a.s. and the theorem is proven.

Remark. Another argument for showing the existence of the random fixed point for a random φ-generalized contraction operator can be given as a corollary of Theorem 4.2.9.

In the case $\max\Big\{\| x - y \|, \ \| x - T(\omega)y \|, \ \| y - T(\omega)y \|,$

$$\frac{1}{2}[\| x - T(\omega)y \| + \| y - T(\omega)x \|]\Big\} = \| x - y \| \quad \text{a.s.},$$

we obtain Theorem 3 due to A. Nowak [314]. The same result was reached by Lee and Padgett who used the successive approximation method [276] to obtain a probabilistic version of a theorem of Boyd and Wong.

A similar result to Nowak's Theorem 4.2.9 for the existence of a random fixed point in a closed random subset $E \subset X$ was obtained by G. Bocşan [53, Th. 1] as follows:

THEOREM 4.2.12. *Let X be a complete separable metric space and let T be a continuous random operator on X.*

If T has a fixed point in the wide sense in the closed random subset E of X, then it has a random fixed point in E.

Moreover, the multivalued mapping $F : \Omega \to 2^X$, defined by

$$F(\omega) = \{x \in E(\omega) : T(\omega, x) = x\},$$

is compactly measurable and there exists a Castaing representation of F, i.e. there exist an (at most) countable set of random fixed points $U = \{u_n\}_{n \in \mathbb{N}}$ such that

$$F(\omega) = \overline{U(\omega)}$$

for all $\omega \in \Omega$.

Proof. Since T has a fixed point in the wide sense in E and since T is continuous, it follows that $F(\omega) \in 2^X$, i.e. F is a multivalued mapping with closed-valued sets.

Let C be an arbitrary compact subset of X, i.e. $C \in K(X)$, and let

$$C(\omega) = E(\omega) \cap C, \quad \omega \in \Omega_0,$$

with

$$\Omega_0 = \{\omega \in \Omega : C(\omega) \neq \emptyset\}.$$

Since E is a random subset of X, it follows that $\Omega_0 \in \mathscr{K}$. Hence, $\omega \to C(\omega)$, $\omega \in \Omega_0$, is a compact-valued mapping and a random set on $(\Omega_0, \mathscr{K} \cap \Omega_0)$, i.e., a measurable multivalued mapping.

Applying Theorem 2.2.8 to the random function

$$f(\omega, x) = d(x, T(\omega, x)), \quad \omega \in \Omega_0, \ x \in X,$$

it follows that the function

$$y(\omega) = \min_{x \in C(\omega)} d(x, T(\omega, x)), \quad \omega \in \Omega_0$$

is $\mathscr{K} \cap \Omega_0$ measurable.

Since

$$\{\omega \in \Omega_0 : F(\omega) \cap C \neq \emptyset\} = \{\omega \in \Omega_0 : F(\omega) \cap C(\omega) \neq \emptyset\} =$$

$$= \{\omega \in \Omega_0 : y(\omega) = 0\} \in \mathscr{K} \cap \Omega_0 \subset \mathscr{K},$$

it follows that F is a compactly measurable multivalued mapping. Since F is a closed-valued mapping and since X is complete, F is a complete-valued mapping. From Theorem 2.2.5, it follows that F has a Castaing representation and the proof of the theorem is complete.

COROLLARY. *Suppose that the conditions of the Theorem are satisfied but P is a complete σ-finite measure. Then T has a random fixed point in E and the multivalued mapping F is a closed random set which admits a Castaing representation.*

Proof. It suffices to prove that $\{\omega \in \Omega : E(\omega) \cap C \neq \emptyset\} \in \mathscr{K}$ for every closed subset $C \subset X$. Since P is a complete σ-finite measure, $\mathscr{K}$ is a Šuslin family [cf. 433, p. 864]. Since X is a metric Šuslin space, it follows (see [433, Th. 4.2 (g)]) that the measurability of multivalued maps are equivalent with compact measurability. Since F is compactly measurable, it is hence measurable, i.e.. it is a random set which admits a Castaing representation.

Remark. Theorem 4.2.1 can be augmented by the conclusion that $F(\omega) = \{x \in E(\omega) : T(\omega, x) = x\}$ is a compact random subset of X which admits a Castaing representation.

Indeed, $F(\omega)$ is a compact set and F is compact measurable. According to Theorem 2.2.5, the mapping F has a Castaing representation. Since F is a compact-valued mapping, Theorem 2.2.5 implies that F is measurable, i.e. F is a compact random set.

The above results may be used to establish the existence, uniqueness and measurability of solutions of random operator equations. Since this is one of the most important applications of both random fixed point theorems and random operators, we shall give it a separate coverage.

4.3. RANDOM FIXED POINT THEOREMS FOR MULTIVALUED RANDOM OPERATORS

The study of random differential and random integral equations with multivalued right-hand side are based on the random fixed point theorem for multivalued mappings. More general problems, including stochastic nonlinear programing and stochastic optimal control problems are solved with the aid of the theory of random multivalued mappings.

In this paragraph, we intend to present the most significant results regarding the existence of random fixed points for multivalued random operators.

Let X be a Polish space and let $(\Omega, \mathscr{K}, \mu)$ be a σ-finite measure space. We shall use the words "stochastic" and "random" interchangeably also when μ is not a probability measure. In the absence of any measure, we assume that $(\Omega, \mathscr{K})$ is a measurable space with σ-algebra $\mathscr{K}$.

A first result was given by S. Itoh [237] by generalizing a theorem due to S. B. Nadler [305]. This result is also of interest owing to its contributions to the concepts of measurability and measurable selectors of multivalued mappings.

THEOREM 4.3.1. *Let (X, d) be a Polish space and let $(\Omega, \mathscr{K})$ be a measurable space. Let $T : \Omega \times X \to CB(X)$ be a multivalued random operator such that, for every $\omega \in \Omega$, $T(\omega,.)$ is a $k(\omega)$-contraction (i. e. there exists a measurable function $k : \Omega \to [0,1]$ such that $D(F(\omega, x), F(\omega, y)) \leqslant k(\omega)d(x, y)$, where D is the Hausdorff metric on $CB(X)$ induced by d). Then T has a random fixed point, i. e. there exists a measurable $x : \Omega \to X$ such that for any $\omega \in \Omega$, $x(\omega) \in T(\omega, x(\omega))$.*

Proof. If $A_1 = \{\omega \in \Omega : 0 < k(\omega)\}$ and $A_2 = \Omega - A_1$, then $A_1, A_2 \in \mathscr{K}$. Let $v_0 : A_1 \to X$ be a measurable mapping. From Proposition 2.2.2, it follows that the mapping $T(., v_0(.)) : A_1 \to CB(X)$ is weakly measurable (in the sense of Himmelberg), hence there exists a measurable selector $v_1 : A_1 \to X$ of $(F., v_0(.))$ by Kuratowski and Ryll-Nardzewski (Th. 2.2.4). Then, from Proposition 2.2.3, it follows that there exists a measurable selector $v_2 : A_1 \to X$ of $T(., v_1(.))$ such that for any $\omega \in \Omega$,

$$d(v_1(\omega), v_2(\omega) \leqslant D(T(\omega, v_0(\omega)), T(\omega, v_1(\omega))) + k(\omega).$$

From Proposition 2.2.3, it follows that there exists a measurable selector $v_3 : A_1 \to X$ of $T(., v_2(.))$ such that, for any $\omega \in \Omega$,

$$d(v_2(\omega), v_3(\omega)) \leqslant D(T(\omega, v_1(\omega)), T(\omega, v_2(\omega))) + [k(\omega)]^2.$$

Applying the induction procedure, we can choose a sequence of measurable mappings $v_n : A_1 \to X$ such that, for each $\omega \in \Omega$,

$$v_n(\omega) \in T(\omega, v_{n-1}(\omega))$$

and

$$d(v_n(\omega), v_{n+1}(\omega)) \leqslant D(F(\omega, v_{n-1}(\omega)), F(\omega, v_n(\omega))) + \\ + [k(\omega)]^n, \quad n \in \mathbb{N}.$$

Let $t \in A_1$ be arbitrary fixed. For any n, one has

$$d(v_n(\omega), v_{n+1}(\omega)) \leqslant D(T(\omega, v_{n-1}(\omega)), T(\omega, v_n(\omega))) + [k(\omega)]^n \leqslant \\ \leqslant k(\omega)d(v_{n-1}(\omega), v_n(\omega)) + [k(\omega)]^n \leqslant k(\omega)\{D(T(\omega, v_{n-2}(\omega)), \\ T(\omega, v_{n-1}(\omega))) + [k(\omega)]^{n-1}\} + k(\omega)^n \leqslant \\ \leqslant k(\omega)^n d(v_{n-2}(\omega), v_{n-1}(\omega)) + 2k(\omega)^n \leqslant \dots \leqslant \\ \leqslant k(\omega)^n d(v_0(\omega), v_1(\omega)) + nk(\omega)^n.$$

Thus, for every $n \leqslant m$,

$$d(v_n(\omega), v_{m+1}(\omega)) \leqslant \sum_{i=1}^{m} d(v_i(\omega), v_{i+1}(\omega)) = \\ = \sum_{i=n}^{m} k(\omega)^i d(v_0(\omega), v_1(\omega)) + \sum_{i=n}^{m} ik(\omega)^i.$$

Since $0 < k(\omega) < 1$, $\{v_n(\omega)\}_{n \in \mathbb{N}}$ is a Cauchy sequence in X, and hence it converges to some $v(\omega) \in X$, it follows that, for any n,

$$d(v(\omega), T(\omega, v(\omega))) \leqslant d(v(\omega), v_n(\omega)) + \\ + d(v_n(\omega), T(\omega, v(\omega))) \leqslant d(v(\omega), v_n(\omega)) + \\ + D(T(\omega, v_{n-1}(\omega)), T(\omega, v(\omega))) \leqslant d(v(\omega), v_n(\omega)) + \\ + k(\omega)d(v_{n-1}(\omega), v(\omega)).$$

Hence, $d(v(\omega), T(\omega, v(\omega)) = 0$, and, since $T(\omega, v(\omega)) \in CB(X)$ it follows that $v(\omega) \in T(\omega, v(\omega))$. Since $v : A_1 \to X$ is the pointwise limit of measurable mappings $\{v_n\}_{n \in \mathbb{N}}$, it also follows that v is measurable.

On the other hand, for $\omega \in A_2$ and for every $x, y \in X$,

$$D(T(\omega, x), T(\omega, y)) \leqslant k(\omega)\, d(x, y) = 0.$$

Hence, there exists a measurable mapping $T_0 : A_2 \to CB(X)$ such that $T(\omega, x) = T_0(\omega)$ for all $\omega \in A_2$ and $x \in X$. According to Kuratowski and Ryll-Nardzewski (Th. 2.2.4), there exists a measurable selector $w : A_2 \to X$ of T_0. Defining $u : \Omega \to X$ as

$$u(\omega) = \begin{cases} v(\omega) & \text{if } \omega \in A_1 \\ w(\omega) & \text{if } \omega \in A_2 \end{cases},$$

it follows that u is measurable and $u(\omega) \in T(\omega, u(\omega))$ for every $\omega \in \Omega$.

A multivalued version of Nowak's theorem 4.2.9 is

THEOREM 4.3.2. *Let $(\Omega, \mathscr{K}, P)$ be a complete probability space and let (X, d) be an analytic subset of a Polish space. Suppose that $T : \Omega \times X$ is a multivalued random operator and that, for each $\omega \in \Omega$, $T(\omega,.) : (X, d) \to (CB(X), D)$ is continuous. If $F(\omega) = \{x \in X : x \in T(\omega, x)\} \neq \emptyset$, then there exists an X-valued random variable y such that $y(\omega) \in T(\omega, y(\omega))$ a.s.*

Proof. The multivalued mapping F is a closed-valued mapping and the measurability of $T(., x)$ implies (cf. Theorem 2.2.1) that $\omega \to d(x, T(\omega, x))$ is a measurable function of ω for each $x \in X$. On the other hand, since $|d(x, A) - d(x, B)| \leqslant D(A, B)$ for $x \in X$ and $A, B \in CB(X)$, the function $d(x, T(\omega, x))$ is continuous in x for each $\omega \in \Omega$. Hence (cf. Theorem 2.1.4), the function $d(x, T(\omega, x))$ is $\mathscr{K} \times \mathscr{B}_X$-measurable on $\Omega \times X$. Consequently, F is a $\mathscr{K} \times \mathscr{B}_X$-measurable subset of $\mathscr{K} \times X$.

From Theorem 2.2.6, it follows that there exists a measurable selector $y : \Omega \to X$ such that $y(\omega) \in T(\omega, y(\omega))$ a.s.

In order to give a generalization of Theorems 4.3.1, 4.3.2 and a generalization of Boyd and Wong's [71] result to a multivalued nonlinear random contraction, we have to introduce some results from ref. [13]:

DEFINITION 4.3.1. A mapping $F : X \to \mathscr{P}(X)$ ($\mathscr{P}(X)$ = power set of X) is said to be a *shrinking diamater mapping* if there exists a function $f : [0, \infty) \to [0, \infty)$ satisfying:

(i) $f(0) = 0 = f(0+)$;

and

(ii) $\operatorname{diam} F(x) \leqslant f(d(x, F(x)))$ whenever $F(x) \neq \emptyset$.

We may assume that $f(t) \geqslant t$ for $t \geqslant 0$. Obviously, any function $F : X \to X$ is a shrinking diameter mapping.

DEFINITION 4.3.2. A *contractive function* is a function $\varphi : [0, \infty) \to [0, \infty)$ satisfying:

(i) $\varphi(t) < t$ for $t > 0$;

and

(ii) $\limsup_{u \to t+} \varphi(u) \leqslant \varphi(t)$ for $t \geqslant 0$.

A *measurable contractive function* is a measurable function $h : \Omega \times [0, \infty) \to [0, \infty)$ such that $k(\omega, .) : [0, \infty) \to [0, \infty)$ is a contractive function for each $\omega \in \Omega$.

A generalization of a Boyd and Wong [71] theorem to multivalued mappings is given by

LEMMA 4.3.1 [13, Th. 3.7]. *Suppose X is a complete metric space, $F : X \to CB(X)$ is a shrinking diameter mapping and φ is a contractive function such that $D(F(x), F(y)) \leqslant \varphi(d(x, y))$ for $x, y \in X$. Then there exists an $\bar{x} \in X$ such that $\bar{x} \in F(\bar{x})$.*

A random version is given by:

THEOREM 4.3.3 [13, Th. 3.8]. *Let X be a complete metric space, let $(\Omega, \mathscr{K}, P)$ be a complete probability space and let $T : \Omega \times X \to CB(X)$ be a multivalued random operator. Suppose that for almost every $\omega \in \Omega$, one of the following conditions holds:*

(A) There exists a constant k_ω such that $0 \leqslant k_\omega < 1$ and

$$D(T(\omega, x), T(\omega, y)) \leqslant k_\omega d(x, y) \text{ for } x, y \in X,$$

(B) $T(\omega,.) : X \to CB(X)$ is a shrinking diameter mapping and k_ω is a contractive function such that $D(T(\omega, x), T(\omega, y)) \leqslant k_\omega(d(x, y))$ for $x, y \in X$.

Then there exists a random fixed point for T, i. e. there exists a measurable selector $y(\omega) \in T(\omega, y(\omega))$ a.s.

Proof. For almost every $\omega \in \Omega$, one has $F(\omega) \neq \emptyset$ and $F(\omega,.) : (X, d) \to (CB(X), D)$ is continuous. Thus, the theorem follows from Theorem 4.3.2.

Remark. Under the hypothesis in the above theorem, conditions (A) and (B) are only pointwise requirements in $\omega \in \Omega$, but have no measurability restrictions.

In order to relax the condition of a complete measure space in the above results to the condition of a measurable space $(\Omega, \mathscr{K})$, C. T. Andrus and T. Nishiura [13] gave some bounds on the convergent sequences from Nadler's fixed point theorem [305].

LEMMA 4.3.2 [13, Th. 31]. *Let $F : X \to CB(X)$ be a multivalued contraction (i.e. $D(F(x), F(y)) \leqslant kd(x, y)$ for $0 \leqslant k < 1$ and $x, y \in X$). Let $x_0 \in X$, δ_0 and k_0 be such that $d(x_0, F(x_0)) < \delta_0$ and $k < k_0 < 1$. Then there exists an $\bar{x} \in X$ such that $\bar{x} \in F(\bar{x})$ and $d(x_0, \bar{x}) < \delta_0(1 - k_0)^{-1}$.*

In order to improve a result due to Racotch [13] and to obtain some bounds, they introduce the *α-function* as a decreasing function $\alpha : [0, \infty) \to [0, 1]$, $\alpha(0) = 1$ and $\alpha(t) < 1$ for $t > 0$.

LEMMA 4.3.3 [13, Prop. 3.3]. *Let $t(\eta)$ be the extended real number associated to an α-function and let $\eta > 0$ be the right endpoint of the interval $[t : \eta \geqslant 1 - \alpha(t)]$. Then $\lim_{\eta \to 0+} t(\eta) = 0$, and if $\bar{t}(1 - \alpha(\bar{t})) < \eta^2$, then $\bar{t} \leqslant$* max $\{\eta, t(\eta)\}$.

LEMMA 4.3.4 [13, Th. 3.5]. *Suppose that $F : X \to CB(X)$ is a shrinking diameter mapping and that α is an α-function such that*

$$D(F(x), F(y)) \leqslant \alpha(d(x, y))\, d(x, y) \text{ for } x, y \in X.$$

Let $x_0 \in X$. Then there exists an $\bar{x} \in X$ such that $\bar{x} \in F(\bar{x})$ and

$$d(x_0, \bar{x})\, [1 - \alpha(d(x_0, \bar{x}) -)] \leqslant d(x_0, F(x_0)).$$

If $d(x_0, F(x_0)) < a^2$, $a > 0$, then

$$d(x_0, \bar{x}) \leqslant 2 \max \{a, t(a)\} \text{ and } d(x_0, \bar{x}) \leqslant \frac{a^2}{1 - \alpha(a^2)}.$$

DEFINITION 4.3.3. We say that a multivalued mapping $T : \Omega \times X \to CB(X)$ has the pointwise estimable fixed-point property if, for each $\omega \in \Omega$, there exists a function $E_\omega : [0, \delta_\omega] \to \to [0, \infty)$, $\delta_\omega > 0$, satisfying

(i) $E_\omega(0) = 0 = E_\omega(0+)$,

(ii) for each $x \in X$ with $d(x, T(\omega, x)) < \delta_\omega$ there exists an $\bar{x} \in X$ such that $\bar{x} \in T(\omega, \bar{x})$ and $d(x, \bar{x}) \leqslant E_\omega(d(x, T(\omega, x)))$.

THEOREM 4.3.4 [13, Th. 4.3]. *Let $(\Omega, \mathscr{K})$ be a measurable space, let X be a complete metric space and let $T : \Omega \times X \to \to CB(X)$ be a multivalued random operator. Suppose that $T(\omega, .) : (X, d) \to (CB(X), D)$ is continuous for each $\omega \in \Omega$ and that T has the pointwise estimable fixed-point property. If $F(\omega) = \{x \in X : x \in T(\omega, x) \neq \emptyset$ for each $\omega \in \Omega\}$, then there exists at least one random fixed point of T.*

Proof. First, we prove that F is a weakly measurable closed-valued mapping. To this purpose, we shall prove that for each nonempty open set $G \subseteq X$, we have $\Gamma(G) = F^{-1}(G)$, where

$$\Gamma(G) = \bigcup_{m=2}^{\infty} \bigcap_{j=1}^{\infty} \bigcup_{r=1}^{\infty} \{\omega \in \Omega : B(\xi_r, j^{-1}) \cap T(\omega, \xi_r) \neq \emptyset \wedge B(\xi_r, j^{-1}) \cap$$
$$\cap\, G_m \neq \emptyset\},$$

where $G_m = \{x \in G : d(x, X - G) \geqslant m^{-1}\}$, $m \geqslant 2$, $\{\xi_r\}_{r \in \mathbb{N}}$, is a dense subset of G, while $B(\xi_r, s)$ is the open ball with center ξ_r and radius in the metric of X.

From the continuity of $T(\omega, .)$, for each $\omega \in \Omega$, one concludes that for every $\omega \in F^{-1}(G)$ there exists an $\bar{x} \in X$ such that $\bar{x} \in F(\omega, \bar{x})$ and $\bar{x} \in G$. Choose $m \geqslant 2$ so that $\bar{x} \in G_m$. Let $j \geqslant 1$ and $\xi_{r_m} \to \bar{x}$ as $n \to \infty$. Then $D(T(\omega, \xi_{r_n}), T(\omega, \bar{x})) \to 0$. Also, $d(\xi_{r_n}, T(\omega, \xi_{r_n})) \leqslant d(\xi_{r_n}, \bar{x}) + d(\bar{x}, T(\omega, \xi_{r_n})) \leqslant d(\xi_{r_n}, \bar{x}) + D(T(\omega, \bar{x}), T(\omega, \xi_{r_n}))$. Hence, for n sufficiently large, we have $d(\xi_{r_n}, T(\omega, \xi_{s_n})) < j^{-1}$ and $d(\xi_{r_m}, G_m) \leqslant d(\xi_{r_n}, \bar{x}) < j^{-1}$. Thus, $\omega \in \Gamma(G)$, i.e., $F^{-1}(G) \subset \Gamma(G)$.

On the other hand, the pointwise estimable fixed point property for every $\omega \in \Gamma(G)$ implies that there exists a function $E_\omega : [0, \delta_\omega) \to (0, \infty)$ such that (i) and (ii) of Definition 4.3.3 hold. Let $m \geqslant 2$ be such that

$$\omega \in \bigcap_{j=1}^{\infty} \bigcup_{r=1}^{\infty} \{\omega \in \Omega : B(\xi_r, j^{-1}) \cap T(\omega, \xi_r) \neq \emptyset \wedge B(\xi_r, j^{-1}) \cap \cap G_m \neq \emptyset\}.$$

There exist $\delta > 0$ such that $\delta < \delta_\omega$ and $E_\omega(t) < (2m^2)^{-1}$ for $t < \delta$. Choose j such that $j^{-1} < \delta$ and $j > 2m^2$. Then there exists an r such that $B(\xi_r, j^{-1}) \cap T(\omega, \xi_r) \neq \emptyset$ and $B(\xi_r, j^{-1}) \cap \cap G_m \neq \emptyset$. Hence, $d(\xi_r, T(\omega, \xi_r)) < j^{-1} < \delta$ and $d(\xi_r, X - G) \geqslant \geqslant m^{-1} - j^{-1}$. There exists an $\bar{x} \in X$ such that $\bar{x} \in T(\omega, \bar{x})$ and $d(\xi_r, \bar{x}) \leqslant E_\omega(d(\xi_r, T(\omega, \xi_r))) < (2m^2)^{-1}$. So, $d(\bar{x}, X - G) \geqslant \geqslant m^{-1} - j^{-1} - (2m^2)^{-1} \geqslant (m-1)/m^2 > 0$. Hence, $\bar{x} \in G$ and thus $\omega \in F^{-1}(G)$.

Since $\Gamma(G)$ is obviously measurable, it follows that $F : \Omega \to \to \mathscr{P}(X)$ is a weakly measurable closed-valued mapping. According to the theorem of Kuratowski and Ryll-Nardzewski (Th. 2.2.4), it follows that there exists a measurable selector $y : \Omega \to X$ such that $y(\omega) \in T(\omega, y(\omega))$.

THEOREM 4.3.5 [13, Th. 4.4]. *Let $(\Omega, \mathscr{K})$ be a measurable space, let X be a complete metric space and let $T : \Omega \times X \to CB(X)$ be a multivalued random operator. Suppose that, for each $\omega \in \Omega$, one of the following conditions holds:*

(A) *There exists a constant k_ω such that $0 \leqslant k_\omega < 1$ and*

$$D(T(\omega, x), T(\omega, y)) \leqslant k_\omega d(x, y) \text{ for } x, y \in X;$$

(B) *$T(\omega, .) : X \to CB(X)$ is a shrinking diameter mapping and α_ω is an α-function such that $D(T(\omega, x), T(\omega, y)) \leqslant \leqslant \alpha_\omega(d(x, y))\, d(x, y)$ for $x, y \in X$.*

Then there exists a measurable mapping $y : \Omega \to X$ such that $y(\omega) \in T(\omega, y(\omega))$ for each $\omega \in \Omega$.

Proof. The multivalued random operator T verifies the conditions of Theorem 4.3.4, as follows from Lemmas 4.3.2,

4.3.3 and 4.3.4. Since $F(\omega) \neq \emptyset$, for each $\omega \in \Omega$, the conclusion follows from Theorem 4.3.4.

Remark. The above theorem is a generalization of Theorem 4.3.1, since k_ω of (A) is not measurable.

A reasoning similar to that used in [237] yields

THEOREM 4.3.6 [13, Th. 4.6]. *Let $(\Omega, \mathscr{K})$ be a measurable space, let X be a complete metric space and let $T : \Omega \times X \to$ $\to CB(X)$ be a multivalued random operator. Suppose that Ω_A and Ω_B are measurable partitions of Ω such that:*

(a) *for each $\omega \in \Omega_A$, either (A) or (B) of Theorem 4.3.5 holds;*

(b) *$T(\omega,.) : X \to CB(X)$ is a shrinking diameter mapping for each $\omega \in \Omega_B$ and $k : \Omega_B \times [0, \infty) \to [0, \infty)$ is a measurable contractive function such that $D(T(\omega, x), T(\omega, y)) \leqslant k(\omega, d(x, y))$ for $\omega \in \Omega_B$ and $x, y \in X$. Then there exists a measurable mapping $y : \Omega \to X$ such that $y(\omega) \in T(\omega, y(\omega))$ for each $\omega \in \Omega$.*

Another generalization of Itoh's result [237] is given by N. S. Papageorgiou [336] for multivalued continuous (in Hausdorff metric) mappings.

THEOREM 4.3.7. *Let $(\Omega, \mathscr{K})$ be a measurable space, let (X, d) be a complete separable metric space and let $T : \Omega \times X \to CB(X)$ be a continuous multivalued random operator. If $T(\omega, .)$ is a $k(\omega)$-generalized contraction for all $\omega \in \Omega$, i.e., if*

$$\text{(c)}\ D(T(\omega, x), T(\omega, y)) \leqslant k(\omega) \max \left\{ d(x, y), \frac{1}{2}[d(x, T(\omega, x)) + \right.$$
$$\left. + d(y, T(\omega, y))], \frac{1}{2}[d(x, T(\omega, y)) + d(y, T(\omega, x))] \right\},$$

where $k : \Omega \to (0,1)$ is measurable, then T has at least one random fixed point.

Proof. Let $g : \Omega \times X \to \mathbb{R}^+$ be the function

$g(\omega, x) = d(x, T(\omega, x))$.

From the measurability of $T(., x)$, for all $x \in X$, and from Theorem 2.2.1, there follows the measurability of $g(., x)$. The continuity of $T(\omega, .)$ for all $\omega \in \Omega$ and Lemma 2.2.2 follow from the continuity of $g(\omega, .)$ for all $\omega \in \Omega$, and hence $g : \Omega \times X$ is a Carathéodory function. Therefore, if $u : \Omega \to X$ is a measurable mapping, then $\omega \to g(\omega, u(\omega))$ is a measurable function. Now, let $s_0 : \Omega \to X$ be a measurable mapping; then, according to Kuratowski and Ryll-Nardzewski (Th. 2.2.4), there exists a measurable selector $s_1 : \Omega \to X$ for the multivalued mapping $F(., s_0(.)) : \Omega \to CB(X)$. Using Lemma 2.2.2, there exists a measurable selector s_2 for $F(.,s_1(.)):\Omega\to CB(X)$ and

$$d(s_1(\omega),\ s_2(\omega)) \leqslant \frac{1}{\sqrt{k(\omega)}} D(T(\omega,\ s_0(\omega)),\ T(\omega,\ s_1(\omega))).$$

In this way, we can inductively define a sequence of measurable mappings $s_n:\Omega\to X$ such that

(a) $s_n(\omega)\in T(\omega,\ s_{n-1}(\omega))$ for $\omega\in\Omega$,

(b) $d(s_n(\omega), s_{n+1}(\omega)) \leqslant \dfrac{1}{\sqrt{k(\omega)}} D(T(\omega,\ s_{n-1}(\omega)),\ T(\omega,\ s_n(\omega))),$

$\forall\ n\in\mathbb{N}$.

Now, we get

$$\begin{aligned} d(s_n(\omega),\ s_{n+1}(\omega)) &\leqslant \frac{1}{\sqrt{k(\omega)}} D(T(\omega,\ s_{n-1}(\omega)),\ T(\omega,\ s_n(\omega))) \leqslant \\ &\leqslant \sqrt{k(\omega)} \max\Big\{ d(s_{n-1}(\omega),\ s_n(\omega)), \frac{1}{2}[d(s_{n-1}(\omega),\ T(\omega, s_{n-1}(\omega))) + \\ &+ d(s_n(\omega),\ T(\omega,\ s_n(\omega)))], \frac{1}{2}[d(s_{n-1}(\omega),\ T(\omega, s_n(\omega)))) + \\ &+ d(s_n(\omega),\ T(\omega,\ s_{n-1}(\omega))))]\Big\} \leqslant \\ &\leqslant \sqrt{k(\omega)} \max\Big\{ d(s_{n-1}(\omega),\ s_n(\omega)), \frac{1}{2}[d(s_{n-1}(\omega),\ s_n(\omega)) + \\ &+ d(s_n(\omega),\ s_{n+1}(\omega))], \frac{1}{2}[d(s_{n-1}(\omega),\ s_{n+1}(\omega))]\Big\} \leqslant \\ &\sqrt{k(\omega)}\ d(s_{n-1}(\omega),\ s_n(\omega)) \leqslant \ldots \leqslant (\sqrt{k(\omega)})^n d(s_0(\omega),\ s_1(\omega)). \end{aligned}$$

Moreover, $\{s_n(\omega)\}_{n\in\mathbb{N}\cup\{0\}}$ is a Cauchy sequence since, for all $0<m<n$,

$$\begin{aligned} d(s_m(\omega),\ s_n(\omega)) &\leqslant \sum_{j=m}^{n-1} d(s_j(\omega),\ s_{j+1}(\omega)) \leqslant \\ &\leqslant \sum_{j=m}^{n-1} (\sqrt{k(\omega)})^j d(s_0(\omega),\ s_1(\omega)) = d(s_0(\omega),\ s_1(\omega)) \sum_{j=m}^{n-1} (\sqrt{k(\omega)})^j. \end{aligned}$$

Hence, $d(s_m(\omega),\ s_n(\omega))\to 0$ as $m\to\infty$, since $\sqrt{k(\omega)}<1$ for all $\omega\in\Omega$. From the completeness of X, we deduce that there exists $s:\Omega\to X$, $s_n(\omega)\to s(\omega)$ for all $\omega\in\Omega$. Then, we can also state :

$$d(s(\omega),\ T(\omega, s(\omega))) \leqslant d(s(\omega),\ T(\omega, s_n(\omega))) +$$
$$+ D(T(\omega, s(\omega)),\ T(\omega, s_n(\omega))) \leqslant d(s(\omega), s_{n+1}(\omega)) +$$
$$+ k(\omega) \max\Big\{d(s(\omega), s_n(\omega)),\ \frac{1}{2}[d(s(\omega),\ T(\omega,\ s(\omega))) +$$
$$+ d(s_n(\omega),\ T(\omega, s_n(\omega)))],\ \frac{1}{2}[d(s(\omega), T(\omega,\ s_n(\omega))) +$$
$$+ d(s_n(\omega),\ T(\omega, s(\omega)))]\Big\} \leqslant d(s(\omega),\ s_{n+1}(\omega)) +$$
$$+ k(\omega) \max\Big\{d(s(\omega), s_n(\omega)),\ \frac{1}{2}[d(s(\omega),\ T(\omega, s(\omega))) +$$
$$+ d(s_n(\omega),\ s_{n+1}(\omega))],\ \frac{1}{2}[d(s(\omega),\ s_{n+1}(\omega)) +$$
$$+ d(s_n(\omega),\ T(\omega,\ s(\omega)))]\Big\}.$$

For $n \to \infty$, we get that

$$d(s(\omega),\ T(\omega, s(\omega))) \leqslant \frac{k(\omega)}{2} d(s(\omega),\ T(\omega, s(\omega))),$$

and, since $k(\omega) < 1$,

$$d(s(\omega),\ T(\omega,\ s(\omega))) = 0,$$

i.e.,

$$s(\omega) \in T(\omega, s(\omega)).$$

Remark. The above result holds for the $k(\omega)$-generalized contraction of the following type :

(c') $D(T(\omega, x),\ T(\omega, y)) \leqslant k(\omega) \max\{d(x, y),\ d(x,\ T(\omega,\ x)),$
$d(y,\ T(\omega, y)),\ d(x,\ T(\omega, y)),\ d(y,\ T(\omega,\ x))\},$

where $k : \Omega \to (0, 1)$ is measurable (cf. [179] and [336, Th. 2.2]).

The previous result holds for a pair of multivalued random operators.

THEOREM 4.3.8. *Let* $S, T : \Omega \times X \to CB(X)$ *be two multivalued continuous random operators such that*

(D) $D(S(\omega, x),\ T(\omega, y)) \leqslant k(\omega) \max\Big\{d(x, y), \frac{1}{2}[d(x,\ S(\omega,x)) +$
$+ d(y,\ T(\omega, y))],\ \frac{1}{2}[d(x,\ T(\omega, y)) + d(y,\ S(\omega,\ x))]\Big\},$

where $k:\Omega\to(0,1)$ is a measurable function. Then S and T have the same nonempty sets of random fixed points and if s is one such point, then

$$S(\omega, s(\omega)) = T(\omega, s(\omega)) \text{ for all } \omega\in\Omega.$$

Proof. In a way similar to that used in the previous theorem for a measurable mapping $s_0:\Omega\to X$, we can define a sequence of measurable mappings $s_n:\Omega\to X$, $n\in\mathbb{N}$, such that

(a) $s_n(\omega)\in S(\omega, s_{n-1}(\omega))$,

(b) $s_{n+1}(\omega)\in T(\omega, s_n(\omega))$,

(c) $d(s_n(\omega), s_{n+1}(\omega)) \leqslant \frac{1}{\sqrt{k(\omega)}} D(S(\omega, s_{n-1}(\omega)), T(\omega, s_n(\omega)))$.

Then, one gets

$$d(s_n(\omega), s_{n+1}(\omega)) \leqslant \sqrt{k(\omega)}\max\Big\{d(s_{n-1}(\omega), s_n(\omega)), \frac{1}{2}[d(s_{n-1}(\omega),$$

$$S(\omega,s_{n-1}(\omega)))+d(s_n(\omega),T(\omega,s_n(\omega)))], \frac{1}{2}[d(s_{n-1}(\omega),T(\omega,s_n(\omega))+$$

$$d(s_n(\omega), S(\omega, s_{n-1}(\omega)))]\Big\} \leqslant$$

$$\leqslant \sqrt{k(\omega)}\max\Big\{d(s_{n-1}(\omega), s_n(\omega)), \frac{1}{2}[d(s_{n-1}(\omega),$$

$$s_n(\omega)) + d(s_n(\omega), s_{n+1}(\omega))], \frac{1}{2}[d(s_{n-1}(\omega), s_{n+1}(\omega))]\Big\} \leqslant$$

$$\leqslant \sqrt{k(\omega)}d(s_{n-1}(\omega), s_n(\omega)) \leqslant \ldots \leqslant (\sqrt{k(\omega)})^n d(s_0(\omega), s_1(\omega)).$$

It follows that $\{s_n(\omega)\}_{n\in\mathbb{N}}$ is a Cauchy sequence for all $\omega\in\Omega$ and there exists a measurable mapping $s:\Omega\to X$, $s=\lim\limits_{n\to\infty} s_n$.
On the other hand,

$$d(s(\omega), S(\omega, s(\omega))) \leqslant d(s(\omega), T(\omega, s_n(\omega))) + D(S(\omega, s(\omega)),$$

$$T(\omega, s_n(\omega))) \leqslant d(s(\omega), s_{n+1}(\omega)) + k(\omega)\max\Big\{d(s(\omega), s_n(\omega)),$$

$$\frac{1}{2}[d(s(\omega), S(\omega, s(\omega)))+d(s_n(\omega), T(\omega, s_n(\omega)))],$$

$$\frac{1}{2}[d(s(\omega), T(\omega, s_n(\omega))) + d(s_n(\omega), S(\omega, s(\omega)))]\Big\} \leqslant$$

$$\leqslant d(s(\omega),\ s_{n+1}(\omega)) + k(\omega) \max\Big\{ d(s(\omega),\ s_n(\omega)),$$

$$\frac{1}{2}\,[d(s(\omega),\ S(\omega,\ s(\omega))) + d(s_n(\omega),\ s_{n+1}(\omega))],$$

$$\frac{1}{2}\,[d(s(\omega),\ s_{n+1}(\omega)) + d(s_n(\omega),\ S(\omega,\ s(\omega))]\Big\}.$$

Letting $n \to \infty$, we get that

$$d(s(\omega),\ S(\omega,\ s(\omega))) \leqslant k(\omega)\,\frac{1}{2}\,d(s(\omega),\ S(\omega,\ s(\omega))) \text{ for all } \omega \in \Omega.$$

Since $k(\omega) < 1$, we conclude that $s(\omega) \in S(\omega, s(\omega))$ for all $\omega \in \Omega$. Hence, the set of selectors of S, $\mathscr{S}(S) = \{s : \Omega \to X$ measurable, $s(\omega) \in S(\omega, s(\omega))$ for all $\omega \in \Omega\} \neq \emptyset$.

We now prove that $\mathscr{S}(S) = \mathscr{S}(T) = \mathscr{S}$ and, for $s \in \mathscr{S}$, $S(\omega, s(\omega)) = T(\omega, s(\omega))$ for all $\omega \in \Omega$.

Indeed, let $s : \Omega \to X$, $s \in \mathscr{S}(S)$. Then using the properties of the Hausdorff metric, we have

$$d(s(\omega),\ T(\omega,\ s(\omega))) \leqslant D(S(\omega, s(\omega)),\ T(\omega,\ s(\omega))) \leqslant$$

$$\leqslant k(\omega)\ d(s(\omega), T(\omega,\ s(\omega))).$$

Since $k(\omega) < 1$ for all $\omega \in \Omega$, we conclude that

$$d(s(\omega),\ T(\omega,\ s(\omega))) = 0 \text{ for all } \omega \in \Omega,$$

i.e., $s(\omega) \in T(\omega, s(\omega))$ for all $\omega \in \Omega$ and $\mathscr{S}(S) \subseteq \mathscr{S}(T)$.

Interchanging the roles of S and T, we conclude from the above argument that $\mathscr{S}(T) \subseteq \mathscr{S}(S)$, and hence $\mathscr{S}(S) = \mathscr{S}(T) = \mathscr{S}$. Now, for every $s \in \mathscr{S}$, we get

$$D(S(\omega,\ s(\omega)),\ T(\omega,\ s(\omega))) \leqslant k(\omega)\,\frac{1}{2}\,[d(s(\omega),\ S(\omega,\ s(\omega))) +$$

$$+ d(s(\omega),\ T(\omega,\ s(\omega)))] = 0$$

such that $S(\omega, s(\omega)) = T(\omega, s(\omega))$ for all $\omega \in \Omega$.

COROLLARY. *Let* $T_n : \Omega \times X \to CB(X)$, $n \in \mathbb{N}$, *be a sequence of multivalued continuous random operators for which* $\{T_1, T_n\}$ *satisfy relation* (D) *for all* $n \in \mathbb{N}$. *Then* $\{T_n\}_{n\in\mathbb{N}}$ *has a common random fixed point and for such a point* s, $T_1(\omega, s(\omega)) = T_n(\omega, s(\omega))$ *for all* $\omega \in \Omega$ *and* $n \in \mathbb{N}$.

Remark. Theorem 4.3.8 and the above corollary also hold when relation (D) is replaced by

$$(D') \quad D(F_1(\omega, x), F_n(\omega, y)) \leqslant k(\omega) \max\{d(x, y), d(x, T_1(\omega, x)), d(y, F_n(\omega, y)), d(x, F_n(\omega, y)), d(y, T_1(\omega, x))\}$$

for all $n \in \mathbb{N}$, where $k : \Omega \to (0,1)$ is a measurable function.

Another result similar to the deterministic case is given by

PROPOSITION 4.3.1. *Let $T : \Omega \times X \to CB(X)$ be a multivalued continuous randon operator for which there exist $a, b, c, d, e : \Omega \to X$ measurable mappings such that $a(\omega) + b(\omega) + c(\omega) + d(\omega) + e(\omega) < 1$ for all $\omega \in \Omega$ and*

$$D(T(\omega, x), T(\omega, y)) \leqslant a(\omega)\, d(x, y) + b(\omega)\, d(x, T(\omega, x)) + c(\omega)\, d(y, T(\omega, y)) + d(\omega)\, d(x, T(\omega, y)) + e(\omega)\, d(y, T(\omega, x)).$$

Then, T has a random fixed point.

THEOREM 4.3.9. *Let X be a separable σ-compact metrizable (not necessarily complete) space with two metrics d_1, d_2 on it and let $T : \Omega \times X \to CB((X, d_1)) \cap CB((X, d_2))$ be a multivalued random operator. Suppose that:*

(i) *$d_2(x, y) \leqslant d_1(x, y)$ for all $(x,y) \in X \times X$,*

(ii) *(X, d_2) is a complete metric space,*

(iii) *$D_1(T(\omega, x), T(\omega, y)) \leqslant k(\omega)\, d_1(x, y)$, where $k : \Omega \to (0,1)$ is measurable,*

(iv) *$T(\omega, .)$ is d_1-continuous for all $\omega \in \Omega$.*

Then T has a fixed random point.

In refs. [238], [241], [251] almost all known fixed point theorems in the Banach space were extended to multivalued random operators. The following is a random fixed point theorem for a multivalued condensing mapping of S. Itoh [242].

THEOREM 4.3.10. *Let $(\Omega, \mathscr{K})$ be a measurable space, let X be a separable closed convex subset of a Banach space and let $T : \Omega \times X \to CK(X)$ be a continuous α-condensing random operator. Suppose that for any $\omega \in \Omega$, $T(\omega, x)$ is bounded. Then T has a random fixed point.*

Proof. Let $F : \Omega \to K(X)$ be the mapping $F(\omega) = \{x \in X, x \in F(\omega, x)\}$ which is well defined by a fixed point theorem due to Petryshyn and Fitzpatrick [342]. Let $\{x_n\}_{n \in \mathbb{N}}$ be a countable dense subset of X. For each $C \in 2^X$, let $C_n = \Big\{x \in X : d(x, C) <$

$< \dfrac{1}{n}\Big\}$ and let

$$L(C) = \bigcap_{n=1}^{\infty} \bigcup_{x_i \in C_n} \left\{ \omega \in \Omega : d(x_i, F(\omega, x_i)) < \frac{2}{n} \right\}.$$

We shall show that $F^{-1}(C) = L(C)$. Indeed, for each $\omega \in F^{-1}(C)$, there exists an $x \in C$ with $x \in T(\omega, x)$. Since $T(\omega, .)$ is continuous, it is continuous from X into the metric space $(K(X), D)$ (cf. Th. 2.2.9). Then for each $n \in \mathbb{N}$, there exists an x_{i_n}, such that $\|x_{i_n} - x\| < \dfrac{1}{n}$ and $D(T(\omega, x_{i_n}), T(\omega, x)) < \dfrac{1}{n}$. Also,

$$d(x_{i_n}, T(\omega, x_{i_n})) \leqslant \|x_{i_n} - x\| + d(x, T(\omega, x_{i_n})) \leqslant \|x_{i_n} - x\| + \\ + D(T(\omega, x), T(\omega, x_{i_n})) < \frac{2}{n}$$

and since $x_{i_n} \in C_n$, it follows that $\omega \in L(C)$. Conversely, if $\omega \in L(C)$, then for each $n \in \mathbb{N}$ we can take $x_{i_n} \in C_n$ for which $d(x_{i_n}, T(\omega, x_{i_n})) < \dfrac{2}{n}$. Since $\{x_{i_n}\}_{n \in \mathbb{N}}$ is precompact, we may assume that $x_{i_n} \to x \in C$ as $n \to \infty$. For each n, choose $y_n \in T(\omega, x_{i_n})$ such that $\|x_{i_n} - y_n\| = d(x_{i_n}, T(\omega, x_{i_n}))$. Then $y_n \to x$. Since $T(\omega, .)$ is upper semicontinuous, $x \in T(\omega, x)$ and hence $\omega \in F^{-1}(C)$. Therefore, $T^{-1}(C) = L(C)$. According to Theorem 2.2.1, $\left\{\omega \in \Omega : d(x, T(\omega, x)) < \dfrac{2}{n}\right\} \in \mathscr{K}$ for any $x \in X$, and hence $F^{-1}(C) = L(C)$ and F is measurable. Any measurable selector of F is random fixed point of T.

As a corollary, we give the following Bohnenblust- and Karlin-type result [242].

COROLLARY. *Let X be a compact (or separable and closed) convex subset of a Banach space, let $T : \Omega \times X \to CK(X)$ be a continuous compact random operator. Then T has a random fixed point.*

The general theorem for obtaining random fixed points using deterministic fixed points were given by Engl [132], [138]. The case of multivalued random operators with stochastic domain was included. The continuity assumption is essential in the Carathéodory condition in order to obtain the joint measurability (see [138, Example 15]). Nevertheless, Engl [138] generalized the ideas contained in the proof by considering an auxiliary mapping and obtained a complete result of Bohnenblust and Karlin type for u.s.c. compact multivalued random operators.

The probabilistic analogue of Kakutani's fixed-point theorem had been obtained by R. Kannan [249]. A probabilistic analogue of the Bohnenblust and Karlin fixed point theorem was obtained by D. D. Angh and S. A. Thanh [14] as follows:

THEOREM 4.3.11. *Let* $(\Omega, \mathcal{K}, P)$ *be a probability space, let* C *be a nonempty (not necessarily closed) convex subset of a separable Banach space* X. *Let* $T : \Omega \times C \to CC(X)$ *be an u.s.c. multivalued random operator such that* $T(\Omega \times C) \in K(C)$. *Then* T *has a random fixed point.*

We prefer to present the proof of Engl's result [138, Th. 16] in paragraph 4, herein.

Adopting Engl's method, S. Itoh [241] proved random fixed point theorems for u.s.c. condensing multivalued random operators.

THEOREM 4.3.12 [241, Th. 3.1]. *Let* $\mathcal{K}$ *be a Šouslin family and let* X *be a nonempty closed convex subset of a separable Banach space* Y. *Let* $T : \Omega \times X \to CK(Y)$ *be a mapping satisfying the following conditions:*

(i) *For any* $\omega \in \Omega$, $T(\omega, X)$ *is bounded and* $T(\omega, bdX) \subset X$, *where* bdX *is the boundary of* X;

(ii) *For any* $\omega \in \Omega$, $T(\omega, .)$ *is u.s.c. and condensing;*

(iii) *For any* $x \in X$, $T(., x)$ *is weakly measurable in the sense of Himmelberg.*

Then there exists a random fixed point of T.

Proof. Let $\{x_n\}_{n \in \mathbb{N}}$ be a countable dense set of X, and let us define $F : \Omega \times X \to CK(Y)$ as in Proposition 2.2.4; then F is $\mathcal{K} \times \mathcal{B}_X$-measurable. It is easy to see that the mapping $v : \Omega \times \times X \to X$ defined by $v(\omega, x) = x$, $\omega \in \Omega$, $x \in X$ is $\mathcal{K} \times \mathcal{B}_X$-measurable. Using Lemma 2.2.3, one can show that $f(\omega, x) = = d(v(\omega, x), F(\omega, x))$, $\omega \in \Omega$, $x \in X$ is a $\mathcal{K} \times \mathcal{B}_X$-measurable function on $\Omega \times X$. Defining $G(\omega) = \{x \in X : x \in F(\omega, x)\}$, $\omega \in \Omega$, then, for any $\omega \in \Omega$, $G(\omega)$ is nonempty and compact according to Petryshyn and Fitzpatrick. One also has:

$$\operatorname{Gr} G = \{(\omega, x) \in \Omega \times X,\ x \in G(\omega)\} = \{(\omega, x) \in \Omega \times X :$$

$$f(\omega, x) = 0\} \in \mathcal{K} \times \mathcal{B}_X.$$

Then there exists a measurable selector $u(\omega) \in G(\omega)$ (cf. [433, Th. 42]) and hence $u \in F(\omega, u(\omega)) \subset T(\omega, u(\omega))$.

COROLLARY [241]. *Let* $(\Omega, \mathcal{K}, \mu)$ *be a complete* σ*-finite measure space and let* Y, X *and* $T : \Omega \times X \to CK(Y)$ *be as in*

Theorem 4.3.11. *Then there exists a measurable selector such that* $u(\omega) \in T(\omega, u(\omega))$ *for* μ*-a.e. (all)* $\omega \in \Omega$.

Regarding the nonexpansive mapping, we give

THEOREM 4.3.13 [251, Th. 6]. *Let* $(\Omega, \mathscr{K}, P)$ *be a complete probability space and let* X *be a separable Banach space satisfying Opial's condition. Let* $C \in WK(X)$ *and let* $T : \Omega \times C \to K(C)$ *and*

$$D(T(\omega, x), T(\omega, y)) \leqslant \|x - y\| \text{ for } \omega \in \Omega \text{ and } x, y \in C.$$

Then T *has a random fixed point.*

Proof. Since T is compact-valued for each ω and since T is nonexpansive, it follows from Theorem 2.2.11 that T is separable. Now, we shall prove that T satisfies the hypotheses of Theorem 2.2.12. For each $\omega \in \Omega$ the operator T satisfies the hypotheses of Theorem 2.2.16 and hence there exists an $x_0 \in T(\omega, x_0)$ for each $\omega \in \Omega$. Now, we can prove that $(I - T)^{-1}$ is closed-graph continuous, i.e., given that $y_n \to y$ and $x_n \to x$, where $x_n \in (I - T)^{-1}y_n$, we only have to show that $y \in (I - T)x$. Clearly, $x_n \in C$ implies that $x \in C$. Also, $y_n \in \in (I - T)x_n$ implies that $y_n = x_n - v_n$, $v_n \in Tx_n$. Since T is nonexpansive with respect to the Hausdorff metric, we can find $v'_n \in Tx$ such that

$$\|v_n - v'_n\| \leqslant \|x_n - x\|.$$

Thus,

$$\liminf \|x_n - x\| \geqslant \liminf \|v_n - v'_n\| \geqslant \liminf \|x_n - y_n - v'_n\|$$

Since T is compact-valued and $y_n \to y$, passig to subsequences, we have $v'_n \to v \in Tx$ and thus,

$$\liminf \|x_n - x\| \geqslant \liminf \|x_n - y - v\|.$$

From Opial's condition, it follows that $y + v = x$. Thus, $y = x - v \in x - Tx$. Hence, Theorem 2.2.12 implies that $(I - T)^{-1}$ is a multivalued random operator. Thus, there exists a random solution of the random operator equation $x(\omega) - - T(\omega, x) = 0$.

Remark. The same result is valid for a weakly compact star-shaped subset C (see [242, Th.3.4]).

We shall conclude this paragraph with several common fixed point theorems for singlevalued random operators and multivalued random operators, as obtained by S. Itoh in [242].

It is known that a random operator $f : \Omega \times X \to X$ is said to *commute* with a multivalued random operator $T : \Omega \times X \to 2^X$ if, for each $\omega \in \Omega$, $f(\omega, T(\omega, x)) \subset T(\omega, f(\omega, x))$.

A mapping $F : X \to 2^X$ is called *asymptotically regular* if, for each $x \in X$, there exists a sequence $\{x_n\}_{n \in \mathbb{N}}$ such that $x_0 = x$, $x_{n+1} \in F(x_n)$ and $d(x_n, x_{n+1}) \to 0$ as $n \to \infty$. If $f : X \to X$ is singlevalued, f is called asymptotically regular if for any $x \in X$, $d(f^n(x), f^{n+1}(x)) \to 0$ as $n \to \infty$.

THEOREM 4.3.14. *Let X be a separable bounded closed convex subset of a Banach space, let $f : \Omega \times X \to X$ be a continuous asyptotically regular random operator, and let $T : \Omega \times X \to$ $\to CK(X)$ be a continuous condensing multivalued random operator. Suppose that f and T commute. Then there exists a common random fixed point y of f and T, i.e., for each $\omega \in \Omega$, $f(\omega, y(\omega)) =$ $= y(\omega) \in T(\omega, y(\omega))$.*

Proof. According to Theorem 4.3.10, T has a random fixed point y_1. Then the mapping $y_2 : \Omega \to X$, defined by $y_2(\omega) =$ $= f(\omega, y_1(\omega))$, is measurable. On the other hand, since f commutes with T, y_2 is a random fixed point of T. By induction, the sequences $\{y_n\}_{n \in \mathbb{N}}$ of mappings y_n: $\Omega \to X$, for which $y_{n+1}(\omega) =$ $= f(\omega, y_n(\omega))$, $\omega \in \Omega$, $n \in \mathbb{N}$, are random fixed points of T. Since T is condensing, for any $\omega \in \Omega$, $\{y_n(\omega)\}_{n \in \mathbb{N}}$ is precompact. For each n, define $F_n : \Omega \to K(X)$ by

$$F_n(\omega) = \mathrm{cl}\,\{y_i,\ i \geqslant n\},$$

and define $F : \Omega \to K(X)$ by

$$F(\omega) = \bigcap_{n=1}^{\infty} F_n(\omega);$$

then F is measurable (see Th. 2.2.2) and hence there exists a measurable selector y of F. Now we prove that y is a common random fixed point of f and T. Let $\omega \in \Omega$ be a fixed element; then there exists $\{y_{n_k}(\omega)\}_{k \in \mathbb{N}} \subset \{y_n(\omega)\}_{n \in \mathbb{N}}$ such that $y_{n_k}(\omega) \to y(\omega)$ as $k \to \infty$. From the upper semicontinuity of $T(\omega, .)$, $y(\omega) \in$ $\in T(\omega, y(\omega))$, since $f(\omega, .)$ is asymptotically regular, $y_{n+1}(\omega) \to y(\omega)$ and hence $f(\omega, y(\omega)) = y(\omega)$.

THEOREM 4.3.15. *Let X be a weakly compact star-shaped subset of a separable Banach space satisfying Opial's condition. Let $f : \Omega \times X \to X$ be a nonexpansive and asymptotically regular random operator and let $T : \Omega \times X \to K(X)$ be a nonexpansive random operator. Suppose that f commutes with T. Then there exists a common random fixed point of f and T.*

Proof. From the Remark to Theorem 4.3.13, it follows that T has a random fixed point y_1. As in the proof of the above theorem, there exists a sequence $\{y_n\}_{n\in\mathbb{N}}$ of random fixed points of T such that $y_{n+1}(\omega) = f(\omega, y_n(\omega))$ for $\omega \in \Omega$, $n \in \mathbb{N}$. Define a sequence of mappings $F_n : \Omega \to WK(X)$ and a mapping $F : \Omega \to WK(X)$ as in the proof of the above theorem. Then F is w^*-measurable and has a measurable selector y. We shall now prove that y is a common random fixed point of f and T. For any $\omega \in \Omega$, there exists $\{y_{n_k}(\omega)\}_{k\in\mathbb{N}} \subset \{y_n(\omega)\}_{n\in\mathbb{N}}$ such that $y_{n_k} \xrightarrow{w} y(\omega)$. Since $f(\omega,.)$ is asymptotically regular, $y_{n_{k+1}} \xrightarrow{w} y(\omega)$. On the other hand, $I - f(\omega,.)$ and $I - T(\omega,.)$ are demiclosed (cf. Opial [324] and Lami Dozo [274]), and hence $f(\omega, y(\omega)) = y(\omega) \in T(\omega, y(\omega))$ and the theorem is proven.

For nonexpansive random operators in Hilbert spaces, the following result was given in [238]:

THEOREM 4.3.16. *Let X be a bounded closed convex subset of a separable Hilbert space, let $f : \Omega \times X \to X$ be a nonexpansive random operator and let $T : \Omega \times X \to CK(X)$ be a nonexpansive multivalued random operator. If f and T commute, then there exists a common random fixed point of T and f.*

Proof. Let $P : \Omega \times X \to X$ be the nonexpansive random operator defined by $P(\omega, x) = P_{M(\omega)}(x)$, where $M(\omega) = \{x \in X;\ f(\omega, x) = x\}$. Since a Hilbert space is strictly convex, then from a fixed point theorem due to Browder [75], for each $\omega \in \Omega$, $M(\omega)$ is nonempty, closed and convex. Define a mapping $F : \Omega \times X \to CK(X)$ by $F(\omega, x) = T(\omega, P(\omega, x))$. Then, according to Proposition 2.2.2, $F(., x)$ is measurable for every $x \in X$; hence, F is a nonexpansive random operator. By virtue of Theorem 4.3.13, F has a random fixed point y. For each fixed point $\omega \in \Omega$,

$$f(\omega, F(\omega, y(\omega))) = f(\omega, T(\omega, P(\omega, y(\omega)))) \subset T(\omega, f(\omega, P(\omega, y(\omega)))) =$$

$$= T(\omega, P(\omega, y(\omega))) = F(\omega, y(\omega)).$$

Thus, $P(\omega, F(\omega, y(\omega))) \subset F(\omega, y(\omega))$ (cf. [238]), and, in particular $P(\omega, y(\omega)) \in F(\omega, y(\omega))$. It follows that the measurable mapping $y : \Omega \to X$, defined by $y(\omega) = P(\omega, y(\omega))$, is a common random fixed point of T and f.

COROLLARY. *If X is a bounded closed convex subset of a separable Hilbert space, $f, g : \Omega \times X \to X$ are nonexpansive singlevalued random operators. If f and g commute, then there exists a random fixed point of g and f.*

4.4. FIXED POINT THEOREMS FOR RANDOM SINGLEVALUED AND MULTIVALUED OPERATORS ON STOCHASTIC DOMAINS

A. T. Bharucha-Reid mentions in most of his works on the random fixed point theorem that the domain of a random operator may be stochastic. A definition and a rigurous study of the classes of singlevalued and multivalued random operators is due to H. W. Engl [132], [133], [134], [136], [137], [138]. He obtained a quite general result which ensures a stochastic analogue to every fixed point theorem regarding continuous operators on separable metric spaces. A random fixed point theorem for only upper semicontinuous multivalued random operators is also given in refs. [132], [138]. Important deterministic fixed point theorems as the Browder-Göhde-Klee-Kirk Theorem for nonexpansive mappings or respectively Kakutani and Bohnenblust and Karlin are generalized into the stochastic case.

Interesting results were also obtained by A. Nowak [313], [314], G. Bocşan [51], [55] and G. Bocşan, Gh. Constantin and V. Radu [58], [61] [62]. Some of the above-mentioned results will be presented in the sequel.

4.4.1. *Random Fixed Point Theorems for Random Operators with Stochastic Domains*

Since the random operator is a measurable mapping, it seems reasonable to allow the domain of $T(\omega,.)$ to vary measurably with ω. This requirement is imposed by the definition of random fixed points themselves whose existence must be ensured.

DEFINITION 4.4.1. Let X be a real separable Banach space and let $(\Omega, \mathscr{K}, \mu)$ be a σ-finite measure space. A multivalued mapping $C:\Omega \to \{A : A \subseteq X \land A \neq \emptyset\}$ is said to be *measurable* if for all open $D \subseteq X \{\omega \in \Omega : C(\omega) \cap D \neq \emptyset\} \in \mathscr{K}$. Such a multivalued mapping is also called a random subset of X (if $C(\omega)$ is closed for all $\omega \in \Omega$, then C is called a closed *random subset*).

We call C *separable* if C is measurable and if there exists a countable set $Z \subseteq X$ such that for all $\omega \in \Omega$, $\mathrm{cl}(Z \cap C(\omega)) = C(\omega)$.

We shall use the words *stochastic* and *random* interchangeably also when μ is not a probability measure.

DEFINITION 4.4.2. Let $C:\Omega \to 2^X$. A mapping $T:\mathrm{Gr}C \to X$ will be called *random operator with stochastic domain* C (i.e. the domain of each realization $T(\omega, .)$ depends on ω), if C is measurable and if for all open $D \subseteq X$ and all $x \in X$, $\{\omega \in \Omega : x \in C(\omega) \wedge T(\omega, x) \in D\} \in \mathscr{K}$.

A mapping $x : \Omega \to X$ will be called *wide-sense fixed point of* T if, for all $\omega \in \Omega$, $x(\omega) \in C(\omega)$ and $x(\omega) = T(\omega, x(\omega))$; it will be called *random fixed point of* T if, for μ almost all $\omega \in \Omega$, $x(\omega) \in C(\omega)$ and $x(\omega) = T(\omega, x(\omega))$ and x is a measurable function of ω.

We now state and prove a composite theorem due to Hanš [204] for the case of deterministic domains, and to Engl [132] for the case of stochastic domains. This theorem is used in the study of the solutions of random equations.

THEOREM 4.4.1. *Let X be a separable Banach space, let $C : \Omega \to 2^X$ be separable and let T be a random operator with stochastic domain C. Let $x : \Omega \to X$ be measurable such that for all $\omega \in \Omega$, $x(\omega) \in C(\omega)$. Furthermore, let $T(\omega, .)$ be continuous on $C(\omega)$ for all $\omega \in \Omega$. Then $\omega \to T(\omega, x(\omega))$ is measurable.*

Proof. For all $\omega \in \Omega$, $i, n \in \mathbb{N}$, we define by induction

$$A_{i,n}(\omega) = \begin{cases} \left\{z \in C(\omega) : \|z - z_i\| < \dfrac{1}{n}\right\} - (A_{1,n} \cup \ldots \cup A_{i-1,n}) & \text{if } z_i \in C(\omega), \\ \emptyset & \text{if } z_i \notin C(\omega) \end{cases}$$

where $\{z_i\}_{i \in N}$ is a set which can be taken as the set Z appearing in Definition 4.4.1. Then, for all $\omega \in \Omega$, $z \in C(\omega)$, $n \in \mathbb{N}$, there exists exactly one $i \in \mathbb{N}$ such that $z \in A_{i,n}(\omega)$. Hence, one can define $T_n(\omega, x(\omega)) = T(\omega, z_i)$, for all $\omega \in \Omega$, where i is the unique number such that $x(\omega) \in A_{i,n}(\omega)$. Now, for all $n \in \mathbb{N}$ and for open $D \subseteq X$,

$$\{\omega \in \Omega : T_n(\omega, x(\omega)) \in D\} =$$

$$= \bigcup_{i=1}^{\infty} (\{\omega \in \Omega : z_i \in C(\omega) \wedge T(\omega, z_i) \in D\} \cap$$

$$\cap \{\omega \in \Omega : x(\omega) \in A_{i,n}(\omega)\}).$$

For $i, n \in \mathbb{N}$, let us define

$$B_{i,n} = \left\{\omega \in \Omega : z_i \in C(\omega) \wedge \|x(\omega) - z_i\| \leqslant \frac{1}{n}\right\} =$$

$$= \left\{\omega \in \Omega : \|x(\omega) - z_i\| \leqslant \frac{1}{n}\right\} \cap$$

$$\cap \bigcap_{m \in \mathbb{N}} \left\{\omega \in \Omega : C(\omega) \cap \left\{x \in X : \|x - z_i\| < \frac{1}{m}\right\} \neq \emptyset\right\} \in \mathscr{K},$$

and note that because of the closedness of $C(\omega) : z_i \in C(\omega) \Leftrightarrow$, for all $m \in \mathbb{N}$, $C(\omega) \cap \left\{x \in X : \|x - z_i\| < \frac{1}{m}\right\} \neq \emptyset$. Then,

$$\{\omega \in \Omega : x(\omega) \in A_{i,n}(\omega)\} = B_{i,n} \setminus (B_{1,n} \cup \ldots \cup B_{i-1,n}) \in \mathscr{K}.$$

Therefore, $\{\omega \in \Omega : T_n(\omega, x(\omega)) \in D\} \in \mathscr{K}$, which means that $y_n : \Omega \to X$ as defined by $\omega \to T_n(\omega, x(\omega))$ is measurable. Hence, $f \circ y_n$ is measurable for every $f \in X^*$. Since all $T(\omega, .)$ are continuous, then, for all $\omega \in \Omega$, $\lim_{n\to\infty} T_n(\omega, x(\omega)) = T(\omega, x(\omega))$. Therefore, for all $f \in X^*$, $f(T(., x(.))) = \lim_{n\to\infty} (f \circ y_n)$ is measurable, which yields the weak-measurability of $\omega \to T(\omega, x(\omega))$. This fact together with the remark that in a separable Banach space the strong random variable and weak random variable are equivalent implies the measurability of $\omega \to T(\omega, x(\omega))$.

As a direct application we give the following

COROLLARY (Stochastic version of Banach's fixed point theorem [135]). *Let $C : \Omega \to 2^X$ be separable, let $K : \Omega \to [0, 1)$ and let T be a random operator with stochastic domain C such that for all $\omega \in \Omega$, $T(\omega, .)(C(\omega)) \subseteq C(\omega)$ and $\|T(\omega, x) - T(\omega, y)\| \leqslant k(\omega)\|x - y\|$ for all $x,y \in C(\omega)$. Then T has a random fixed point.*

Proof. Let $x_0 : \Omega \to X$ be a measurable selector of C which exists according to the Kuratowski and Ryll-Nardzewski selection theorem (Th. 2.2.4.). Let $x_n(\omega) = T(\omega, x_{n-1}(\omega))$, $n \in \mathbb{N}$. According to the above theorem, every x_n is measurable and $x_n(\omega) \to x(\omega)$ as $n \to \infty$ for all $\omega \in \Omega$, where $x(\omega)$ is the unique fixed point of $T(\omega,.)$; x is measurable as the limit of measurable mappings.

Remarks. (1) The above stochastic version of Banach's fixed point theorem could be also derived (as other classical theorems are) from the general theorems given below. In spite of this fact, Engl [135] gave a specifically tailored proof for the nonexpansive mapping obtaining a stochastic analogue to the Browder-Gohde-Klee-Kirk theorem. The Chebyshef centers method should also be highlighted [133], [135], [360].

(2) The separability condition for the domain-function C is not restrictive since if Ω is countable, then every measurable C is separable. Reference [138, Prop. 4] also shows that every measurable mapping $C : \Omega \to CC(X)$ with solid values (i.e., for all $\omega \in \Omega$ int$C(\omega) \neq \emptyset$) is separable; in particular, if $x_0 : \Omega \to X$ and $r : \Omega \to \mathbb{R}_0^+$ are measurable, then $C : \Omega \to CC(X)$, defined by $C(\omega) = \{x \in X : \|x - x_0(\omega)\| \leqslant r(\omega)\}$, is measurable and if $r(\omega) > 0$ for all $\omega \in \Omega$, then C is separable.

As an application of the existence of a fixed point and weak convergence of Mann's iteration procedure to a fixed point of a nonexpansive mapping [see Theorem 2.2.15] and of Theorem 4.4.1, we give

THEOREM 4.4.2 [134]. *Let X be a reflexive separable Opial space and let $C : \Omega \to CC(X)$ be a separable and uniformly bounded mapping (i.e., there exists an $r > 0$ such that for all $\omega \in \Omega$ and $x \in C(\omega)$, $\|x\| \leqslant r$). If T is a random operator with stochastic domain C such that, for all $\omega \in \Omega$, $T(\omega, .)$ is a nonexpansive self-mapping of $C(\omega)$ (i.e., for all $\omega \in \Omega$, and all $x,y \in C(\omega)$, $\| T(\omega, x) - T(\omega, y) \| \leqslant \| x - y \|$), then T has a random fixed point.*

Proof. (a) Let us first assume that $(\Omega, \mathscr{K}, \mu)$ is complete. Let $x_0 : \Omega \to X$ be measurable with $x_0(\omega) \in C(\omega)$ for all $\omega \in \Omega$; according to the Kuratowski and Ryll-Nardzewski selection theorem, such an x_0 exists. We define, for all $n \in \mathbb{N}$ and $\omega \in \Omega$, $x_{n+1}(\omega) = \frac{1}{2}(x_n(\omega) + T(\omega, x_n(\omega)))$. From Theorem 4.4.1 and from the properties of Banach-space-valued random variables, it follows that each $x_n : \Omega \to X$ is measurable. We define $\check{C} : \Omega \to 2^X$ as

$$\check{C}(\omega) = \{x \in C(\omega) : \limsup \|x_n(\omega) - x\| = \inf_{y \in C(\omega)} \|x_n(\omega) - y\|\}.$$

According to Corollary 2.2.1, there exists an $x : \Omega \to X$ with $\check{C}(\omega) = \{x(\omega)\}$ and $T(\omega, x(\omega)) = x(\omega)$ for all $\omega \in \Omega$. By Proposition 2.2.6, $\check{C}$ and hence x too are measurable and so x is a random fixed point of T.

(b) If $(\Omega, \mathscr{K}, \mu)$ is not complete, we denote by $(\Omega, \mathscr{K}^*, \mu^*)$ the Lebesgue completion of $(\Omega, \mathscr{K}, \mu)$. Since the assumptions of the theorem are also fulfilled for $(\Omega, \mathscr{K}^*, \mu^*)$, we can consider x^* as a fixed point of T which exists as in (a) but with respect to the complete measure space. Let $\{B_i\}_{i \in \mathbb{N}}$ be a countable generator of the σ-algebra on $K_r = \{x \in X : \|x\| \leqslant r\}$ which is generated by the (norm)-open sets. Then, for all $n \in \mathbb{N}$, $x^{*-1}(B_n) = A_n \cup L_n$ with $A_n \in \mathscr{K}$, $L_n \subseteq N_n$, $N_n \in \mathscr{K}$ with $\mu(N_n) = 0$. Let $N = \bigcup_{n \in \mathbb{N}} N_n$ and define $x : \Omega \to K_r$ by

$$\omega \to \begin{cases} x^*(\omega) & \text{if } \omega \notin N \\ \bar{x} & \text{if } \omega \in N \end{cases},$$

where $\bar{x}$ is an arbitrary but fixed element in K_r. Then x is a random fixed point of T.

Remark. This result improves Theorem 12 of ref. [135] since X is not necessarily a uniformly convex space.

In order to prove the general stochastic fixed point theorem, which assumes the existence of a random fixed point of a continuous random operator with stochastic domain under rather general conditions provided that the corresponding deterministic fixed point problem is solvable, we give

DEFINITION 4.4.3. Let $C : \Omega \to 2^X$ be measurable and let T be a random operator with stochastic domain C. We recall that T is *continuous* if $T(\omega, .)$ is continuous on $C(\omega)$ for all $\omega \in \Omega$.

DEFINITION 4.4.4 [136]. Let $C : \Omega \to 2^X$ be measurable with convex and solid (i.e. int $C(\omega) \geqslant \emptyset$) values, let T be a continuous random operator with stochastic domain C, let $x_0 : \Omega \to X$ be a measurable selector of int$C$ (the condition for the existence is given in Theorem 2.2.13 or is of course true if $\bigcap_{\omega \in \Omega} C(\omega) \neq \emptyset$ or if it is verified directly). We define :

$$b : \Omega \times X \to \overline{\mathbb{R}}$$

by $(\omega, x) \to \sup \{t \geqslant 0 : x_0(\omega) + t(x - x_0(\omega)) \in C(\omega)\}$;

$$a : \Omega \times X \to X$$

$$\text{by } (\omega, x) \to \begin{cases} x & \text{if } x \in C(\omega) \\ x_0(\omega) + b(\omega, x)(x - x_0(\omega)) & \text{if } x \notin C(\omega) \end{cases}$$

(here, $b(\omega, x) < \infty$)

$$N : \Omega \times X \to \mathbb{R} \text{ by } (\omega, x) \to \|a(\omega, x) - T(\omega, a(\omega, x))\| + \\ + \|x - a(\omega, x)\|.$$

LEMMA 4.4.1. (i) $N(\omega, x) = 0 \Leftrightarrow a \in C(\omega)$ and $T(\omega, x) = x$.
(ii) *For all* $x \in X$, $b(., x)$ *is measurable.*
(iii) *For all* $x \in X$, $a(., x)$ *is measurable.*
(iv) $N : \Omega \times X \to \mathbb{R}$ *is jointly measurable (i.e. measurable as a function on the measurable space* $(\Omega \times X, \mathscr{K} \times \mathscr{B}_X)$).

Proof. (i) This follows immediately from Definition 4.4.4.
(ii) Let $Z = \{x_n\}_{n \in \mathbb{N}}$ be dense in X, let $x \in X$ be fixed but arbitrary. Then, for all $r > 0$, $[b(., x)]^{-1}([r, +\infty]) = \{\omega \in \Omega :$

$$x_0(\omega) + r(x - x_0(\omega)) \in C(\omega)\} = \bigcap_{n \in \mathbb{N}} \bigcup_{m \in \mathbb{N}} \left(\left\{ \omega \in \Omega : \|(1-r)x_0(\omega) - \right.\right.$$

$$\left. - z_m\| \leqslant \frac{1}{n} \right\} \cap \left\{ \omega \in \Omega : C(\omega) \cap \left(rx + \left\{ z \in X : \|z - z_m\| < \frac{1}{n} \right\} \right) \neq \right.$$

$$\left. \neq \emptyset \right\} \in \mathscr{K}.$$

For $r \leqslant 0$, $[b(.\,,x)]^{-1}([r,\infty]) = \Omega$. So $b(.\,,x)$ is measurable.

(iii) We have

$$a(\omega, x) = x\chi_{C(\omega)}(x) + [x_0(\omega) + b(\omega, x)(x - x_0(\omega))](1 - \chi_{C(\omega)}(x)),$$

where $\chi_.$ is the indicator function. Let $x \in X$ be fixed but arbitrary; then, for all $r \in \mathbb{R}$,

$$[\chi_{C(.)}(x)]^{-1}((-\infty, r]) = \begin{cases} \Omega & \text{if } r \geqslant 1 \\ \emptyset & \text{if } r < 0 \\ \Omega - \{\omega \in \Omega : C(\omega) \cap \{x\} \neq \emptyset\} & \text{otherwise} \end{cases} \in \mathscr{K}$$

as $\{\omega \in \Omega : C(\omega) \cap \{x\} \neq \emptyset\} = \bigcap_{n \in \mathbb{N}} \left\{\omega \in \Omega : C(\omega) \cap \left\{y \in X : \|y - x\| < \frac{1}{n}\right\} \neq \emptyset\right\}$ because of the closedness of $C(\omega)$. Hence, $\chi_{C(\omega)}$ is measurable.

Let $f \in X^*$ be arbitrary. Then $f(a(\omega, x)) = f(x).\ \chi_{C(.)}(x) + (1 - \chi_{C(.)}(x))\ (f(x_0(.)) + b(.\,, x)\,f(x - x_0(.)))$ is measurable since $b(.\,,\ x)$ is measurable.

From the connection between weak and strong measurability in separable Banach spaces, it follows also that $a(.\,, x)$ is a measurable mapping.

(iv) From the measurability of $a(.\,, x)$, and from the properties of the sum and norm of Banach space-valued random variables (see [31, Chap. I]) follows the measurability of $N(.\,, x)$. Indeed, from Remark 2, following Theorem 4.4.1, we conclude that C is separable and hence Theorem 4.4.1 is applicable to T and $a(.\,, x)$.

Moreover, $N(\omega, .)$ is continuous for all $\omega \in \Omega$. Indeed, for every $\omega \in \Omega$ fixed but arbitrary, we can suppose $x_0(\omega) = 0$. Then for all $x \in X$, $b(\omega, x) = \sup\ \{t \geqslant 0 : t\,x\,(\omega) \in C(\omega)\} = \sup\,\{s^{-1} > 0 : x(\omega) \in s\ C(\omega)\} = p(\omega, x)^{-1}$, where $p(\omega,\ .)$ is the Minkowski functional of $C(\omega)$. Then, $a(\omega, x) = \min\{1, p(\omega, x)^{-1}\}, x$ is continuous as a function of x and therefore $N(\omega, .)$ is continuous. From these properties and from Theorem 2.1.4, it follows that N is jointly measurable, i.e. measurable as a function on the measurable space $(\Omega \times X, \mathscr{K} \times \mathscr{B}_X)$.

A general stochastic fixed point theorem is given by

THEOREM 4.4.3 [136, Th. 8]. *Let $C : \Omega \to CC(X)$ be measurable with solid values. We assume that there exists a measurable selector of int$C$ (cf. Theorem 2.2.13). Let T be a continuous random operator with stochastic domain C which has a wide-sense fixed point. Then T has a random fixed point.*

Proof. (a) Let $(\Omega, \mathscr{K}, \mu)$ be complete.

We take b, a, N as in Definition 4.4.4 and define $F : \Omega \to 2^X$ by $F(\omega) = \{x \in C(\omega) : T(\omega, x) = x\}$. Then from the assumption about the existence of a wide-sense fixed point, it follows that each $F(\omega) \neq \emptyset$. Since T is continuous, each $F(\omega)$ is closed. For all $\omega \in \Omega$, $F(\omega) = \{x \in X : N(\omega, x) = 0\}$ (see Lemma 4.4.1 (i)). Since N is jointly measurable, $\text{Gr}F = \{(\omega, x) \in \Omega \times X : x \in F(\omega)\} = N^{-1}(\{0\})$ is a measurable subset of $\Omega \times X$. Then F is measurable (cf. Theorem 2.2.1) and from the Kuratowski and Ryll-Nardzewski selection theorem there exists a measurable selector of F, i.e. a random fixed point of T.

(b) If $(\Omega, \mathscr{K}, \mu)$ is not complete, the rest of the proof is as in point (b) of the proof of Theorem 4.4.2.

By combining the above result with the selection Theorem 2.2.13 one gets

THEOREM 4.4.4. *Let X be reflexive, let $C : \Omega \to 2^X$ be measurable and uniformly bounded with convex and solid values, let T be a continuous random operator with stochastic domain C which has a wide-sense fixed point. Then T has a random fixed point.*

Remark. An improved result will be obtained as a corollary of the general result for multivalued random operators with stochastic domain. Interesting applications will be given in Chapter 5.

Before proceeding to the multivalued case, we shall consider several results of the singlevalued case. Several measurability properties obtained for metric spaces will be presented since these are used to prove the stochastic version of the classical Urysohn and Tietze theorem, as well as a general random fixed point. This will allow us to construct a unified view of the above results together with those of Sec. 4.3.

Let X be a complete metric space and let $(\Omega, \mathscr{K}, \mu)$ be a σ-finite and complete measure.

PROPOSITION 4.4.1. *Let $C : \Omega \to 2^X$ be separable, let f be a continuous random function with stochastic domain C and let $a : \Omega \to \mathbb{R}$ be a measurable function. Then the multivalued mapping*

$$F_<(\omega) = \{x \in C(\omega) : f(\omega, x) < a(\omega)\}, \quad \omega \in \Omega$$

is measurable.

Proof. Let $Z \subset X$ be a countable set such that $C(\omega) = \overline{Z \cap C(\omega)}$ for all $\omega \in \text{Dom}\, C$. Let $Y = \bar{Z}$ be a complete separable metric space and let C be a random subset of Y.

Since C is closed-valued, since Y is a complete separable metric space and since μ is σ-finite and complete, from Theorem 2.2.5 it follows that C has a Castaing representation U, i.e. there exists at most one countable set $U = \{u_n\}_{n \in \mathbb{N}}$, each u_n being a measurable selector of C and $C(\omega) = \overline{U(\omega)}$ for all $\omega \in \Omega$.

Then, we have :

$$(*)\ \{\omega \in \Omega : F_<(\omega) \cap D \neq \emptyset\} = \bigcup_{u_n \in U} [\{\omega \in \Omega : f(\omega, u_n(\omega)) < a(\omega) \cap u_n^{-1}(D)\}]$$

for every open subset $D \subseteq Y$. Indeed, let $\omega \in \Omega$ be such that $F_<(\omega) \cap D \neq \emptyset$, i.e., there exists an $x \in C(\omega) \cap D$ and $f(\omega, x) < a(\omega)$. Since D is open and C has a Castaing representation, there exists a sequence $x_n = u_n(\omega)$, $x_n \to x$, $x_n \in D$, $u_n \in U$. From the continuity of $f(\omega,.)$ on $C(\omega)$ and $f(\omega, x) < a(\omega)$, it follows that there exists an n_0 such that $x_{n_0} \in D$ and $f(\omega, x_{n_0}) < a(\omega)$, i.e.

$$\omega \in \{\omega \in \Omega : f(\omega, u_{n_0}(\omega)) < a(\omega)\} \cap u_{n_0}^{-1}(D).$$

Since the converse inclusion is obvious, equality $(*)$ is proven.

On the other hand, by Theorem 4.4.1, the function $\omega \to f(\omega, u(\omega))$ is measurable and, hence,

$$\{\omega \in \Omega : F_<(\omega) \cap D \neq \emptyset\} \in \mathcal{K}.$$

The proof of the proposition is complete.

R. T. Rockafellar [363, Prop. 2.1] proved that if $f : \Omega \times \mathbb{R}^n \to \mathbb{R}$ is a Charathéodory function and if $a : \Omega \to \mathbb{R}$ is a measurable function, then the multivalued function

$$\omega \to \{x \in \mathbb{R}^n : f(\omega, x) \leqslant a(\omega)\},\ \omega \in \Omega$$

is closed-valued and measurable. An improvement on this result is given by

PROPOSITION 4.4.2. *Let $C : \Omega \to 2^X$ be separable, let f be a continuous random function with stochastic domain C and let $a : \Omega \to \mathbb{R}$ be a measurable function. Then the multivalued mapping*

$$F_\leqslant(\omega) = \{x \in C(\omega) : f(\omega, x) \leqslant a(\omega)\},\quad \omega \in \Omega$$

is a closed-valued measurable mapping.

Proof. For a compact subset $E \subset X$ and for each $n \in \mathbb{N}$, let

$$D_n = \left\{x \in X : d(x, E) < \frac{1}{n}\right\}$$

and

$$F_n(\omega) = \left\{ x \in C(\omega) : f(\omega, x) < a(\omega) + \frac{1}{n} \right\}.$$

Since each D_n is an open set of X, it follows from the previous proposition that every F_n is measurable. Let us prove that

$$\{\omega \in \Omega : F_{\leqslant}(\omega) \cap C \neq \emptyset\} = \bigcap_{n \in \mathbb{N}} \{\omega \in \Omega, F_n(\omega) \bigcap D_n \neq \emptyset\}.$$

Consider $\omega \in \bigcap_{n \in \mathbb{N}} \{\omega \in \Omega : F_n(\omega) \cap D_n \neq \emptyset\}$. Hence, for every $n \in \mathbb{N}$, there exists an $x_n \in F_n \cap D_n$. But $x_n \in D_n$ implies that there exists a $y_n \in C$ such that

$$d(x_n, y_n) = \inf_{y \in C} d(x_n, y) = d(x_n, C) < \frac{1}{n} \cdot$$

Since $\{y_n\} \subset C$ and since C is compact, there exists $\{y_{n_k}\} \subset \subset \{y_n\}$ such that $y_{n_k} \to y$ as $k \to \infty$. Then

$$d(x_{n_k}, y) \leqslant d(x_{n_k}, y_{n_k}) + d(y_{n_k}, y) < \frac{1}{n_k} + d(y_{n_k}, y),$$

and hence $x_{n_k} \to y$ as $k \to \infty$. As $x_{n_k} \in F_{n_k}(\omega) \subset C(\omega)$, it follows that $f(\omega, x_{n_k}) < a(\omega) + \dfrac{1}{n_k}$. Since $C(\omega)$ is closed and since $f(\omega,.)$ is continuous on $C(\omega)$, it follows that $y \in C(\omega)$ and that $f(\omega, y) \leqslant a(\omega)$, i.e. $y \in F_{\leqslant}(\omega)$. As $y \in C$, then $F_{\leqslant}(\omega) \cap C \neq \neq \emptyset$ and the inclusion is proved. The converse inclusion is obvious, since $F_{\leqslant}(\omega) \leqslant F_n$ and $C \subseteq D_n$ for all $n \in \mathbb{N}$.

Hence $F_{\leqslant}$ is a compactly-measurable multivalued mapping. It is clear that $F_{\leqslant}(\omega)$ is closed for all $\omega \in \Omega$ and that C can be considered as a closed random subset of the complete separable metric space $Y = \bar{Z}$, where Z is a countable subset of X for which $C(\omega) = \overline{Z \bigcap C(\omega)}$. Thus, the compact measurability of $F_{\leqslant}$ is equivalent with the open-measurability of $F_{\leqslant}$, because μ is σ-finite and complete (cf. Theorem 2.2.1) and, hence, the theorem is proven.

COROLLARY 1. *Let $f : \Omega \times X \to \mathbb{R}$ be a Carathéodory function and let $a : \Omega \to \mathbb{R}$ be a measurable function. If C is a separable measurable multivalued mapping on X, then the multivalued mapping*

$$F(\omega) = \{x \in C(\omega) : f(\omega, x) \leqslant a(\omega)\}, \quad \omega \in \Omega$$

is a closed-valued random subset of X.

Proof. It suffices to apply the above proposition to the restriction g_C of f to C, defined as

$$g(\omega, x) = f(\omega, x), \quad x \in C(\omega).$$

COROLLARY 2. *Let $f : \Omega \times X \to \mathbb{R}$ be a Carathéodory function and let $a : \Omega \to \mathbb{R}$ be a measurable function.*

If the complete metric space X is separable, then the function

$$F(\omega) = \{x \in X : f(\omega, x) \leqslant a(\omega)\}, \quad \omega \in \Omega$$

is a closed random subset of X.

Proof. Since X is separable, the multivalued mapping

$$C(\omega) = X, \omega \in \Omega$$

is a separable random subset of X. Applying Corollary 1, the conclusion is immediate.

COROLLARY 3 (Rockafellar [363]). *If $f : \Omega \times \mathbb{R}^n \to \mathbb{R}$ is a Carathéodory function and $a : \Omega \to \mathbb{R}$ is a measurable function, then the multivalued mapping*

$$F(\omega) = \{x \in \mathbb{R}^n : f(\omega, x) \leqslant a(\omega)\}, \omega \in \Omega$$

is a closed random subset of $\mathbb{R}^n$.

Proof. It is easy to see that this corollary is obtained as a particular case of Corollary 2 for $X = \mathbb{R}^n$.

COROLLARY 4. *Let f be a continuous random function with stochastic domain C and let $a . \Omega \to \mathbb{R}$ be a measurable function. If C is a separable random subset of X, we have:*

(i) $F_{\geqslant}(\omega)=\{x \in C(\omega) : f(\omega, x) \geqslant a(\omega)\}$, $\omega \in \Omega$ *is a closed random subset of X,*

(ii) $F_{>}(\omega) = \{x \in C(\omega) : f(\omega, x) > a(\omega)\}$, $\omega \in \Omega$ *is a measurable multivalued mapping.*

Proof. It is sufficient to apply the above properties to the functions $-f(\omega, x)$ and $-a(\omega)$, respectively.

A stochastic analogue of Urysohn's Lemma and Tietze's theorem is given below.

First note that we shall say that the random subsets C_1 and C_2 of the complete metric space X are disjoint if the sets $C_1(\omega)$ and $C_2(\omega)$ are disjoint for all $\omega \in \Omega$.

PROPOSITION 4.4.3. *Let C_1 and C_2 be the disjoint and closed random subsets of the separable metric space X and let a and b be two measurable functions on Ω such that $a(\omega) < b(\omega)$ for all $\omega \in \Omega$.*

Then there exists a Carathéodory function $f : \Omega \times X \to \mathbb{R}$ *with the properties*

(i) $f(\omega, x) = \begin{cases} a(\omega) \text{ for } x \in C_1(\omega) \\ b(\omega) \text{ for } x \in C_2(\omega) \end{cases}$

(ii) $f(\omega, x) \in [a(\omega), b(\omega)]$, *for* $x \in X$

for each $\omega \in \Omega$.

Proof. Since C_1 and C_2 are measurable, the sets DomC_1 and DomC_2 are $\mathscr{K}$-measurable. Since X is separable and since μ is σ-finite and complete, from Theorem 2.2.1, it follows that the mappings

$\omega \to d(x, C_1(\omega))$, $\omega \in \text{Dom}C_1$ and $\omega \to d(x, C_2(\omega))$, $\omega \in \text{Dom}C_2$ are $\mathscr{K}$-measurable for each $x \in X$.

Putting $d(x, \emptyset) = 1$, it becomes obvious that

$$f(\omega, x) = a(\omega) + \frac{(b(\omega) - a(\omega))d(x, C_1(\omega)),}{d(x, C_1(\omega)) + d(x, C_2(\omega))}, \; \omega \in \Omega, \; x \in X$$

satisfies the required conditions.

THEOREM 4.4.5. *Let f be a continuous random function with stochastic domain C. If C is a separable random subset of the complete separable metric space X, then there exists a Carathéodory function* $\tilde{f} : \Omega \times X \to \mathbb{R}$ *such that $\tilde{f}$ is a continuous extension of f, i.e., for every* $\omega \in \Omega$,

$$\tilde{f}(\omega, x) = f(\omega, x), \; x \in C(\omega).$$

Outline of the proof. First, let f be a bounded function, i.e., there exists $m : \Omega \to [0, \infty)$ such that

$$|f(\omega, x)| \leqslant m(\omega), \; x \in E(\omega), \text{ for all } \omega \in \Omega.$$

Then define by induction the sequences $\{f_n\}$, $\{g_n\}$, $\{C_n\}$ and $\{E_n\}$, $n = 0,1,2,\ldots$ such that

(i) $f_0(\omega, x) = f(\omega, x)$, $f_n(\omega, x) = f_{n-1}(\omega, x) + g_{n-1}(\omega, x)$, $\omega \in \Omega$, $x \in C(\omega)$

(ii) $C_n(\omega) = \left\{x \in C(\omega) : f_n(\omega, x) \leqslant -\left(\frac{2}{3}\right)^n \frac{m(\omega)}{3}\right\}$, $\omega \in \Omega$

$E_n(\omega) = \left\{x \in C(\omega) : f_n(\omega, x) \geqslant \left(\frac{2}{3}\right)^n \frac{m(\omega)}{3}\right\}$, $\omega \in \Omega$

(iii) g_n is the Carathéodory function on $\Omega \times X$ given by Proposition 4.4.3 for the random subsets C_n and E_n of X and the measurable functions $a_n(\omega) = -\left(\frac{2}{3}\right)^n m(\omega)$ and $b_n(\omega) = = \left(\frac{2}{3}\right)^n m(\omega)$.

Proposition 4.4.3 can be used since f_n is a random continuous function with random domain C. Applying then Proposition 4.4.2 and Corollary 4, it follows that C_n and E_n are the closed random subsets of X, and hence X is a separable metric space.

Let $\tilde{f}$ be the random function defined by

$$\tilde{f}(\omega, x) = \sum_{n=0}^{\infty} g_n(\omega, x).$$

Since

$$|f_n(\omega, x)| \leqslant \left(\frac{2}{3}\right)^n m(\omega), \quad x \in C(\omega), \text{ for all } \omega \in \Omega,$$

it follows that

$$|g_n(\omega, x)| \leqslant \left(\frac{2}{3}\right)^n \frac{m(\omega)}{3}, \quad \omega \in \Omega, \quad x \in X.$$

Thus, $\tilde{f}$ is well-defined and $\tilde{f}$ is a Carathéodory function. Moreover,

$$|\tilde{f}(\omega, x)| \leqslant \sum_{n=0}^{\infty} \left(\frac{2}{3}\right)^n \frac{m(\omega)}{3} = m(\omega),$$

and for $x \in C(\omega)$ we have

$$\tilde{f}(\omega, x) = \sum_{n=0}^{\infty} (f_n(\omega, x) - f_{n+1}(\omega, x)) = f_0(\omega, x) = f(\omega, x).$$

Therefore, the function $\tilde{f}$ is the required function.

If f is not bounded, then the proof is reduced to f bounded by considering a homeomorphism of $\mathbb{R}$ on $(-1,1)$.

As a first application of the random version of Tietze's theorem, we give

THEOREM 4.4.6. *Let f be a continuous function with stochastic domain C. If C is a compact and separable random subset of X, then the mapping $a : \Omega \to \mathbb{R}$, defined by*

$$a(\omega) = \min_{x \in C(\omega)} f(\omega, x),$$

is measurable.

Proof. Let Z be a countable set of X such that $C(\omega) = \overline{Z \cap C(\omega)}$ and $Y = \bar{Z}$. By Theorem 4.4.5, there exists an extension g of f which is a Carathéodory function on $\Omega \times Y$, because Y is a separable metric space and C can be considered as a multivalued mapping in Y. Hence,

$$g(\omega, x) = f(\omega, x), \; x \in C(\omega)$$

for all $\omega \in \Omega$.

Thus,

$$a(\omega) = \min_{x \in C(\omega)} g(\omega, x).$$

Another application of the above result is an implicit function theorem for random functions with stochastic domain:

THEOREM 4.4.7. *Let f be a continuous random function with stochastic domain C and let F be the closed random subset of X such that*

$$F(\omega) \subseteq C(\omega) \text{ for all } \omega \in \Omega.$$

If C is separable and $a(\omega) \in f(\omega, F(\omega))$ for all $\omega \in \Omega$, where $a : \Omega \to \mathbb{R}$ is a given measurable function, then there exists a measurable selector b of F such that

$$a(\omega) = f(\omega, b(\omega))$$

for all $\omega \in \Omega$.

Proof. Let Z be a countable subset of X such that $C(\omega) = \overline{Z \cap C(\omega)}$ and let $Y = \bar{Z}$. Since Y is a complete separable metric space and C is a separable random subset of Y, it follows that f can be extended to a Carathéodory function g on $\Omega \times Y$. Moreover, since F is measurable, it is compactly measurable (μ is σ-finite and complete).

The rest follows by applying Theorem 2.2.5 (ii) to Y, $\mathbb{R}$, F and a.

Remark. The above results also hold in a more general case when the condition of separability of C is replaced by the following condition [51]:

(c) "C is a closed random subset of X and there exists a Castaing representation U of C such that the mapping $\omega \to f(\omega, u(\omega))$, $u \in U$, is measurable".

This condition is satisfied, for example, in the following case:

1) If X is a complete separable metric space and $C \subset X$ is a separable random subset of X, then condition (c) is satisfied for every continuous random mapping T with stochastic domain C.

Likewise, condition (c) is satisfied if C is an arbitrary closed random subset of X and T is the restriction to C of a Carathéodory mapping on $\Omega \times X$.

2) If C is a closed random subset of a complete separable metric space X, then from Theorem 4.4.5 it follows that condition (c) is satisfied if the mappings $\omega \to f(\omega, u(\omega))$ are measurable for every measurable selector of C.

An application of this condition is the new version of the general stochastic fixed point theorem given below.

THEOREM 4.4.8. *Let X be a complete separable metric space and let T be a continuous random operator with stochastic domain C.*

If condition (c) is satisfied for C and T and if T has a wide-sense fixed point, then T has a random fixed point.

Moreover, the multivalued mapping F defined by

$$F(\omega) = \{x \in C(\omega) : T(\omega, x) = x\}, \quad \omega \in \Omega$$

is a nonempty, closed random subset of X and there exists an (at most) countable set of random fixed point U such that

$$F(\omega) = U(\omega) \text{ for all } \omega \in \Omega.$$

Proof. Since $C(\omega)$ is closed and T has a wide-sense fixed point, $F(\omega)$ is nonempty and closed for each $\omega \in \Omega$.

Since C is measurable, the set

$$\Omega_x = \{\omega \in \Omega : x \in C(\omega)\}$$

is $\mathscr{K}$-measurable for every $x \in X$ and since $\omega \to T(\omega, x)$, $\omega \in \Omega_x$, are $\mathscr{K}$-measurable for every $x \in X$, it follows that

$$f(\omega, x) = d(T(\omega, x), x), \quad \omega \in \Omega, \ x \in C(\omega)$$

is a continuous random function with random domain C.

Now, since C and f satisfy condition (c), Proposition 4.4.2, with the separability condition of C replaced by condition (c), implies that

$$F(\omega) = \{x \in C(\omega) : f(\omega, x) \leqslant 0\}$$

is a nonempty closed random subset of X. Hence, F has a Castaing representation U and each $u \in U$ is a random fixed point of T.

Remark. From the first part of the above remark, we obtain the general stochastic fixed point theorem of Engl [132, Corollary 5.7] and Theorem 4.2.12.

The application of random operators satisfying condition (c) on a separable Hilbert space is illustrated below with some results obtained by Pazy [337] and Brézis and Browder.

Let T be a random operator with stochastic domain C and let $R(\omega) = \text{cl}\{y \in H : \ y = x - T(\omega, x), \quad x \in C(\omega)\} = \text{cl}[I - T(\omega,.)]$. Hence, one obtains a multivalued mapping $\omega \to R(\omega)$, $\omega \in \Omega$.

PROPOSITION 4.4.4. *Let T be a random operator on the closed random subset $C \subseteq H$. If T satisfies condition* (c), *then R is a closed random subset of H. If $C(\omega) = H$ for all $\omega \in \Omega$, then R is also convex.*

Proof. Let U be a Castaing representation of C and, for each $u \in U$, let

$$v(\omega) = u(\omega) - T(\omega, u(\omega)), \ \omega \in \Omega.$$

We show that the (at most) countable set $V = \{v\}$ is a Castaing representation of R.

Indeed, we observe that each v is a selector of R, and since T satisfies condition (c), v is measurable. Thus, V is an (at most) countable set of measurable selectors of R.

It remains to show that $R(\omega) = \overline{V(\omega)}$ for all $\omega \in \Omega$. To show this, let ω be fixed in Ω, but arbitrary, and let $x \in R(\omega)$. There exists a sequence $\{y_n\}_{n \in \mathbb{N}}$, $y_n = x_n - T(\omega, x_n)$, $x_n \in C(\omega)$ such that $y_n \to x$. Since $x_n \in C(\omega) = \overline{U}(\omega)$, it follows that there exists $\{u_m^{(n)}\} \subset U$ such that $u_m^{(n)}(\omega) \to x_n(\omega)$ and, from the continuity of T, it follows that

$$v_m^{(n)}(\omega) = u_m^{(n)}(\omega) - T(\omega, u_m^{(n)}(\omega)) \to x_n - T(\omega, x_n) = y_n.$$

Since $y_n \to y$, there exists $\{v_k^{(i)}\} \subset \{v_m^{(n)}\}$ such that $v_k^{(i)}(\omega) \xrightarrow[k,i]{} y$.

Therefore, V is a Castaing representation of R, i.e., R is closed random subset of H.

Now, since T is nonexpansive and $C(\omega) = H$, it follows that $T(\omega,.)$ is nonexpansive on H for each $\omega \in \Omega$. Hence, applying Lemma 2.2.7, $R(\omega)$ is a convex subset of H for each $\omega \in \Omega$.

The following result is a stochastic version of Pazy's fixed point theorem.

PROPOSITION 4.4.5. *Let T be a random nonexpansive operator on a convex closed random subset C of H. Then T has a random fixed point if and only if there exists a $y_0 \in \mathscr{S}(C)$ such that the sequence $\{\| T^n(\omega, y_0(\omega)) \|\}_{n \in \mathbb{N}}$ is bounded for all $\omega \in \Omega$.*

Proof. If $y \in \mathscr{S}(C)$ is a random fixed point of T, then $y(\omega)$ is a fixed point of $T(\omega,.)$ in $C(\omega)$, and hence $T^n(\omega, y(\omega)) = y(\omega)$ is bounded.

Now, if, for $y_0 \in \mathscr{S}(C)$, the sequence $\{T^n(\omega, y_0(\omega))\}_{n\in \mathbb{N}}$ is bounded, then Lemma 2.2.7 with $x_0 = y_0(\omega)$ implies that the operator T has a fixed point in the wide sense. Since T satisfies condition (c), T has a random fixed point.

COROLLARY 1. *Let T be a nonexpansive random operator on H. Then T has a random fixed point if and only if there exists an H-random variable y such that the sequence $\{\| T^n(\omega,y(\omega)) \|\}_{n\in \mathbb{N}}$ is bounded for all $\omega \in \Omega$.*

Proof. T is defined on the convex closed random subset $C(\omega) = H$; hence, T satisfies condition (c) and the corollary follows from the previous theorem.

COROLLARY 2. *Let T be a nonexpansive random operator on a bounded convex closed random subset $C \subset H$. Then T has a random fixed point.*

Proof. Since C is bounded, the sequence $\{\| T^n(\omega, y_0(\omega)) \|\}_{n\in \mathbb{N}}$ is bounded for every $y_0 \in \mathscr{S}(C)$ and for all $\omega \in \Omega$.

Remark. Corollary 1 is a stochastic version of a theorem due to Browder and Petryshyn, while Corollary 2 is a stochastic version of a theorem due to Kirk and Browder.

Let us now consider a random nonlinear ergodic theorem which extends a theorem of Brézis and Browder. We recall that ergodicity establishes the general averaging process convergence

$$A_n(\omega, x) = \sum_{m=0}^{\infty} a_{n,m}\, T^m(\omega, x),\ \omega \in \Omega,\ x \in C(\omega),$$

$0 \leqslant a_{n,m}(\omega) \leqslant 1$, $\sum_{m=0}^{\infty} a_{n,m}(\omega) = 1$, $\omega \in \Omega$, ($a_{n,m}$ are real random variables)

to the random fixed point of T.

PROPOSITION 4.4.6. *Let T be a nonexpansive random operator on the bounded convex closed random subset $C \subset H$.*

If, for all $\omega \in \Omega$,

(i) $a_{n,m}(\omega) \to 0$ *for each* m,

(ii) $\sum_{m=0}^{\infty} (a_{n,m+1}(\omega) - a_{n,m}(\omega)) \underset{n}{\longrightarrow} 0$,

then for every $y \in \mathscr{S}(C)$, the sequence $A_n(\omega, y(\omega))$ is weakly convergent to a random fixed point of T, for each $\omega \in \Omega$.

Proof. According to Corollary 2, T has a random fixed point. Applying now the nonlinear ergodic theorem due to Brézis and Browder for every ω, leads to the statement of the proposition.

4.4.2. *Random Fixed Point Theorems for Multivalued Random Operators with Stochastic Domains*

In this section, we present a general stochastic fixed point theorem [132], [133] for continuous multivalued random operators which improves upon the results of the foregoing section. We shall examine the case of upper semicontinuous multivalued random operators, which give a stochastic version of the Kakutani-Bohnenblust-Karlin Theorem [132], [133] [137]. Nowak's results offering some generalization of Engl's results are also presented.

DEFINITION 4.4.5. Let $C : \Omega \to 2^X$. A mapping $T : \mathrm{Gr}C \to 2^X$ is called *multivalued random operator with stochastic domain* C if C is measurable and if for all $x \in X$ and open $D \subseteq X$, $\{\omega \in \Omega : x \in C(\omega) \wedge T(\omega, x) \cap D \neq \emptyset\} \in \mathscr{K}$. A mapping $x : \Omega \to X$ will be called *wide-sense fixed point of* T if for all $\omega \in \Omega$, $x(\omega) \in C(\omega)$ and $x(\omega) \in T(\omega, x(\omega))$. It will be called *random fixed point of* T if for μ-almost all $\omega \in \Omega$, $x(\omega) \in C(\omega)$ and $x(\omega) \in T(\omega, x(\omega))$ and x is a measurable mapping of ω.

Remark 4.4.1. A mapping $T : \mathrm{Gr}C \to X$ is a random operator with stochastic domain C (Def. 4.4.2) if and only if $\tilde{T} : \mathrm{Gr}C \to 2^X$, defined by $\tilde{T}(\omega, x) = \{T(\omega, x)\}$, is a multivalued random operator with stochastic domain C (Def. 4.4.5). In this case, the notions of wide-sense and of random fixed point, as given in Def. 4.4.2 and Def. 4.4.5, respectively, coincide. Also, the continuity of T in the sense of Def. 4.4.3 coincides with the continuity of $T : \mathrm{Gr}C \to CB(X)$ as a mapping from $(C(\omega), \|.\|)$ into $[CB(X), D)$ for all $\omega \in \Omega$ (i.e., in the Hausdorff metric).

The main result for continuous multivalued random operators is given by

THEOREM 4.4.9 [132, Th. 5.6]. *Let* $C : \Omega \to 2^X$ *be separable and let* $T : \mathrm{Gr}C \to CB(X)$ *be a continuous (in the Hausdorff metric) multivalued random operator with stochastic domain* C *which has a wide-sense fixed point. Then,* T *has an (at most) countable set of random fixed points* $\{x_n\}_{n \in \mathbb{N}}$ *such that for* μ*-almost all* $\omega \in \Omega$, $\{x_n(\omega)\}_{n \in \mathbb{N}}$ *is dense in the set of all fixed points of* $T(\omega, .)$.

Proof. (a) We assume first that $(\Omega, \mathscr{K}, \mu)$ is complete and define $N : \mathrm{Gr}C \to \mathbb{R}$ as $N(\omega, x) = d(x, T(\omega, x))$. Let $\omega \in \Omega$ and $x, y \in C(\omega)$ be arbitrary but fixed. Then, we have

$$|N(\omega, x) - N(\omega, y)| \leqslant D(T(\omega, x), T(\omega, y)) + d(x, y).$$

From the continuity of T, it follows that each $N(\omega.,)$ is continuous on $C(\omega)$.

Since T is a multivalued random operator, it follows that, for every x and $r \in [0, \infty)$ chosen arbitrary but fixed,

$$\{\omega \in \Omega : x \in C(\omega) \wedge N(\omega, x) < r\} =$$
$$= \{\omega \in \Omega : x \in C(\omega) \wedge T(\omega, x) \cap \{y \in X : \|y - x\| <$$
$$< r\} \neq \emptyset\} \in \mathscr{K}.$$

From the continuity of each $N(\omega,.)$ and from the fact that GrC is a measurable set of $\mathscr{K} \times \mathscr{B}_X$ (cf. Theorem 2.2.1), we have

$$\{(\omega, x) \in \mathrm{Gr}C : N(\omega, x) \leqslant r\} =$$
$$= \{(\omega, x) \in \Omega \times X : x \in C(\omega) \wedge N(\omega, x) \leqslant r\} =$$
$$= \bigcap_{n \in \mathbb{N}} \bigcap_{z \in Z} \left[\left(\left\{\omega \in \Omega : z \in C(\omega) \wedge N(\omega, z) <\right.\right.\right.$$
$$\left.\left.\left. < z + \frac{1}{n}\right\} \times \left\{x \in X : \|x - z\| < \frac{1}{n}\right\}\right) \cap \mathrm{Gr}C\right] \in \mathscr{K} \times \mathscr{B}_X,$$

where Z is a countable set as it appears in the definition of the separability of C.

Hence, N is a measurable function from the measurable space (GrC, Gr$C \cap \mathscr{K} \times \mathscr{B}_X$) into $\mathbb{R}$. In particular, $\{(\omega, x) \in \mathrm{Gr}C : N(\omega, x) = 0\} \in \mathscr{K} \times \mathscr{B}_X$.

Let us now define

$$F : \Omega \to \{A : A \subseteq X\} \text{ by } F(\omega) = \{x \in X : x \in C(\omega) \cap T(\omega, x)\}.$$

Since T has a wide-sense fixed point, each $F(\omega)$ is nonempty. Choose $\omega \in \Omega$ arbitrary but fixed and let $\{x_n\}_{n \in \mathbb{N}} \in F(\omega)$ be such that $x_m \to x \in X$ as $m \to \infty$. Thus, $x \in C(\omega)$; since

$$d(x, T(\omega, x)) \leqslant \|x - x_m\| + d(x_m, T(\omega, x_m)) +$$
$$+ D(T(\omega, x_m), T(\omega, x))$$

for all $m \in \mathbb{N}$ and the continuity of T, it follows that $x \in F(\omega)$. Hence, every $F(\omega)$ is closed, i.e., $F(\Omega) \subseteq 2^X$ and

$$\mathrm{Gr}F = \{(\omega, x) \in \Omega \times X : x \in F(\omega)\} =$$
$$= \{(\omega, x) \in \mathrm{Gr}C : N(\omega, x) = 0\} \in \mathscr{K} \times \mathscr{B}_X.$$

From Theorem 2.2.1, it follows that F is measurable and then the Theorem of Kuratowski and Ryll-Nardzewski (Theorem

2.2.4) implies that there exists a measurable selector of F. The statement regarding the countable dense family of random fixed points follows from Theorem 2.2.5 applied to F.

(b) Let now $(\Omega, \mathscr{K}, \mu)$ be not necessarily complete and let $(\Omega, \mathscr{K}^*, \mu^*)$ denote the completion of $(\Omega, \mathscr{K}, \mu)$. Since $\mathscr{K} \subseteq \mathscr{K}^*$, the assumptions of the theorem are also fulfilled with $(\Omega, \mathscr{K}^*, \mu^*)$ instead of $(\Omega, \mathscr{K}, \mu)$. According to part (a) of this proof, there exists a countable set $\{x_m^*\}_{m\in\mathbb{N}}$ of $\mathscr{K}^*$-measurable mappings from Ω to X such that, for all $\omega \in \Omega$, $\{x_m^*\}_{m\in\mathbb{N}}$ is a dense subset of $F(\omega)$. Let $\mathscr{D}$ be a countable generator of $\mathscr{B}_X$ which exists because of the separability of X. Then, for all $D \in \mathscr{D}$ and $m \in \mathbb{N}$, $x_m^{*-1}(d) = A_{D,m} \cup L_{D,m}$ with $A_{D,m} \in \mathscr{K}$, $L_{D,m} \subseteq N_{D,m}$, $N_{D,m} \in \mathscr{K}$ and $\mu(N_{D,m}) = 0$. Let $M = \bigcup_{m\in\mathbb{N}} \bigcup_{D\in\mathscr{D}} N_{D,m}$ and define for every $m \in \mathbb{N}$

$$x_m(\omega) = \begin{cases} x_m^*(\omega) & \omega \notin M \\ 0 & \omega \in M \end{cases}.$$

Then, every x_m is $\mathscr{K}$-measurable and, for all $\omega \in \Omega - M$, $\{x_m(\omega)\}_{m\in\mathbb{N}}$ is a dense subset of $F(\omega)$. Since $\mu(M) = 0$, the proof of the theorem is complete.

COROLLARY 4.4.1. *Let $C : \Omega \to 2^X$ be separable and let T: $\mathrm{Gr}C \to X$ be a continuous random operator with stochastic domain C such that $\{x \in C(\omega) : T(\omega, x) = x\} \neq \emptyset$ for all $\omega \in \Omega$. Then T has a countable set of random fixed points $\{x_m\}_{m\in\mathbb{N}}$ such that, for μ-almost all $\omega \in \Omega$, $\{x_n(\omega)\}$ is dense in the set of fixed points of $T(\omega,.)$.*

The proof follows immediately from the above theorem and Remark 4.4.1.

Remark 4.4.2. The above corollary is an improvement on Theorems 4.4.3 and 4.4.4.

The proof of Theorem 4.4.9 implies that "μ-almost all" can be replaced in the conclusion and all its consequences by "all" if $(\Omega, \mathscr{K}, \mu)$ is complete, i.e., this refers also to the meaning of "random fixed point". Hence, either new or at least more general random fixed point theorems follow immediately. For example :

COROLLARY 4.4.2 (stochastic version of Banach, Schauder and Krasnosel'skiĭ-type theorems). *Let $C : \Omega \to CC(X) \cap CB(X)$ be separable, let $k : \Omega \to [0, 1)$ and let T and S be random operators with stochastic domain C such that for all $\omega \in \Omega$*

(i) *for all $x \in C(\omega)$, $T(\omega, x) + S(\omega, x) \in C(\omega)$,*

(ii) *for all $x, y \in C(\omega)$, $\|S(\omega, x) - S(\omega, y)\| \leqslant \leqslant k(\omega)\|x - y\|$,*

(iii) *$T(\omega,.)$ is completely continuous.*
Then, there exists a countable set of measurable mappings $x_m : \Omega \to X$ such that for μ-almost all $\omega \in \Omega$ and all $m \in \mathbb{N}$, $x_m(\omega) \in C(\omega)$ and $T(\omega, x_m(\omega)) + S(\omega, x_m(\omega)) = x_m(\omega)$. Furthermore, $\{x_m(\omega)\}_{m\in\mathbb{N}}$ is dense in the set of fixed points of $T(\omega,.) + S(\omega,.)$ for μ-almost all $\omega \in \Omega$.

The proof is a direct consequence of the above corollary and of Reinermann's results [359].

For $T \equiv 0$, it is sufficient to assume $C : \Omega \to 2^X$.

COROLLARY 4.4.3. *Let $C : \Omega \to 2^X$ be separable and let $T : \mathrm{Gr}C \to CB(X)$ be a multivalued random operator with stochastic domain C such that, for all $\omega \in \Omega$, there exists a $k(\omega) \in [0, 1)$ for all $x, y \in C(\omega)$ with $D(T(\omega, x), T(\omega, y)) \leqslant k(\omega)\|x - y\|$. Furthemore, let us assume that $T(\omega,x) \subseteq C(\omega)$ for all $\omega \in \Omega$ and $x \in bdC(\omega)$. Then, the conclusions of Theorem 4.4.9 hold.*

Remark. This result is a random version of Assad and Kirk's theorem and of Itoh's theorem [237], when $C = X$ is a Polish space.

A more general result is given by

COROLLARY 4.4.4. *Let X be a separable Opial space, let $C : \Omega \to 2^X$ be separable such that, for all $\omega \in \Omega$, $C(\omega)$ is convex and weakly compact, and let $T : \mathrm{Gr}C \to CB(X)$ be a multivalued random operator with stochastic domain C such that, for all $(\omega, x) \in \mathrm{Gr}C$, $T(\omega, x)$ is compact and, for all $\omega \in \Omega$ and all $x \in bdC(\omega)$, $T(\omega, x) \subseteq C(\omega)$. Furthermore, we assume that for all $\omega \in \Omega$ and all $x, y \in C(\omega)$, $D(T(\omega, x), T(\omega, y)) \leqslant \|x - y\|$. Then, the conclusions of Theorem 4.4.9 hold.*

The proof follows from Theorems 2.2.17 and 4.4.9.

In what follows, we shall need

DEFINITION 4.4.6. Let $C : \Omega \to 2^X$ be measurable and let $T : \mathrm{Gr}C \to 2^X$ be a multivalued random operator with stochastic domain C. T will be called "u.s.c" ("compact") if for all $\omega \in \Omega$, $T(\omega,.)$ is u.s.c. (compact) in the sense of Definition 2.2.4.

It is known that in the deterministic fixed point theory of multivalued operators, upper semicontinuous operators seem to be of more interest than continuous operators. But in spite of this fact, the proof of the above theorem cannot be simply transferred to u.s.c. mappings. Indeed, as is shown in Example 2.2.1, the fact that a mapping of two variables which is u.s.c. in one and measurable in other variable need not be true to be jointly measurable. Also, for some connections between some definitions of u.s.c., see Definition 2.2.4 and Lemma 2.2.1, since unfortunately the definition of u.s.c. vary in the literature.

Nevertheless, Engl [132, Th. 5.23 ; 138, Th. 16] generalizes the idea contained in the proof of Theorem 4.4.9 by considering, instead of T, an auxiliary mapping H which will be defined and studied in

PROPOSITION 4.4.7. *Let* $C : \Omega \to 2^X$ *be separable and let* $T : \mathrm{Gr}C \to CC(X)$ *be a compact multivalued random operator with stochastic domain* C (in the sense of Def. 4.4.6). *Define* $H : \mathrm{Gr}C \to CC(X)$ *as*

$$H(\omega, x) = \bigcap_{n \in N} \mathrm{cl\,conv} \bigcup_{\substack{z \in Z \cap C(\omega) \\ \|z-x\| < 1/n}} T(\omega, z),$$

where Z *is a countable set as it appeared in Definition* 4.4.1. *Then* :

(1) *For all* $(\omega, x) \in \mathrm{Gr}C$, $\emptyset \neq H(\omega, x) \subseteq T(\omega, x)$.
(2) *For all* $\omega \in \Omega$ *and* $x \in C(\omega) \cap Z$, $H(\omega, x) = T(\omega, x)$.
(3) H *is (jointly) measurable (i. e., for all open* $D \subseteq X$,

$$\{(\omega, x) \in \mathrm{Gr}H : H(\omega, x) \cap D \neq \emptyset\} \in \mathscr{K} \times \mathscr{B}_X).$$

(4) *For all* $\omega \in \Omega$, $H(\omega,.)$ *is compact.*

Proof. Let $\omega \in \Omega$ be arbitrary but fixed. Hence, we shall omit writing ω when dealing with the proof of (1), (2) and (4).

(1) Let L be a finite subset of $\mathbb{N}$, $m = \max L$, $x \in C$. As T is compact, the sets $\mathrm{cl\,conv} \bigcup_{\substack{\|z-x\| < 1/n \\ z \in Z \cap C}} T(z)$ are closed subsets of a compact set which is independent of n ; this is a consequence of Mazur's theorem in the closed convex hull of a compact set. Since

$$\emptyset \neq \mathrm{cl\,conv} \bigcup_{\substack{z \in Z \cap C \\ \|z-x\| < 1/m}} T(z) \subseteq \bigcap_{n \in L} \mathrm{cl\ conv} \bigcup_{\substack{z \in Z \cap C \\ \|z-x\| < 1/n}} T(z),$$

the family $\{\mathrm{cl\ conv} \bigcup_{\substack{z \in Z \cap C \\ \|z-x\| < 1/n}} T(z)\}_{n \in \mathbb{N}}$ has the finite intersection property ; therefore, $H(x) \neq \emptyset$.

Let $\varepsilon > 0$ be arbitrary but fixed and choose $n \in \mathbb{N}$ such that, for all $z \in Z \cap C$ with $\|z - x\| < 1/n$, $\sup_{y \in T(z)} d(y, T(x)) < \varepsilon$, which is possible according to Lemma 2.2.1, (3). Then, for all such z, $T(z) \subseteq (T(x))_\varepsilon = \{y \in X : d(y, T(x)) < \varepsilon\}$ and considering the convexity of $(T(x))_\varepsilon$, it follows that $H(x) \subseteq \mathrm{cl\ conv} \bigcup_{\substack{z \in Z \cap C \\ \|z-x\| < 1/n}} T(z) \subseteq \mathrm{cl}(T(x))_\varepsilon$. As ε was arbitrary, $H(x) \subseteq T(x)$.

(2) Let $x \in C \cap Z$. Then $T(x) \subseteq \bigcup_{\substack{z \in C \cap Z \\ \|z-x\| < 1/n}} T(z)$ for all $n \in \mathbb{N}$,

and therefore, $T(x) \subseteq H(x)$ which together with (1) implies that $T(x) = H(x)$.

(4) Let $\{x_m\}_{m\in\mathbb{N}} \in C^{\mathbb{N}}$ with $x_m \to x$ (since C is closed $x \in C$) and let $y_m \in H(x_m)$ with $y_m \to y$. Let $n \in \mathbb{N}$ be arbitrary but fixed and choose $s \in \mathbb{N}$ such that $\|x_m - x\| < \dfrac{1}{2n}$ if $m \geqslant s$. For all such m, one has

$$y_m \in H(x_m) \subseteq \text{cl conv} \bigcup_{\substack{z\in Z\cap C\\ \|z-x_m\|<1/2n}} T(z) \subseteq \text{cl conv} \bigcup_{\substack{z\in Z\cap C\\ \|z-x_i\|<1/n}} T(z).$$

Therefore, $y \in \text{cl conv} \bigcup_{z\in Z\cap C} T(z)$; since n was arbitrary, $y \in H(x)$. Considering (1) and the compactness of T, it follows that $\text{cl} \bigcup_{x\in C} H(x)$ is compact and hence H is compact too.

(3). Let for all $n \in \mathbb{N}$, $T_n : \text{Gr}C \to \{\emptyset \neq A \subseteq X\}$ be given by

$$T_n(\omega, x) = \bigcup_{\substack{z\in Z\cap C(\omega)\\ \|z-x\|<1/n}} T(\omega, z)$$

and let $D \subseteq X$ be open. Then

$$\{(\omega, z) \in \text{Gr}C : T_n(\omega, x) \cap D \neq \emptyset\} = \bigcup_{z\in Z} [\{(\omega, x) \in \text{Gr}C : \|z -$$
$$- x\| < 1/n\} \cap (\{\omega \in \Omega : z \in C(\omega) \wedge T(\omega, z) \cap D \neq \emptyset\} \times X)] =$$
$$= \bigcup_{z\in Z} [(\Omega \times \{x \in X : \|z - x\| < 1/n\}) \cap \text{Gr}C \cap$$
$$\cap (\{\omega \in \Omega : z \in C(\omega) \wedge T(\omega, z) \cap D \neq \emptyset\} \times X)] \in \mathscr{K} \times \mathscr{B}_X,$$

since T is a multivalued random operator and $\text{Gr}C \in \mathscr{K} \times \mathscr{B}_X$ (cf. Theorem 2.2.1). Therefore, T_n is measurable as a multivalued mapping on the measurable space $(\text{Gr}C, (\mathscr{K} \cap \mathscr{B}_X) \cap \text{Gr}C)$ and so is $\text{cl}\, T_n$ (cf. Proposition 2.2.1). Applying now Theorem 2.2.3, one gets that $\text{cl conv}\, T_n = \text{conv cl}\, T_n$ is measurable (on $\text{Gr}C$). As all $\text{cl conv}\, T_n(\omega, x)$ are compact (see Mazur's theorem), one can apply Theorem 2.2.2 and get that $H = \bigcap_{n\in\mathbb{N}} \text{cl conv}\, T_n$ is measurable as a mapping on $(\text{Gr}C, (\mathscr{K} \times \mathscr{B}_X) \cap \text{Gr}C)$. This means that $\{(\omega, x) \in \text{Gr}C : H(\omega, x) \cap D \neq \emptyset\} \in \mathscr{K} \times \mathscr{B}_X$.

We give now a general theorem which is a stochastic analogue to Bohnenblust-Karlin-Kakutani's theorem.

THEOREM 4.4.10 [132, Th. 5.23; 138, Th. 16]. *Let $C : \Omega \to CC(X)$ be separable and let $T : \text{Gr}C \to CC(X)$ be a compact multivalued random operator with stochastic domain C such that, for every $\omega \in \Omega$ and every $x \in \text{bd}C(\omega)$, $T(\omega, x) \subseteq C(\omega)$. Then, T has a random fixed point.*

Proof. Let $(\Omega, \mathscr{K}, \mu)$ be a complete set and let Z be a countable set as in Definition 4.4.1. Let us define H as in Proposition 4.4.7. It is now known that H is measurable and the mapping $\tilde{x} : \mathrm{Gr}C \to X$, defined by $\tilde{x}(\omega, x) = x$, is also measurable. Using Lemma 2.2.3 with measurable space $(\mathrm{Gr}C, (\mathscr{K} \times \mathscr{B}_X) \cap \mathrm{Gr}C)$, we obtain that $N : \mathrm{Gr}C \to X$, defined by $(\omega, x) \to d(\tilde{x}(\omega, x), H(\omega, x))$, is jointly measurable.

Define now $F : \Omega \to \{A : A \subseteq X\}$ as $F(\omega) = \{x \in C(\omega) : x \in H(\omega, x)\}$. Rothe's variant of Theorem 2.2.18 ([132, Th. 5.12]) implies that $F(\omega) \neq \emptyset$ for all $\omega \in \Omega$. Let $\omega \in \Omega$ be arbitrary but fixed and let $\{x_m\}_{m \in \mathbb{N}} \in F(\omega)^{\mathbb{N}}$ with $x_m \to x \in C(\omega)$. For every $m \in \mathbb{N}$, $x_m \in H(\omega, x_m)$, and since $H(\omega, .)$ is compact, it follows from Lemma 2.2.1, 2) with $\{y_m\} = \{x_m\}$ that $x \in H(\omega, x)$. Hence, $x \in F(\omega)$, i.e., $F(\omega)$ is closed. Now, the measurability of N implies that

$$\mathrm{Gr}F = \{(\omega, x) \in \Omega \times X : x \in F(\omega)\} =$$
$$\{(\omega, x) \in \mathrm{Gr}C : N(\omega, x) = 0\} \in \mathscr{K} \times \mathscr{B}_X.$$

Then, from Theorem 2.2.1, it follows that F is measurable. The Theorem of Kuratowski and Ryll-Nardzewski (Theorem 2.2.4) implies that there exists a measurable selector x of F, i.e., $x(\omega) \in C(\omega)$ and $x(\omega) \in H(\omega, x(\omega)) \subseteq T(\omega, x)$ for all $\omega \in \Omega$.

If $(\Omega, \mathscr{K}, \mu)$ is not complete, the reasoning is as in point (b) of the proof of Theorem 4.4.2.

Remark 4.4.3. (1) It is not known whether a similar conclusion about a countable dense set of random fixed points as in Theorem 4.4.9 can be drawn. Examples are known when $\mathrm{Fix}\ (H(\omega, .)) \nsubseteq \mathrm{Fix}\ T(\omega, .)$.

(2) Throughout the above proof, separability plays an important rôle since a measurability theory for singlevalued or multivalued mappings in non-separable spaces leads to pathological conclusions. For example, the sum of two measurable singlevalued mappings with values in a non-separable Banach space need not be measurable [31, p. 18].

A. Nowak [315] considered an even larger class of multivalued mappings with stochastic domain.

DEFINITION 4.4.7. Let (X, d) be a metric space and let $(\Omega, \mathscr{K}, P)$ be a probability space. Let C be a multivalued mapping from Ω to X, whose graph $\mathrm{Gr}C$ is $\mathscr{K} \times \mathscr{B}_X$-measurable and let $T : \mathrm{Gr}C \to \mathscr{P}(X)$. C is called the *stochastic domain* of T.

THEOREM 4.4.11. *Let (X, d) be a Polish space and let T be a multivalued mapping with stochastic domain C. If $\emptyset \neq F(\omega) =$ $= \{x \in X : x \in C(\omega) \cap T(\omega, x)\}$ a.s. and if the graph of T is*

$\mathscr{K} \times \mathscr{B}_X \times \mathscr{B}_X$-measurable, then T has a random fixed point, i.e., there exists a measurable mapping $y : \Omega \to X$ such that $y(\omega) \in C(\omega) \cap T(\omega, y(\omega))$ a.s.

Proof. There exists an $A \in \mathscr{K}$ with $P(A) = 1$, such that $F(\omega) \neq \emptyset$ for all $\omega \in A$. Let $\mathscr{A} = A \cap \mathscr{K}$ be the trace σ-field on A. We shall prove that $F : A \to \mathscr{P}(X)$ has a measurable graph. Indeed,

$$\mathrm{Gr}F = \{(\omega, x) \in A \times X : x \in C(\omega) \cap T(\omega, x)\} =$$

$$= \mathrm{pr}_{A \times X}\{(A \times V) \cap \mathrm{Gr}T\},$$

where $V = \{(x, y) \in X \times X : x = y\}$. Since the graph of T is measurable, $(A \times V) \cap \mathrm{Gr}\, T \in \mathscr{A} \times \mathscr{B}_X$. It is known (see Kuratowski [270, 39, IV]) that the projection of a Borel subset of V is a Borel subset of X, as a one-to-one continuous image. It follows that for every $E \in \mathscr{A} \times \mathscr{B}_V$, $\mathrm{pr}_{A \times X} E \in \mathscr{A} \times \mathscr{B}_X$ and $\mathrm{Gr}F \in \mathscr{A} \times \mathscr{B}_X$.

There exists $z : A \to X$ such that $z(\omega) \in F(\omega)$ for all $\omega \in A$ and $z^{-1}(B) \in \mathscr{A}_P$ for all $B \in \mathscr{B}_X$, where $\mathscr{A}_P$ is the completion of $\mathscr{A}$ (see Sainte-Beuve [373] and Remark 2.2.1). Also, it is known that there exist $A_1 \in \mathscr{A}$ with $P(A_1) = 1$ and a measurable mapping $y : A \to X$ such that $z(\omega) = y(\omega)$ for all $\omega \in A_1$. For $\omega \in \Omega - A$, we define $y(\omega) = x_0$, where x_0 is a point of X. The mapping y is a random fixed point of T.

Remarks. (1) If $T(\omega,.)$ has a fixed point, then $F(\omega) \neq \emptyset$ and the sufficient conditions for the measurability of the graph T are given by Theorem 2.2.14.

(2) Theorem 4.4.11 generalizes Theorem 4.4.9.

A random version of Nadler's theorem is given by

COROLLARY 1 [315, Th. 4]. *Let (X, d) be a Polish space and let C be a separable multivalued mapping from Ω to X. Let T be a closed-valued mapping with stochastic domain C, such that, for all $(\omega, x) \in \mathrm{Gr}\, C$, $T(\omega, x) \subset C(\omega)$. We assume that T is measurable in ω and contractive in x, i.e., for all $\omega \in \Omega$ there exists $k(\omega) \in [0, 1)$ such that, for all $x, y \in C(\omega)$, $D(T(\omega, x), T(\omega, y)) \leqslant k(\omega)\, d(x, y)$. Then, T has a random fixed point.*

Proof. For every $\omega \in \Omega$, $T(\omega,.)$ has a fixed point in $C(\omega)$ (see Nadler [306]). $T(\omega,.)$ being continuous in the generalized Hausdorff metric, we have that $\mathrm{Gr}\, T$ is $\mathscr{K} \times \mathscr{B}_X \times \mathscr{B}_X$-measurable (cf. Theorem 2.2.14). According to Theorem 4.4.11, T has a random fixed point.

COROLLARY 2. *Let X be a Banach space and let C be a separable multivalued mapping from Ω to X with convex values. Let T*

be a compact-valued mapping from GrC *to* X, *measurable in* ω *and satisfying* $T(\omega, x) \subset C(\omega)$ *for all* $(\omega, x) \in$ GrC. *We assume that there exists a function* $\varphi : \Omega \times [0, \infty) \to [0,\infty)$ *such that* $\varphi(\omega, 0) = 0$, $\varphi(\omega, t) < t$ *for all* $\omega \in \Omega$, $t > 0$, *and* $D(T(\omega, x), T(\omega, y)) \leqslant \varphi(\omega, d(x, y))$ *for all* $\omega \in \Omega$, $x, y \in C(\omega)$. *Then,* T *has a random fixed point.*

A random version of Fan-Kakutani's theorem is given by

THEOREM 4.4.12 [315, Th. 6]. *Let* X *be a Fréchet (i.e., locally convex, complete, metric) space and let* C *a separable multivalued mapping from* Ω *to* X *with compact convex values. Let* T *be a closed convex-valued multivalued mapping with stochastic domain* C. *We assume that* T *is measurable in* ω, *upper semicontinuous in* x *(in the sense of Def.* 2.2.4), *and that* $T(\omega, x) \subset C(\omega)$ *for all* $(\omega, x) \in$ GrC. *Then,* T *has a random fixed point.*

Proof. For every ω, the set

$$\bigcup_{x \in C(\omega)} T(\omega, x)$$

is compact (see Berge [22, Chap. VI, §1, Th. 3). Hence T is a compact multivalued random operator in the terminology of Engl. Then, there exists a multivalued mapping H from GrC to X satisfying the following conditions:

(a) for each $(\omega, x) \in$ GrC, $H(\omega, x)$ is a compact convex subset of $T(\omega, x)$,

(b) for every $\omega \in \Omega$, $H(\omega,.)$ is u.s.c. on $C(\omega)$,

(c) H is jointly measurable.

This result formulated for a real Banach space (see Proposition 4.4.7) with the obvious changes can be extended for the Fréchet-space case.

For every $\omega \in \Omega$, $H(\omega,.)$ has a fixed point in $C(\omega)$ (cf. Ky Fan [147]).

Since H is closed-valued and measurable, H has a measurable graph. According to Theorem 4.4.11, H has a random fixed point, and hence so has T.

Chapter 5

Applications of the fixed point theory to random operator equations

Given the limited space of this book, it is difficult to encompass the whole importance and the large number of applications of the theory of random fixed points. Hence, only four main directions will be investigated with the aim of identifying those elements that justify the intention of the chapter.

We shall present the most significant results regarding the applications of some fixed point theorems to probabilistic structures. A close correlation between the probabilistic structure results and those for random operator equations will be established. Since the latter are the kernel of probabilistic functional analysis, their treatment will fill the main body of this chapter.

Several classes of random differential and integral equations as studied by A. T. Bharucha-Reid [31], [32], H. Engl [132], [136], [138], A. Nowak [318], [319], J. R. Kannan [247], [250], Gh. Bocşan, Gh. Constantin and V. Radu [52], [58], [60] will be considered.

The last part of this chapter will describe some applications of the random contraction mapping theorem to solve questions from stochastic approximation and experience theory. Some of the results obtained by O. Hanš [205], [207] and M. Driml [125], introducing a new approach to the stochastic approximation theory, will be selectively presented (some more concrete applications of these will be given in Chap. 7).

5.1. SOME APPLICATIONS OF THE FIXED POINT THEOREMS TO PROBABILISTIC STRUCTURES

The connections between either metric spaces or normed spaces and probabilistic structures allow for some applications to several classes of random operator equations.

As an illustration of the fixed point theorems applied in probabilistic inner product spaces, let us consider the existence and uniqueness of solution of Urysohn's integral equation on the $L^2(G)$ space, where G is a bounded closed set of $\mathbb{R}^2$, and $L^2(G)$ is a real linear space of all Lebesgue square integrable functions.

(We shall use the results and definitions given by Shih-sen Chang [98], [99]).

It is easy to verify that if $(E, \langle.,.\rangle)$ is a real Hilbert space and if $(E, \mathscr{F}, \mathrm{Min})$ is the probabilistic inner product space induced by $(,\langle.,.\rangle)$ $(\mathscr{F}_{p,q}(t) = H_0(t - \langle p, q\rangle), \forall p, q \in E)$, then

$$\lim_{n\to\infty} \|p_n - p\|_E = 0 \Leftrightarrow \lim_{n\to\infty} (\sup\{t \in \mathbb{R} : F_{x_n-x,x_n-x}(t) < 1\})^{1/2} = 0.$$

Let $K(s, t, u)$, $s, t \in G$, $u \in \mathbb{R}$ be a continuous function in u satisfying the following conditions:

$$|K(s, t, u)| \leqslant r(s, t)(a + b|u|), \quad s, t \in G, \ u \in \mathbb{R},$$

where $a,b > 0$ and

$$\int_G \int_G |r(s, t)|^2 \, ds \, dt < \infty.$$

Then the operator T defined by

$$(1) \quad Tx(s) = y(s) + \int_G K(s, t, x(t)) \, dt$$

is continuous from $L^2(G)$ to $L^2(G)$.

THEOREM 5.1.1. *Consider the integral equation of Urysohn type on* $L^2(G)$

$$(2) \quad x(s) = y(s) + \int_G K(s, t, x(t)) \, dt,$$

where the function $K(s, t, u)$, $s, t \in G$, $u \in \mathbb{R}$ *is continuous in* u *satisfying the above conditions and also*

$$\sum_{n=0}^{\infty} \|T^n x - T^n u\|_{L^2(G)} < \infty, \quad \forall\, x(s), u(s) \in L^2(G).$$

Then, for a given $y(s) \in L^2(G)$, *Eq.* (1) *has a unique solution* $\bar{x}(s) \in L^2(G)$ *and for any given* $x_0(s) \in L^2(G)$ *the iterative sequence*

$$x_{n+1}(s) = y(s) + \int_G K(s, t, x_n(t)) \, dt, \quad n = 0, 1, 2, \ldots$$

converges in norm $\|\cdot\|_{L^2(G)}$ *to* $\bar{x}(s)$.

Several applications to other classes of equations on probabilistic structure can be considered. For example, let us study the existence and uniqueness of solution of the following kind of nonlinear Volterra integral equation

$$(3)\quad x(t) = x_0(t) + \int_0^t K(t, s, x(s))\, ds, \quad 0 \leqslant t \leqslant a.$$

Let $[0, a]$ be a fixed real interval $(0 < a < \infty)$ and let $(E, \|\cdot\|_E)$ be a real Banach space. Let us denote by $C([0, a], E)$ the linear space of all E-valued continuous functions. We know that $C([0, a], E)$ is a Banach space with norm defined by

$$\|x\|_C = \sup_{0 \leqslant t \leqslant a} \|x(t)\|_E, \qquad \forall\, x \in C([0, a], E).$$

In the sequel, we also denote by $C([0, a] \times [0, a] \times E, E)$ the linear space of all E-valued continuous functions $f(t, s, x)$ defined on $[0, a] \times [0, a] \times E$.

Besides the norm $\|\cdot\|_C$, the space $C([0, a], E)$ can be endowed with another norm $\|\cdot\|_*$ defined as

$$\|x\|_* = \sup_{0 \leqslant t \leqslant a} e^{-bt} \|x(t)\|_E, \qquad \forall\, x(t) \in C([0, a], E),$$

where b is any positive real number. It is clear that the norm $\|\cdot\|_*$ is equivalent to the norm $\|\cdot\|_C$.

If $(C([0, a], E), \mathscr{F}, \mathrm{Min})$ is the induced Menger space, where $F_{x,y}(t) = H_0(t - \|x - y\|_*)$, $\forall\, x, y \in C([0, a], E)$, $t \geqslant 0$, then it is a complete Menger space.

THEOREM 5.1.2. *Let* $(E, \|\cdot\|_E)$, $C([0, a], E)$ *and* $(C([0, a], E), \mathscr{F}, \mathrm{Min})$ *be as stated above. Let the kernel* $K(t, s, x)$ *in* (3) *satisfy the following conditions:*

(i) $K(t, s, x(s)) \in C([0, a] \times [0, a] \times C([0, a], E), E)$ *and*

$$\|K\|_C = \sup_{t, s \in [0, a], x \in E} \|K(t, s, x)\|_E < \infty$$

(ii) *there exists an* $m \in \mathbb{N}$ *such that for all* $x(t), y(t) \in C([0, a], E)$ *and all* $t, s \in [0, a]$

$$\|K(t, s, T^{m-1}x(s)) - K(t, s, T^{m-1}y(s))\|_E \leqslant$$

$$\leqslant L \max_{p, q \in \{x(s), y(s), T^m x(s), T^m y(s)\}} \|p - q\|_E$$

where L *is any positive number whereas the mappings* T *and* T^m

are defined as

$$Tx(t) = x_0(t) + \int_0^t K(t, s, x(s))\mathrm{d}s, \ x_0(t) \in C([0, a], E),$$

$$T^m x(t) = x_0(t) + \int_0^t K(t, s, T^{m-1}x(s))\mathrm{d}s.$$

Suppose further that for any $x(t) \in C([0, a], E)$ *the set* $O_T(x, 0, \infty)$ *is bounded. Then, for any* $x(t) \in C([0, a], E)$, *the sequence*

$$x_n(t) = T^m x_{n-1}(t), \ n \geqslant 2, \ x_1(t) = T^m x(t)$$

converges in norm $\|\cdot\|_*$ *to a solution* $\bar{x}(t) \in C([0, a], E)$ *of equation* (3).

In a similar way [98], we can prove the existence and uniqueness of solution of the following kind of systems of nonlinear Volterra integral equations

$$(4) \quad \begin{cases} x(t) = x_0(t) + \int_0^t K_1(t, s, x(s))\,\mathrm{d}s. \\ y(t) = x_0(t) + \int_0^t K_2(t, s, y(s))\mathrm{d}s. \end{cases}$$

THEOREM 5.1.3. *Let* $(E, \|\cdot\|_E)$, $C([0, a], E)$ *and* $(C([0, a], E), \mathscr{F}, \mathrm{Min})$ *be as in Theorem* 5.1.2. *Let the kernel* $K_i(t, s, x)$, $i = 1,2$ *in* (4) *satisfy the following conditions:*

(i) $K_i(t, s, x) \in C([0, a] \times [0, a] \times E, E)$, $i = 1, 2$;

$$\|K_i\|_C = \sup_{\forall\, t,s \in [0,a], x \in E} \|K_i(t, s, x)\|_E < \infty;$$

(ii) *there exist* $m, n \in \mathbb{N}$ *and* L_1 *such that for all* $x(t), y(t) \in C([0, a], E)$ *and all* $t, s \in [0, a]$ *the following holds*

$$\|K(t, s, T_1^{m-1}x(s)) - K_2(t, s, T_2^{n-1}y(s))\|_E \leqslant$$
$$\leqslant L_1 \max\{\|x(s) - y(s)\|_E, \|x(s) - T_1^m x(s)\|_E, \|y(s) - T_2^n y(s)\|_E\},$$

where the mappings T_i, $i = 1, 2$, *and the mappings* T_i^k, $i = 1, 2$ *are defined as*

$$T_i x(t) = x_0(t) + \int_0^t K_i(t, s, x(s))\mathrm{d}s, \quad x_0(t) \in C([0, a], E),$$

$$T_i^k x(t) = x_0(t) + \int_0^t K_i(t, s, T_i^{k-1}x(s))\mathrm{d}s.$$

Suppose further that there exist some $y_0(t) \in C([0, a], E)$ *such that the sequence* $\{y_i\}_{i \in \mathbb{N}}$ *defined as*

$$(5)\quad y_{2i+1} = T_1^m y_{2i},\ y_{2i+2} = T_{2i+1}^n y_{2i+1},\ i = 0, 1, 2, \ldots,$$

is bounded. Then the system of equations (4) *has a unique common solution* $\bar{x}(t) \in C([0, a], E)$ *and the sequence* (5) *converges in the* (ε, λ)*-topology (hence in the norm* $\|\cdot\|_C$*) to* $\bar{x}(t)$.

The use of other fixed point theorems on probabilistic structures to the study of the existence and uniqueness of random solution to nonlinear random equations is given below (Shih-sen Chang [95], [97]).

THEOREM 5.1.4. *Let* (E, d) *be a complete separable linear metric space and let* $(E, \mathscr{F}, \mathrm{Min})$ *be the induced complete Menger space. Suppose that* $T(.,.) : \Omega \times E \to E$ *is a d-continuous random operator, such that for any* $x \in E$ *and any* $\omega \in \Omega$, $\delta(O_S(x, (\omega), 0, \infty))$ *is d-bounded, and there exist* $m, n \in \mathbb{N}$ *satisfying*

$$(6)\quad P(\{\omega \in \Omega : d(S^n(\omega, x), S^m(\omega, y)) \leqslant$$
$$\leqslant \Phi(\omega, \delta(O_S(x(\omega), y(\omega), 0, \infty)))\}) =$$
$$= P(E_{x,y}) = 1, \quad \forall x, y \in E,$$

where the random oprator $S : \Omega \times E \to E$ *is defined as*

$$(7)\quad S(\omega, x) = T(\omega, x) + y(\omega)$$

while $\Phi : \Omega \times [0, \infty) \to [0, \infty)$ *is a random function satisfying the following conditions :*

(1) *for any fixed* $\omega \in \Omega$, $\Phi(\omega, 0) = 0$ *and* $\Phi(\omega, .)$ *is strictly increasing with respect to* $t \in [0, \infty)$;

(2) *for any given* $\omega \in \Omega$ *and* $t > 0$ $\lim_{n\to\infty} \Phi^{-n}(\omega, t) = \infty$, *where* $\Phi^{-1}(\omega, .)$ *denotes the inverse of* $\Phi(\omega, .)$ *and* $\Phi^{-n}(\omega, .)$ *is the n-th iteration of* $\Phi^{-1}(\omega, .)$.

Then for any given E-valued random variable $\bar{y}(\omega)$, *there exists a unique solution* $x(\omega)$ *of the nonlinear random operator equation*

$$(8)\quad x(\omega) = T(\omega, x(\omega)) + y(\omega),$$

and for any E-valued random variable $x_0(\omega)$, *the sequence*

$$x_n(\omega) = T(\omega, x_{n-1}(\omega)) + \bar{y}(\omega),\ n = 1, 2, \ldots,$$

of E-valued random variables converges a.s. to $\bar{x}(\omega)$.

Proof. From the separability of E and the continuity of $T(\omega, .)$, it follows that

$$G = \bigcap_{x \in E} \bigcap_{y \in E} E_{x,y} \in \mathscr{K}, \text{ and } P(G) = 1.$$

For any $\bar{\omega} \in G$, we have

$$d(S^n(\bar{\omega}, x), S^m(\bar{\omega}, y)) \leqslant \Phi(\bar{\omega}, \delta(O_S(x(\bar{\omega}), y(\bar{\omega}), 0, \infty))), \ \forall\, x, y \in E.$$

Hence, for all $x,y \in E$ and any $t \geqslant 0$,

$$\begin{aligned} F_{S^n(\bar{\omega},x), S^m(\bar{\omega},y)}(t) &= H_0(t - d(S^n(\bar{\omega}, x), S^m(\bar{\omega}, y))) \geqslant \\ &\geqslant H_0(t - \Phi(\bar{\omega}, \delta(O_S(x(\bar{\omega}), y(\bar{\omega}), 0, \infty)))) = \\ &= H_0(\Phi(\bar{\omega}, \Phi^{-1}(\bar{\omega}, t)) - \Phi(\bar{\omega}, \delta(O_S(x(\bar{\omega}), y(\bar{\omega}), 0, \infty)))) = \\ &= H_0(\Phi^{-1}(\bar{\omega}, t) - \delta(O_S(x(\bar{\omega}), y(\bar{\omega}), 0, \infty))) = \\ &= \inf_{p,q \in O_S(x(\bar{\omega}), y(\bar{\omega}), 0, \infty)} F_{p,q}(\Phi^{-1}(\bar{\omega}, t)). \end{aligned}$$

Since $\Phi^{-1}(\bar{\omega}, t)$ satisfies condition (Φ_1) of Th. 3.1, it follows that $S(\bar{\omega}, .)$ is continuous in the (ε, λ)-topology. For any $x \in E$, the orbit $O_S(x, (\bar{\omega}), 0, \infty)$ is probabilistically bounded. Then from Theorem 3.1.42 and its remark, it follows that there exists a unique fixed point $\bar{x}(\bar{\omega}) \in E$ and for any E-valued random variable $x_0(\bar{\omega})$, where $\omega = \bar{\omega}$, the sequence $\{x_n(\bar{\omega})\}_{n \in \mathbb{N}}$ converges in the (ε, λ)-topology (hence in d) to $\bar{x}(\omega)$. Since $x_n(\omega) \to \bar{x}(\omega)$ a.s. and since $\bar{x}(\omega)$ is the unique random fixed point of $S(.,.)$, it follows that $\bar{x}(\omega)$ is the unique solution of Eq. (8).

THEOREM 5.1.5. *Let $(\Omega, \mathscr{K}, P)$ be a complete probability measure space, let $(X, \|.\|)$ be a separable complete Banach space and let Φ be a function satisfying condition (Φ_1) of Th. 3.1. Suppose that $\alpha(.)$, $\beta(.)$ are given X-valued random variables, that $A(.,.)$, $B(.,.) : \Omega \times X \to X$ are any given continuous random operators and that $S(.,.)$ and $T(.,.)$ are the continuous random operators defined as*

$$(9) \quad \begin{cases} S(\omega, .) = A(\omega, .) + \alpha(\omega) \\ T(\omega, .) = B(\omega, .) + \beta(\omega) \end{cases}.$$

If $(E, \mathscr{F}, T_m)$ is an E-space over $(X, \|.\|)$ and if there exist $i, j \in \mathbb{N}$ such that for any $x(\omega)$, $y(\omega) \in E$ the following holds

$$\begin{aligned} &\inf_{p,q \in \{O_S(S^i(\omega), x(\omega), 0, \infty) \cup O_T(T^j(\omega), y(\omega), 0, \infty)\}} F_{p,q}(t) \geqslant \\ &\geqslant \inf_{p,q \in \{O_S(x(\omega), 0, \infty) \cup O_T(y(\omega), 0, \infty)\}} F_{p,q}(\Phi(t)), \end{aligned}$$

then the system of random operator equations (9) has a unique common random solution, and for any $x_0(\omega) \in E$, the sequences $\{S^n(\omega, x_0(\omega))\}_{n \in \mathbb{N} \cup \{0\}}$ and $\{T^n(\omega, x_0(\omega))\}_{n \in \mathbb{N} \cup \{0\}}$ of random variables converge in the (ε, λ)-topology to this random solution.

Proof. The proof follows from Proposition 4.1.2.

THEOREM 5.1.6. *Let* $(\Omega, \mathscr{K}, P)$, $(X, \|\cdot\|)$, $(E, \mathscr{F}, T_m)$, Φ, $B(\omega, .)$, $\beta(.)$ *and* $T(\omega, .)$ *be as in Theorem* 5.1.5. *Suppose there exists* $n, m \in \mathbb{N}$ *such that, for any* $x(\omega), y(\omega) \in E$ *and any* $t \geqslant 0$, *the following holds:*

$$\text{(I)}\quad F_{T^m(\omega,x(\omega)),T^n(\omega,y(\omega))}(t) \geqslant$$
$$\geqslant \inf_{p,q \in \{O_T(x(\omega),0,\infty) \cup O_T(y(\omega),0,\infty)\}} F_{p,q}(\Phi(t)).$$

Then there exists a unique random solution $\overline{x}(\omega) \in E$ *of the following random operator equation*

$$(10)\quad y(\omega) = B(\omega, y(\omega)) + \beta(\omega)$$

and for any $x_0(\omega) \in E$ *the sequence* $\{T^n(\omega, x_0(\omega))\}_{n\in\mathbb{N}}$ *converges in the* (ε, λ)*-topology to this solution* $\overline{x}(\omega)$.

The proof mimics that of Proposition 4.1.3.

COROLLARY. *Let* $(\Omega, \mathscr{K}, P)$, $(X, \|\cdot\|)$, $(E, \mathscr{F}, T_m)$, $\Phi(.)$, $B(\omega, .)$, $\beta(\omega)$ *and* $T(\omega, .)$ *be as in Theorem* 5.1.6. *Suppose that for any* $x(\omega)$, $y(\omega) \in E$ *the following holds:*

$$\text{(II)}\quad F_{T(\omega,x(\omega)),T(\omega,y(\omega))}(t) \geqslant$$
$$\geqslant \inf_{p,q \in \{x(\omega),y(\omega),T(\omega,x(\omega)),T(\omega,y(\omega))\}} F_{p,q}(\Phi(t)) \quad \forall t \geqslant 0.$$

Then the conclusion of Theorem 5.1.6 *still holds.*

Proof. From condition (II) it follows that condition (I) holds. Therefore, the conclusion of the Corollary follows from Theorem 5.1.6.

5.2. SOME EXTREMUM PROBLEMS FOR FUNCTIONALS ON *PM*-SPACES AND STOCHASTIC OPTIMIZATION

First, let us briefly recall a problem of stochastic optimization due to R. Kannan [251]. Let us consider the problem of maximizing the functional $I(\omega, x, u)$ subject to the set of constraints characterized by $0 \in \Phi(\omega, x, u)$. We assume that $x(\omega)$ and $y(\omega)$ are in the range of random correspondence $A : \Omega \to 2^X$, $B : \Omega \to 2^X$, X and Y being separable Banach spaces. An example of the above would be to maximize $I(\omega, x, u) = \int_{t_1}^{t_2} f_0(\omega, t, x(t, \omega),$ $u(t, \omega))\,\mathrm{d}t$ with $\dfrac{\mathrm{d}x}{\mathrm{d}t} = f(\omega, t, x, u)$ and with appropriate constraints on $u(t, \omega) \in U(\omega, t, x)$ together with the boundary conditions.

We assume for the rest of the discussion that the hypotheses on I and Φ are such that, for any $\omega \in \Omega$, the deterministic problem is solvable. In order to prove the randomness of the correspondence associated with every $\omega \in \Omega$, the set of couples (x, u), on which I attains its maximum subject to the constraints, are defined in the following steps:

(1) establish the randomness of the correspondence T_1, which is defined by

$$T_1(\omega) = \{(x, u) : x \in A(\omega),\ u \in B(\omega) \text{ and } 0 \in \Phi(\omega, x, u)\},$$

(2) show that the correspondence $g(\omega)$ given by

$$g(\omega) = \sup_{(x,u) \in T_1(\omega)} I(\omega, x, u)$$

is random;

(3) finally, show that the correspondence

$$T(\omega) = \{(x, u) \in T_1(\omega) : I \text{ attains its maximum at } (\omega, x, u)\}$$

is random.

We shall revert to such practical problems in Chapter 7.

It is known that a nonlinear optimization problem consists in minimizing a nonlinear functional $f : H \to R$, where H is a real Hilbert space. The main difficulty is that we cannot calculate $f(x)$. All we can obtain are observations of the type $f(x) + \varepsilon(x)$, where $\varepsilon(x)$ is a random disturbance. We would like to generate a sequence $\{x_n\}_{n \in \mathbb{N}}$ by methods analogous to the deterministic case, where we can calculate $f(x)$ such that $\{x_n\}_{n \in \mathbb{N}}$ converges in a probabilistic sense to $x^* \in H$, where $f(x)$ attains its minimum. As is shown in a wide variety of practical situations, x^* is not necessarily unique. The problem of minimizing $f(x)$ reduces to a nonlinear operator equation.

Sometimes one considers [155] a family of functionals $(x, \lambda) \to f(x, \lambda)$ defined on a metric space (X, ρ), depending on a parameter λ in a topological space Λ. The dependence of the set of minimum points on λ is investigated. This problem is closely related to the problem of finding the solution of equations of the form $f(p) = p_0$ or $f(p) = p$.

The aim of the following part of this paragraph is to present some results due V. Radu [348] on the case of $\mathscr{D}^+$-valued functionals defined on a complete Menger space $(S, \mathscr{F}, T)$ with a continuous t-norm T. Thus, we shall be able to give several interesting applications of some fixed point theorems on PM-spaces and of Kuratowski's function presented in Chap. 2.

In what follows, the Lévy metric on $\mathscr{D}^+$ will be denoted by L, whereas f will be a functional defined on S with values in $\mathscr{D}^+$.

Let $F^* \in \mathscr{D}^+$ be the function $\sup_{p \in S} f(p)$ and let us say that the *minimum problem for* $f : S \to \mathscr{D}^+$ *is well posed in the generalized sense* (w.p.g.s.) if

(a) there exists a sequence $p_n \in S$ such that $f(p_n) \xrightarrow{L} F^*$,

(b) if $f(p_n) \xrightarrow{L} F^*$, then the set $\{p_n\}_{n \in \mathbb{N}}$ is relatively compact.

By $\mathscr{U}(S)$ we denote the set of all nonempty subsets of S, with the *upper semi-finite topology* (*u.s.f. topology*), i.e., the topology which has as a basis all sets of the form $\{A \in \mathscr{U}(S) : A \subset U, U$ open set in $S\}$.

Let $f : S \times \Lambda \to R$ be a $\mathscr{D}^+$-valued functional defined on the cartesian product $S \times \Lambda$, where S is a complete Menger space and Λ is a topological space. We shall use the following notations:

$$f_\lambda(p) = f(p, \lambda),$$

$$A_a = \{p \in S : L(f(p), F^*) \leqslant a\}, \ a \in (0,1).$$

$$A(\lambda) = \{p' \in S : f_\lambda(p') = \sup_{p \in S} f_\lambda(p)\}.$$

We say that $f_\lambda(p)$ converges to $f_{\lambda_0}(p)$ uniformly in S as $\lambda \to \lambda_0$ if for any $\varepsilon > 0$ there exists a neighbourhood N of λ_0 such that $L(f_\lambda(p), f_{\lambda_0}(p)) < \varepsilon$ for any $p \in S$, $\lambda \in N$; $A_a(\lambda) \xrightarrow{u.s.f.} A(\lambda_0)$ as $(\lambda, a) \to (\lambda_0, 0)$ if for any open set U containing $A(\lambda_0)$, there exists a neighbourhood N of λ_0 and $\delta > 0$ such that $A_a(\lambda) \subset U$ for any $\lambda \in N$, $0 < a < \delta$.

The sets $\{A_a\}_{a \in (0,1)}$ are supposed to be closed subsets of S.

THEOREM 5.2.1. *The minimum problem for f is well posed in the generalized sense if and only if* $\alpha_{A_a} \to H_0$ *as* $a \to 0$.

Proof. It is known (see Th. 1.2.5) that the set $A = \bigcap_a A_a$ is a nonempty compact set.

Let N_k, $k \in \mathbb{N}$ be a neighbourhood of A such that $A \bigcap_{k \in \mathbb{N}} N_k$. Then it is easy to see that for every k there exists an a_k such that $A_{a_k} \subset N_k$. Clearly, we may suppose that $a_k \to 0$ as $k \to \infty$.

Let $p_n \in S$ be such that $f(p_n) \to F^*$. Then there exists an n_k such that $n \geqslant n_k$ implies that $L(f(p_n), F^*) \leqslant a_k$, that is $p_n \in N_k$ for $n \geqslant n_k$, which clearly implies that $\{p_n\}_{n \in \mathbb{N}}$ is relatively compact.

Conversely, let d be a distance which metrizes S and let h_d be the Hausdorff metric on 2^S. It is easy to see that $h_d(A_a, A) = \inf\{\eta : A_a \subset B(A, \eta)\}$, where $B(A, \eta) = \{p \in S : d(p, A) < \eta\}$.

Suppose that $h_d(A_a, A)$ does not converge to zero as $a \to 0$. Then let η_0 be such that for every a' there exists an $a \leqslant a'$ with $h_d(A_a, A) \geqslant \eta_0$. Let $\eta_1 < \eta_0$ and $a'_n = \dfrac{1}{n}$. Then there exists a $p_n \in A_{a_n}$ such that $d(p_n, A) \geqslant \eta_1$. Since

$$L(f(p_n), F^*) \leqslant a_n \leqslant \frac{1}{n}, \text{ then } f(p_n) \to F^*,$$

which says that $\{p_n\}_{n\in\mathbb{N}}$ is relatively compact and its limit points are in A, which contradicts $d(p_n, A) \geqslant \eta_1$. Therefore $h_d(A_a, A) \to 0$, that is A_a converges to A as $a \to 0$.

Since h_d metrizes 2^S, then $F_{A_a,B} \to F_{A,B}$ for each finite subset B of S. Since A is compact, then $\beta_A = H_0$ and it is easy to see that $\beta_{A_a} \to H_0$ which implies that $\chi_{A_a} \to H_0$ and thus $\alpha_{A_a} \to H_0$. The proof is complete.

Let $\Lambda = I$ be a topological space. For every $i \in I$, let $f_i : S \to \to \mathscr{D}^+$, $F_i^* = \sup\limits_{p\in S} f_i(p)$ and for $a \in (0,1)$ let

$$A_a(i) = \{p \in S : L(f_i(p), F_i^*) \leqslant a\}.$$

Suppose also that $A_a(i)$ is closed and bounded.

THEOREM 5.2.2. *Assume that*

(1) $\lim\limits_{i\to i_0} f_i(p) = f_{i_0}(p)$, *uniformly in* S.

(2) *The minimum problem for* f_{i_0} *is w.p.g.s.*

Then

(α) $A(i_0)$ *is a nonempty compact set.*

(β) $A_a(i) \xrightarrow{u.s.f.} A(i_0)$ *as* $a \to 0$ *and* $i \to i_0$.

Proof. (α) follows from (2) and Theorem 5.2.1. In order to prove (β), let $N_{A(i_0)}$ be an open neighbourhood of $A(i_0)$ and let $\eta > 0$ be such that $A_{3\eta}(i_0) \subset A(i_0)$. According to (1), there exists an N_{i_0} such that $i \in N_{i_0}$ implies that $L_i(f_i(p), f_{i_0}(p)) < \eta$ for every $p \in S$. Hence, $L(F_i^*, F_{i_0}^*) \leqslant \eta$. Let $a < \eta$, $i \in N_{i_0}$ and $p \in A_a(i)$. Since $L(F_{i_0}^*, f_{i_0}(p)) \leqslant L(F_{i_0}^*, F_i^*) + L(F_i^*, f_i(p)) + L(f_i(p), f_{i_0}(p)) < 3\eta$, then $A_a(i) \subset N_{A(i_0)}$ and the proof is complete.

PROPOSITION 5.2.1. *Let* $V_n : S \to S$ *be a sequence of mappings uniformly convergent to a continuous mapping* $V : S \to S$. *Let* $A_a^0 = \{p \in S : L(F_{p,V(p)}, H_0) \leqslant a\}$ *and suppose that* $\alpha_{A_a^0} \to H_0$ *as* $a \to 0$. *Then any sequence* $\{p_n\}_{n\in\mathbb{N}}$ *such that* $F_{p_n,V_n(p_n)} \to H_0$ *is relatively compact and all its limit points are fixed points for* V.

Proof. Let $I = \left\{0, 1, \frac{1}{2}, \ldots, \frac{1}{n}, \ \ldots\right\}$ with the relative topology as a subset of $\mathbb{R}$ and define

$$f_i(p) = \begin{cases} F_{p,V(p)} & \text{if } i = 0 \\ F_{p,V_i(p)} & \text{if } i = \frac{1}{n}. \end{cases}$$

As $\alpha_{A_a^0} \to H_0$, then $A = \bigcap_a A_a^0$ is a nonempty compact set. Thus, V has a fixed point which implies that $F_0^* = H_0$. Also, $A_a(0) = A_a^0$ and hence $\alpha_{A_a(0)} \to H_0$ as $a \to 0$. It is known that the mapping $(p, q) \to F_{p,q}$ is uniformly continuous and thus $f_i(p) \to f_0(p)$ uniformly in S. From Theorem 5.2.2, it follows that $A_a(i) \to A(0)$ as $i \to 0$ and $a \to 0$. Let $\{p_n\}_{n\in\mathbb{N}}$ be such that $F_{p_n,V_n(p_n)} \to H_0$. Then $F_{1/n}^* = \sup_{p\in S} f_{1/n}(p) \to H_0$. For each $a \in (0,1)$ there exists an n_a such that $n \geqslant n_a$ implies that

$$L(F_{p_n,V_n(p_n)}, H) \leqslant \frac{a}{2} \quad \text{and} \quad L(F_{1/n}^*, H_0) \leqslant \frac{a}{2}$$

which implies that $L(f_{1/n}(p_n), F_{1/n}^*) \leqslant a$.

Now let N_k be a neighbourhood of $A(0) = \bigcap_k N_k$. Then there exist n_k, a_k such that $A_a(1/n) \subset N_k$ for $a \leqslant a_k$ and $n \geqslant n_k$. Now if $n \geqslant \max\{n_a, n_k\}$, then $L(f_{1/n}(p_n), F_{1/n}^*) \leqslant a$, that is $p_n \in A_a\left(\frac{1}{n}\right)$ and thus $p_n \in N_k$. This implies that $\{p_n\}_{n\in\mathbb{N}}$ is relatively compact and its limit points are in $A(0)$ and hence are fixed for V.

Remark. Let (S, d) be a metric space and let $(S, \mathscr{F}, \mathrm{Min})$ be the PM-space induced by $\mathscr{F}(p, q) = H_{d(p,q)}$. Suppose that (S, d) is complete and that $\psi_i : S \to \mathbb{R}^+$ and $\Phi_i^* = \inf_{p\in S} \psi_i(p)$. If we define $f_i : S \to \mathscr{D}^+$ by $f_i(p) = H_{\psi_i(p)}$, then

(1) $F_i^* = H_{\Phi_i^*}$

(2) $A_a(i) = \{p \in S : \psi_i(p) \leqslant \Phi_i^* + a\}$

(3) $\alpha_{A_a(i)} \to H_0$ if $\alpha(A_a(i)) \to 0$, where $\alpha(A)$ is Kuratowski's number for A.

The following result [155, Th. 1] is a conseauence of Theorem 5.2.2.

Let $\psi_i : (S, d) \to \mathbb{R}^+$ be such that

(i) $\psi_i(p) \to \psi_{i_0}(p)$ uniformly in S;

(ii) ψ_{i_0} is lower semicontinuous;

(iii) the minimum problem for ψ_{i_0} is well posed in the generalized sense.

Then

(a) $A(i_0)$ is nonempty and compact,

(b) $A_a(i) \xrightarrow{u.s.f.} A(i_0)$.

THEOREM 5.2.3. *Suppose that for any $i_0 \in I$, $f_i(p)$ converges to $f_{i_0}(p)$ uniformly on S as $i \to i_0$ and that the minimum problem for f_{i_0} is well posed in the generalized sense. Then the multi-valued mapping $\theta : I \to \mathscr{U}(S)$, $\theta(i) = A(i)$ is continuous.*

Proof. For every i, $A(i)$ is nonempty and compact and $A(i) \cap \subset A_a(i)$. The theorem follows from Theorem 5.2.2.

LEMMA 5.2.1. *The following statements are equivalent:*

(i) $f_i(p_n) \to F_i^* \Rightarrow \{p_n\}_{n\in\mathbb{N}}$ *is convergent;*

(ii) $D_{A_a(i)} \to H_0$ *as* $a \to 0$.

Proof. To prove (i) $\Rightarrow$ (ii), it is easy to see that condition (i) implies that $A(i)$ contains only one point $p = \lim_{n\to\infty} p_n$. Let ε, η be given. Since clearly $\alpha_{A_a(i)} \to H_0$, then there exists an a such that $A_a(i) \subset N_p(\varepsilon, \eta)$, that is $q \in A_a(i)$ implies $F_{p,q}(\varepsilon) > > 1-\eta$. Let $q_1, q_2 \in A_a(i)$; then $F_{q_1,q_2}(2\varepsilon) \geqslant T(F_{q_1,p}(\varepsilon), F_{p,q_2}(\varepsilon)) \geqslant \geqslant T(1-\eta, 1-\eta)$ which implies that $\inf_{q_1,q_2\in A_a(i)} (2\varepsilon) \geqslant T(1-\eta, 1-\eta)$.

Let $x > 0$ and $2\varepsilon < x$. Then $D_{A_a(i)}(x) \geqslant T(1-\eta, 1-\eta)$, which shows that $D_{A_a(i)}(x) \to 1$ and thus $D_{A_a(i)} \to H_0$.

Conversely, suppose that $D_{A_a(i)} \to H_0$ as $a \to 0$. Then it is easy to see that $A(i) = \bigcap_a A_a(i) = \{p\}$. If $f_i(p_n) \to F_i^*$, then $\{p_n\}_{n\in\mathbb{N}}$ is relatively compact and any limit point is in $A(i)$, that is $\{p_n\}_{n\in\mathbb{N}}$, is convergent to p, and the proof is complete.

We say that the *minimum problem for f_i is well posed* if one of the statements in the above lemma is verified.

THEOREM 5.2.4. *Suppose that for any $i_0 \in I$, $f_i(p)$ converges to $f_{i_0}(p)$ on S as $i \to i_0$ and that the minimum problem for f_{i_0} is* *well posed. Then the mapping* $\theta : I \to S$, *where* $\theta(i)$ *is the unique minimum point of* f_i, *is continuous.*

Proof. The uniqueness of the minimum point for f_i follows immediately from the definition of the well-posed problem. From Theorem 5.2.3, we obtain that $i \to A(i)$ is continuous. On the other hand, $\{p\} \to p$ is continuous and thus the mapping θ is continuous.

PROPOSITION 5.2.2. *Let* $V : S \times I \to S$ *be a continuous mapping. If for any* $i_0 \in I$

(i) $V_i(p) = V(p, i) \to V(p, i_0) = V_{i_0}(p)$ *uniformly on* S *as* $i \to i_0$,

(ii) V_{i_0} *is a contraction mapping on the complete Menger space* $(S, \mathscr{F}, T)$, *i.e.*, $F_{V_{i_0}(p), V_{i_0}(q)}(kt) \geqslant F_{p,q}(t)$ *for some* $k \in (0, 1)$, *then there exists a unique mapping* $\theta : I \to S$ *such that* $\theta(i) = V(\theta(i), i)$ *and* θ *is continuous.*

Proof. Consider the functional $f_i(p) = F_{p, V_i(p)}$. Let $F_i = \sup_a D_{A_a(i)}$ and $x, y \in (0, \infty)$. Let $a < a' < y$ and $p, q \in A_a(i)$. Since V_i is contractive, then $F_i^* = H_0$ and thus $L(f_i(p), H_0) \leqslant a$, $L(f_i(q), H_0) \leqslant a$ which implies that

$$H_0(y - a') \leqslant F_{p,V_i(p)}(y - a' + a') + a'$$

and

$$H_0(y - a') \leqslant F_{q,V_i(q)}(y - a' + a') + a',$$

which imply that

$$F_{p,V_i(p)}(y) \geqslant 1 - a' \text{ and } F_{q,V_i(q)}(y) \geqslant 1 - a'.$$

Then

$$F_{p,q}(x + 2y) \geqslant T(F_{p,V_i(p)}(y),\ T(F_{V_i(p),V_i(q)}(x),\ F_{q,V_i(q)}(y))) \geqslant$$
$$\geqslant T(1 - a',\ T(F_{p,q}(x/k_i),\ 1 - a')),$$

where $k_i < 1$ (V_i is contractive).

Therefore, $F_{p,q}(x + 2y) \geqslant T(1 - a', T(D_{A_a(i)}(x/k_i), 1 - a'))$.

If $y' > y$, then $D_{A_a(i)}(x + 2y') \geqslant T(D_{A_a(i)}(x/k_i),\ T(1 - a', 1 - a'))$, which implies that

$$F_i(x + 2y') \geqslant T(F_i(x/k_i),\ T(1 - a', 1 - a'))$$

for each a', and thus $F_i(x + 2y') \geqslant F_i(x/k_i)$ for every $y' > 0$.

Suppose that x is a continuity point of F_i. Then

$$F_i(x) \geqslant F_i(x/k_i) \geqslant F_i\left(\frac{x}{k_i^2}\right) \geqslant \ldots \geqslant F\left(\frac{x}{k_i^n}\right).$$

But $F_i\left(\frac{x}{k_i^n}\right) \to 1$, which says that $F_i(x) = 1$ i.e., $F_i = H_0$.

If $a_1 \leqslant a_2$, then $A_{a_1}(i) \subseteq A_{a_2}(i)$, which implies that $D_{A_{a_1}(i)} \geqslant D_{A_{a_2}(i)}$. Then $\lim\limits_{a\to 0} D_{A_a(i)}(x) = 1$ for every $x > 0$, which says that $D_{A_{a_1}(i)} \to H_0$.

Hence, the minimum problem for f_i is well posed. From Theorem 5.2.4 above, it follows that there exists $\theta : I \to S$ which is continuous, where $\theta(i)$ is the unique point of $A(i)$. Since $F_i^* = H_0$, then $f_i(\theta(i)) = H_0$, that is $F_{\theta(i), V_i(\theta(i))} = H_0$, which implies that $V_i(\theta(i)) = \theta(i)$ or $V(\theta(i), i) = \theta(i)$. Since V_i has a unique fixed point, then θ is unique.

LEMMA 5.2.2. *If $A \subset S$ and $B \subset (A, \varepsilon, \lambda) = \{p : \sup\limits_{q\in A} F_{p,q}(\varepsilon) > > 1 - \lambda\}$, then*

$$\alpha_{B(A,\varepsilon,\lambda)}(x + 2\varepsilon') \geqslant T(1 - \lambda, T(\alpha_A(x), 1 - \lambda))$$

for all $\varepsilon' > \varepsilon$.

Proof. Let $\eta < \alpha_A(x)$. Then there exist $A_1, A_2, \ldots, A_n$ such that $\bigcup\limits_{i=1}^{n} A_i = A$ and $D_{A_i}(x) \geqslant \eta$. Let $B_i = A_i \cup \{p \in B(A, \varepsilon, \lambda) : F_{p,q}(\varepsilon) > 1 - \lambda, q \in A_i\}$. If $p', p'' \in B_i$, then

$$F_{p',p''}(x + 2\varepsilon) \geqslant T(F_{p',q'}(\varepsilon), T(F_{q',q''}(x), F_{q'',p''}(\varepsilon))) \geqslant$$
$$\geqslant T(1 - \lambda, T(D_{A_i}(x), 1 - \lambda)) \geqslant T(1 - \lambda, T(\eta, 1 - \lambda)).$$

Thus,

$$\inf_{p',p''\in B_i} F_{p',p''}(x + 2\varepsilon) \geqslant T(1 - \lambda, T(\eta, 1 - \lambda)) \Rightarrow D_{B_i}(x + 2\varepsilon') \geqslant$$
$$\geqslant T(1 - \lambda, T(\eta, 1 - \lambda)).$$

Therefore,

$$\alpha_{B(A,\varepsilon,\lambda)}(x + 2\varepsilon') \geqslant T(1 - \lambda, T(\eta, 1 - \lambda)), \quad \forall\, \eta < \alpha_A(x),$$

which implies that

$$\alpha_{B(A,\varepsilon,\lambda)}(x + 2\varepsilon') \geqslant T(1 - \lambda, T(\alpha_A(x), 1 - \lambda))$$

for every $\varepsilon' > \varepsilon$.

PROPOSITION 5.2.3. *Let C be a closed and bounded subset of a complete random space S and let $V : C \to S$ be a probabilistic α-condensing mapping. Consider the equation*

$$p + V(p) = q,$$

where q is a given point of S. Let $E = \{p + V(p) : p \in C\}$. Then the multivalued mapping $\theta : E \to \mathcal{U}(C)$, $\theta(q) = \{p \in S : p + V(p) = q\}$

is upper semicontinuous, i.e., it is continuous in the u.s.f. topology of $\mathcal{U}(C)$.

Proof. Let $f_q(p) = F_{p+V(p)-q}$, $q \in E$. Then the minimum problem for f_q is well posed in the generalized sense and $f_q(p) \to f_{q_0}(p)$ uniformly on C as $q \to q_0$ for any $q_0 \in E$. The assertion follows from Theorem 5.2.3.

PROPOSITION 5.2.4. *Let* $(S, \mathcal{F}, T)$ *be a complete RN-space with t-norm* T *of* $\mathcal{H}$*-type, let* Q *be a compact convex subset of* S *and let* $V : Q \times I \to Q$ *be a continuous mapping. If for any* $i_0 \in I$. $V(p, i) = V_i(p) \to V_{i_0}(p)$ *uniformly on* Q *as* $i \to i_0$, *then the mapping* $\theta : I \to \mathcal{U}(Q)$, *with* $\theta(i) = \{p \in Q : V(p, i) = p\}$, *is continuous.*

Proof. It is known that S is a locally convex space and that every continuous self-mapping on Q has fixed point. Let $f_i(p) = F_{p-V_i(p)}$. If Q is compact and convex, then $A(i)$ is nonempty and $F_i^* = H_0$, i.e., $\theta(i) = A(i)$. It remains to show that $f_i(p) \to f_{i_0}(p)$ as $i \to i_0$, uniformly on Q. But this follows from the hypothesis and the uniform continuity of the random norm, and the proof is complete.

Remarks. Regarding the continuity of the mapping θ, we recall some results from [348]:

(1) If $V : S \times I \to S$ is a continuous mapping and if for every $i_0 \in I$, $V_i(p) \to V_{i_0}(p)$ uniformly on S as $i \to i_0$, and if V_{i_0} has a strong unique fixed point p_{i_0} (i.e., V_{i_0} has a unique fixed point p_{i_0} and $\forall \varepsilon$, $\forall \lambda$, $\exists \varepsilon'$, $\exists \lambda'$ such that $F_{p,V(p)}(\varepsilon') > 1 - \lambda' \Rightarrow F_{p,p_{i_0}}(\varepsilon) > 1 - \lambda$), then the mapping $\theta : I \to S, \theta(i) = p_i$, is continuous.

(2) Let S_1, S_2 be two Menger spaces and suppose that S_1 is compact. If $V : S_1 \to S_2$ is a continuous mapping of S_1 onto S_2, then the mapping $\theta : S_2 \to \mathcal{U}(S_1)$, $\theta(q) = \{p \in S_1 : V(p) = q\}$ is continuous.

(3) Let $\varepsilon_n \searrow 0$, $\lambda_n \searrow 0$ and $A_n(q) = \{p \in S_1 : F_{q,V(p)}(\varepsilon_n) > 1 - \lambda_n\}$. If $\alpha_{A_n(q)} \to H_0$ for any $q \in S_2$, then the mapping $\theta : S_2 \to \mathcal{U}(S_1)$, $\theta(q) = \{p \in S_1 : V(p) = q\}$ is continuous.

5.3. APPLICATIONS OF RANDOM FIXED POINT THEOREMS TO RANDOM DIFFERENTIAL AND INTEGRAL EQUATIONS

We give below some applications of random fixed point theorems to different types of equations (random differential equations, random integral equations and functional equations with random parameters), given by H. W. Engl [132], [136], [138], G. Bocşan, G. Constantin and V. Radu [52], [60], [61], [62], [113], S. Itoh [242], A. Nowak [316], [317], [318], A. C. Lee and W. J. Padgett [276], [277].

First, we shall consider a theorem about the existence of a random solution to a nonlinear differential equation with random vector field, subject to random initial conditions. We shall give such a formulation to ensure the existence of a dense set of random solutions.

Let $\mathbb{R}^n$ be endowed with an arbitrary norm $\|\cdot\|_n$ and the Borel σ-algebra; for $a < b$, denote $C_n[a, b] = \{x : [a, b] \to \mathbb{R}^n$, x continuous$\}$ endowed with the norm $\|x\| = \sup_{t \in [a,b]} \|x(t)\|_n$.

A Peano-type existence result for nonlinear differential equations is given by

THEOREM 5.3.1. *Let $U \subseteq \mathbb{R}$ be an open neighbourhood of 0, let $f : U \times \mathbb{R}^n \times \Omega \to \mathbb{R}^n$ be such that $f(.,., \omega)$ is continuous for all $\omega \in \Omega$ and let $f(t, x,.)$ be measurable for all $(t, x) \in U \times \mathbb{R}^n$ ("random vector field" or "random coefficients"). Let $x_0 : \Omega \to \mathbb{R}^n$ be measurable ("random initial conditions"), $K : \Omega \to (0, \infty)$ measurable. We assume that there exists an $\varepsilon > 0$ such that for all $\omega \in \Omega$ and $(t, x) \in U \times \mathbb{R}^n$ with $|t| \leqslant \varepsilon$ and $\|x - x_0(\omega)\|_n \leqslant \varepsilon$, $\|f(t, x, \omega)\|_n \leqslant K(\omega)$ and that there exists a $K_0 > 0$ with $K(\omega) < K_0$ for all $\omega \in \Omega$. Let $d_0 = \min\{\varepsilon, \varepsilon/K_0\}$.*

Then there exists a countable set of functions $x_k : \Omega \times [0, d_0] \to \mathbb{R}^n$ such that, for all $k \in \mathbb{N}$,

(1) $x_k(.,t)$ is measurable for all $t \in [0, d_0]$;

(2) $x_k(\omega,.)$ is continuously differentiable for all $\omega \in \Omega$;

(3) For μ-almost all $\omega \in \Omega$

$$\frac{\mathrm{d}x_k(\omega, t)}{\mathrm{d}t} = f(t, x_k(\omega, t), \omega) \quad \textit{and} \quad x_k(\omega, 0) = x_0(\omega);$$

(4) For all $\omega \in \Omega$ and $t \in [0, d_0]$

$$\| x_k(\omega, t) - x_0(\omega)\|_n \leqslant \min\{\varepsilon, \varepsilon.K(\omega)\}.$$

Furthermore,

(5) For μ-almost all $\omega_0 \in \Omega$, every solution on $[0, d_0]$ of the deterministic initial value problem

$$\frac{\mathrm{d}x(t)}{\mathrm{d}t} = f(t, x(t), \omega_0), \quad x(0) = x_0(\omega_0)$$

is the uniform limit of a sequence of elements of the set $\{x_1(\omega_0,.), x_2(\omega_0,.), x_3(\omega_0,.), \ldots\}$.

If $(\Omega, \mathscr{K}, \mu)$ is complete, then "μ-almost all" can be replaced by "all".

Proof. Without loss of generality (cf. the proof of Theorem 4.4.2), let, for all $\omega \in \Omega$, $d(\omega) = \min\{\varepsilon, \varepsilon/K(\omega)\}$, $C(\omega) = \{x \in C_n[0, d_0]$ and for all $t \in [0, d_0]$, $\|x(t) - x_0(\omega)\|_n \leqslant d(\omega)K(\omega)\}$. Then define $\tilde{T}: \Omega \times C_n[0, d_0] \to C_n[0, d_0]$ as

$$\tilde{T}(\omega, x)(t) = x_0(\omega) + \int_0^t f(s, x(s), \omega)\, ds,$$

$$T = \tilde{T}|_{\{(\omega,x):\omega\in\Omega,\ x\in C(\omega)\}=\mathrm{Gr}C}.$$

It is easy to see that for every fixed but arbitrary $x \in X = \mathbb{R}^n$, the mapping $\omega \to d(x, C(\omega)) = \max\{0, \|x - x_0(\omega)\| - d(\omega, K(\omega))\}$ is measurable. According to Theorem 2.2.1, C is measurable, and from Remark 2, Sec. 4.4, it follows that it is separable. Hence, for all $\omega \in \Omega$, $x \in C_n[0, d_0]$ and $t \in [0, d_0]$,

$\omega \to x_0(\omega) + \int_0^t f(s, x(s), \omega)\, ds$ is measurable as the limit of a sequence of finite sums of measurable functions. Together with Lemma 2.2.5, this implies that for all open $D \subset C_n[0, d_0]$ and $x \in C_n[0, d_0]$ $\{\omega \in \Omega : x \in C(\omega) \wedge T(\omega, x) \in D\} = \Big[\bigcap_{n\in N}\Big\{\omega \in \Omega : C(\omega) \cap$

$$\cap \left\{y \in C_n[0, d_0] : \|y - x\| < \frac{1}{n}\right\} \neq \emptyset\Big\}\Big] \cap \Big\{\omega \in \Omega : x_0(\omega) +$$

$$+ \int_0^t f(s, x(s), \omega)\, ds \in D\Big\} \in \mathscr{K}.$$

Therefore, T is a random operator with stochastic domain C and every $C(\omega)$ is bounded, closed, convex and solid. It is well known (cf. e.g. [411, Th. 6.2.2 p. 45]) that, for every $\omega \in \Omega$, $T(\omega,.)$ is a completely continuous self-mapping of $C(\omega)$. The conclusions of the theorem follow now from Corollary 4.4.2 (with $S = 0$) and Lemma 2.2.5.

Similar results can be obtained for the case when f is defined on a random subset of $\mathbb{R} \times \mathbb{R}^n$ and the initial time t_0 is a random variable [52].

Let $(\Omega, \mathscr{K}, \mu)$ be a σ-finite and complete measure space. Consider that the following are given:

(a) the random variables $t_0 : \Omega \to \mathbb{R}$, $a : \Omega \to (0, \infty)$ and $r : \Omega \to (0, \infty)$;

(b) the closed random interval $I_a \subset \mathbb{R}$, $\omega \to I_a(\omega) = \{t \in R : |t - t_0(\omega)| \leqslant a(\omega)\}$, $\omega \in \Omega$ and the closed random sphere $S_r \subseteq \mathbb{R}^n$, $\omega \to S_r(\omega) = \{x \in \mathbb{R}^n : \|x - x_0(\omega)\|_n \leqslant r(\omega)\}$, where $x_0 : \Omega \to \mathbb{R}^n$ is a random variable;

(c) the continuous random function f with random domain $\omega \to I_a(\omega) \times S_r(\omega)$, $\omega \in \Omega$, i.e.,

$$f : \{(t, x, \omega) \in \mathbb{R} \times \mathbb{R}^n \times \Omega : (t, x) \in I_a(\omega) \times S_r(\omega)\} \to \mathbb{R}^n$$

such that $(t, x) \to f(t, x, \omega)$ is continuous for every $\omega \in \Omega$ and $\omega \to f(t, x, \omega)$ is $\mathscr{K}$-measurable on $\{\omega \in \Omega : (t, x) \in I_a(\omega) \times \times S_r(\omega)\}$ i.e., $\{\omega \in \Omega : t \in I_a(\omega),\ x \in S_r(\omega),\ f(t, x, \omega) \in D\} \in \mathscr{K}$ for every open set $D \subset \mathbb{R}^n$ and every $(t, x) \in \mathbb{R} \times \mathbb{R}^n$.

DEFINITION 5.3.1. We say that $x : \{(t, \omega) \in \mathbb{R} \times \Omega : t \in I_d(\omega)\} = J_d \to \mathbb{R}^n$ is a random solution of the differential equation

$$(1) \quad \begin{cases} \dfrac{dx}{dt} = f(t, x, \omega) \\ x(t_0(\omega), \omega) = x_0(\omega) \end{cases}$$

if the following conditions are satisfied:

(i) x is a random operator such that, a.s. on J_d, $x(t, \omega) \in \in S_r(\omega)$;

(ii) x satisfies (1) a.s. on J_d and t_0 is a selector of J_d.

THEOREM 5.3.2. *Suppose that Equation* (1) *is true and that f satisfies* (a), (b) *and* (c). *Then there exists a random solution of* (1) *on the closed random interval $\omega \to I_d(\omega)$, where*

$$d(\omega) = \min\left\{a(\omega), \frac{r(\omega)}{m(\omega)}\right\}, \quad \omega \in \Omega$$

and

$$m(\omega) = \max\{\|f(t, x, \omega)\|_n,\ (t, x) \in I_a(\omega) \times S_r(\omega)\}.$$

Before proceeding with the proof of this theorem, we shall first consider several lemmas which are of interest in themselves.

Let X be the linear space of the bounded continuous functions $x = (x_1, \ldots, x_n)$ of $\mathbb{R}$ into $\mathbb{R}^n$ such that $\lim\limits_{t \to \pm\infty} x_j(t) < \infty$ for every $j = 1, 2, \ldots, n$. For $x \in X$, we put $\|x\| = \sup\limits_{t \in \mathbb{R}} \|x(t)\|_n$.

LEMMA 5.3.1. (i) *The linear space X is a separable Banach space;*

(ii) *$x : \Omega \times \mathbb{R} \to \mathbb{R}^n$ with trajectories in X is a continuous random function if the mapping $\omega \to x(\omega)$ is an X-valued random variable.*

Proof. (i) Denoting $\overline{\mathbb{R}} = \mathbb{R} \cup \{-\infty, \infty\}$, one can see that $\overline{\mathbb{R}}$ is a compact metric space with Baire's *limitative metric*. Hence,

$C(\overline{\mathbb{R}}, \mathbb{R}^n)$, the space of the continuous functions on $\overline{\mathbb{R}}$ endowed with the max-norm, is a separable Banach space. Since $\mathbb{R}$ is dense in $\overline{\mathbb{R}}$, the mapping $x \to \bar{x}$ is an isometric isomorphism between X and $C(\overline{\mathbb{R}}, \mathbb{R}^n)$, where $\bar{x}$ is the unique continuous extension of x to $\overline{\mathbb{R}}$.

(ii) If $x : \Omega \times \mathbb{R} \to \mathbb{R}^n$ is a continuous random function, then for every $x_0 \in X$, $r > 0$ and $t \in \mathbb{R}$ the mapping

$$\omega \to \| x(\omega, t) - x_0(t) \|_n, \quad \omega \in \Omega,$$

is measurable. Therefore, if Z is a dense countable set in $\mathbb{R}$, it follows that

$$\sup_{t \in Z} \| x(\omega, t) - x_0(t) \|_n = \| x_\omega - x_0 \|, \quad \omega \in \Omega$$

is measurable.

Thus,

$$\{\omega \in \Omega, \| x_\omega - x_0 \| < r\} \in \mathscr{K}$$

i.e., $\omega \to x_\omega$ is Price-measurable and since X is separable it is also Borel-measurable (see Ch. 2).

Conversely, we suppose that $\omega \to x_\omega$ is an X-valued random function and for $t \in \mathbb{R}$ let f_t^j be the linear continuous functionals defined on X by

$$f_t^j(x) = x_j(t),$$

where $x = (x_1, \ldots, x_n) \in X$, $j = 1, 2, \ldots, n$.

Since X is separable, it follows that $\omega \to f_t^j(x_\omega) = x_j(\omega, t)$ is measurable for every j, hence $\omega \to x(\omega, t)$ is measurable.

Now, for two random variables a and b such that $a(\omega) < b(\omega)$ for $\omega \in \Omega$, let I be the closed random interval $I(\omega) = [a(\omega), b(\omega)]$. The multivalued mapping $\omega \to I(\omega)$ is a compact random subset of $\mathbb{R}$. Let $X(\omega)$ be the set of the continuous functions $x : \mathbb{R} \to \mathbb{R}^n$ such that $x(t) = x(a(\omega))$ for $t < a(\omega)$ and $x(t) = x(b(\omega))$ for $t > b(\omega)$.

LEMMA 5.3.2. (i) *The mapping* $\varphi : \Omega \times X \to X$ *defined by* $\varphi(\omega, x) = x\chi_{I(\omega)} + x(a(\omega))\chi_{(-\infty, a(\omega))} + x(b(\omega))\,\chi_{(b(\omega), \infty)}$ *is a linear continuous random operator* (χ_A *is the indicator of* A).

(ii) *the multivalued mappings* $\omega \to X(\omega)$ *and* $\omega \to E(\omega)$ *defined by*

$$E(\omega) = \{x \in X(\omega) : \| x - x_0(\omega) \| \leqslant r(\omega)\}, \quad \omega \in \Omega$$

are closed random subsets of X, *where* x_0 *is a measurable selector of* $\omega \to X(\omega)$ *and* r *is a random variable.*

Proof. (i) Let x be fixed in X. Since a and b are measurable it follows from Lemma 5.3.1 (ii) that $\omega \to \varphi(\omega, x)$, $\omega \in \Omega$, is measurable.

Since the linearity of $\varphi(\omega,.)$ is obvious, it follows that φ is a linear random operator. The continuity of $\varphi(\omega,.)$ follows from

$$\|(\varphi, x)\| = \|x|_{I(\omega)}\| \leqslant \|x\|.$$

(ii) Since $X(\omega)$ is a closed subset of X, the measurability of $\omega \to X(\omega)$ is sufficiently shown (see Theorem 2.2.5) by finding a Castaing representation of $\omega \to X(\omega)$. To do this, let Z be a dense countable subset of X and for each $z \in Z$ let

$$\varphi_z(\omega) = \varphi(\omega, z), \ \omega \in \Omega.$$

Since $\varphi(\omega, z) \in X(\omega)$, $\varphi(\omega, x) = x$ for $x \in X(\omega)$, and since φ is a continuous random operator on X, it is easy to see that the set $U = \{\varphi_z : z \in Z\}$ is a Castaing representation of $\omega \to X(\omega)$.

The measurability of E follows from applying Proposition 4.4.2 to the Carathéodory function $f(\omega, x) = \|x - x_0(\omega)\|$, $\omega \in \Omega$, $x \in X$.

Let $\tilde{X}(\omega) = C(I(\omega), \mathbb{R}^n)$ be the Banach space of the continuous function on $I(\omega)$ endowed with the norm

$$\|x\|_J = \max_{t \in I(\omega)} \|x(t)\|_n.$$

Denoting by $\psi_\omega(x)$ the restriction of $x \in X(\omega)$ to $I(\omega)$, it becomes obvious that ψ_ω is an isometric isomorphism between $X(\omega)$ and $\tilde{X}(\omega)$.

The following lemma shows that the set $\mathscr{Y}$ of the continuous random functions defined on the closed random interval I can be identified with the set $\mathscr{X}$ of the measurable selectors of the closed random subset $\omega \to X(\omega)$ of X.

LEMMA 5.3.3. *The mapping* $\psi : \mathscr{Y} \to \mathscr{X}$ *defined by*

$$(\psi(x))(\omega) = \psi_\omega^{-1}(x_\omega), \ \omega \in \Omega$$

is bijective, where $x_\omega : I(\omega) \to \mathbb{R}^n$ *is given by* $x_\omega(t) = x(\omega, t)$.

Proof. Let us first prove that ψ is well defined, i.e., ψ is $\mathscr{X}$-valued. Since $(\psi(x))(\omega) \in X$ and $\psi(x)$ is a selector of $\omega \to X(\omega)$, it is sufficient to prove the measurability of $\omega \to (\psi(x))(\omega)$. To do this, let $x \in \mathscr{Y}$, and $t \in \mathbb{R}$. Since I is a closed random interval, the set $\Omega_t = \{\omega \in \Omega : t \in I(\omega)\}$ is $\mathscr{K}$-measurable.

For every open subset of $\mathbb{R}^n$, we have

$$\{\omega \in \Omega : ((\psi(x))(\omega))(t) \in G\} = \{\omega \in \Omega_t : (\psi_\omega^{-1}(x_\omega))(t) \in G\} \cup$$
$$\cup \{\omega \notin \Omega_t : (\psi_\omega^{-1}(x))(t) \in G\}.$$

But $\omega \in \Omega_t$ implies that $t \in I(\omega)$, hence $(\psi_\omega^{-1}(x_\omega))(t) = x(\omega, t)$. Since $x \in \mathscr{Y}$ and $\Omega_t \in \mathscr{K}$, it follows that

$$\{\omega \in \Omega_t : (\psi_\omega^{-1}(x_\omega))\,(t) \in G\} = \{\omega \in \Omega_t : x(t, \omega) \in G\} \in \mathscr{K}.$$

For $\omega \in \Omega_t$, i.e., $t \in I(\omega)$, we have

$$(\psi_\omega^{-1}(x_\omega))(t) = x(\omega, a(\omega)) \text{ for } t < a(\omega)$$

and

$$(\psi_\omega^{-1}(x_\omega))(t) = x(\omega, b(\omega)) \text{ for } t > b(\omega).$$

Hence (see Theorem 4.4.1),

$$\{\omega \in \Omega_t : (\psi_\omega^{-1}(x_\omega))(t) \in G\} = \{\omega \notin \Omega_t : x(\omega, a(\omega)) \in G \text{ or }$$
$$x(\omega, b(\omega)) \in G\} \in \mathscr{K}$$

since a and b are measurable selectors of I, $x \in \mathscr{Y}$ and I is a separable random subset of $\mathbb{R}$.

Thus, for each $t \in \mathbb{R}$, the mapping $\omega \to ((\psi(x))\,(\omega)\,(t)$ is measurable, and using Lemma 5.3.1 (ii), it follows that the mapping $\omega \to (\psi(x))(\omega)$ is an X-valued measurable mapping. Hence, ψ is $\mathscr{X}$-valued.

The injectivity of ψ follows from the injectivity of ψ_ω. If $y \in \mathscr{X}$, then from Lemma 5.3.1 (ii) it follows that the function x defined by

$$x(\omega, t) = (y(\omega))(t), \quad \omega \in \Omega, \quad t \in I(\omega)$$

is in $\mathscr{Y}$ and $\psi(x) = y$.

LEMMA 5.3.4. *Let $f : \mathbb{R} \times \Omega \to \mathbb{R}^n$ be a continuous random function and let $t_0 : \Omega \to \mathbb{R}$ be a measurable selector of the closed random interval I.*

Then the function $i : \operatorname{Gr} I \to \mathbb{R}^n$ given by

$$i(t, \omega) = \int_{t_0(\omega)}^{t} f(s, \omega)\, ds, \quad \omega \in \Omega, \quad t \in I(\omega)$$

is a continuous random function with random domain I.

Proof. Let t be fixed in $\mathbb{R}$; then $\xi_t : \Omega_t \to \mathbb{R}$, $\xi_t(\omega) = t$ is a measurable selector of $I|_{\Omega_t}$, the restriction of I to $\Omega_t \in \mathscr{K}$. Therefore, $i(t,.)$ is $\mathscr{K} \cap \Omega_t = \mathscr{K}_t$-measurable as the limit of a sequence of finite sums of $\mathscr{K}_t$-measurable functions.

LEMMA 5.3.5. *Let $g : \mathbb{R} \times \mathbb{R}^n \times \Omega \to \mathbb{R}^n$ be a continuous random function, let t_0 be a measurable selector of the closed random interval I and let $x_0 : \Omega \to \mathbb{R}^n$ be a random vector. For $x \in X$,*

$\omega \in \Omega$ and $t \in I(\omega)$, let

$$s(t, x, \omega) = \int_{t_0(\omega)}^{t} g(u, x(u), \omega)\, du.$$

Then:

(i) *for every $\omega \in \Omega$, the mapping $x \to s_\omega(., x)$ of X into $\tilde{X}(\omega)$ is continuous,*

(ii) *for every x, the mapping $(t, \omega) \to s_x(t, \omega)$ belongs to $\mathcal{Y}$,*

(iii) *the mapping $T : \Omega \times X \to X$ given by*

$T(\omega, x) = x_0(\omega) + (\psi(s_x))(\omega)$,

is a continuous random operator and $T(., x)$ is a measurable selector of $\omega \to X(\omega)$ for every $x \in X$ (here $x_0(\omega)$ is considered as a constant on $\mathbb{R}$ into $\mathbb{R}^n$).

Proof. (i) follows from [129, §3.6.3]. (ii) follows from Lemma 5.3.4 (iii). From (ii) and Lemma 5.3.3, it follows that T is a random operator and $\omega \to T(\omega, x)$ is a measurable selector of $\omega \to X(\omega)$. By (i) and

$$\| T(\omega, x) - T(\omega, y)\| = \| \psi_\omega^{-1}(s_{x,\omega} - s_{y,\omega})\| = \| s_\omega(x,.) - s_\omega(y,.)\|_\omega,$$

we obtain the continuity of T.

Proof of Theorem 5.3.2. Let us first observe that $m(.)$ is a random variable and that $I_d(\omega)$ is a closed random interval.

Let g be a continuous extension of f to $\mathbb{R} \times \mathbb{R}^n$. The existence of this extension follows from Theorem 4.4.5 since the random set $\omega \to I_a(\omega) \times S_r(\omega)$ is separable and thus the operator T from Lemma 5.3.5 is defineable. Moreover, T has a wide-sense fixed point in the closed random subset $E \subset X$ (see Lemma 5.2.2) with $I(\omega) = I_d(\omega) \subset I_a(\omega)$. Indeed,

$$(1)\ (\psi_\omega(T(\omega, x)))(t) = x_0(\omega) + \int_{t_0(\omega)}^{t} g(u, x(u), \omega)\, du,\ t \in I_d(\omega).$$

Let $\omega \in \Omega$ be fixed and let $y \in \tilde{X}(\omega)$. Since $x = \psi_\omega^{-1}(y) \in X(\omega)$ and since $x(s) = y(s)$ for $s \in I_d(\omega)$, from (1) we obtain:

$$(2)\ (S_\omega y)(t) = (\psi_\omega(T(\omega, \psi_\omega^{-1}(y))))(t) = x_0(\omega) + $$
$$+ \int_{t_0(\omega)}^{t} f(s, y(s), \omega) ds,\ t \in I_d(\omega).$$

Hence, the operator $S_\omega : \tilde{X}(\omega) \to \tilde{X}(\omega)$ defined by (2) is the operator associated to the deterministic equation obtained when ω

is fixed. From the deterministic results (see, e.g. [129, § 3.6.3]), S has a fixed point y_ω in the sphere $\{y \in \tilde{X}(\omega) : \|y - x_0(\omega)\|_\omega \leqslant \leqslant r(\omega)\}$, i.e.,

(3) $\psi_\omega(T(\omega, \psi_\omega^{-1}(y_\omega))) = S_\omega(y_\omega) = y_\omega$

and

(4) $\|y_\omega - x_0(\omega)\| \leqslant r(\omega)$.

Denoting $x_\omega = \psi_\omega^{-1}(y_\omega)$, we have $x_\omega \in X(\omega)$ and using (3) and (4) we obtain

$T(\omega, x_\omega) = x_\omega$

and

$\|x_\omega - x_0(\omega)\| = \|\psi_\omega^{-1}(y_\omega - x_0(\omega))\| = \|y_\omega - x_0(\omega)\|_\omega \leqslant r(\omega)$.

Thus, $\omega \to x_\omega$ is a wide-sense fixed point of T in E, and therefore (see Theorem 4.4.3) T has a random fixed point in E, i.e., there exists a measurable mapping $\xi : \Omega \to X$ such that

(5) $T(\omega, \xi(\omega)) = \xi(\omega), \quad \xi(\omega) \in E(\omega), \quad \omega \in \Omega.$

If we put

$x(t, \omega) = (\psi_\omega(\xi(\omega)))(t), \quad t \in I_d(\omega)$

then Lemma 5.3.3 and (5) imply that $x \in \mathscr{Y}$ and that

(6) $x(t, \omega) \in S_r(\omega), \quad t \in I_d(\omega)$.

From (5) and (1), we obtain for $t \in I_d(\omega)$ that

(7) $x(t, \omega) = \psi_\omega(T(\omega, \xi(\omega)))(t) = x_0(\omega) +$

$$+ \int_{t_0(\omega)}^{t} g(s, (\xi(\omega))(s), \omega)\, \mathrm{d}s.$$

But $s \in I_d(\omega)$ and hence (b) implies that

$(\xi(\omega))(s) = (\psi_\omega(\xi(\omega)))(s) = x(s, \omega) \in S_r(\omega)$.

Since g is an extension of f, equation (7) implies that

$$x(, \omega) = x_0(\omega) + \int_{t_0(\omega)}^{t} f(s, x(s, \omega), \omega)\, \mathrm{d}s, \quad t \in I_d(\omega)$$

for every $\omega \in \Omega$. The proof is complete.

Let us consider next a stochastic analogue of Ambrosetti's theorem [12, Th. 3.1] as follows. Let a, r, t_0, x_0, I_a, S_r be as in Theorem 5.3.2.

THEOREM 5.3.3. *Let X be a separable real Banach space and let f be a random operator with stochastic domain* $\omega \to I_a(\omega) \times S_r(\omega)$, $\omega \in \Omega$, *i.e.*, $f : \{(t, x, \omega) \in \mathbb{R} \times X \times \Omega : (t, x) \in I_a(\omega) \times S_r(\omega)\} \to X$ *having the following properties:*

(i) *For every* $\omega \in \Omega$, $(t, x) \to f(t, x, \omega)$ *is uniformly continuous;*

(ii) *For every* $\omega \in \Omega$, $t \in I_a$, $x \to f(t, x, \omega)$ *is* $h(\omega)$*-set-Lipschitz, i.e., for any bounded set* $A \subseteq X$, $\alpha(f(\omega)A) \leqslant h(\omega)\alpha(A)$ *where* $\sup \{h(\omega), \omega \in \Omega\} < \infty$;

(iii) $\omega \to f(t, x, \omega)$ *is* $\mathscr{K}$*-measurable on* $\Omega_{(t,x)}$ *for each* (t, x), *where* $\Omega_{(t,x)} = \{\omega \in \Omega : (t, x) \in I_a(\omega) \times S_r(\omega)\} \in \mathscr{K}$;

(iv) $m = \sup \{\|f(t, x, \omega)\| : t \in I_a(\omega), \ x \in S_r(\omega), \ \omega \in \Omega\} < \infty$.

Then there exists at least one random solution of Eq. (1) *on the closed random interval* $\omega \to I_d(\omega)$, *where*

$$d(\omega) = \min \left\{ d'(\omega), a(\omega), \frac{r(\omega)}{m(\omega)} \right\},$$

$$m(\omega) = \sup \{\|f(t, x, \omega)\|, \ (t, x) \in I_a(\omega) \times S_r(\omega)\}$$

and $d'(\omega)$ *is a real random variable such that* $d'(\omega)h(\omega) < 1$ *for all* $\omega \in \Omega$.

Proof. The proof is similar to that for the previous theorem. Let us consider $Y(\omega) = C(I_d(\omega), X)$ with $d < \infty$, as the Banach space of all continuous functions $x : \mathrm{Gr}(I_d) \to X$ with the norm $|||x|||_\omega = \sup \{\|x(t, \omega)\|_\omega, \ t \in I_d(\omega)\}$. Consider also $Z = C_b(\mathbb{R}, X)$ as the space of all bounded continuous functions $z : \mathbb{R} \to X$ for which $\lim_{t \to t_0} z(t)$ exists. It can be proved that $Y(\omega)$ and Z are separable Banach spaces.

Indeed, if K is a compact space with a countable base and if X is a uniform metrizable space with a countable base, then $C(K, X)$ with the uniform convergence is a topological space with a countable base.

The separability of Z can be proved as in Lemma 5.3.1 (i). A similar characterization of the generalized random process $x : \mathbb{R} \times \Omega \to X$ (with trajectories in Z) is given by : x is a continuous X-valued random mapping if the mapping $\omega \xrightarrow{\tilde{x}} x_\omega(\,.\,) \in Z$ is a Z-valued random variable.

Indeed, if $\omega \xrightarrow{\tilde{x}} x_\omega(\,.\,)$ is a Z-valued random variable and since the projection $\pi_t : Z \to X$ defined by $\pi_t(y) = y(t)$ is a continuous mapping, then we deduce that, for every open set $G_X \subset X$, we have $A = \{y \in Z : y(t) \in G_X\} \in \mathscr{B}_Z$, where $\{\omega \in \Omega : x_t(\omega) \in G_X\} = \tilde{x}^{-1}(A) \in \mathscr{K}$. Therefore, $x_t(\,.\,)$ is an X-valued random variable for every t.

Conversely, suppose that $x : \mathbb{R} \times \Omega \to X$ is a continuous X-valued random function. Let x^0 be a fixed point of Z and let $S = \{y \in Z : \|y - x^0\| \leqslant d\}$; then $\tilde{x}^{-1}(S) = \{\omega : x_\omega(t) \in S\} = \bigcap_{q \in Q} \{\omega : \|x_\omega(q) - x^0(q)\| \leqslant d\} \in \mathscr{K}$, where Q is the set of rational numbers. Since Z is a separable Banach space, it follows that $\tilde{x}^{-1}(\mathscr{B}_Z) \subset \mathscr{K}$ and $\tilde{x}$ is a Z-valued random variable.

Let a and b be two random variables such that $a(\omega) < b(\omega)$. Then for all $\omega \in \Omega$, $I(\omega) = [a(\omega), b(\omega)]$,

$Z(\omega) = \{x \in Z : x(t) = x(a(\omega))$ for $t \leqslant a(\omega)$, $x(t) = x(b(\omega))$ for $t \geqslant b(\omega)\}$ and the mapping $\varphi : Z \times \Omega \to Z$ defined by

$$\varphi(x, \omega) = x\chi_{I(\omega)} + x(a(\omega))\chi_{(-\infty, a(\omega))} + x(b(\omega))\,\chi_{(b(\omega), \infty)},$$

where χ_A is the indicator of A.

As in Lemmas 5.3.2 and 5.3.3, the multivalued mappings $\omega \to Z(\omega)$ and $\omega \to E(\omega)$

$$E(\omega) = \{x \in Z(\omega) : \|x - x_0(\omega)\| \leqslant r(\omega)\},\ \omega \in \Omega,$$

are closed subsets of Z and x_0 is a measurable selector of $\omega \to Z(\omega)$.

Let $\psi_\omega : Z(\omega) \to Y(\omega)$ be the restriction of $x \in Z(\omega)$ to $I(\omega)$. Then the mapping $\psi(x)(\omega) = \psi_\omega^{-1}(\tilde{x}_\omega)$, $\omega \in \Omega$, is a bijection between $Y(\omega)$ and the set of all measurable selectors of $\omega \to Z(\omega)$.

Let $v : \mathbb{R} \times \Omega \to X$ be a continuous X-valued random function and let $t_0 : \Omega \to \mathbb{R}$ be a measurable selector of the closed random interval I. Then the function $w : \operatorname{Gr} I \to X$ defined by

$$w(t, \omega) = \int_{t_0(\omega)}^{t} v(s, \omega)\, \mathrm{d}s,\ \omega \in \Omega,\ t \in I(\omega)$$

is a continuous X-valued random function with random domain I.

Let $g : \mathbb{R} \times X \times \Omega \to X$ be a uniform continuous X-valued random function and let $x : \Omega \to X$ be an X-valued random variable. For $z \in Z$, $\omega \in \Omega$ and $t \in I(\omega)$, let

$$u(t, z, \omega) = \int_{t_0(\omega)}^{t} g(s, z(s), \omega)\, \mathrm{d}s;$$

then :

(a) $z \to u_\omega(z, .)$ is a continuous mapping on Z into $Y(\omega)$ for every $\omega \in \Omega$,

(b) $(t, \omega) \to u_z(t, \omega)$ is in $Y(\omega)$ for every $z \in Z$,

(c) the mapping $T : Z \times \Omega \to Z$ given by

$T(z, \omega) = x_0(\omega) + (\psi(u_z))(\omega)$

is a continuous random operator and $T(z,.)$ is a measurable selector of $\omega \to Z(\omega)$, for every $z \in Z$.

As in Proposition 2.2.5, one can prove that $m(.)$ is a random variable and, hence, that $I_a(.)$ is a closed random interval. Since the random set $\omega \to I_a(\omega) \times S_r(\omega)$ is a separable, we can consider a continuous extension g of f to $\mathbb{R} \times X$ from Theorem 4.4.5. Hence, the operator T is defineable and for $I(\omega) = I_d(\omega) \subset I_a(\omega)$, we have

$$(\psi_\omega(T(z, \omega)))(t) = x_0(\omega) + \int_{t_0(\omega)}^{t} g(s, z(s), \omega)\mathrm{d}s, \; t \in I_d(\omega).$$

Then, for $\omega \in \Omega$ fixed and for $y \in Y(\omega)$,

$$(\psi_\omega(T(\psi_\omega^{-1}(y), \omega)))(t) = x_0(\omega) + \int_{t_0(\omega)}^{t} f(s, z(s), \omega)\mathrm{d}s = (A_\omega y)(t).$$

The operator $A_\omega : Y(\omega) \to Y(\omega)$ is the operator associated to the deterministic equation obtained for ω fixed. Ambrosetti's theorem [12] implies that A_ω has a fixed point x_ω in the set $\{y \in Y(\omega) : \|y - x_0(\omega)\| \leqslant r(\omega)\}$ and that $\omega \to z_\omega = \psi_\omega^{-1}(y_\omega) \in \in Y(\omega)$. Since $T(z_\omega, \omega) = z_\omega$ and $\|z_\omega - x_0(\omega)\| \leqslant r(\omega)$, it follows that $\omega \to z_\omega$ is a wide-sense fixed point of T in E. Then there exists a random fixed point $\xi : \Omega \to Z$ in E (see Theorem 4.4.3) and

$z(t, \omega) = (\psi_\omega(\xi(\omega)))(t), \; t \in I_d(\omega)$

is a random solution of equation (1).

If $x(t, \omega)$ were a function of $t \in I = \{t : |t - t_0| \leqslant a\}$ with values in the space $L^2(\Omega, \mathscr{K}, P)$, then we would obtain [12]:

PROPOSITION 5.3.1. *Consider the random differential equation*

$$(8) \quad \frac{\mathrm{d}x}{\mathrm{d}t} = f(t, x), \quad x_0(t_0, \omega) = x_0(\omega),$$

where $x(t, \omega)$ is a second-order random process continuous on I, whereas

$$f(.,.) : I \times S \to L^2(\Omega, \mathscr{K}, P),$$

$$S = \{x \in L^2(\Omega, \mathscr{K}, P) : \|x - x_0\|_{L^2} \leqslant b\}.$$

If f is a uniform continuous function on $I \times S$ and if, for all $t \in I$, f is an $h(\omega)$-set-Lipschitz operator related to x, with the module h independent of t, and if there exists a number d such that $dh < 1$, then equation (8) *has a random solution* $x(t, \omega)$ *on* $I' = \{t : |t - t_0| < d'\}$, *where* $d' = \min\left\{d, a, \dfrac{b}{M}\right\}$, $M = \sup_{(t,x) \in I \times S} \|f(t, x)\|_{L^2(\Omega, \mathscr{X}, P)}$.

Another stochastic version of Ambrosetti's theorem is given by S. Itoh in [242, Th. 4.3].

Consider a random delay-time differential equation

$$(9) \quad \frac{\mathrm{d}x(t, \omega)}{\mathrm{d}t} = f(t, x(t, \omega), x(t - h, \omega), \omega), \quad h > 0$$

with the condition

$$(9') \quad x(t, \omega) = x_0(t, \omega), \quad t \in [t_0 - h, t_0].$$

THEOREM 5.3.4. *Let* $D = \{(t, x, y) : t_0 \leqslant t \leqslant a,\ x, y \in \mathbb{R}\}$ *and* $f : D \times \Omega \to \mathbb{R}$ *satisfy the conditions*

(i) $(t, x, y) \to f(t, x, y, \omega)$ *is continuous on* D, *a.s.*,

(ii) *there exists a positive random variable* $L : \Omega \to (0, \infty)$ *such that*

$$|f(t, x, y, \omega) - f(t, u, v, \omega)| \leqslant L(\omega)\,(|x - u| + |y - v|), \text{ a.s.},$$

(iii) $t \to x_0(t, \omega)$ *is a continuous function on* $[t_0 - h, t_0]$, *a.s.*

Then there exists a unique solution of Eq. (9) *with condition* (9').

Proof. As in the proof of Theorem 5.3.3, we define the operator

$$(T(x, \omega))(t) = \begin{cases} x_0(t, \omega), & t_0 - h \leqslant t \leqslant t_0, \\ x_0(t_0, \omega) + \displaystyle\int_{t_0}^{t} f(s, x(s), x(s - h), \omega)\mathrm{d}s, & t_0 \leqslant t \leqslant a, \end{cases}$$

which is a random operator on $X = C[a, b]$ with the max-norm.

Let ω be a fixed point in Ω and choose $m(\omega) > 0$ such that $m(\omega) > 2L(\omega)$. Define on X the metric

$$d(x, y) = \max_{t \in [t_0 - h, a]} (\mathrm{e}^{-\alpha(t - t_0)}|x(t) - y(t)|),$$

which introduces an equivalent topology with the norm topology. It follows from the deterministic case that $T(.,\omega) : X \to$

$\to X$ is a contraction operator, and hence T is a continuous random operator which has a unique fixed point in the wide sense. According to Theorem 4.2.12, T has a random fixed point which is a random solution of (9).

A stochastic version of Darboux's problem can also be given [113]. To do this, let $x_0, y_0 \in \mathbb{R}$, $d \in (0, \infty)$, $I = [x_0, x_0 + d]$, $J = [y_0, y_0 + d]$ and $D = I \times J \times \mathbb{R}^3$. Consider the random functions $f : D \times \Omega \to \mathbb{R}$, $\varphi : I \times \Omega \to \mathbb{R}$ and $\psi : J \times \Omega \to \mathbb{R}$ satisfying the conditions:

(i) $(x, y, u, v, w) \xrightarrow{f_\omega} f(x, y, u, v, w, \omega)$ is a continuous mapping on D, for every $\omega \in \Omega$, and $\omega \to f(x, y, u, v, w)$ is a measurable mapping, for every fixed element $(x, y, u, v, w) \in D$.

(ii) there exists a nonnegative real-valued right continuous random function $\Phi : [0, \infty) \times \Omega \to \mathbb{R}$ such that $\Phi_\omega(.)$ is a nondecreasing function for every $\omega \in \Omega$ and there exists a number $a \in (0, 1)$ such that $at < \Phi_\omega(t) < t$ for every $\omega \in \Omega$, $t > 0$ and

$$|f(x, y, u_1, v_1, w_1, \omega) - f(x, y, u_2, v_2, w_2, \omega)| \leqslant$$

$$\leqslant \Phi(\max \{|u_2 - u_1|, |v_2 - v_1|, |w_2 - w_1|\}, \omega)$$

or all (x, y, u_1, v_1, w_1), $(x, y, u_2, v_2, w_2) \in D$ and every $\omega \in \Omega$;

(iii) the mappings $x \xrightarrow{\varphi} \varphi(x, \omega)$, $y \xrightarrow{\psi_\omega} \psi(y, \omega)$ are of class C^1 on I and J, respectively; for every $\omega \in \Omega$, $\omega \to \varphi(x, \omega)$, $\omega \to \psi(y, \omega)$ are measurable and

$$\varphi(x_0, \omega) = \psi(y_0, \omega), \quad \omega \in \Omega.$$

A continuous random function $u : I \times J \times \Omega \to \mathbb{R}$ is called a random solution of Darboux's problem

$$\text{(I)} \quad \frac{\partial^2 u}{\partial x \partial y} = f\left(x, y, u, \frac{\partial u}{\partial x}, \frac{\partial u}{\partial y}, \omega\right) = 0$$

$$\text{(II)} \quad u(x, y_0, \omega) = \varphi(x, \omega), \quad u(x_0, y, \omega) = \psi(y, \omega)$$

if $u(.,.,\omega) \in C^1(I \times J)$, there exists $\dfrac{\partial^2 u}{\partial x \partial y} \in C(I \times J)$ which verifies (I) and (II) for all $\omega \in \Omega$.

THEOREM 5.3.5. *The random Darboux problem* (I) *and* (II) *for which the functions* f, φ, ψ *satisfy conditions* (i) – (iii) *has a unique random solution.*

Proof. The random Darboux problem is equivalent with the random system of integral equations

$$(10)\quad \begin{cases} u(x, y, \omega) = \psi(y, \omega) + \int\limits_{x_0}^{x} v(t, y, \omega)\,dt \\ v(x, y, \omega) = \varphi'(x, \omega) + \int\limits_{y_0}^{y} f(x, s, u(x, s, \omega), v(x, s, \omega), \\ \qquad w(x, s, \omega), \omega)\,ds \\ w(x, y, \omega) = \psi'(y, \omega) + \int\limits_{x_0}^{x} f(t, y, u(t, y, \omega), v(t, y, \omega), \\ \qquad w(t, y, \omega), \omega)\,dt \end{cases}$$

for every $\omega \in \Omega$, where

$$v = \frac{\partial u}{\partial x}, \quad w = \frac{\partial u}{\partial y}.$$

Hence, we first define a convenient metric function space X and a convenient operator A on X.

Let $X = \{(u, v, w) : u, v, w \in C(I \times J)$ and $u(x_0, y) = \psi(y)$, $v(x, y_0) = \varphi'(x)$, $w(x_0, y) = \psi'(y)\}$ and consider the metric

$$d((u_1, v_1, w_1), (u_2, v_2, w_2)) = \max_{(x,y)\in I\times J} \{|u_2 - u_1|\, e^{-\tau(x-x_0+y-y_0)},$$
$$|v_2 - v_1|\, e^{-\tau(x-x_0+y-y_0)}, \ |w_2 - w_1|\, e^{-\tau(x-x_0+y-y_0)}\},$$

where $\tau > 0$.

It is easy to see that (X, d) is a complete metric space and that the operator $A_\omega : X \to X$ given by

$$A_\omega(u_\omega(x, y), v_\omega(x, y), w_\omega(x, y)) =$$
$$= \left(\psi_\omega(y) + \int\limits_{x_0}^{x} v_\omega(t, y)\right) dt, \ \varphi'_\omega(x) +$$
$$+ \int\limits_{y_0}^{y} f_\omega(x, s, u_\omega(x, s), v_\omega(x, s), w_\omega(x, s))\,ds, \ \psi'_\omega(y) +$$
$$+ \int\limits_{x_0}^{x} f_\omega(t, y, u_\omega(t, y), v_\omega(t, y), w_\omega(t, y))\,dt$$

for all $x, y \in I$ is a Φ-contraction, i.e.,

$$d(A(u_1, v_1, w_1), A(u_2, v_2, w_2)) \leqslant \Phi(d((u_1, v_1, w_1), (u_2, v_2, w_2)))$$

for all (u_1, v_1, w_1), $(u_2, v_2, w_2) \in X$.

Using one of Browder's results [77], we conclude that there exists a unique fixed point of A. Since X is a complete separable metric space and A is a continuous random operator on X with a wide-sense fixed point in $E = X$, it follows that A has a random fixed point in $E = X$. Therefore, the random Darboux problem has a unique random solution. Moreover, the random solution is obtained by the successive approximation method.

A similar result can be proved for the following random Darboux problem:

(I′) $$\frac{\partial^2 u(x, y, \omega)}{\partial x \partial y} = f(x, y, u(g(x, y)), \omega)$$

(II′) $u(x, 0, \omega) = 0$, $x \in [0, a]$ and $u(0, y, \omega) = 0$, $y \in [0, b]$ for all $\omega \in \Omega$, where

(i′) $(x, y, u) \to f(x, y, u, \omega)$ is a continuous mapping on $[0, a] \times [0, b] \times \mathbb{R}$, for every $\omega \in \Omega$, and the mapping $\omega \to f(x, y, u, \omega)$ is measurable for every fixed element (x, y, u);

(ii′) $|f(x, y, u_1, \omega) - f(x, y, u_2, \omega)| \leqslant \Phi(|u_2 - u_1|, \omega)$ for all (x, y, u_1), $(x, y, u_2) \in [0, a] \times [0, b] \times \mathbb{R}$ and $\omega \in \Omega$, where Φ is as in (ii);

(iii′) $g \in C([0, a] \times [0, b], [0, a] \times [0, b])$.

Problems (I′) and (II′) are equivalent with the following integral equation of Volterra type:

$$u(x, y, \omega) = \int_0^x \int_0^y f(s, t, u(g(s, t)), \omega) \, ds \, dt.$$

The operator

$$A_\omega : C([0, a] \times [0, b]) \to C([0, a] \times [0, b]),$$

defined by

$$(A_\omega u)(x, y) = \int_0^x \int_0^y f_\omega(s, t, u(g(s, t)), \omega) \, ds \, dt,$$

is a Φ-contraction in the metric space $X = (C([0, a] \times [0, b]), d_\omega)$, where

$$d_\omega(u, v) = \max_{(x,y) \in [0,a] \times [0,b]} |u(x, y) - v(x, y)| e^{-\tau(x+y)}, \quad \tau > 0.$$

Hence, we state

THEOREM 5.3.6. *The random Darboux problem* (I′) *and* (II′) *with the functions f and g satisfying* (i′)-(iii′) *has a unique random solution.*

It seems that the theory of the generalized random differential and integral equations with the multivalued right-hand side is still under development. The first results on such random equations were obtained by C. Castaing [87], [88] and Phan Van Chuong [343]. We give below some of A. Nowak's [318] results on the existence of absolutely continuous solutions depending measurably on a random parameter.

To do this, we shall study the *generalized random* differential equation

(11) $\dot{x}(t) \in \varphi(\omega, t, x(t)), \; t \in [t_0, t] = T$

with the initial condition

(11′) $x(t_0) = y(\omega)$,

where $\varphi : \Omega \times T \times \mathbb{R}^n \to 2^{\mathbb{R}^n}$ and $y : \Omega \to \mathbb{R}^n$ are given mappings.

Consider that $(\Omega, \mathscr{K}, P)$ is a probability space, $\mathscr{L}_T$ is the Lebesgue σ-algebra on T and $L(T)$ is the space of all integrable functions $u : T \to \mathbb{R}^n$ endowed with the norm $\| U \| = \int_T \| u(t) \|_n \, dt$.

A mapping $x : \Omega \times T \to \mathbb{R}^n$ is called a random solution of (11)-(11′) if it is measurable in ω, absolutely continuous in t, and if for P-almost all $\omega \in \Omega$, the condition

$$\frac{dx(\omega, t)}{dt} \in \varphi(\omega, t, x(\omega, t))$$

holds a.e. in T with $x(\omega, t_0) = y(\omega)$.

Let $\varphi : \Omega \times T \to 2^{\mathbb{R}^n}$ be a multivalued mapping and let Φ be a new multivalued mapping associated to φ by the formula

$$\Phi(\omega) = \{x \in L(T) : x(t) \in \varphi(\omega, t), \text{ a.e. in } T\}.$$

LEMMA 5.3.6. *Let $(\Omega, \mathscr{K}, P)$ be a complete probability space and let $\varphi : \Omega \times T \to 2^{\mathbb{R}^n}$ satisfy the following conditions:*

(i) *φ is measurable with respect to the product σ-algebra $\mathscr{K} \times \mathscr{L}_T$,*

(ii) *for any* $\omega \in \Omega$, *there exists an integrable function* $a_\omega : T \to \mathbb{R}^n$ *such that* $|\varphi(\omega, t)| \leqslant a_\omega(t)$ *a.e. in* T.

Then, for each $\omega \in \Omega$, $\Phi(\omega)$ *is a nonempty closed subset of* $L(T)$ *and* Φ *is a measurable multivalued mapping.*

Proof. Let $\omega \in \Omega$ be arbitrary but fixed. The Kuratowski and Ryll-Nardzewski selection theorem implies that there exists a measurable function $x : T \to \mathbb{R}^n$ such that $x(t) \in \varphi(\omega, t)$ for all $t \in T$. According to condition (ii), x is integrable, and thus $\Phi(\omega) \neq \emptyset$. Let the function $g : \Omega \times L(T) \to \mathbb{R}^n$ be given by

$$g(\omega, x) = \int_T d(x(t), \varphi(\omega, t))\, dt, \quad \omega \in \Omega, \quad x \in L(T),$$

where $d(a, B) = \inf\{|a - b|, \ b \in B\}$, $a \in \mathbb{R}^n$, $B \subset \mathbb{R}^n$. It follows from (i) that for any measurable function $x : T \to \mathbb{R}^n$, the mapping $(\omega, t) \to d(x(t), \varphi(\omega, t))$ is also measurable. Hence g is well defined and measurable in ω. We shall prove that y is continuous in x. Indeed, for fixed $\omega \in \Omega$ and for any $x, x' \in$ $\in L(T)$,

$$\|g(\omega, x) - g(\omega, x')\|_n \leqslant \int_T \|d(x(t), \varphi(\omega, t)) - d(x'(t), \varphi(\omega, t))\|_n\, dt \leqslant$$

$$\leqslant \int_T \|x(t) - x'(t)\|_n\, dt = \|x - y\|.$$

Using the continuity of g in x, we conclude that, for every $\omega \in \Omega$, $\Phi(\omega)$ is a closed subset of $L(T)$. Since $\Phi(\omega) = \{x \in L(T) :$ $: g(\omega, x) = 0\}$ and g being measurable in the first variable and continuous in the second, g is product-measurable and

$$\operatorname{Gr} \Phi = \{(\omega, x) \in \Omega \times L(T) : g(\omega, x) = 0\} \in \mathscr{K} \times \mathscr{B}_{L(T)}.$$

The measurability of Φ follows from Theorem 2.2.1 since $(\Omega, \mathscr{K}, P)$ is complete.

Remark. The above lemma holds if T is a Lebesgue measurable subset of $\mathbb{R}^n$.

We give now a stochastic version of a result due to Himmelberg and Van Vleck [213].

THEOREM 5.3.7. *Let* $(\Omega, \mathscr{K}, P)$ *be a complete probability space and let* $y : \Omega \to \mathbb{R}^n$ *and* $\varphi : \Omega \times T \times \mathbb{R}^n \to 2^{\mathbb{R}^n}$ *be given such that:*

(i) *for every* $z \in \mathbb{R}^n$ *the multivalued mapping* $\varphi(.,z)$ *is* $\mathscr{K} \times \mathscr{L}_T$-*measurable;*

(ii) y *is an* $\mathbb{R}^n$-*valued variable;*

(iii) *for each* $\omega \in \Omega$, *there exist two integrable functions* $a_\omega, b_\omega : T \to \mathbb{R}$ *such that*

$$\|\varphi(\omega, t, x)\|_n \leqslant a_\omega(t)$$

$$D(\varphi(\omega, t, z), \varphi(\omega, t, v)) \leqslant b_\omega(t)\|z - v\|_n$$

for all $t \in T$, $z, v \in \mathbb{R}^n$, *where* D *is the generalized Hausdorff metric on* $2^{\mathbb{R}^n}$.

Then the generalized Cauchy problem (11)-(11′) *has at least one random solution.*

Proof. Let Φ be a multivalued mapping on $\Omega \times L(T)$ defined by $\Phi(\omega, x) = \{u \in L(T) : u(t) \in \varphi(\omega, t, y(\omega) + I(x)(t))$, a.e. in $T\}$, where the operator I is given by

$$I(x)(t) = \int_{t_0}^{t} x(s)\,\mathrm{d}s, \quad x \in L(T), \quad t \in T.$$

It is easy to see that if $V : \Omega \to L(T)$ is a random fixed point of Φ, then

$$x(\omega, t) = y(\omega) + I(V(\omega))\,(t), \quad \omega \in \Omega, \; t \in T$$

is a random solution of (11) — (11′).

To prove that Φ has a random fixed point, we choose an arbitrary but fixed $x \in L(T)$. Then the mapping

$$(\omega, t) \to y(\omega) + I(x)\,(t)$$

is measurable. Since φ is measurable in (ω, t) and satisfies the Lipschitz condition in the third variable, the multivalued mapping

$$(\omega, t) \to \varphi(\omega, t, y(\omega) + I(x)\,(t))$$

is also $\mathscr{K} \times \mathscr{L}_T$-measurable (see Proposition 2.2.2). According to the previous lemma, Φ has nonempty closed subsets of $L(T)$ and Φ is measurable in ω.

For a fixed $\omega \in \Omega$, we define an equivalent norm in $L(T)$

$$\|x\|_\omega = \int_T \exp\,(-\lambda c_\omega(t)\|x(t))\|_n\,\mathrm{d}t,$$

where

$$c_\omega(t) = \int_{t_0}^{t} b_\omega(s)\mathrm{d}s, \quad t \in T,$$

and $\lambda > 1$ is a constant. According to a result due to Himmelberg and Van Vleck ([213, Prop. 2]),

$$D_\omega(\Phi(\omega, x), \varphi(\omega, u)) \leqslant \frac{1}{\lambda} \|x - u\|_\omega, \quad x, y \in L(T),$$

where D_ω is the generalized Hausdorff metric on $2^{L(T)}$ induced by the norm $\|\cdot\|_\omega$. From Theorem 4.3.1 (or Theorem 4.4.9) it follows that Φ has a random fixed point and thus the proof is complete.

THEOREM 5.3.8. *Let assumptions* (i)-(ii) *of the previous theorem as well as the following conditions be satisfied:*

(iv) *the multivalued mapping* $\varphi(\omega, t, .)$ *is closed for any* $(\omega, t) \in \in \Omega \times T$;

(v) *there exist functions* $a, b : \Omega \times T \to [0, \infty)$ *measurable in* ω *and integrable in* t *such that*

$$\|\varphi(\omega, t, z)\|_n \leqslant a(\omega, t) + b(\omega, t)\|z\|_n$$

for all $\omega \in \Omega$, $t \in T$, $z \in \mathbb{R}^n$;

(vi) $\varphi(\omega, t, z)$ *is convex for every* $(\omega, t, z) \in \Omega \times T \times \mathbb{R}^n$.

Then, problem (11) — (11′) *admits a radom solution.*

Next, we consider some applications of random fixed point theorems to some classes of random nonlinear inegral equations.

We start with a result due to H. Engl [136, Th. 21] regarding a nonlinear Volterra-Urysohn integral equation with random kernels and a random free term ($(\Omega, \mathscr{K}, \mu)$ is a σ-finite measure space):

THEOREM 5.3.9. *Let* $x_0 : \Omega \times [0,1] \to \mathbb{R}^n$ *be such that* $x_0(\omega, .)$ *is continuous for all* $\omega \in \Omega$ *and* $x_0(., t)$ *is measurable for all* $t \in [0,1]$. *Let* $K : [0,1]^2 \times \mathbb{R}^n \times \Omega \to \mathbb{R}^n$ *be such that* $K(., ., ., \omega)$ *is continuous for all* $\omega \in \Omega$ *and* $K(t, s, x, .)$ *is measurable for all* $(t, s, x) \in [0,1]^2 \times \mathbb{R}^n$. *Let also* $f : [0,1] \times \mathbb{R}^n \times \Omega \to \mathbb{R}^n$ *be such that* $f(., x(.), \omega)$ *is Riemann-integrable for all* $\omega \in \Omega$ *and for* $x \in C_n[0,1]$ *and* $f(t, x, .)$ *is measurable for all* $(t, x) \in [0,1] \times \mathbb{R}^n$. *Then assume that there exists a function* $L : \Omega \to [0,1)$ *such that, for all* $\omega \in \Omega$, $x, y \in \mathbb{R}^n$, *and* $t \in [0,1]$, $\|f(t, x, \omega) - f(t, y, \omega)\|_n \leqslant \leqslant L(\omega)\|x - y\|_n$.

Let $r : \Omega \to (0, \infty)$ *be measurable, and also assume that there exist two functions* $m_1, m_2 : \Omega \to (0, \infty)$ *such that* $\|f(s, x(s), \omega)\|_n \leqslant \leqslant m_1(\omega)$ *and* $\|K(t, s, x(s), \omega)\|_n \leqslant m_2(\omega)$ *for all* $\omega \in \Omega$, $(t, s) \in \in [0, 1]^2$ *and* $x \in C_n[0, 1]$ *with* $\|x - x_0(\omega, .)\| \leqslant r(\omega)$. *Furthermore, assume that* $\inf_{\omega \in \Omega} \dfrac{r(\omega) - m_2(\omega)}{m_1(\omega)} \geqslant 1$. *Then, there exists an*

$x : \Omega \times [0, 1] \to \mathbb{R}^n$ *such that*

(i) $x(., t)$ *is measurable for all* $t \in [0, 1]$;

(ii) $x(\omega, .)$ *is continuous for all* $\omega \in \Omega$;

(iii) *for* μ*-almost all* $\omega \in \Omega$ *and all* $t \in [0, 1]$

$$x(\omega, t) = x_0(\omega, t) + \int_0^1 K(t, s, x(\omega, s), \omega)\, ds + \int_0^t f(s, x(s, \omega), \omega)\, ds$$

(iv) $\|x(\omega, .) - x_0(\omega, .)\| \leqslant r(\omega)$ *for* μ*-almost all* $\omega \in \Omega$.

Proof. Without loss of generality, consider that $(\Omega, \mathscr{K}, \mu)$ is complete. Let $C(\omega) = \{x \in C_n[0, 1] : \|x - x_0(\omega, .)\| \leqslant r(\omega)$ for all $\omega \in \Omega\}$. Let also $\tilde{T} : \Omega \times C_n[0, 1] \to C_n[0, 1]$ be defined by

$$\tilde{T}(\omega, x)(t) = x_0(\omega, t) + \int_0^1 K(t, s, x(s), \omega)\, ds + \int_0^t f(s, x(s), \omega)\, ds$$

and let $T = \tilde{T}|_{\mathrm{Gr}C}$. As in Theorem 5.3.1, we can prove that C is a measurable and that T is a random operator with stochastic domain C.

Let $\omega \in \Omega$, $x \in C(\omega)$; then $\|T(\omega, x) - x_0(\omega, .)\| \leqslant m_2(\omega) + m_1(\omega) \leqslant r(\omega)$. Therefore, $T(\omega, .)(C(\omega)) \subseteq C(\omega)$ for all $\omega \in \Omega$. Using the assumptions in the Theorem and some results from [264, Th. 31, p. 33], it follows that for every $\omega \in \Omega$, $T(\omega, .)$ is the sum of a completely continuous operator and a contraction with Lipschitz constant $L(\omega) < 1$. Therefore, $\{x \in C(\omega) : T(\omega, x) = x\} \neq \emptyset$ for all $\omega \in \Omega$. Since for all $\omega \in \Omega$, $\omega \to x_0(\omega, .)$ is a selector of $\operatorname{int} C(\omega)$, Theorem 4.4.3 and Corollary 4.4.1 hold and the proof is complete.

A stochastic version of the nonlinear contraction mapping theorem in the sense of Boyd and Wong [71], together with some applications to random operator equations, were given by C. H. Lee and W. J. Padgett [276].

The concept of admissibility of function spaces was used and we select some of their associated results.

Let $y_j(\omega)$, $j = 1, 2, \ldots, n$ be second-order real-valued random variables on a probability space $(\Omega, \mathscr{K}, P)$, that is, $y_j(\omega) \in L^2(\Omega, \mathscr{K}, P)$. The collection of all n-component random vectors $\bar{y}(\omega) = (y_1(\omega), \ldots, y_n(\omega))$ constitutes a linear vector space if all equivalent random vectors are identified. Denote the norm of $\bar{y}$ by

$$\|\bar{y}\|_{L_2^n} = \max_{1 \leqslant j \leqslant n} \|y_i\|_{L_2} = \max_{1 \leqslant j \leqslant n} \left(\int_\Omega |y_j|^2\, dP \right)^{1/2}$$

Then the space of all n-component random vectors $\bar{y}$ with second-order components and norm given by $\|\cdot\|_{L_2^n}$ is a separable Banach space denoted by $L_2^n(\Omega, \mathscr{K}, P)$.

Let S be a locally compact metric space with metric d defined on $S \times S$ and let μ be a complete σ-finite measure on the Borel subsets of S. Also, assume that S is a countable union of compact subsets $\{C_n\}$, with $\mu(C_n) < \infty$, such that $C_1 \subset C_2 \subset \ldots$ and that, for any other compact set in S, there exists a C_i which contains it.

Let $C(S, L_2^n(\Omega, \mathscr{K}, P))$ be the space of all continuous functions from S into $L_2^n(\Omega, \mathscr{K}, P)$ with the topology of uniform convergence on compacta. This space is a locally convex space whose topology is defined by a countable family of seminorms given by

$$\|x(t, \omega)\|_j = \sup_{t \in C_j} \|x(t, \omega)\|_{L_2^n}, j = 1, 2, \ldots$$

It is known that this topology is metrizable and that the resulting metric space is a separable Fréchet space.

DEFINITION 5.3.2. Let B and D be Banach spaces. The pair (B, D) is said to be *a.s. admissible* with respect to a random operator $U(\omega)$ if $U(\omega)B \subset D$ a.s.

The continuity of a linear operator $U(\omega)$ from B into D follows from the continuity of $U(\omega)$ from $C(S, L_2^n(\Omega, \mathscr{K}, P))$ into itself if $B, D \subset C(S, L_2^n(\Omega, \mathscr{K}, P))$ are Banach spaces stronger than $C(S, L_2^n(\Omega, \mathscr{K}, P))$ and if (B, D) is a.s. admissible with respect to $U(\omega)$.

We recall that a random operator $T(\omega)$ on a Banach space X with domain $\mathscr{D}(T(\omega))$ is said to be a *random nonlinear contraction operator* if there exists a nonnegative real-valued continuous random function $\varphi(\omega)$ satisfying $\varphi(\omega)(r) < r$ a.s. for $r > 0$, $\varphi(\omega)(0) = 0$ *a.s.*, such that

$$\|T(\omega)x_1 - T(\omega)x_2\| \leqslant \varphi(\omega)(\|x_1 - x_2\|) \text{ a.s.}$$

for all $x_1, x_2 \in \mathscr{D}(T(\omega))$.

A result contained in Theorem 4.2.11 or Theorem 4.2.6 is given by

PROPOSITION 5.3.2 ([276, Theorem 3.1]). *Let $A(\omega)$ be a random nonlinear contraction operator from a separable Banach space X into itself. Then there exists an X-valued random variable $\xi(\omega)$ which is the unique random fixed point of $A(\omega)$. Moreover, if for any $\omega \in \Omega$, $x \in X$, and $n = 1, 2, \ldots$, we put $A^1(\omega)x = A(\omega)x$ and $A^{n+1}(\omega)x = A(\omega)[A^n(\omega)x]$, then all successive approximations $\{A^n(\omega)x\}$ converge to $\xi(\omega)$ a.s. as $n \to \infty$.*

We investigate the existence, uniqueness and measurablility of the random solutions of the random nonlinear integral equation :

$$(12)\quad x(t, \omega) = h(t, \omega) + \int_S K(t, s, \omega) f(s, x(s, \omega))\, d\mu(s),$$

where S is a locally compact metric space, μ is a complete σ-finite measure defined on the Borel subsets of S, and the integral is a Bochner integral.

We assume the following conditions concerning the random kernel $K(t, s, \omega) : S \times S \times \Omega \to \mathbb{R}$:

(1) $K(t, s, \omega) : S \times S \to L_\infty(\Omega, \mathscr{K}, P)$ such that $|||K(t, s, \omega)||| \cdot$

$\|x(s, \omega)\|_{L_2^n}$

is μ-integrable with respect to $s \in S$ for each $t \in S$ and $x \in C(S, L_2^n(\Omega, \mathscr{K}, P)$, where, for each $(t, s) \in S \times S$,

$$|||K(t, s, \omega)||| = \|K(t, s, \omega)\|_{L_\infty(\Omega, \mathscr{K}, P)} = P\text{-ess sup } |K(t, s, \omega)|$$

is the norm in $L_\infty(\Omega, \mathscr{K}, P)$,

(2) $K(t, s, \omega)$ is continuous in $t \in S$ from S into $L_\infty(\Omega, \mathscr{K}, P)$ for every $s \in S$; $K(t, s, \omega)$ is continuous in $s \in S$ from S into $L_\infty(\Omega, \mathscr{K}, P)$ for every $t \in S$;
and

(3) there exists a positive real-valued function H defined μ-a.e. on S such that $H(s)\|x(s, \omega)\|_{L_2^n}$ is μ-integrable for $x \in C(S, L_2^n(\Omega, \mathscr{K}, P))$ and such that, for every $t, s \in S$,

$$|||K(t, u, \omega) - K(s, u, \omega)||| \cdot \|x(u, \omega)\|_{L_2^n} \leqslant H(u) \|x(u, \omega)\|_{L_2^n}$$

μ — a.e.

Then, for every $(t, s) \in S \times S$, we have $K(t, s, \omega)\, x(s, \omega) \in L_2^n(\Omega, \mathscr{K}, P)$.

Let us denote the random integral operator $T(\omega)$ on $C(S, L_2^n(\Omega, \mathscr{K}, P))$ by

$$(13)\qquad [T(\omega)x](t, \omega) = \int_S K(t, s, \omega)\, x(s, \omega)\, d\mu(s),$$

where the integral is a Bochner integral. The Bochner integral is well defined and for every $t \in S$ and $T(\omega)$, $[T(\omega)x](t, \omega) \in L_2^n(\Omega, \mathscr{K}, P)$ is a continuous linear operator from $C(S, L_2^n(\Omega, \mathscr{K}, P))$ into itself.

THEOREM 5.3.10 [276, Th. 4.1]. *Consider the stochastic integral equation* (12) *subject to the following conditions*:

(i) *B and D are Banach spaces stronger than $C(S, L_2^n(\Omega, \mathscr{K}, P))$ such that (B, D) is a.s. admissible with respect to the integral operator defined by equation* (13):

(ii) *$x(t, \omega) \to f(t, x(t, \omega))$ is an operator from the set*

$$Q(\rho) = \{x(t, \omega) : x(t, \omega) \in D, \|x(t, \omega)\|_D \leqslant \rho\}$$

into the space B satisfying

$$\|f(t, x(t, \omega)) - f(t, y(t, \omega))\|_B \leqslant \varphi(\omega)(\|x(t, \omega) - y(t, \omega)\|_D) \ a.s.$$

for $x(t, \omega)$, $y(t, \omega) \in Q(\rho)$;
and

(iii) *$h(t, \omega) \in \Omega$.*

Then there exists a unique random solution of (12) *in $Q(\rho)$, provided $c(\omega) < 1$ a.s. and $\|h(t, \omega)\|_D + c(\omega)\|f(t, 0)\|_B \leqslant \rho(1 - c(\omega))$ a.s. where $c(\omega)$ is the norm of $T(\omega)$.*

Proof. Define the operator $U(\omega)$ from $Q(\rho)$ into D as

$$[U(\omega)x](t, \omega) = h(t, \omega) + \int_S K(t, s, \omega) f(s, x(t, \omega)) \, d\mu(s).$$

Using the conditions of the theorem, one gets

$$\begin{aligned} \|[U(\omega)x](t, \omega)\|_D &\leqslant \|h(t, \omega)\|_D + c(\omega)\|f(t, x(t, \omega))\|_B \leqslant \|h(t, \omega)\|_D + \\ &+ c(\omega)\|f(t, 0)\|_B + c(\omega)\varphi(\omega)(\|x(t, \omega)\|_D) \leqslant \rho(1 - c(\omega)) + \\ &+ c(\omega)\varphi(\omega)(\rho) \leqslant \rho(1 - c(\omega)) + c(\omega)\rho = \rho \text{ a.s.} \end{aligned}$$

Hence $[U(\omega)x](t, \omega) \in Q(\rho)$.

For $x(t, \omega)$, $y(t, \omega) \in Q(\rho)$, use condition (ii) to get

$$\begin{aligned} \|[U(\omega)x](t, \omega) - [U(\omega)y](t, \omega)\|_D &= \Big\|\int_S K(t, s, \omega)[f(s, x(s, \omega)) - \\ &- f(s, y(s, \omega))] \, d\mu(s)\Big\|_D \leqslant c(\omega)\|f(t, x(t, \omega)) - f(t, y(t, \omega))\|_B \leqslant \\ &\leqslant c(\omega)\varphi(\omega)(\|x(t, \omega) - y(t, \omega)\|_D) \leqslant \varphi(\omega)(\|x(t, \omega) - y(t, \omega)\|_D) \\ &\text{a.s.,} \end{aligned}$$

since $c \leqslant 1$ a.s. Thus, U is a random nonlinear contraction operator on $Q(\rho)$.

Recalling Proposition 5.3.2, we conclude that there exists a unique X-valued random variable $x(t, \omega) \in Q(\rho)$ which is the fixed point of $U(\omega)$; that is, $x(t, \omega)$ is the unique random solution of Eq. (12).

Various useful corollaries to the general existence Theorem 5.3.10 may be easily obtained by choosing B and D to be particular Banach spaces in $C(S, L_2^n(\Omega, \mathscr{K}, P))$. For example, the space $BC(S, L_2^n(\Omega, \mathscr{K}, P))$ of all bounded continuous functions from S into $L_2^n(\Omega, \mathscr{K}, P)$ may be chosen as either or both of these spaces.

To conclude this section, we shall consider a random Hammerstein equation in an L_2-space as treated by Engl [136].

THEOREM 5.3.11. *Let* $(\Omega, \mathscr{K}, \mu)$ *be a* σ*-finite measure space and let* $k:[0,1]^2 \times \Omega \to \mathbb{R}$ *be jointly measurable such that* $\int_0^1\int_0^1 k(t, s, \omega)^2\, dt\, ds < 1$ *for all* $\omega \in \Omega$. *Let* $g:[0,1] \times \mathbb{R} \times \Omega \to \mathbb{R}$ *be jointly measurable such that* $g(t, ., \omega)$ *is continuous for all* $\omega \in \Omega$ *and* $t \in [0,1]$ *and* $|g(t, x, \omega) - x| \leqslant \sum_{i=1}^{n(\omega)} g_i(t, \omega)|x|^{1-b_i(\omega)} + g_0(t, \omega)$ *for all* $t \in [0,1]$ *and* $x \in \mathbb{R}$, *where* $g_0(., \omega) \in L_2[0,1]$, $g_i(., \omega) \in L_{2/b_i(\omega)}[0,1]$ *with* $0 < b_i(\omega) < 1$, $i = 1, 2, \ldots, n(\omega)$, $\omega \in \Omega$. *Let* $x_0 : \Omega \to L_2[0,1]$ *which, for* μ*-almost all* $\omega \in \Omega$, *fulfills*

$$x(\omega,.) = x_0(\omega,.) + \int_0^1 k(., s, \omega)\, g(s, x(\omega, s), \omega)\, ds$$

("=" *in the usual sense of equality of equivalence classes of functions*).

Proof. Let $T : \Omega \times L_2[0,1] \to L_2[0,1]$ be defined as

$$(\omega, x) \to x_0(\omega) + K(\omega, N(\omega, x)),$$

where

$$N : \Omega \times L_2[0,1] \to L_2[0,1], \text{ by } (\omega, x) \to (t \to g(t, x(t), \omega))$$

$$K : \Omega \times L_2[0,1] \to L_2[0,1] \text{ by } (\omega, x) \to \left(t \to \int_0^1 k(t, s, \omega)\, x(s)\, ds\right).$$

It has been shown [264, p. 30] that $N(\omega, .)$ really maps into $L_2[0,1]$, that it maps bounded sets into bounded sets and that it is continuous for all $\omega \in \Omega$. Since $K(\omega, .)$ is completely continuous for all $\omega \in \Omega$, $T(\omega, .)$ is also completely continuous.

Let $y \in L_2[0,1]$ be arbitrary. Using the assumptions of the theorem, we get that, for all $x \in L_2[0,1]$,

$$\omega \to \langle T(\omega, x), y \rangle = \int_0^1 \left(x_0(\omega) + \int_0^1 k(t, s, \omega)\, g(s, x(s), \omega)\, ds\right) y(t)\, dt$$

is measurable; therefore, $T(., x)$ is measurable for all $x \in L_2[0,1]$ (cf. [31, p. 16]), and hence T is a random operator. Hence, we conclude that for all $\omega \in \Omega$

$$\{x \in L_2[0,1] : T(\omega, x) = x\} \neq \emptyset.$$

Indeed [308, p. 773], $K(\omega, N(\omega, .))$ is quasibounded for all $\omega \in \Omega$ with quasinorm $\|K(\omega, .)\| \leqslant \left(\int_0^1\int_0^1 k(t, s, \omega)^2 \, dt \, ds\right)^{1/2} < 1.$

It also follows [165, p. 64] that the concept of quasiboundedness is used and it can be proved that for every $\omega \in \Omega$, there exists an $x \in L_2[0,1]$, with $T(\omega, x) = x$.

Using Theorem 4.4.3, we complete the proof.

5.4. APPLICATIONS OF RANDOM FIXED POINT THEOREMS TO STOCHASTIC APPROXIMATION

Random contraction mapping theorems are of fundamental importance in the theory of random equations, since they can be used to establish the existence, uniqueness, and measurability of solutions of random operator equations. They have been used in the study of random linear and nonlinear integral equations [25], [27], [31]. They have also been used to establish the existence of random solution of the Fredholm integral equations as obtained by Boyce [69] in his studies of differential equations with random boundary conditions. Random contraction mappings have been employed in the study of stochastic approximation procedures (cf. Gardner [157], Hanš [205] and Wasan [435]). An application of Theorem 4.1.2 may be found in [325] and [326], where an algorithm for the identification of a random linear discrete-time system described by a random difference equation is obtained. In what follows, we refer to the approach of Hanš and Špaček [208], Hanš [207], Driml and Hanš [125] and the Czech school of stochastic approximation which is different from the direction taken by Robbins and Monro [362].

It is known [207] that an important question in the theory of random operator equation is the relationship between the random solution of the random operator equation and the solution of the corresponding deterministic operator equation. More precisely, one discusses the case when the random equation

(1) $T[\omega, x(.)] = y(.)$

is such that the Bochner integrals

$$(2) \quad \int_{\Omega} y(\omega)\, d\mu(\omega) = y \text{ and } \int_{\Omega} T(\omega, x)\, d\mu(\omega) = S(x)$$

exist for every $x \in X$ (X is a separable Banach space). Let us assume that the solution of the deterministic operator $S(u) = y$ is equal to z. The question arises whether the expected value of the random solution of the random operator equation (1) exists, and if so, whether it is equal to the deterministic solution z. There are examples showing cases in which the answer is positive but it is also easy to give examples where the answer is negative.

However, there are many practical problems in which we are interested in the solution of the deterministic operator equation corresponding to the random operator equation, rather than in the expected value or even in the probability distribution of the random solution of the random equation. It is a fact that the deterministic or, at least, the estimation of the fixed point of the expected value of random operators is the aim of such practical problems as are, for example, different tests in biology and many distribution-free estimates.

The expected value of the random operator under examination is unknown to the experimentator, who can only make independent experiments. This fact leads to the assumption that we have a sequence of independent, equally distributed random operators, whose fixed point of common expected value we want to approximate. In the case of independent trials, we must distinguish between two different cases : In the first case we may determine the fixed point of each realization. In the second case, we must above all choose our arbitrary but fixed point $x \in X$ and only then may we obtain the value of the realization for this previously chosen x, i.e., in technical terms, we can make experiments on fixed, previously chosen levels alone. Hence, in several cases, we cannot assume individual random fixed points, because we cannot obtain them. This is the case of all "destructive" experiments.

Hence, the following problem is posed : given a sequence of independent equally distributed random operators, satisfying certain conditions, we seek a method to estimate the fixed point of the common expected value of the given random operators. This method is sometimes required to ask for only one value of each random operator for a fixed, previously chosen level alone.

First, we give an interesting application of the random contraction mapping theorem using strong laws of large numbers for Banach space-valued random variables.

LEMMA 5.4.1 [205, Th. 2]. *Let* $(\Omega, \mathscr{K}, P)$ *be a probability space, let* (X, d) *be a separable metric space, let* c *be a function on* Ω, *let* $\varphi, f_1, f_2, \ldots$ *be a sequence of* X*-valued random variables and let* $T_1, T_2, \ldots$ *be a sequence of continuous random operators on* $\Omega \times X$ *into* X *so that the following conditions are satisfied:*

$0 \leqslant c(\omega) < 1$ *for every* $\omega \in \Omega$;

$P(\{\omega : \lim_{n\to\infty} d(f_n(\omega), \varphi(\omega)) = 0\}) = 1$;

$P(\{\omega : \lim_{n\to\infty} d(T_n(\omega, \varphi(\omega)), \varphi(\omega)) = 0\}) = 1$;

$P(\{\omega : d(T_n(\omega, x), T_n(\omega, f_n(\omega))) \leqslant c(\omega)\ d(x, f_n(\omega))\}) = 1$;

for every $n = 1, 2, \ldots$ *and* $x \in X$. *Let us choose the* X*-valued random variable* V_1 *arbitrary and define, for every* $n = 1, 2, \ldots$, *the mapping* V_{n+1} *of the space* Ω *into the space* X, *for every* $\omega \in \Omega$, *by the formula*

$V_{n+1}(\omega) = T_n(\omega, V_n(\omega))$.

Then, $V_1, V_2, \ldots$ *is a sequence of* X*-valued random variables which converges a.s. to the generalized random variable* φ.

Proof. According to the substitution theorem (see Th. 4.4.1), $\{V_n\}_{n\neq \mathbb{N}}$ is a sequence of X-valued random variables. Let us denote

$$E = \{\omega : \lim_{n\to\infty} f_n(\omega) = \varphi(\omega)\} \cap \{\omega : \lim_{n\to\infty} T_n(\omega, \varphi(\omega)) = \varphi(\omega)\} \cap$$

$$\cap \bigcap_{n=1}^{\infty} \bigcap_{x\in H} \{\omega : d(T_n(\omega, x), T_n(\omega, f_n(\omega))) \leqslant c(\omega)\ d(x, f_n(\omega))\},$$

where H is a countable dense set in the space X. Since $E \in \mathscr{K}$, $P(E) = 1$ and the random operators are continuous, we obtain

$$E \subset \bigcap_{n=1}^{\infty} \bigcap_{x\in X} \{\omega : d(T_n, (\omega, x), T_n(\omega, f_n(\omega))) \leqslant c(\omega)\ d(x, f_n(\omega))\}.$$

Let $\omega \in E$ be fixed but arbitrary. Then, we have

$$d(V_{n+1}(\omega), \varphi(\omega)) = d(T_n(\omega, V_n(\omega)), \varphi(\omega)) \leqslant d(T_n(\omega, V_n(\omega)),$$
$$T_n(\omega, f_n(\omega))) + d(T_n(\omega, f_n(\omega)), T_n(\omega, \varphi(\omega))) + d(T_n(\omega, \varphi(\omega)),$$
$$, \varphi(\omega)) \leqslant c(\omega)\ d(V_n(\omega), f_n(\omega)) + c(\omega)\ d(f_n(\omega), \varphi(\omega)) + d(T_n(\omega),$$
$$, \varphi(\omega)), \varphi(\omega)) \leqslant c(\omega)\ d(V_n(\omega), \varphi(\omega)) + 2c(\omega) d(f_n(\omega), \varphi(\omega)) +$$
$$+ d(T_n(\omega, \varphi(\omega)), \varphi(\omega)).$$

Consider an arbitrary $\varepsilon > 0$ and some chosen $\omega \in E$. Then there exists an $n_0 \in N$ such that for every $n \geqslant n_0$

$$c(\omega)\, d(f_n(\omega), \varphi(\omega)) \leqslant \frac{1 - c(\omega)}{8}\, \varepsilon$$

and

$$d(T_n(\omega, \varphi(\omega)), \varphi(\omega)) \leqslant \frac{1 - c(\omega)}{4}\, \varepsilon.$$

Thus, for every $n \geqslant n_0$,

$$d(V_{n+1}(\omega), \varphi(\omega)) \leqslant c(\omega)\, \mathrm{d}(V_n(\omega), \varphi(\omega)) + \frac{1 - c(\omega)}{2}\, \varepsilon$$

and then, by repeated induction, for every $k \in \mathbb{N}$,

$$d(V_{n_0+k}(\omega), \varphi(\omega)) \leqslant c^k(\omega)\, d(V_{n_0}(\omega), \varphi(\omega)) + \frac{1 - c(\omega)}{2}\, \varepsilon \sum_{i=1}^{k} c^{i-1}(\omega).$$

Since $0 \leqslant c(\omega) < 1$ and $\varepsilon > 0$, for every $V_1(\omega)$ and for every n_0, there exists a positive integer n_1 such that, for every $k \geqslant n_1$,

$$c^k(\omega)\, \mathrm{d}(V_{n_0}(\omega), \varphi(\omega)) \leqslant \frac{\varepsilon}{2},$$

and hence, for the above fixed $\omega \in E$, the inequality

$$d(V_{n_0}(\omega), \varphi(\omega)) \leqslant \frac{\varepsilon}{2}$$

holds for every $n \geqslant n_0 + n_1$. Let us remark that the integers n_0 and n_1 depend on the choice of the element $\omega \in E$. Thus, at least for $\omega \in E$, the sequence $\{V_n(\omega)\}_{n \in \mathbb{N}}$ converges to the point $\varphi(\omega)$. Since the conditions of Theorem 2.2.1 are satisfied, the proof of the assertion of our lemma is complete.

In order to solve the above problem, we give below some results due to Hanš [205]. First, recall that a random operator T is called local contraction in the X-valued random variable f (in the sense of Hanš) if there exists $c : \Omega \to R$, $c(\omega) < 1$, for every ω, such that

$$d(T(\omega, x), T(\omega, f(\omega))) \leqslant c(\omega)\, d(x, f(\omega))$$

holds for every $x \in X$ and $\omega \in \Omega$.

LEMMA 5.4.2. *Let $(\Omega, \mathscr{K}, P)$ be a probability space, let X be a separable Banach space and let $V_1, V_2, \ldots$ be a sequence of independent and equally distributed X-valued random variables such that* $\int_\Omega \| V_1(\omega) \| \mathrm{d}P(\omega) < \infty$.

Then the sequence $U_1, U_2, \ldots$ *of X-valued random variables, defined for every* $n \in \mathbb{N}$ *and every* $\omega \in \Omega$ *as*

$$U_n(\omega) = \frac{1}{n} \sum_{k=1}^{n} V_k(\omega),$$

converges strongly a.s. to the expected value $E(V_1)$ *of the X-valued random variable* V_1.

THEOREM 5.4.1. *Let* $(\Omega, \mathscr{K}, P)$ *be a probability space, let X be a separable Banach space and let* $T_1, T_2, \ldots$ *be a sequence of weakly independent, weakly equally distributed continuous random operators on* $\Omega \times X$ *into X satisfying the following conditions:*

(i) *for every* $x \in X$, *there exists the Bochner integral*

$$S(x) = \int_{\Omega} T_1(\omega, x) \, \mathrm{d}P(\omega);$$

(ii) *there exists a number* $c < 1$ *such that for every pair* $x, y \in X$ *and every* $n \in \mathbb{N}$,

$$P(\{\omega : \| T_n(\omega, x) - T_n(\omega, y) \| \leqslant c\|x - y\|\}) = 1.$$

Let us choose an arbitrary X-valued random variable V_1 *and, for every* $n \in \mathbb{N}$ *and every* $\omega \in \Omega$, *let us define the mapping* V_{n+1} *of the space* Ω *into X as*

$$V_{n+1}(\omega) = S_n(\omega, V_n(\omega)),$$

where the mapping S_n *on* $\Omega \times X$ *into X is defined for every* $(\omega, x) \in \Omega \times X$ *as*

$$S_n(\omega, x) = \frac{1}{n} \sum_{i=1}^{n} T_i(\omega, x).$$

Then there exists a unique point $x_0 \in X$ *such that*

$$S(x_0) = x_0$$

and $V_1, V_2, \ldots$ *is a sequence of X-valued random variables converging strongly a.s. to the fixed point* x_0.

Proof. Since both the random operator S and the random operators $S_1, S_2, \ldots$ are a.s. uniform local contractions with the same constant c, there exists a sequence $\varphi, f_1, f_2, \ldots$ of X-valued random variables and $P(\{\omega : \varphi(\omega) = x_0\}) = 1$. According to Lemma 5.4.2, the sequence $S_1(\cdot, x), S_2(\cdot, x), \ldots$ of X-valued random variables converges strongly a.s. to the point $S(x)$ for very $x \in X$. Since for every $n \in \mathbb{N}$, f_n is a random fixed point, then, for every

$$\omega \in \bigcap_{i=1}^{\infty} \bigcap_{x \in X} \{\omega : \| T_i(\omega, x) - T_i(\omega, x_0) \| \leqslant c \| x - x_0 \|\},$$

one can write the estimate

$$\|f_n(\omega) - x_0\| \leqslant \|f_n(\omega) - S_n(\omega, x_0)\| + \|S_n(\omega, x_0) - x_0\| \leqslant$$
$$\leqslant c\|f_n(\omega) - x_0\| + \|S_n(\omega, x_0) - x_0\|$$

and, hence,

$$\|f_n(\omega) - x_0\| \leqslant \frac{1}{1-c} \|S_n(\omega, x_0) - x_0\|.$$

Therefore,

$$P(\{\omega : \lim_{n\to\infty} f_n(\omega) = \varphi(\omega)\}) = 1$$

and, thus, all conditions of Lemma 5.4.1 are fulfilled and the assertion holds.

Remark. In the above theorem we have assumed that almost all realizations of each random operator are uniform contractions with the same constant c. A similar result (see [205, Th. 5]) would hold assuming the existence and uniqueness of the common expected value and that almost all realizations of each random operator are only local contractions in this fixed point.

The following theorem [207], [250] can be considered to be one of the basic theorems of the experience theory.

THEOREM 5.4.2. *Let $(\Omega, \mathscr{K}, P)$ be a probability space and let X be a separable Banach space. Let T be a generalized stochastic process mapping the cartesian product space $\Omega \times [0, \infty) \times X$ into the space X and almost surely continuous with respect to both arguments ($t \geqslant 0$ and $x \in X$) simultaneously. Let there be an element $\hat{x} \in X$ and a real-valued random variable c, and assume that the following assumptions be satisfied:*

(i) $P(\{\omega : c(\omega) < 1\}) = 1$;

(ii) $P\left(\left\{\omega : \lim_{t\to\infty} \left\| t^{-1} \int_0^t T(\omega, s, \hat{x})\, ds - \hat{x} \right\| = 0\right\}\right) = 1$;

(iii) $P(\{\omega : \| T(\omega, t, x) - T(\omega, t, y) \| \leqslant c(\omega) \| x - y \|\}) = 1$

for every $t \geqslant 0$ and every two elements $x, y \in X$. Further, let

$x_t(.)$ be the solution of the random operator equation

(3) $\xi_0(.) = T(., 0, \xi_0(.))$;

$$\xi_t(.) = t^{-1}\int_0^t T(., x, \xi_s(.))\, ds \quad for \quad t > 0.$$

Then, for every $t \geqslant 0$, x_t is a generalized random variable and, hence,

(4) $P(\{\omega : x_t(\omega) \text{ is continuous in } t\}) = 1$

and

(5) $P(\{\omega : \lim\limits_{t\to\infty} \| x_t(\omega) - \hat{x} \| = 0\}) = 1.$

Proof. Let us denote by C_∞ the space of all continuous mappings f of the space $[0, \infty)$ into the space X such that the relation $\lim\limits_{t\to\infty} \|f(t) - \hat{x}\| = 0$ holds. Introducing the distance function $d : C_\infty \times C_\infty \to \mathbb{R}$ by $d(f, g) = \sup\limits_{t\geqslant 0} \|f(t) - g(t)\|$, the space C_∞ becomes a separable metric space whose σ-algebra of all Borel subsets is the σ-algebra generated by the class of sets $\{f : f(t) \in B\}$, with t running over $[0, \infty)$ and B over $\mathscr{B}_X$. Also, consider the operator S defined over the cartesian product space $\Omega \times C_\infty$ by $(S(\omega, f))\,(0) = T(\omega, 0, f(0))$ and by

$$(S(\omega, f))\,(t) = t^{-1}\int_0^t T(\omega, s, f(s))\, ds, \quad f \in C_\infty, \quad t > 0$$

for every $\omega \in E$, where E equals the intersection over all $t \geqslant 0$, $x \in X$, $y \in X$ of the sets occurring in (i), (ii), (iii) and the set of those $\omega \in \Omega$ for which $T(\omega, t, x)$ is continuous in both t and x simultaneously. For every $\omega \in \Omega - E$, $f \in C_\infty$ and $t \in [0, \infty)$, put $(S(\omega, f))\,(t) = \hat{x}$.

First let us prove that the mapping S maps the cartesian product space $\Omega \times C_\infty$ into the space C_∞. Choose arbitrary $\omega \in E$, $f \in C_\infty$ and $\delta > 0$. There exists a real number t_0 such that, for every $t \geqslant t_0$, $\|f(t) - x\| \leqslant \dfrac{\delta}{3}$ simultaneously with $\left\| t^{-1}\int_0^t T(\omega, s, \hat{x})\, ds - \hat{x} \right\| \leqslant \dfrac{\delta}{3}$. Then, one can write

$$\|(S(\omega, f))(t)-\hat{x}\| \leqslant t^{-1}\int_0^t \left\| T(\omega, s, f(s)) - T(\omega, s, \hat{x}) \right\| ds +$$

$$+ \|(S(\omega, h))(t) - \hat{x}\| \leqslant c(\omega)\left(\frac{t_0}{t}\right) d(f, h) +$$

$$+ c(\omega)\left(1 - \frac{t_0}{t}\right)\left(\frac{\delta}{3}\right) + \frac{\delta}{3} \leqslant \delta,$$

for every $t \geqslant t_0 \, d(f, h)\dfrac{3}{\delta}$, where h denotes the function for which $h(t) = \hat{x}$ for every $t \in [0, \infty)$.

Since the $\Omega - E$ part is trivial, we have just proved that S maps the cartesian product space $\Omega \times C_\infty$ into C_∞. Now, let us prove that S is a contraction mapping. Thus, let us take arbitrarily $f \in C_\infty$, $g \in C_\infty$ and $\omega \in E$. Using (iii), one gets

$$d(S(\omega, f), S(\omega, g)) \leqslant c(\omega)\, d(f, g).$$

Since the mapping S is a random contraction mapping of $\Omega \times C_\infty$ into the complete separable metric space C_∞, one can apply Theorem 4.1.2 and obtain a C_∞-valued random variable φ such that $S(\omega, \varphi(\omega)) = \varphi(\omega)$ holds a.s. However, we have $E \in \mathscr{K}$, $P(E) = 1$; hence, if we define x_t for every $\omega \in \Omega$ and every $t \in [0, \infty)$ by $x_t(\omega) = (\varphi(\omega))\,(t)$, all the assertions in the theorem are immediate.

A generalization of the foregoing theorem for almost surely linear bounded random operators is

THEOREM 5.4.3. *Let T be a generalized stochastic process mapping the cartesian product space $\Omega \times [0, \infty) \times X$ into the space X and almost surely continuous with respect to both the arguments $t \in [0, \infty)$ and $x \in X$ simultaneously. Denote by $T_{t_1, t_2, \ldots, t_n}$ the mapping of the cartesian product space $\Omega \times X$ into the space X, defined for every $\omega \in \Omega$, $x \in X$, and $j = 1, 2, \ldots$ by $T_{t_1}(\omega, x) = T(\omega, t_1, x)$ and $T_{t_1, t_2, \ldots, t_{j+1}}(\omega, x) = T(\omega, t_{j+1}, T_{t_1, \ldots, t_j}(\omega, x))$. Let an element $\hat{x} \in X$ be such that condition* (ii) *as well as the following conditions are satisfied:*

(a) *$T(\omega, t, \alpha x + \beta y) = \alpha T(\omega, t, x) + \beta T(\omega, t, y)$ holds with probability one for every $t \in [0, \infty)$, $x \in X$, $y \in X$, $\alpha \in \mathbb{R}$ and $\beta \in \mathbb{R}$ separately;*

(b) $\sup_{t \geqslant 0} \| T(\omega, t, .)\| < \infty$ *with probability one;*

(c) $$P\left(\bigcup_{m=1}^{\infty}\bigcup_{n=1}^{\infty}\bigcap_{x\in X}\bigcup_{y\in X}\left\{\omega:\|T_{t_1,\ldots,t_n}(\omega, x-y)\|\leqslant \left(1-\frac{1}{m}\right)\|x-y\|\right\}\right)=1.$$

Further, let $x_t(.)$ be the solution of the random operator equation (3). *Then x_t is, for every $t\in[0, \infty)$, a generalized random variable, the solution is unique in the "almost sure" sense, and* (4) *and* (5) *hold.*

Another result on stochastic approximation for continuous processes is given by

THEOREM 5.4.4 [126]. *Let $f(\omega, t, x)$ be a random function such that*

(i) $P(\{\omega : f(\omega, t, x)$ *is continuous with respect to t and x simultaneously*$\}) = 1$;

(ii) *for every* t,
$P(\{\omega : x_1 < x_2$ *implies that* $f(\omega, t, x_1) < f(\omega, t, x_2),\ x_1, x_2 \in \mathbb{R}\}) = 1$;

(iii) *there exists a function $r(x)$ such that*

$$P\left(\left\{\omega : t^{-1}\int_0^t f(\omega, s, x)\,ds = r(x) \text{ for every } x\right\}\right) = 1.$$

Then the solution of the random differential equation

$$\dot{x}(\omega, t) = -\frac{1}{t}f(\omega, t, x)$$

exists with probability one and further

$$P(\{\omega : \lim_{t\to\infty} d(x(\omega, t), S) = 0\}) = 1,$$

where $S = \{x : r(x) = 0\}$ and d is the distance on the real line.

Remark. The above results are partial answers to a more general problem [250] posed as follows: consider the random differential equation

$$\dot{x}_\eta(t, \omega) = \eta f(x_\eta(t, \omega), \omega, t),\ x_\eta(0) = x_0$$

and the averaged equation

$$\dot{y}_\eta(t) = \eta E(f),\ y_\eta(0) = x_0.$$

Then find sufficient conditions of f in order that $y_\eta(t)$ is a good approximation to the random function $x_\eta(t, \omega)$ as $\eta \to 0$.

Other stochastic approximations and some problems in nonlinear random operator equations together with selected topics will be presented in Chap. 6.

Chapter 6

Some analysis problems and methods for random equations

Random equations arise in a great variety of theoretical and practical problems in a very natural and intuitively clear way. It is easy to pose stochastic problems in applied science and engineering and it is also clear that they are more realistic than their deterministic counterparts. However, their analysis is difficult, and more cumbersome than the analysis of deterministic cases. The applications of the random equation theory are of fundamental importance in the formulation and analysis of various classes of operator equations that arise in the physical, biological, social engineering, and technical sciences. This may partly explain why this theory is at present a very active area of mathematical research.

Methods for solving random equations must establish not only existence and uniqueness, but also the measurability of solutions. In most applications of random operator equations, one is interested in statistical properties (e.g., probability distribution, mean and variance of solution, etc.) and hence it is necessary that (exact or approximate) solutions should be formulated so as to permit the study of their statistical properties. These are important aspects where the study of random operator equations assumes a new nontrivial dimension and departs from the deterministic theory.

According to Nashed and Engl [310], the questions related with random operator equations have soft and hard aspects. The "soft" aspects concern wide-sense solutions of random operator equations. Basically, the "hard" aspects of random operator equations are three :

i) measure-theoretic aspects : random operators, existence of measurable solutions, properties of the set of all measurable solutions as a subset of the set of all wide-sense solutions;

ii) probabilistic and statistical aspects : probability distributions and properties of functionals of random solutions, such as moments ;

iii) "experience theory" and modeling: estimates for the difference between the mean-value of solutions of the random equation and the (deterministic) solution of the "averaged equation"; in other words: How well does a deterministic model approximate the stochastic problem?

In order to summarize the principal options the researcher has when dealing with probabilistic functional analysis, following the lead of the Prague School, we shall recall the four fields of interest [141]:

1) The study of measurability of adjoints, inverses and generalized inverses for linear random operators; applications of these results to the development of random analogues of inverse and implicit function theorems: Hanš [206], Bharucha-Reid [31], Nashed and Salehi [309], Engl and Nashed [310], [141], Bocşan [54].

2) Measurability results for solutions of nonlinear random analogues of classical fixed point theorems; Andrus and Nishiura [13], Bharucha-Reid [31], [32], Bocşan [52], [53], [55], Bocşan, Constantin and Radu [61], Engl [132], [136], [138], Itoh [239], [240], [242], Joshi [243], [244], Kannan [247], [248], [251], [255], Kannan and Salehi [253], [254], Kravvaritis [267], [268], Lee and Padgett [276], [280], Mukherjea [299], [301], Nowak [313], [314], [317], and others.

3) Constructive and approximation schemes for random operator equations: [35], Bharucha-Reid [31], [37], Bharucha-Reid and Kannan [36], Bharucha-Reid and Römisch [39], Engl and Nashed [140], Engl and Römisch [142], [143], [144], Lee and Padgett [277], [278], [279], Padgett [335], Padgett and Rao [334], Radu [351], Römisch [365], [367], and others.

4) Applications of random differential and integral equations and adaptation of the theory to specific models in applications: Bharucha-Reid [31], Engl [132], [136], [139], Morse [296], Nowak [317], Römisch [366], Soong [412], Tsokos and Padgett [431].

We shall summarize now the most significant results regarding the solution of random operator equations. Beside the direct approach based on the measurability theorem for multivalued mappings, we shall outline the three main directions employed in the approximation theory of random equation solutions. In the third paragraph, we shall selectively present some important results regarding the use of some special classes of operators in solving random equations.

6.1. SOME EXAMPLES ON THE USE OF THE SELECTION THEOREM APPROACH

It is known [141] that two basic general approaches have been used for establishing the measurability of solutions of operator equations: *the limit theorem approach* and *the selection theorem approach* (see Chap. 2).

Besides the general results presented in Chapter 2, we shall give here some results and concrete examples due to A. Nowak [313]. The selection theorem approach is applied in the case when an approximation scheme is not known, most frequently when the solution is not unique. The key step in this technique is to establish the measurability of the multivalued map which takes each $\omega \in \Omega$ into the solution set of the realization of the equation corresponding to this ω.

A sufficient condition for the measurable dependence on ω of the solution of the equation $F(\omega, x) = 0$ is given by

THEOREM 6.1.1. *Let X be an analytic subset of a Polish space, let $\mathscr{B}_X$ be the Borel σ-algebra on X and let $(\Omega, \mathscr{K}, P)$ be a probability space. If $F^{-1}(0) \in \mathscr{K} \times \mathscr{B}_X$ and $E(\omega) = \{x \in X : F(\omega, x) = 0\} \neq \emptyset$ a.s., then the equation $F(\omega, x) = 0$ has a random solution ($F: \Omega \times X \to Y$, Y is a linear metric space).*

Proof. There exists an $A \in \mathscr{K}$ such that $P(A) = 1$ and $E(\omega) \neq \emptyset$ for all $\omega \in A$. Let $\mathscr{A}$ be the trace σ-algebra on A. We examine the measurability of the graph of the multivalued mapping E

$$\mathrm{Gr}E = \{(\omega, x) \in A \times X : x \in E(\omega)\} = F^{-1}(0) \cap A \times X \in \mathscr{A} \times \mathscr{B}_X.$$

According to the measurable selection theorem, there exists $y : A \to X$ such that $y(\omega) \in E(\omega)$ for all $\omega \in A$, and $y^{-1}(B) \in \mathscr{A}_c$ for all $B \in \mathscr{B}_X$, where $\mathscr{A}_c$ is the completion of $\mathscr{A}$ (see [373, Corollary 1]). Now, there exist $A_1 \in \mathscr{A}$ with $P(A_1) = 1$ and a measurable mapping $x : A \to X$ such that $x(\omega) = y(\omega)$ for all $\omega \in A_1$. An extension of x to a measurable mapping from Ω to X is a random solution of the given equation.

Remark. The above result still holds in the case of a stochastic domain.

Example 1. Let $G : \Omega \times [0,1] \times \mathbb{R} \times \mathbb{R} \to \mathbb{R}$ and $K : \Omega \times [0,1] \times \mathbb{R} \times \mathbb{R} \to \mathbb{R}$ be two functions such that for all $\omega \in \Omega$, $G(\omega, ., .,.)$ and $K(\omega, ., ., .)$ are continuous, for all $(t, s, r) \in [0,1] \times [0,1] \times \mathbb{R}$, $K(., t, s, r)$ is measurable and for all $(t, s, r) \in [0,1] \times \mathbb{R} \times \mathbb{R}$, $G(., t, s, r)$ is measurable. If for almost all $\omega \in \Omega$ the random integral equation

$$G(\omega, t, x(t), \int_0^1 K(\omega, t, s, x(s))\, \mathrm{d}s) = 0, \quad t \in [0,1],$$

has a continuous solution x, then there exists a Carathéodory mapping $h : \Omega \times [0,1] \to \mathbb{R}$ such that, for almost all ω and all $t \in [0,1]$,

$$G(\omega, t, h(\omega, t), \int_0^1 K(\omega, t, s, h(\omega, s))\, \mathrm{d}s) = 0.$$

The proof follows from Theorem 6.1.1. Let $X = Y = C[0,1]$ with sup norm. Define the operator $F : \Omega \times X \to Y$ by

$$F(\omega, x)(t) = G(\omega, t, x(t), \int_0^1 K(\omega, t, s, x(s))\, \mathrm{d}s).$$

Under the above assumptions, F is measurable in ω and continuous in x. For almost all $\omega \in \Omega$, the equation $F(\omega, x) = 0$ has a solution. Then it has a random solution y. The random function $h(\omega, t) = y(\omega)(t)$ is the desired solution of the random integral equation.

Example 2. Let $G : \Omega \times [0,1] \times \mathbb{R} \times \mathbb{R} \to \mathbb{R}$ and $x_0 : \Omega \to \mathbb{R}$ be two functions such that, for all $(t, s, r) \in [0,1] \times \mathbb{R} \times \mathbb{R}$, $G(., t, s, r)$ is measurable for all $\omega \in \Omega$, $G(\omega, ., ., .)$ is continuous and x_0 is measurable. If for almost all $\omega \in \Omega$ the initial-value problem

$$(1)\quad G(\omega, t, x(t), x'(t)) = 0, \; t \in [0,1]$$
$$x(0) = x_0(\omega)$$

has a continuously differentiable solution x, then there exists a random function $h : \Omega \times [0,1] \to \mathbb{R}$ such that

(i) for all $t \in [0,1]$, $h(., t)$ is measurable,
(ii) for all $\omega \in \Omega$, $h(\omega, .)$ is continuously differentiable,
(iii) for almost all $\omega \in \Omega$ and all $t \in [0,1]$

$$G(\omega, t, h(\omega, t), \frac{\mathrm{d}}{\mathrm{d}t} h(\omega, t)) = 0, \quad h(\omega, 0) = x_0(\omega).$$

First we give a useful lemma:

Let T be a non-empty set, let $(W, \|.\|_W)$ be a normed linear space and let $(X, \|.\|_X)$ be a normed linear space of functions $x : T \to W$. We assume that the norm of X is not weaker than *the sup norm. If* $y : \Omega \to X$ *is measurable, then the function* $h(\omega, t) = y(\omega)(t)$ *is measurable in* ω.

To prove this assertion, we consider X to be the linear space of all continuously differentiable functions $x : [0,1] \to \mathbb{R}$ with the norm

$$\|x\|_X = \sup_{t \in [0,1]} (|x(t)| + |x'(t)|),$$

(X is a separable Banach space). Let $Y = C[0,1] \times \mathbb{R}$, and

$$(*) \quad \|(x, r)\|_Y = \sup_{t \in [0,1]} |x(t)| + |r|.$$

In order to apply Theorem 6.1.1, we associate to (1) a random operator $F : \Omega \times X \to Y$ defined as

$$F(\omega, x) = (G(\omega, ., x(.), x'(.)), x(0) - x_0(\omega)).$$

The operator F is measurable in ω and continuous in x. Since we assumed that, for almost all $\omega \in \Omega$, the equation $F(\omega, x) = 0$ has a solution, it follows that it has a random solution y. The function $h(\omega, t) = y(\omega)(t)$ satisfies (i)—(iii).

Example 3. Let $H : \Omega \times [0,1] \times \mathbb{R} \to \mathbb{R}$ and $g : \Omega \times [0,1] \to [0,1]$ be two functions such that they are measurable in ω and continuous in other variables. Suppose that, for almost all $\omega \in \Omega$, the functional equation with random parameter ω

$$(2) \quad x(t) = H(\omega, t, x[g(\omega, t)]), \quad t \in [0,1],$$

has a continuous solution $x : [0,1] \to \mathbb{R}$ and that the additional condition

$$(2') \quad x(0) = \eta(\omega)$$

is satisfied, where $\eta : \Omega \to \mathbb{R}$ is a given measurable function such that $\eta(\omega) = H(\omega, 0, \eta(\infty))$ for all $\omega \in \Omega$. Then there exists a Carathéodory mapping $h : \Omega \times [0,1] \to \mathbb{R}$ satisfying

$$h(\omega, t) = H(\omega, t, h[\omega, g(\omega, t)]), \; h(\omega, 0) = \eta(\omega)$$

for almost all $\omega \in \Omega$ and all $t \in [0,1]$.

This example can be considered a corollary of Theorem 6.1.1 by taking $X = C[0,1]$ and $Y = C[0,1] \times \mathbb{R}$. Let $\|x\| = \sup_{t \in [0,1[} |x(t)|$ be the norm on X and let the norm defined by (*) be on Y. Then the operator $F : \Omega \times X \to Y$, defined as

$$F(\omega, x) = (x(.) - H(\omega, ., x[g(\omega, .)]), x(0) - \eta(\omega)),$$

is measurable in ω and continuous in x. Since for almost all $\omega \in \Omega$ the equation $F(\omega, x) = 0$ has a solution, then it has a random solution y.

The function $h(\omega, t) = y(\omega)(t)$ is the desired function.

6.2. APPROXIMATE SOLUTION OF RANDOM EQUATIONS

Recently, the study of approximate methods for solving random equations has grown into an area of active research. This area of research, which utilizes current results from numerical functional analysis, theory of stochastic processes, operator theory, measure theory, has led to the study and development of methods for obtaining approximate solutions of various classes to random equations. We note that approximation methods play two roles in the treatment of random operator equations. Clearly, such methods are needed when explicit solutions are difficult to obtain. They also provide an effective approach for establishing measurability of solutions. It should be pointed out that most of the equations are not necessarily uniquely solvable. A natural approach is to consider these equations in the framework of generalized inverses, thereby studying least-squares and projectional solutions.

The contributions brought by A. T. Bharucha-Reid [31], [35], [32], [38], [93], [39], [36], Nashed and Engl [310], [141], Engl and Römisch [142], [143], [144], Römisch [364], [365], [367], Lee and Padgett [277], [278], [279], V. Radu [351], and others opened several ways to the development of this area. In this section we shall attempt to sketch some of their results.

6.2.1. *Iterative Methods*

Random variants of iterative methods for fixed point problems involving operators that fulfill contraction conditions were first given in refs. [31], [36], [35], [39]. Results of this type were used in [38] and [351] to approximate the solution of random equations.

Consider the random analogue of the *iterative method of steepest descent* [310], [141] for the least-square solutions of the random equation

(1) $T(\omega)x = y(\omega)$,

where $T : \Omega \times X \to Y$ is a random bounded linear operator on Hilbert spaces and $y : \Omega \to Y$ is a random variable such that $y(\omega) \in D(T^+(\omega))$.

It is known [310] that $N(T)$ is a closed subspace, and thus

$$X = N(T) \oplus N(T)^\perp = N(T) \oplus \overline{R(T^*)}$$

$$Y = \overline{R(T)} = R(T)^\perp = \overline{R(T)} + \overline{R(T)^*}$$

Of course, these orthogonal decompositions induce orthogonal (equivalently : self-adjoint) projectors P and Q. The generalized inverse relative to these orthogonal complements (equivalently : projectors) is called the *Moore-Penrose inverse* in the Hilbert space. T^+ can be characterized in this case by the following extremal (least-square) property

(2) $\begin{cases} \{u = T^+y \text{ for } y \in R(T) + R(T)^\perp \text{ minimizes } \| Tx - y \| \\ \text{over } x \in D(T) \text{ and has the smallest norm among all the} \\ \text{other minimizers.} \end{cases}$

LEMMA 6.2.1 [310]. *The following sets are identical:*

(3) $\{u : \inf_{x \in D(T)} \| Tx - y \| = \| Tu - y \|\}$,

(3′) $\{u : Tu = Qy\}$,

(3″) $\{u : T^*(Tu - y) = 0\}$.

In the case of the bounded operator, (3″) can be written as $\{u : T^*Tu = T^*y\}$, such that the set of all least-square solutions coincides with the set of all solutions of the normal equations $T^*Tu = T^*y$. The generalized inverse T^+ can thus be characterized by the property that T^+y is the minimal-norm element in any of the sets in Lemma 6.2.1.

Let $U_y(\omega)$ denote the set of all least-square solutions of (1). Then $U_y(\omega) = T^+(\omega)y(\omega) + N(T(\omega))$ is a closed linear manifold for all ω. Let $P_U : \Omega \times X \to X$ be the metric projector onto U_y. Then

$$P_U(x) = T^+y + P_N x,$$

where P_N is the orthogonal projector of X onto $N(T)$. Note that $P_U(z)$ is the unique element in U_y which minimizes $\| x - z \|$ over $x \in U_y$. Consider now iterative methods for computing

$$P_U(\omega)z(\omega),$$

where z is a given random variable. In particular, the best approximate solution $T^+(\omega)y(\omega)$ can be computed by taking $z = 0$. Let

$$J(\omega, x) = \frac{1}{2} \| T(\omega)x - y(\omega) \|^2.$$

The method of steepest descent [307] for minimizing $J(\omega, x)$ is defined by the sequence

(4) $x_{n+1}(\omega) = x_n(\omega) - \alpha_n \operatorname{grad} J(\omega, x_n(\omega))$,

where α_n is chosen to minimize $J(\omega, x_{n+1}(\omega))$, thus indicating that in general α_n depends also on ω. It follows that

$$(5)\ r_n(\omega) = \operatorname{grad} J(\omega, x) = T^*(\omega)T(\omega)x - T^*(\omega)y(\omega)$$

and

$$(6)\ \alpha_n(\omega) = \frac{\| r_n\omega) \|^2}{\| T(\omega)r_n(\omega) \|^2}.$$

It follows from Remarks 2.2.3 and 2.2.4 that r_n is a random operator and α_n is a random variable. Thus, the iteration scheme (4) generates random variables if the initial approximation x_0 is a random variable. Together with convergence results for the deterministic case (see [307]), this proves

THEOREM 6.2.1. *Let $x_0 : \Omega \to X$ be measurable. Then the method of steepest descent defined by* (4), (5) *and* (6) *generates random approximants and converges to*

$$P_U(\omega)x_0(\omega) = T^+(\omega)y(\omega) + P_N(\omega)x_0(\omega).$$

For other methods, see [310]. An illustration of the applicability of this method to random linear boundary-value problems is given in [35]. Let us now present a general method for obtaining approximate solutions of random operator equations given by V. Radu [351]. Both an iterative process and an approximation for the random solution by a sequence of simpler random equations are given.

Let X and Y be separable Banach spaces and consider the random operator equations

$$(1.\text{n})\ T_n(\omega, x) = y_n(\omega), \quad n \in \mathbb{N},$$

where $T_n : \Omega \times X \to Y$, $T_n(\omega, 0) = 0$ is a random operator which is continuous for each $\omega \in \Omega$, the support of the complete probability measure space $(\Omega, \mathscr{K}, P)$.

We are concerned with two problems. Firstly, to obtain the solution of (1.n) by an iterative process and, secondly, to give conditions under which the solutions of (1.n), $n \in \mathbb{N}$, approximate the solution of $(1.\infty)$ if y_n and T_n approximate in a given sense y_∞ and T_∞. The iterative process will be chosen under the form

$$(2.\text{n})\ x_n^{j+1}(\omega) = x_n^j(\omega) + \tau(\omega), L(\omega, y_n(\omega)) - T_n(\omega, x_n^j(\omega)),$$

where $L : \Omega \times Y \to X$ is a random operator which is bounded, linear and invertible for each $\omega \in \Omega$, whereas $\tau : \Omega \to (0, \infty)$ is a random variable.

Consider the operator $U_n = I_X - \tau LT_n$,

$U_n(\omega, x) = x - \tau(\omega) L(\omega, T_n(\omega, x))$, $U_n : \Omega \times X \to X$.

Suppose that U_n is random contraction, i.e.,

$\| U_n(\omega, x) - U_n(\omega, y) \| \leqslant q_n(\omega) \| x - y \|$ a.s.,

where $q_n : \Omega \to R$ is a random variable, $q_n(\omega) \in [0,1)$ a.s. Then, for each Y-valued random variable y_n, Eq. (1.n) has a unique random solution x_n which is obtained by the iterative process (2.n) with an arbitrary X-valued random variable x_n^0.

According to the previous assumptions, U_n is a continuous random operator. Since $L(\omega, .)$ is linear and invertible, it follows that (1.n) is equivalent to

(1′.n) $x = U_n(\omega, x) + \tau(\omega)L(\omega, y_n(\omega))$.

As U_n is a random contraction, then (see Sec. 4.1) Eq. (1′n) has a unique random solution $x_n(\omega)$ which is obtained by use of the method of successive approximations, and this leads to the iterative process (2.n).

Let us consider now the conditions under which the random solutions of (1.n), $n \in \mathbb{N}$, approximate the solution of $(1, \infty)$.

THEOREM 6.2.2. *Let* τ *and* L *be as above. If*

(i) $T_n(\omega, .)$ *is linear for each* ω, $n \in \mathbb{N}$;

(ii) $\| T_n(\omega, x) - T_\infty(\omega, x) \| \leqslant \varepsilon_n(\omega) \| x \|$, *a.s. for all* $x \in X$, *where* $\varepsilon_n(\omega) \to 0$ *a.s. for* $n \to \infty$;

(iii) U_∞ *is a random strict contraction*;

(iv) $\| y_n(\omega) - y_\infty(\omega) \| \to 0$ *a.s.*,

then

(v) *Equation* (1.n) *has a unique random solution* x_n *for* n *sufficiently large and for each Y-valued random variable* y_n;

(vi) x_n *is obtained by the iterative process* (2.n);

(vii) $\| x_n(\omega) \| \leqslant \dfrac{\tau(\omega) \| L(\omega, y_n(\omega)) \|}{1 - q_n(\omega)}$ *a.s.*;

(viii) $\| x_n(\omega) - x_\infty(\omega) \| \leqslant \dfrac{\tau(\omega) L(\omega, .)}{1 - q_\infty(\omega)} (\varepsilon_n(\omega) \| x_n(\omega) \| + \| y_n(\omega) - y_\infty(\omega) \|)$,

where $0 \leqslant q_n(\omega) \leqslant q(\omega) < 1$ *a.s.*;

(ix) x_n *converges a.s. to* x_∞.

Proof. Consider $\Omega_0 \in \mathscr{K}$ with $P(\Omega_0) = 1$, such that (i) – (iv) hold for each $\omega \in \Omega$. Let ω be fixed in Ω_0 and let $n < \infty$. If

$q_\infty(\omega)$ is the contraction constant of $U_\infty(\omega,.)$, then from (ii) we have

$$\|U_n(\omega, x)\| \leqslant [q_\infty(\omega) + \tau(\omega)\|L(\omega,.)\|\varepsilon_n(\omega)]\|x\|.$$

Since $\varepsilon_n(\omega) \to 0$ and $q_\infty(\omega) < 1$, then for n sufficiently large

$$q_n(\omega) = q_\infty(\omega) + \tau(\omega)\ \|L(\omega,.)\|\varepsilon_n(\omega) \leqslant q(\omega) < 1,$$

that is $U_n(.)$ is a strict contraction. Hence, (v) and (vi) hold.

Let $y_n(\omega) \in Y$. Then

$$\|x_n(\omega) - x_\infty(\omega)\| \leqslant \|U_n(\omega, x_n(\omega)) - U_\infty(\omega, x_n(\omega))\| + \\ + \|U_\infty(\omega, x_n(\omega)) - U_\infty(\omega, x_\infty(\omega))\| + \tau(\omega)\|L(\omega, y_n(\omega) - \\ - y_\infty(\omega))\| \leqslant \varepsilon(\omega)\|L(\omega,.)\|\varepsilon_n(\omega)\|x_n\| + q_\infty(\omega)\|x_n(\omega) - \\ - x_\infty(\omega)\| + \tau(\omega)\ \|L(\omega, y_n(\omega) - y_\infty(\omega))\|,$$

which clearly implies (viii).

If $y_\infty(\omega) = 0$, then $x_\infty(\omega) = 0$ and thus

$$\|x_n(\omega)\| \leqslant [\tau(\omega)\|L(\omega,.)\varepsilon_n(\omega) + q_\infty(\omega)]\|x_n(\omega)\| + \\ + \tau(\omega)\|L(\omega, y_n(\omega))\|,$$

that is

$$(1 - q_n(\omega))\|x_n(\omega)\| \leqslant \tau(\omega)\|L(\omega, y_n(\omega))\|$$

and hence (vii) holds also.

Since $q_n(\omega) \leqslant q(\omega)$, then (iv) implies that $\{\|x_n(\omega)\|\}$ is bounded. Thus, (ix) follows from (viii).

THEOREM 6.2.3. *Let U_n be as above. Suppose that*

(i) $q_n(\omega) \leqslant q(\omega) < 1$ *a.s. for* $n \in \mathbb{N}$;

(ii) $\|y_n(\omega) - y_\infty(\omega)\| \to 0$ *a.s.*;

(iii) $\|T_n(\omega, x_\infty(\omega)) - T_\infty(\omega, x_\infty(\omega))\| \to 0$ *a.s.*

Then

(iv) $$\|x_n(\omega) - x_\infty(\omega)\| \leqslant \frac{\tau(\omega)\|L(\omega,.)\|}{1 - q(\omega)}[T_n(\omega, x_\infty(\omega)) - \\ - T_\infty(\omega, x_\infty(\omega))\| + \|y_n(\omega) - y_\infty(\omega)\|];$$

(v) *If $T_n(\omega)$ is linear and if $x_n^0(\omega) = \tau(\omega)\, L(\omega, y_n(\omega))$, then*

$$\|x_n(\omega) - x_n^j(\omega)\| \leqslant \frac{q_n^{j+1}(\omega)\|x_n^0(\omega)\|}{1 - q_n(\omega)}\ a.s.$$

Proof. Let $\Omega_0 \in \mathscr{K}$, $P(\Omega_0) = 1$ and $\omega \in \Omega_0$. Then

$$\|x_n(\omega) - x_\infty(\omega)\| = \|U_n(\omega, x_n(\omega)) + \tau(\omega)L(\omega, y_n(\omega)) - \\ - U_\infty(\omega, x_\infty(\omega)) - \tau(\omega)L(\omega, y_\infty(\omega))\| \leqslant \\ \leqslant \|U_n(\omega, x_n(\omega)) - U_n(\omega, x_\infty(\omega))\| + \|U_n(\omega, x_\infty(\omega)) -$$

$$- U_\infty(\omega, x_\infty(\omega))\| + \tau(\omega)\|L(\omega,.)\| \|y_n(\omega) - y_\infty(\omega)\| \leqslant$$
$$\leqslant q_n(\omega)\|x_n(\omega) - x_\infty(\omega)\| + \tau(\omega)\|L(\omega,.)\| \| T_n(\omega, x_\infty(\omega)) -$$
$$- T_\infty(\omega, x_\infty(\omega))\| + \tau(\omega)\|L(\omega,.)\| \|y_n(\omega) - y_\infty(\omega)\|$$

and, hence, one obtains (iv).

On the other hand,

$$\|x_n(\omega) - x_n^j(\omega)\| = \|U_n x_n(\omega) - U_n x_n^{j-1}(\omega)\| \leqslant q(\omega)\|x_n(\omega) -$$
$$- x_n^{j-1}(\omega)\| \leqslant q_n^j(\omega)\|x_n(\omega) - x_n^0(\omega)\| \leqslant q_n^j(\omega)\|x_n(\omega) -$$
$$- x_n^j(\omega)\| + q_n^j(\omega)\|x_n^j(\omega) - x_n^0(\omega)\| \leqslant q_n^j(\omega)\|x_n(\omega) -$$
$$- x_n^0(\omega)\| + q_n^j(\omega)\frac{1 - q_n^j(\omega)}{1 - q_n(\omega)}\|x_n^1(\omega) - x_n^0(\omega)\| \cdot$$

Thus,

$$\|x_n(\omega) - x_n^j(\omega)\| \leqslant \frac{q_n^j(\omega)}{1 - q_n(\omega)}\|U_n(\omega, x_n^0(\omega))\| \leqslant$$
$$\leqslant \frac{q_n^{j+1}}{1 - q_n(\omega)}\|x_n^0(\omega)\|$$

and the proof of the theorem is complete.

Remark. The deterministic problem is well known (see, for example, [156], [351]).

Let us now consider some sufficient conditions for the existence and uniqueness of a random solution, using the method suggested by S. Olech [322] for the deterministic case.

Let X be a separable Banach space and let $(\Omega, \mathscr{K}, P)$ be a complete probability measure space. If a and r are positive random variables, then setting

$$I_a(\omega) = [0, a(\omega)],\ S_r(\omega) = \{x \in X,\ \|x\| < r(\omega)\}$$

we can see that I_a are closed multivalued mappings.

We shall use also the following notation:

$$I_a = \{(\omega, t) \in \Omega \times \mathbb{R},\ t \in I_a(\omega)\},$$

$$D_{ar} = (\{\omega, t, x) \in \Omega \times \mathbb{R} \times X,\ (t, x) \in I_a(\omega) \times S_r(\omega)\}$$

and similarly for d.

Let $f : D_{ar} \to X$ be such that

(F_1) $f(\omega, .,.)$ is continuous on $I_a(\omega) \times S_r(\omega)$;

(F_2) f is a random operator with stochastic domain;

(F_3) there exists a random variable $M : \Omega \to (0, \infty)$ such that $\|f(\omega, t, x,\| \leqslant M(\omega)$;

(F_4) $\|f(\omega, t, x) - f(\omega, t, y)\| \leqslant k(\omega, t, \|x-y\|),\quad \forall t, x, y \in D_{ar}$;

where $k :\{(\omega, t, u) \in \Omega \times \mathbb{R} \times [0, \infty),\ t \in I_a(\omega)\} = K_a$.

Let also $k : K_a \to [0, \infty)$ be such that

(k_1) $k(.,.,u)$ is a random operator for each $u \in [0, \infty)$;

(k_2) $k(\omega, t, .)$ is continuous for almost all $\omega \in \Omega$ and each $t \in I_a(\omega)$;

(k_3) $k(\omega, . u)$ is Lebesgue-measurable for almost all $\omega \in \Omega$ and $u \in [0, \infty)$;

(k_4) for each bounded subset $B(\omega) \subset I_a(\omega) \in [0, \infty)$, there exists $m : I_a(\omega) \to [0, \infty)$ which is Lebesgue-measurable on $I_a(\omega)$ and integrable on $[s, a(\omega))$ for every $s > 0$, a.s.

(k_5) for each $s \in (0, a(\omega))$, the identically zero function is the only absolutely continuous solution of the equation

$$u' = k(\omega, t, u) \text{ a.s. on } [0, s]$$

which satisfies

$$u(0) = D_+u(0) = 0 \text{ for a.s. } \omega \in \Omega.$$

THEOREM 6.2.4. *There exists a random solution of the random differential equation*

$$\frac{dx}{dt} = f(\omega, t, x)$$

$$x(\omega, 0) = 0$$

on I_d, *where* $d(\omega) = \min\left\{a(\omega), \frac{r(\omega)}{M(\omega)}\right\}$ *and is unique.*

Proof. Let $\Omega_0 = \{\omega \in \Omega : (k_1) - (k_5), (F_3), (F_4) \text{ holds}\}$. Clearly, $P(\Omega_0) = 1$. Let $\omega \in \Omega_0$ be a fixed point. Then one can construct [322] a sequence $\{q_n\}_{n\in\mathbb{N}}$ as follows:

$$q_n(0) = 0,$$

$$q_n(\omega, t) = q_n\left(\omega, \frac{j\,d(\omega)}{n}\right) + \int_{\frac{j d(\omega)}{n}}^{t} f\left(\omega, u, q\left(\omega, \frac{j\,d(\omega)}{n}\right)\right) du$$

for $t \in \left[\frac{j\,d(\omega)}{n}, \frac{(j+1)\,d(\omega)}{n}\right]$.

As proved by Olech [322], $\{q_n\}_{n\in\mathbb{N}}$ converges uniformly to the unique solution of the equation

$$\begin{cases} \dfrac{dq}{dt} = f(\omega, t, q), \quad t \in I_d(\omega), \quad q(t) \in S_r(\omega), \\ q(0) = 0. \end{cases}$$

It is easy to see that $\Omega_t = \{\omega, t \in I_d(\omega)\} = \{\omega, t \leqslant d(\omega)\} \in \mathscr{K}$. Let $\mathscr{K}_t = \mathscr{K} \cap \Omega_t$. Since f is a continuous random operator, $q_n(., t) : \Omega_t \to X$ is $\mathscr{K}_t$-measurable and therefore $q_n(., t)$ is

$\mathscr{K}_t$-measurable. If

$$x(\omega, t) = \begin{cases} q(\omega, t), & \omega \in \Omega_0. \\ 0, & \omega \in \Omega - \Omega_0, \end{cases}$$

then x is a random solution of the equation and the theorem is proven.

6.2.2. *Projection Methods*

An extensive literature is available regarding projectional methods for deterministic equations (e.g., Galerkin's method, Ritz' method, least-square method, etc.) [265], [338].

In the case of random operator equations, the projectional schemes will largely depend on the random parameter ω. This is even unavoidable in the case when the operator is not onto and one wants to find least-square solutions. Thus, randomness is introduced into the projectional scheme *via* the random operator.

Let X and Y be separable Hilbert spaces. H. Engl and M. Nashed [140], [141] suggested a general methodology for adapting projection methods to random operator equations in Hilbert spaces. Other general concepts and convergence results for approximate solutions of nonlinear random operator equations or equations involving P-compact random operators have applications in projection methods [39], [142], [143], [144], [365].

Several results [140], [310] will be presented below.

DEFINITION 6.2.1. A sequence of triples $(X_n, Y_n; Q_n)$ will be called a *stochastic projectional scheme* if $X_n : \Omega \to 2^X$ and $Y_n : \Omega \to 2^Y$ are measurable and if, for all $n \in \mathbb{N}$ and all $\omega \in \Omega$, $X_n(\omega)$, $Y_n(\omega)$ are closed subspaces of X and Y, respectively, and $Q_n(\omega)$ is the orthogonal projector onto $Y_n(\omega)$. For $T : \Omega \times X \to Y$ a linear random operator, define

(1) $T_n(\omega, x) = Q_n(\omega, T(\omega, x))$.

From Lemma 2.2.5 and Remark 2.2.4, it follows that Q_n and T_n are continuous random operators. The new aspects considered in the random analogues of the projection methods are the measurability and the statistical properties of the approximants.

LEMMA 6.2.2 [140]. *Let $(X_n, Y_n; Q_n)$ be a stochastic projectional scheme, let $y : \Omega \to Y$ be measurable, let $T : \Omega \times X \to Y$ be a bounded linear random operator, and let T_n be defined by* (1). *Assume that, for some $k \in \mathbb{N}$,*

(2) $T_k(\omega, x) = Q_k(\omega, y(\omega))$

is solvable in $x_k(\omega)$ *for all* $\omega \in \Omega$. *Then there exists a measurable function* $x_k : \Omega \to X$ *such that, for all* $\omega \in \Omega$, $x_k(\omega) \in X_k(\omega)$ *and* $T_k(\omega, x_k(\omega)) = Q_k(\omega, y(\omega))$.

Proof. T_k is a continuous random operator; because of Remark 2.2.4, $\omega \to Q_k(\omega, y(\omega))$ is measurable. Thus, $(\omega, x) \to$ $\to x + Q_k(\omega, y(\omega)) - T_k(\omega, x)$ is a continuous random operator. Denote its fixed point set by $F(\omega)$. It follows from Theorem 4.4.9 that F is measurable. Considering also Theorem 2.2.1, it follows that $F \cap X_k$ is measurable. As, by the above assumptions, $F(\omega) \cap X_k(\omega) \neq \emptyset$ for all $\omega \in \Omega$, from the Kuratowski and Ryll-Nardzewski selection theorem (see Theorem 2.2.4) it follows that there exists a measurable selector x_k for X_k.

Let $\Phi_n : \Omega \to X$ and $\psi_n : \Omega \to Y$ be sequences of measurable functions such that, for all $\omega \in \Omega$, $\{\Phi_n(\omega)\}_{n \in \mathbb{N}}$ and $\{\psi_n(\omega)\}_{n \in \mathbb{N}}$ are linearly independent and complete in X and Y, respectively. We shall keep these functions fixed and shall use them to construct stochastic projectional schemes which are suitable for treating different kinds of equations. Ordinarily, one will choose $\{\Phi_n\}_{n \in \mathbb{N}}$ and $\{\psi_n\}_{n \in \mathbb{N}}$ independent of ω. However, projectional schemes for treating random operator equations will in general be stochastic. Hence we can as well permit from the outset the basis functions to be random.

In general, deterministic results are used to establish the existence and convergence of approximate solutions whereas the measurability of these approximate solutions will follow from the above lemma.

We shall use $\{\Phi_n\}_{n \in \mathbb{N}}$ and $\{\psi_n\}_{n \in \mathbb{N}}$ in various ways to construct stochastic projectional schemes for finding least-square solutions of linear operator equations of first and second kind (i.e., in the case of non-closed, respectively closed range). Four typical results will be listed. In the closed-range case, one states

THEOREM 6.2.5. *Let* $T : \Omega \times X \to Y$ *be a bounded linear random operator such that, for all* $\omega \in \Omega$, $T(\omega)$ *is one-to-one and onto,* $y : \Omega \to Y$ *is measurable. For all* $\omega \in \Omega$ *and* $n \in \mathbb{N}$, *let* $X_n(\omega) =$ $= \operatorname{span}\{\Phi_1(\omega), \ldots, \Phi_n(\omega)\}$, $Y_n(\omega) = \operatorname{span}\{\psi_1(\omega), \ldots, \psi_n(\omega)\}$ *with orthogonal projector* $Q_n(\omega)$. *Then* $(X_n, Y_n; Q_n)$ *is a stochastic projectional scheme. Let* T_n *be defined by* (1) *and assume that there exists* $c(\omega) > 0$ *and* $n(\omega) \in \mathbb{N}$ *such that, for all* $n \geqslant n(\omega)$ *and* $x_n \in X_n(\omega)$,

(3) $\|T_n(\omega, x_n)\| \geqslant c(\omega)\|x_n\|$.

Then the measurable functions $x, x_1, \ldots : \Omega \to X$ *exist and have the following properties:*

(1) *For all* $n \in \mathbb{N}$ *and* $\omega \in \Omega$, $x_n(\omega) \in X_n(\omega)$.

(2) *For all* $\omega \in \Omega$ *and* $n \geqslant n(\omega)$, $T_n(\omega, x_n) = Q_n(\omega, y(\omega))$.

(3) *For all* $\omega \in \Omega$, $x(\omega)$ *is the unique solution of* $T(\omega, x) = y(\omega)$.
(4) *For all* $\omega \in \Omega$ *and* $n \in \mathbb{N}$, $x_n(\omega)$ *is the best-approximate solution of* $T_n(\omega)P_n(\omega, x) = Q_n(\omega, y(\omega))$ *in* X, *where* $P_n(\omega)$ *is the orthogonal projector onto* $X_n(\omega)$.
(5) *For all* $\omega \in \Omega$, $x_n(\omega)$ *is the best-approximate solution of* $T_n(\omega, x) = Q_n(\omega, y(\omega))$ *in* $X_n(\omega)$.
(6) *For all* $\omega \in \Omega$, $\{x_n(\omega)\}_{n \in \mathbb{N}}$ *converges to* $x(\omega)$.

Remark. If we choose $\psi_n(\omega) = T(\omega, \Phi_n(\omega))$, then Theorem 6.2.5 becomes a stochastic version of the least-square method.
The following result deals with best-approximate solutions of not necessarily (uniquely) solvable equations of the second kind :

THEOREM 6.2.6. *Let* $T : \Omega \times X \to Y$ *be a bounded linear random operator with closed ranges and let* $y : \Omega \to Y$ *be measurable. For all* $\omega \in \Omega$ *and* $n \in \mathbb{N}$, *let* $X_n(\omega) = \text{span}\{T^*(\omega)\Phi_1(\omega), \ldots, \ldots T^*(\omega)\Phi_n(\omega)\}$, $Y_n(\omega) = T(\omega, X_n(\omega))$ *with orthogonal projector* $Q_n(\omega)$. *Then* $(X_n, Y_n; Q_n)$ *is a stochastic projectional scheme. For all* $\omega \in \Omega$ *and* $n \in \mathbb{N}$, $T_n(\omega, x) = Q_n(\omega, y(\omega))$ *has a unique solution* $x_n(\omega)$ *in* $X_n(\omega)$, *and* $\{x_n(\omega)\}_{n \in \mathbb{N}}$ *converges to* $x(\omega)$, *the best-approximate solution of* $T(\omega, x) = y(\omega)$. *All* x_n *and* x *are measurable.*

Consider now equations involving a linear operator with non-closed range. It is well known from the deterministic situation that one cannot expect a projection method to yield a sequence of approximates that converges to a (best-approximate) solution in this case. All that can be expected is that the residual tends to zero (or to its infimum in the case of measurable equations) [265].

THEOREM 6.2.7. *Let* T *be a (not necessarily bounded) linear random operator from* $\Omega \times X$ *into* Y *such that, for all* $\omega \in \Omega$, $T(\omega)$ *is one-to-one and densely defined*, $R(T(\omega))$ *is dense in* Y *and* $y : \Omega \to Y$ *is measurable. Let* $(X_n, Y_n; Q_n)$ *be a stochastic projectional scheme such that, for all* $\omega \in \Omega$ *and* $n \in \mathbb{N}$, $X_n(\omega) \subseteq \subseteq D(T(\omega))$ *and* $T(\omega, X_n(\omega))$ *is closed in* Y. *Let* T_n *be defined by* (1). *By* $P_n(\omega)$ *we denote the orthogonal projector onto* $T(\omega, X_n(\omega))$. *Assume also that, for all* $b \in Y$ *and* $\omega \in \Omega$,

(4) $\lim_{n \to \infty} \| P_n(\omega, b) - b \| = 0$

and that

(5) $\lim_{n \to \infty} \sup \{\| P_n(\omega) - Q_n(\omega) \| : \omega \in \Omega\} < 1$.

Furthermore, suppose that either all $X_n(\omega)$ *are independent of* ω *and finite-dimensional or that* T *is bounded. Then there exists an* $n_0 \in \mathbb{N}$ *(independent of* ω*) such that, for all* $n \geq n_0$, *there exists a measurable function* $x_n : \Omega \to X$ *with the property that* $x_n(\omega)$ *is the unique solution of* $T_n(\omega, x) = Q_n(\omega, y(\omega))$ *in* $X_n(\omega)$ *for all* $\omega \in \Omega$. *For all* $\omega \in \Omega$,

(6) $\lim_{n\to\infty} \| T(\omega, x_n(\omega)) - y(\omega) \| = 0.$

In the case of the linear random operator equation of the first kind with non-trivial null space, we have

THEOREM 6.2.8. *Let* $T : \Omega \times X \to Y$ *be a bounded linear random operator and let* $y : \Omega \to Y$ *be measurable. For all* $\omega \in \Omega$ *and* $n \in \mathbb{N}$, *let* $X_n(\omega) = \text{span}\{T^*(\omega)T(\omega)\Phi_1(\omega), \ldots, T^*(\omega)T(\omega)\Phi_n(\omega)\}$, *let* $Y_n(\omega) = T(\omega, X_n(\omega))$ *with orthogonal projector* $Q_n(\omega)$ *and let* $Q(\omega)$ *be the orthogonal projector onto* $\overline{R(T(\omega))}$. *Then* $(X_n, Y_n; Q_n)$ *is a stochastic projectional scheme. For all* $\omega \in \Omega$ *and* $n \in \mathbb{N}$, $T_n(\omega, x) = Q_n(\omega, y(\omega))$ *has a unique solution* $x_n(\omega)$ *in* $X_n(\omega)$. *Each* x_n *is measurable. For all* $\omega \in \Omega$, $\lim_{n\to\infty} \| T(\omega, x_n(\omega)) - Q(\omega, y(\omega)) \| = 0$ *and*

$$\lim_{n\to\infty} \| T(\omega, x_n(\omega)) - y(\omega) \| = \inf_{x\in X} \| T(\omega, x) - y(\omega) \|.$$

Remark. Lemma 6.2.2 allows the projectional equations to have non-unique solutions; moreover, it does not depend on the linearity, but only on the continuity of the operator involved, such that generalizations to nonlinear problems are possible. Let us now recall A. T. Bharucha-Reid and W. Römisch's results [39] and their projectional schemes as an approach to proving weak compactness of approximate "solution measures", i.e., of probability distributions of approximate random solution. This approach (as in [37], [143]) provides useful tools for establishing the weak compactness and convergence of solution measures associated with random operator equations in Banach spaces.

Let X be a separable Banach space and let $\mathcal{M}(X)$ be the set of all probability measures on $(X, \mathcal{B}_X)$ equipped with the topology of weak convergence (see e.g. [40]). Let us call $(X_n, P_n)_{n\in\mathbb{N}}$ a *projectional scheme* for X, where $X_n \subset X$, $n \in \mathbb{N}$, is a sequence of monotonically increasing finite-dimensional subspaces, and $P_n : X \to X_n$, $n \in \mathbb{N}$, are linear continuous projections such that $\lim_{n\to\infty} \| P_n x - x \| = 0$ for all $x \in X$.

It is known [345] that a sequence $\{\mu_n\}_{n\in\mathbb{N}} \subset \mathcal{M}(X)$ is *weakly compact* (i.e., relatively compact with respect to weak convergence in $\mathcal{M}(X)$) if and only if $\{\mu_n\}_{n\in\mathbb{N}}$ is *tight*, i.e., for all $\varepsilon > 0$, there exists a compact subset $K_\varepsilon \subset X$ such that

$$\inf_{n\in\mathbb{N}} \mu_n(K_\varepsilon) \geq 1 - \varepsilon.$$

The sequence $\{\mu_n\}_{n\in\mathbb{N}}$ is called *uniformly bounded* if, for all $\varepsilon > 0$, there exists a constant $c_\varepsilon > 0$ such that

$$\inf_{n\in\mathbb{N}} \mu_n(\{x \in X : \|x\| < c_\varepsilon\}) \geqslant 1 - \varepsilon.$$

It is called *flatly concentrated* if, for all $\varepsilon > 0$ and $\delta > 0$, there exists a finite-dimensional subspace L of X such that

$$\inf_{n\in\mathbb{N}} \mu_n(\{x \in X : \mathrm{d}(x, L) \leqslant \delta\}) \geqslant 1 - \varepsilon.$$

The following characterizations of tightness are also given [39]: "$\{\mu_n\}_{n\in\mathbb{N}} \subset \mathcal{M}(X)$ is tight if and only if $\{\mu_n\}_{n\in\mathbb{N}}$ is uniformly bounded and flatly concentrated".
"$\{\mu_n\}_{n\in\mathbb{N}} \subset \mathcal{M}(X)$ is tight if and only if $\{\mu_n\}_{n\in\mathbb{N}}$ is uniformly bounded and, for all $\varepsilon > 0$ and $\delta > 0$, there exists an $n_0(\varepsilon, \delta) \in \mathbb{N}$ such that, for every $m \geqslant n_0$, $\inf_{n\in\mathbb{N}} \mu_n(\{x \in X : \mathrm{d}(x, X_n) \leqslant \delta\}) \geqslant \geqslant 1 - \varepsilon$ holds, where $(X_n, P_n)_{n\in\mathbb{N}}$ is a given projectional scheme for X".

Let $(X_n, P_n)_{n\in\mathbb{N}}$ be a projectional scheme for X, let $T : \Omega \times \times X \to X$ and $T_n : \Omega \times X \to X_n$, $n \in \mathbb{N}$, be linear continuous random operators and let $y : \Omega \to X$ and $y_n : \Omega \to X_n$ be random variables defined on $(\Omega, \mathcal{K}, P)$.

Let us consider the random operator equation

(7) $x = T(\omega, x) + y(\omega)$

and its approximations

(8) $x = T_n(\omega, x) + y_n(\omega), \quad n \in \mathbb{N}.$

The following problem is of interest: "For all $n \in \mathbb{N}$, let $x_n : \Omega \to X_n$ be a random solution of (8) for the index n. Assume that the sequence $\{D(y_n)\}_{n\in\mathbb{N}} \subset \mathcal{M}(X)$ is tight. Find the conditions for T_n, $n \in \mathbb{N}$, that guarantee that the sequence $\{D(x_n)\}_{n\in\mathbb{N}}$ of solution measures forms a tight subset of $\mathcal{M}(X)$".

Recall that $D(x)$ denotes the probability distribution of any X-valued random variable $x : (\Omega, \mathcal{K}, P) \to X$, i.e., $D(x) \in \mathcal{M}(X)$ defined as $D(x)(B) = P(x^{-1}(B))$, $B \in \mathcal{B}_X$.

It is clear that under the above assumptions the use of well-known random fixed point theorems from Chapter 4 yields the existence of a random solution x_n of (8) under the assumption that (8) is deterministically solvable.

An important result relative to the above problem is

THEOREM 6.2.9 ([39], Th. 3.3). *Let T, T_n, y_n, $n \in \mathbb{N}$, be as above and suppose that for P-almost all $\omega \in \Omega$*

$I - T(\omega, .)$ *is injective,*

and

$$\lim_{n\to\infty} \| T_n(\omega, .) - T(\omega, .)\| = 0$$

(*here* $\|\cdot\|$ *denotes the usual operator norm*). *For all* $n \in \mathbb{N}$, *let* x_n *be a random solution of* (8). *Then*

(1) $\{(D(x)_n\}_{n\in\mathbb{N}}$ *is uniformly bounded if* $\{D(y_n)\}_{n\in\mathbb{N}}$ *is uniformly bounded,*

(2) $\{D(x_n)\}_{n\in\mathbb{N}}$ *is tight if* $\{D(y_n)\}_{n\in\mathbb{N}}$ *is tight.*

Another type of approximation procedure is based on the kernel approximation of random Fredholm equations of the second kind. Consider the following equation :

$$(9)\quad x(t) = \int_0^1 K(\omega, t, s)\, x(s)\, ds + y(\omega, t),\ t \in [0, 1].$$

Several assumptions have to be considered :

(a) $K : \Omega\times[0,1]\times[0,1]\to \mathbb{R}$ is a $\mathscr{K}\times\mathscr{B}([0,1])\times\mathscr{B}[0,1])$-measurable stochastic process such that, for all $\omega\in\Omega$, $K(\omega,.,.)$ is an element of $\mathbb{Z} = L_2([0,1]\times(0,1))$;

(b) $y : \Omega\times[0,1]\to\mathbb{R}$ is a $\mathscr{K}\times([0,1])$-measurable stochastic process such that, for all $\omega\in\Omega$, $y(\omega,.)\in L_2(0,1)$;

(c) $X = L_2(0,1)$, $T:\Omega\times X\to X$ is defined as

$$[T(\omega, x)](t) = \int_0^1 K(\omega, t, s)x(s)\, ds,\quad t\in[0,1],\ \omega\in\Omega,\quad x\in X.$$

Since T is a linear completely continuous operator and $y(K)$ can be viewed as an X-valued (Z-valued) random variable, we can consider the following X projectional scheme $(X_n, P_n)_{n\in\mathbb{N}}$: $X_n = \text{span}\ \{\Phi_1, \ldots, \Phi_n\}$, $n\in\mathbb{N}$, $\{\Phi_m\}_{m\in\mathbb{N}}$ is an orthonormal basis of X and P_n is the orthogonal projection from X onto X_n.

Suppose that the sequences $(K_n)_{n\in\mathbb{N}}$ and $(y_n)_{n\in\mathbb{N}}$ of random kernels and random forcing functions, respectively, are such that, for each $n\in\mathbb{N}$,

(10) K_n and y_n satisfy (a) and (b), respectively,

and

$$K_n(\omega, ., .)\ \text{span}\ \{\Phi_i(.)\Phi_j(.),\quad i, j = 1, \ldots, n\},$$

(11) $y_n(\omega)\in X_n$ for all $\omega\in\Omega$.

Each kernel K_n is a so-called [31, p. 150] "random degenerate kernel".

Linear continuous random operators $T_n : \Omega\times X\to X_n$, are similarly defined for all $n\in\mathbb{N}$:

$$[T_n(\omega, x)]\,(t) = \int_0^1 K_n(\omega, t, s)x(s)\, ds,\quad t\in[0,1],\ \omega\in\Omega,\ x\in X,$$

and hence the random Fredholm equations are stated as

(12) $x = T_n(\omega, x) + y_n(\omega), \quad \omega \in \Omega, \quad n \in \mathbb{N}.$

There follows

THEOREM 6.2.10 [39]. *Let K, K_n and y_n, $n \in \mathbb{N}$, satisfy* (a), (10) *and* (11). *Assume that*

(13) $I - T(\omega, .)$ *is injective for P-almost all* $\omega \in \Omega$,

(14) $\lim_{n\to\infty} \int_0^1\int_0^1 [K_n(\omega, t, s) - K(\omega, t, s)]^2 \, ds \, dt = 0$ *a.s.*

(15) $\{D(y_n)\}_{n\in\mathbb{N}}$ *is tight.*

For all $n \in \mathbb{N}$, let x_n be a random solution of (12). *Then $\{D(x_n)\}_{n\in\mathbb{N}}$ is tight.*

THEOREM 6.2.11 [39]. *Let K_n and y_n, $n \in \mathbb{N}$, fulfill* (10), (11) *and assume that the sequences of distributions $\{D(K_n)\}_{n\in\mathbb{N}}$ and $\{D(y_n)\}_{n\in\mathbb{N}}$ are tight. For all $n \in \mathbb{N}$, let x_n be a random solution of* (12) *such that $\{D(x_n)\}_{n\in\mathbb{N}}$ is uniformly bounded. Then $\{D(x_n)\}_{n\in\mathbb{N}}$ is tight.*

It can be shown that every weak limit of probability distributions of random solutions $\{x_n\}_{n\in\mathbb{N}}$ of (8) is the distribution of an X-valued random variable on $(\Omega, \mathscr{K}, P)$ that is the so-called D-solution of (9) which was introduced in [143].

Some results about the convergence in distribution (i.e., $x_n \xrightarrow{D} x$) of approximate solutions of random operator equations are given in [143]. A sequence of X-valued random variables *converges in distribution to x* if $\{D(x_n)\}_{n\in\mathbb{N}}$ converges weakly to $D(x)$ (i.e., $\rightarrow \int_X f(s)\, dD(x_n)(s) \rightarrow \int_X f(s)\, dD(x)(s)$ for all continuous bounded real-valued functions f on X). More precisely, one considers a probability space $(\Omega, \mathscr{K}, P)$, a Polish space Z, a separable Banach space X and the mappings T, T_n, $n \in \mathbb{N}$, from $Z \times X$ into X that are measurable from the Borel-σ-algebra on $Z \times X$ into that on X. Furthermore, let $z : \Omega \rightarrow Z$ and $y : \Omega \rightarrow X$ be measurable with respect to the Borel-σ-algebras on Z and X, respectively. Consider the equation

(16) $x(\omega) = T(z(\omega), x(\omega)) + y(\omega), \quad \omega \in \Omega,$

with unknown $x : \Omega \rightarrow X$ and its approximations

(17) $x_n(\omega) = T_n(z_n(\omega), x_n(\omega)) + y_n(\omega), \quad \omega \in \Omega, \quad n \in \mathbb{N},$

where the random variables $\{z_n\}_{n\in\mathbb{N}}$ and $\{y_n\}_{n\in\mathbb{N}}$ converge in distribution to z and y, respectively. Then $\{T_n\}_{n\in\mathbb{N}}$ will be required to converge to T in a suitable way. Note that T and T_n are deterministic operators and hence the conditions regarding the way $\{T_n\}_{n\in\mathbb{N}}$ approximate T stem from the purely deterministic operator approximation theory. The randomness is present through the random variable z approximated by $\{z_n\}_{n\in\mathbb{N}}$ independently of the approximation of the operator. Note that since T and T_n are assumed to be jointly measurable, the operators $(\omega, x) \to T(z(\omega), x)$ and $(\omega, x) \to T_n(z_n(\omega), x)$ are random operators.

DEFINITION 6.2.2 [143]. We call x a "*D-solution*" of (16) if x is measurable and if there exist measurable maps $\bar{y} : \Omega \to X$ and $\bar{z} : \Omega \to Z$ with $D(\bar{z}) = D(z)$ and $D(\bar{y}) = D(y)$ such that

$$(18)\quad D(x) = D(T(\bar{z}, x) + \bar{y})$$

holds.

In this context, (16) need not hold almost surely; it is only required that both sides of (16) have the same distribution. This concept is related to Ershov's concept of a solution of a stochastic equation [145].

Note that the stochastic inputs z and y can be modified provided that their distributions remain unchanged. Note also that if $D(\bar{z}) = D(z)$, then $T(z, x)$ and $T(\bar{z}, x)$ need not have the same distribution [142, §5]. Hence, if one chooses $\{z_n\}_{n\in\mathbb{N}}$ such that $z_n \xrightarrow{D} z$, then, by definition, $z_n \xrightarrow{D} \bar{z}$ for all $\bar{z}$ with $D(\bar{z}) = D(z)$. However, since $T(\bar{z}, x)$ and $T(z, x)$ need not have the same distribution, one can not even expect that the limit x is a solution of the equation

$$(19)\quad D(x) = D(T(z, x) + y)$$

but only of (18) with suitable $\bar{z}$, $\bar{y}$. The original z, y are in no way distinguished from any other random variables with the same distribution.

Thus, one can conclude that if the concept of D-solution is used, then one can regard (16) in the following way: consider the distributions $\mathscr{Z}$ and $\mathscr{Y}$ given on Z and X. Then find the random variables $\bar{z}$, $\bar{y}$ and x such that $\bar{z}$ and $\bar{y}$ have the given distributions $\mathscr{Z}$ and $\mathscr{Y}$, respectively, and such that (18) holds. This is achieved by approximating $\mathscr{Z}$ and $\mathscr{Y}$ with the random variables $\{z_n\}_{n\in\mathbb{N}}$ and $\{y_n\}_{n\in\mathbb{N}}$ converging in distribution and by calculating the random solutions $\{x_n\}_{n\in\mathbb{N}}$ of the resulting Eq. (17). It can be proven that $(z_n, x_n, y_n)_{n\in\mathbb{N}}$ has a subsequence converging (jointly) in distribution to a $Z\times X\times X$-valued random variable (z, x, y) whose components will be such that (19) is fulfilled. More precisely, consider

THEOREM 6.2.12 [143, Th. 2.11]. *Let* T, $\{T_n\}_{n\in\mathbb{N}}$, z, $\{z_n\}_{n\in\mathbb{N}}$, y, $\{y_n\}_{n\in\mathbb{N}}$ *be as above and let* x_n *be a random solution of* (17) *for the index* n, *for all* $n \in \mathbb{N}$.

(a) *Assume that*

(20) $\{T_n(\bar{z},.)\}_{n\in\mathbb{N}}$ *is collectively compact for all* $\bar{z} \in Z$ (i.e., *for all bounded* $C \subseteq X$, $\bigcup_{n\in\mathbb{N}}\{T_n(\bar{z}, C)\}$ *is compact*;

(21) *for all bounded* $C \subseteq X$ *and compact* $K \subseteq Z$, $\{T_n(\cdot, x)\}_{\substack{n\in\mathbb{N}\\ x\in C}}$ *is equi-continuous on* K;

(22) $\{x_n\}_{n\in\mathbb{N}}$ *is D-bounded* (*i. e., for all* $\varepsilon > 0$, *there exists a bounded Borel set* $C_\varepsilon \subseteq X$ *such that* $D(x_n)(C_\varepsilon) > 1 - \varepsilon$ *holds for all* $n \in \mathbb{N}$) *and* $\{z_n\}_{n\in\mathbb{N}}$ *and* $\{y_n\}_{n\in\mathbb{N}}$ *are tight. Then* $\{x_n\}_{n\in\mathbb{N}}$ *is tight.*

(b) *Assume that* (20), (21) *and* (22) *hold, that* $z_n \xrightarrow{D} z$ *and* $y_n \xrightarrow{D} y$, *and that*

(23) $T_n \to T$ (*i. e., for all* $\{(z_n, x_n)\}_{n\in\mathbb{N}} \subset Z \times X$ *converging to* (z, x), $T_n(z_n, x_n) \to T(z, x)$).

Then, for every subsequence $\{x_{n_k}\}_{k\in\mathbb{N}}$ *of* $\{x_n\}_{n\in\mathbb{N}}$ (*especially for* $\{x_n\}_{n\in\mathbb{N}}$ *itself*), *there exists an X-valued random variable* x *which is a D-solution of* (16) *and for which* $x_{n_{k_j}} \xrightarrow{D} x$ *holds for a suitable subsequence* $\{x_{n_{k_j}}\}_{j\in\mathbb{N}}$ *of* $\{x_{n_k}\}_{k\in\mathbb{N}}$.

Furthermore, if all D-solutions of (16) *have the same distribution, then* $\{D(x_n)\}_{n\in\mathbb{N}}$ *converges weakly to this distribution.*

Remark. The D-solutions appear as limits (with respect to the convergence in distribution) of random solutions, which would also justify the use of the term *weak solution*, which is used in a similar context for stochastic differential equations (see [159], p. 357).

Some applications to a random equation under Peano conditions and to a random Hammerstein integral equation and its quadrature approximation are given in the last part of ref. [143].

The above results were extended [144] to cover the convergence in distribution of approximate solution of random operator equations and to obtain convergence results for approximate solutions of equations between spaces of probability measures. In order to achieve this, an approximation concept was developed [143] for equations in spaces of probability measures ("stochastic equations").

Consider a random equation of the form

(24) $T(z(\omega), x) = y(\omega)$,

and its approximations

(25) $T_n(z_n(\omega), x) = y_n(\omega), \quad \omega \in \Omega_n, \quad n \in \mathbb{N}$,

where T is a mapping from $Z \times X$ into Y, T_n from $Z \times X_n$ into Y, and X, X_n, Y, Z are metric spaces with $X_n \subseteq X$, whereas z, y and z_n, y_n, $n \in \mathbb{N}$, are random variables (defined on possibly different probability spaces $(\Omega, \mathscr{K}, P)$, $(\Omega_n, \mathscr{K}_n, P_n)$, $n \in \mathbb{N}$).

Assume that the joint probability distributions $D(y_n, z_n)$ converge weakly to $D(y, z)$ in the space of probability measures on $Y \times Z$ and look for the conditions on T and T_n, $n \in \mathbb{N}$, that imply weak convergence of $D(x_n)_{n\in\mathbb{N}}$ to a "solution" of (24). Here, $\{x_n\}_{n\in\mathbb{N}}$ is a sequence of X_n-valued random variables defined on $(\Omega_n, \mathscr{K}_n, P_n)$ that are almost surely solutions of (25). The weak limit of $\{D(x_n)\}_{n\in\mathbb{N}}$ is a probability measure defined on X which is a solution of (24) in the weak sense, i.e. in the sense of

DEFINITION 6.2.3. A probability measure $\mu_X \in \mathscr{M}(X)$ is called a *weak solution* of (24) if there exists a probability measure $\mu \in \mathscr{M}(Z \times X)$ such that

$$\mu T^{-1} = D(y)$$

and

$$\mu_X = \mu P_X^{-1}, \quad D(z) = \mu P_Z^{-1},$$

where P_X and P_Z are the coordinate projectors from $Z \times X$ onto X and Z, respectively.

Note that the coordinate projectors are Borel-measurable and that μP_X^{-1} and μP_Z^{-1} are the so-called marginals of $\mu \in \mathscr{M}(Z \times X)$.

This solution concept for (24) is justified by Ershov's concepts (see, e.g., [145]).

The distribution of a D-solution of (24) is a weak solution of (24) since $D(T(\bar{z}(.), x(.)) = D(\bar{z}, x)T^{-1}$ and $D(x)$ and $D(z) = D(\bar{z})$ are the marginals of $D(\bar{z}, x)$. However, a weak solution need not be the distribution of a D-solution. Hence, the weak solution concept is a strict generalization of the D-solution.

In order to examine a basic convergence result, let us now establish the terminology. For a metric space S, let $\mathscr{M}(S)$ be the set of all probability measures defined on $(S, \mathscr{B}_S)$ and equipped with the topology of weak convergence. If S_0 is a Borel subset of S, then $\mathscr{B}_{S_0} = \{B \cap S_0, \ B \in \mathscr{B}_S\}$ and μ is a probability measure on $(S_0, \mathscr{B}_{S_0})$, and $\mu^e \in \mathscr{M}(S)$, defined as $\mu^e(B) =$

$= \mu(B \cap S_0)$ for all $B \in \mathscr{B}_S$ and denoting the "canonical extension" of μ.

A sequence $\{\mu_n\}_{n\in\mathbb{N}} \subset \mathscr{M}(S)$ is called "w-bounded" if, for all $\varepsilon > 0$, there exists a bounded Borel set $B_\varepsilon \subseteq S$ such that $\inf_{n\in\mathbb{N}} \mu_n(B_\varepsilon) \geqslant 1 - \varepsilon$. If $\mu_n \in \mathscr{M}(S_n)$, $n \in \mathbb{N}$, then $\{\mu_n\}_{n\in\mathbb{N}}$ is called "w-bounded" whenever $\{\mu_n^e\}_{n\in\mathbb{N}}$ has this property in $\mathscr{M}(S)$.

Let S and S' be metric spaces, let $S_n \subseteq S$ for $n \in \mathbb{N}$ and let $T : S \to S'$, $T_n : S_n \to S'$ for $n \in \mathbb{N}$.

$\{T_n\}_{n\in\mathbb{N}}$ is said to "converge discretely" to T if

(a) $d(s, S_n) = \inf\{d(s, \bar{s}) : \bar{s} \in S_n\} \to 0$, for all $s \in S$, where d is the metric in S,

and

(b) for all $s \in S$, $s_n \in S_n$, $n \in \mathbb{N}$, with $s_n \to s$ in S,

we have

$T_n s_n \to Ts$ in S'.

Also, T is called "regular" if $T^{-1}(K) \cap B$ is relatively compact in S for each bounded $B \subseteq S$ and compact $K \subseteq S'$.

$\{T_n\}_{n\in\mathbb{N}}$ is called "A-regular" if for all $n_1 < n_2 < n_3 < \ldots$ $\ldots \in \mathbb{N}$ and all $s_{n_k} \in S_{n_k}$, $k \in \mathbb{N}$, the sequence $\{s_{n_k}\}_{k\in\mathbb{N}}$ is relatively compact in S if $\{s_{n_k}\}_{k\in\mathbb{N}}$ is bounded in S and if $\{T_{n_k}s_{n_k}\}_{k\in\mathbb{N}}$ is relatively compact in S'.

Equation (24) is transformed into the so-called "standard stochastic equation" using Ershov's idea [146, p. 606], i.e., considering the induced stochastic equation

(26) $\tilde{T} : Z \times X \to Y \times Z$, $\tilde{T}(z, x) = (T(z, x), x)$,

$(z, x) \in Z \times X$

(27) $\mu\tilde{T}^{-1} = D(y, z)$.

The mapping T is Borel-measurable from $\mathscr{B}_{Z\times X}$ to $\mathscr{B}_{Y\times Z}$ and T is injective if the mapping $T(z, .) : X \to Y$ is injective for each $z \in Z$.

If $\mu \in \mathscr{M}(Z \times X)$ is a solution of (27), then μP_X^{-1} is a weak solution of (24).

Now consider that, for all $n \in \mathbb{N}$, Borel subsets X_n of X, the mappings $T_n : Z \times X_n \to Y$ that are Borel-measurable from $\mathscr{B}_{Z\times X_n}$ to B_Y and the random variables y_n and z_n with values in Y and Z, respectively (defined on some probability space $(\Omega_n, \mathscr{K}_n, P_n)$) are given. Consider also the "approximate" random operator equations (25) and define mappings $\tilde{T} : Z \times$ $\times X_n \to Y \times Z$, $n \in \mathbb{N}$, as in (26). Assume that the random solutions $x_n : \Omega_n \to X_n$ of the approximate equations (26) are given for all $n \in \mathbb{N}$. Then one is interested to find sufficient conditions on T and T_n that imply the weak convergence of the sequence $\{D(x_n)\}_{n\in\mathbb{N}}$ to a weak solution of (24) if $\{D(y_n, z_n)\}_{n\in\mathbb{N}}$ converges weakly to $D(y, z)$. We have

THEOREM 6.2.13 [144, Th. 4.6]. *Let* T, $\{T_n\}$, z, $\{z_n\}$, y, $\{y_n\}$, *be as above and let* x_n *be a random solution of* (25) *for all* $n \in \mathbb{N}$. *Assume that*

(28) *for all* $z \in Z$, *each* $T_n(z, .)$ *is regular and* $\{T_n(z, .)\}_{n\in\mathbb{N}}$ *is* A-*regular*;

(29) *for all bounded* $B \subset X$ *and compact* $K \subset Z$, $\{T_n(., x) : x \in B \cap X_n,\ n \in \mathbb{N}\}$ *is equicontinuous on* K;

(30) $\{T_n\}_{n\in\mathbb{N}}$ *converges discretely to* T *(jointly in both variables)*;

(31) $\{D(y_n, z_n)\}_{n\in\mathbb{N}}$ *converges weakly to* $D(y, z)$;

(32) $\{D(x_n)\}_{n\in\mathbb{N}}$ *is* w-*bounded.*

Then $\{D(x_n)\}_{n\in\mathbb{N}}$ is discretely w-compact (i.e., every subsequence of a sequence contains a further subsequence that is w-convergent), and every limit of a weakly convergent subsequence is a weak solution of (24).

If, furthermore, $T(\tilde{z}, .)$ is injective for all $\tilde{z} \in Z$, then $\{D(x_n)\}_{n\in\mathbb{N}}$ converges weakly to a weak solution of (24).

The applications of this result to random Fredholm integral equations of the second kind and to a random nonlinear elliptic boundary-value problem (a Galerkin scheme) can be found in the last part of ref. [144].

6.2.3. *Approximate Solution of Random Nonlinear Equations by Random Contractors*

The concept of contractors was introduced by M. Altman [11] as a tool for solving deterministic operator equations in Banach spaces. The natural random contractor of a random nonlinear operator and its use in obtaining existence, uniqueness and approximation of random solutions to random nonlinear operator equations has been introduced by Lee-Padgett [277], [278], [279], [334], [335]. In the following lines, we shall present some of their definitions and properties and then illustrate their applicability to some classes of random equations.

Let X and Y be separable Banach spaces and let $T : \Omega \times \mathscr{D}(T) \to Y$ be a nonlinear random operator. Let $\Gamma(x, .) : \Omega \times Y \to X$ be a bounded linear operator associated with $x \in X$.

DEFINITION 6.2.4. We say that $T(\omega)$ has a *random contractor* $\Gamma(x, .)$ at $x \in \mathscr{D}(T) \subset X$ if there exists a positive random variable $q(\omega)$, $0 < q(\omega) < 1$ a.s., and a constant $s > 0$ such that

(1) $\| T(\omega)[x + \Gamma(x, \omega)y] - T(\omega)x - y \| \leqslant q(\omega) \|y\|$ a.s.,

where $y \in Y$ and $\|y\| \leqslant s$.

From this definition, it follows that

$$\|y\| - \| T(\omega)[x + \Gamma(x, \omega)y] - T(\omega)x \| \leqslant$$
$$\leqslant \| T(\omega)[x + \Gamma(x, \omega)y] - T(\omega)x - y \| \leqslant$$
$$\leqslant q(\omega) \|y\| \text{ a.s.}$$

and, thus,

$$\|1 - q(\omega)y\| \leqslant \| T(\omega)[x + \Gamma(x, \omega)y] - T(\omega)x \| \text{ a.s.}$$

Hence, the random contractor $\Gamma(x, \omega)$ is a one-to-one mapping.

Let $T : \Omega \times \mathscr{D}(T) \subseteq \Omega \times X \to Y$ be a nonlinear random operator having a random contractor $\Gamma(x, .)$ in a neighbourhood $S(x_0, r) = \{x \in \mathscr{D}(T) \subset X : \|x - x_0\| \leqslant r\}$, where $x_0(\omega) = x_0$ a.s. is a given approximation solution of the random operator equation

(2) $T(\omega)x = 0.$

To solve (2) we consider the iteration procedure

(3) $x_{n+1}(\omega) = x_n(\omega) - \Gamma(x_n(\omega), \omega)T(\omega)x_n(\omega), \quad n = 0,1,2,\ldots$

and then we have

THEOREM 6.2.14 [277]. *Suppose that there exist two positive random variables $q(w)$, $0 < q(\omega) < 1$ a.s., and $b(\omega)$, and two positive numbers r and s such that the following conditions hold:*

(i) $\|\Gamma(x, \omega)\| \leqslant b(\omega)$ *a.s. for* $x \in S(x_0; r)$;

(ii) $\|T(\omega)x_0\| \leqslant s$ *a.s.*;

(iii) $b(\omega)s(1 - q(\omega))^{-1} \leqslant r$ *a.s.*;

and

(iv) $T(\omega)$ *is an a.s. closed operator on* $S(x_0; r)$.

Then there exists a random solution $\bar{x}(\omega) \in S(x_0; r)$ of equation (2), *and the sequence of X-valued random variables $x_n(\omega)$ defined by* (3) *converges to $\bar{x}(\omega)$ a.s., provided that $T(\omega)x(\omega)$ is a Y-valued random variable with $x(\omega)$ an X-valued random variable.*

The random solution $\bar{x}$ is unique in $S(x_0; r)$ if, in addition, condition (1) is satisfied for all $y \in Y$ such that $\|\Gamma(x, \omega)\| \leqslant 2r$ a.s. and $\Gamma(x, \omega)$ is onto, where $x \in S(x_0 ;r)$.

A random operator T is said to have : (i) a *bounded random contractor* $\Gamma(x,.)$ if there exists a positive random variable $b(\omega)$ such that $\| \Gamma(x, \omega) \| \leqslant b(\omega)$ a.s. for all $x \in \mathscr{D}(T)$; (ii) a *uniform random contractor* at $x_0 \in \mathscr{D}(T) \subset X$ if

$$\| T(\omega)[x + \Gamma(x_0, \omega)y] - T(\omega)x - y \| \leqslant q(\omega) \|y\| \text{ a.s.}$$

for all $x \in S(x_0 ; r)$ and $\|y\| \leqslant s$.

These ideas, together with the following modification of the iteration procedure (3) :

$$x_{n+1}(\omega) = x_n(\omega) - \Gamma(x_0, \omega)\, T(\omega)x_n(\omega), \quad n = 0, 1, 2, \ldots,$$

yields that the above theorem remains true if we replace the random contractor by the uniform random contractor $\Gamma(x_0, \omega)$.

Considering the random operator equation

$$(4) \quad T(\omega)x = \xi(\omega)$$

and assuming that the nonlinear random operator T has a random contractor $\Gamma(x, \omega)$ in some certain region, it results that the random operator $T(\omega)x - \xi(\omega)$ will have the same random contractor $\Gamma(x, \omega)$.

If we now suppose that an a.s. closed random operator T has a bounded random contractor $\Gamma(x, \omega)$ such that condition (1) holds for all $y \in Y$ and $\| \Gamma(x, \omega) \| \leqslant b(\omega)$ a.s. for all $x \in \mathscr{D}(T)$, then the random operator equation (4) has a random solution. The random solution is unique if the random contractor $\Gamma(x, \omega)$ is onto.

A random contractor $\Gamma(x, \omega)$ of a nonlinear random operator T is said to be *regular* at $x_0 \in D(T) \subset X$ if

(i) $\| T(\omega)[x_0 + \Gamma(x_0, \omega)y] - T(\omega)x_0 - y \| \leqslant q(\omega) \|y\|$ a.s. for all $y \in Y$

(ii) $\mathscr{D}(T) = \Gamma(x_0, .)Y$ a.s.

If the nonlinear random operator T has a regular random contractor, then there exists a unique random solution of the random operator equation (2).

A more general type of random contractor is given by

DEFINITION 6.2.5 [278]. The random operator $T : \Omega \times \mathscr{D}(T) \subset \Omega \times X \to Y$ is said to have a *random contractor* $\Gamma(x, \omega)$ at $x \in \mathscr{D}(T)$ *with random nonlinear majorant* $Q(., \omega)$ if, for $x \in \mathscr{D}(T)$ and $y, z \in Y$ with $x + \Gamma(x, \omega)(y - z) \in \mathscr{D}(T)$ a.s., the following inequality is satisfied :

$$(5) \quad \| T(\omega)[x - \Gamma(x, \omega)(y - z)] - T(\omega)\, x - (y - z) \| \leqslant$$
$$\leqslant Q(\|y - z\|, \omega) \text{ a.s.},$$

where $Q(.,\omega)$ is an a.s. nondecreasing continuous (real-valued) random function satisfying $Q(t,\omega) \geqslant 0$ a.s. for $t \geqslant 0$ and $Q(0,\omega) = 0$ a.s. and $\Gamma(x,.) : \Omega \times Y \to X$ is a bounded linear random operator associated with $x \in X$.

Inequality (5) can be replaced by

$$(5') \quad \| T(\omega)[x + \Gamma(x,\omega)y] - T(\omega)x - y \| \leqslant Q(\|y\|,\omega) \text{ a.s.}$$

for all $x \in \mathscr{D}(T)$ and y such that $x + \Gamma(x,\omega)y \in \mathscr{D}(T)$ a.s.

We assume that $T(\omega)$ has a contractor $\Gamma(x,\omega)$ with the random nonlinear majorant $Q(.,\omega)$ in a neighbourhood $S(x_0, r) = \{x \in \mathscr{D}(T) : \|x - x_0\| \leqslant r\}$, where $x_0(\omega) = x_0$ a.s. is a given approximate solution of (2). Let us define the iteration procedure

$$(6) \quad x_{n+1}(\omega) = x_n(\omega) - \Gamma(x_n(\omega),\omega)\,[T(\omega)x_n(\omega)],$$

$$n = 0,1,2,\ldots$$

$$(7) \quad t_{n+1}(\omega) = t_n(\omega) + Q(t_n(\omega) - t_{n-1}(\omega),\omega), \quad t_0(\omega) = 0 \text{ a.s.},$$

$$n = 1,2,\ldots$$

and assume that the sequence of random variables $\{t_n(\omega)\}$ converges a.s. to some random variable $\bar{t}(\omega) > 0$ a.s.

THEOREM 6.2.15 [278]. *Suppose that the positive random variables $b(\omega)$ and $t_1(\omega)$ exist such that the following conditions hold:*

(i) $\|T(\omega, x_0)\| \leqslant t_1(\omega)$ *a.s.;*

(ii) $\|\Gamma(x,\omega)\| \leqslant b(\omega)$ *a.s. for* $x \in S(x_0, b(\omega)\bar{t}(\omega))$, *where* $b(\omega)\bar{t}(\omega) \leqslant r$ *a.s.;*

(iii) $T(\omega)$ *is an a.s. closed random operator on* $S(x_0, b(\omega)\bar{t}(\omega))$. *Then there exists a random solution* $\bar{x}(\omega)$ *in* $S(x_0, b(\omega)\bar{t}(\omega))$ *a.s. for equation* (2) *and the sequence of* X*-valued random variables defined by* (6) *converges a.s. to* $\bar{x}(\omega)$, *provided that* $T(\omega)x(\omega)$ *is a* Y*-valued random variable whenever* $x(\omega)$ *is an* X*-valued random variable. Further, for every* $n = 0,1,2,\ldots,$

$$\|\bar{x}(\omega) - x_n(\omega)\| \leqslant b(\omega)\,[\bar{t}(\omega) - t_n(\omega)] \text{ a.s.}$$

The random solution $\bar{x}(\omega)$ *is unique if, in addition,* $\Gamma(x,\omega)$ *is onto for* $x \in S(x_0, b(\omega)\bar{x}(\omega))$ *a.s. and* $Q(t,\omega) < t$ *a.s. for all* $t > 0$.

A uniform random contractor at $x_0 \in \mathscr{D}(T)$ with *random nonlinear majorant* $Q(.,\omega)$ can also be defined and then the above result remains true.

THEOREM 6.2.16. *Suppose that the a.s. closed random nonlinear operator* T *has a random contractor* $\Gamma(x,\omega)$ *with random nonlinear majorant function* $Q(.,\omega)$ *such that* (5′) *holds for all* $y \in Y$, *where*

$Q(0, \omega) = 0$ a.s., $Q(t, \omega) < t$ a.s. for all $t > 0$ and $\|\Gamma(x, \omega)\| \leqslant b(\omega)$ for all $x \in \mathscr{D}(T)$. Then the random operator equation $T(\omega)x = u(\omega)$ has a random solution, and the random solution is unique if $\Gamma(x, \omega)$ is onto.

Let us now consider the special case of random quadratic majorant functions $Q(t, \omega) = a(\omega)t^2 + b(\omega)t$, where $a(\omega) > 0$ a.s. and $b(\omega) \geqslant 0$ a.s. Then the random contractor inequality becomes

$$(8) \quad \|T(\omega)[x + \Gamma(x, \omega)y] - T(\omega)x - y\| \leqslant a(\omega)\|y\|^2 + b(\omega)\|y\| \text{ a.s.,}$$

whenever $x \in \mathscr{D}(T)$ and $x + \Gamma(x, \omega) \in \mathscr{D}(T)$ a.s. Let $t_0(\omega) = \eta > 0$ a.s. and let us define as before

$$t_{n+1}(\omega) = t_n(\omega) + Q(t_n(\omega) - t_{n-1}(\omega), \omega), \quad n = 1, 2, \ldots .$$

THEOREM 6.2.17. *Suppose that $x_0(\omega) = x_0 \in \mathscr{D}(T)$, $S(x_0, r) \subset \mathscr{D}(T)$, $\|T(\omega)x_0\| \leqslant \eta$ a.s., and $q(\omega) = a(\omega)\eta + b(\omega) < 1$ a.s. If $T(\omega)$ has a random contractor $\Gamma(x, \omega)$ satisfying (8) for $x \in S(x_0, r)$ and if $\|\Gamma(x, \omega)\| \leqslant b(\omega)$ a.s., then all $x_n(\omega)$ defined by the iteration procedure (6) lie in $S(x_0, r)$ a.s. and the sequence of random variables $\{x_n(\omega)\}_{n \in \mathbb{N}}$ converges to a random solution $\bar{x}(\omega)$ of $T(\omega)x = 0$ and the error estimate*

$$\|\bar{x}(\omega) - x_n(\omega)\| \leqslant b(\omega)\eta q^n(\omega)\,[1 - q(\omega)]^{-1}, \quad n = 0, 1, 2, \ldots$$

holds, provided $t_n(\omega) \to \bar{t}(\omega) = \eta[1 - q(\omega)]^{-1}$ a.s., $b(\omega)\bar{t}(\omega) \leqslant r$ a.s., and $T(\omega)x(\omega)$ is a Y-valued random variable whenever $x(\omega)$ is an X-valued random variable.

Lee and Padgett introduced [279] more general random contractors by allowing different random possible contractors at each iteration step of the approximating procedure.

DEFINITION 6.2.6. Let $T : \Omega \times \mathscr{D}(T) \to \Omega \times X \to Y$ be a random nonlinear operator and let $Q_n(t, \omega) \geqslant 0$ a.s. for $t \geqslant 0$ be a continuous real-valued random function for each $n = 0, 1, \ldots$. A sequence of random operators $\Gamma_n(\omega, .) : \Omega \times S \to \mathscr{L}(X, Y)$ ($S = S(x_0, r) = \{x \in X : \|x - x_0\| \leqslant r$, for $r > 0\}$), is said to be *a random step-contractor* (*or random s-contractor*) for the random operator T with random s-majorant $\{Q_n(\,.\,, \omega)\}$ if the inequalities

$$(9) \quad \|T(\omega)[x + \Gamma_n(\omega, x)y] - T(\omega)(x) - y\| \leqslant Q_n(\|y\|, \omega)$$

are satisfied for $x \in S$, $y \in Y$ whenever $x + \Gamma_n(\omega, x)y \in S$ a.s., $n = 0, 1, 2, \ldots$. If $Q_n(.,\omega) = Q(., \omega)$ for all $n = 0, 1, 2, \ldots$, then we say that $\Gamma_n(\omega, .)$ is a random s-contractor for T with uniform random majorant T.

A random s-contractor $\{\Gamma_n(\omega, .)\}$ for T is said to be bounded if there exists a set of random variables $\{b_n(\omega)\}$ such that $\|\Gamma_n(\omega, x)\| \leqslant b_n(\omega)$ a.s. for $x \in S$, $n = 0,1,2, \ldots$.

In order to solve Eq. (2), let us consider the iterative procedure

$$(10)\quad x_{n+1}(\omega) = x_n(\omega) - \Gamma_n(\omega, x_n(\omega))\,[T(\omega)x_n(\omega)],$$
$$n = 0,1,2, \ldots,$$

where $\{\Gamma_n(\omega,.)\}$ is a random s-contractor with a random linear s-majorant $Q_n(.,\omega)$ given by

$$(11)\quad Q_n(t, \omega) = q_n(\omega)t,\ 0 < q_n(\omega) < 1 \text{ a.s.},\ n = 0,1,2, \ldots.$$

We have

THEOREM 6.2.18 [279]. *Suppose that* $T: \Omega \times S \to Y$ *is a random nonlinear closed operator and that* $\{\Gamma_n(\omega, x)\}$ *is a bounded random s-contractor for* $T(\omega)$ *satisfying* (9) *with* $\{Q_n(\cdot, \omega)\}$ *given by* (11). *Let* $x_0 \in X$ *be chosen such that* $\|T(\omega)x_0\| \leqslant \eta$ *a.s. and suppose that*

$$\tilde{q}_n(\omega) = \prod_{i=0}^{n} q_i(\omega)$$

and $b_n(\omega)$ *is such that*

$$\lim_{n\to\infty} q_n(\omega) = 0 \ \text{ a.s. and }\ \bar{t}(\omega) = \sum_{n=0}^{\infty} b_n(\omega)\tilde{q}_{n-1}(\omega),$$

with $\tilde{q}_{n-1}(\omega) = 1$ *a.s. If* $t(\omega)\eta \leqslant r$ *a.s., then all* $x_n(\omega)$ *lie in* S *and the sequence* $\{x_n(\omega)\}$ *defined by* (10) *converges a.s. to a random solution* $\bar{x}(\omega)$ *of* (2), *provided that* $T(\omega)x(\omega)$ *is a* Y*-valued random variable whenever* $x(\omega)$ *is an* X*-valued random variable. Further, the following error bound holds:*

$$\|\tilde{x}(\omega) - x_n(\omega)\| \leqslant \eta \sum_{i=n}^{\infty} b_i(\omega)\tilde{q}_{i-1}(\omega),\ n = 0,1,2, \ldots.$$

The random solution $x(\omega)$ *is unique in* $S(x_0, r)$ *if, in addition, there exists an integer* $m > 0$ *such that*

$$\|T(\omega)[x + \Gamma_m(\omega, x)y] - T(\omega)x - y\| \leqslant q_m(\omega)\|y\| \ \text{a.s.}$$

holds for all $y \in Y$ *with* $\|\Gamma_m(\omega, x)\| \leqslant 2r$ *a.s. and* $\Gamma_m(\omega, x)$ *is onto when* $x \in S(x_0, r)$.

A similar result is true if the random s-contractor is replaced by a uniform random s-contractor.

For the case of a random nonlinear s-majorant, suppose that $Q_n(t, \omega) \geqslant 0$ a.s. for $t \geqslant 0$, $Q_n(0, \omega) = 0$ a.s. and $Q_n(t, \omega)$ is a.s.

a nondecreasing function, $n = 0, 1, 2, \ldots$ and consider the iterative procedure

$$(12)\quad s_{n+1}(\omega) = s_n(\omega) + b_n(\omega)[(s_n(\omega) - s_{n-1}(\omega))/b_{n-1}(\omega), \omega],$$

for $n = 1, 2, \ldots$, $s_0(\omega) = 0$ a.s., $s_1(\omega) = 0$ a.s., and $s_2(\omega) = = \eta b_1(\omega)$ a.s.

Suppose further that $s_n(\omega)$ converges a.s. and denote

$$\lim_{n\to\infty} s_n(\omega) = \bar{s}(\omega) \text{ a.s.}$$

Then the following is true:

THEOREM 6.2.19. *Assume that $T: \Omega \times S \to Y$ is a random closed nonlinear operator and that $\{\Gamma_n(\omega, .)\}$ is a bounded random s-contractor for T satisfying* (9) *with $\{Q_n(., \omega)\}$ satisfying the above assumptions. Assume that x_0 is such that $\| T(\omega)x_0 \| \leqslant \eta$. Assume also that the sequence $\{s_n(\omega)\}$ defined by* (12) *is a.s. convergent. Then all $x_n(\omega)$ defined by* (10) *lie in $S(x_0, r)$ and the sequence $\{x_n(\omega)\}$ converges a.s. to a random solution $\bar{x}(\omega)$ of* (2), *provided that $T(\omega)x(\omega)$ is a Y-valued random variable whenever $x(\omega)$ is an X-valued random variable. Also, the following error estimate holds:*

$$\| \tilde{x}(\omega) - x_n(\omega) \| \leqslant \bar{s}(\omega) - s_n(\omega) \; a.s., \quad n = 0, 1, 2, \ldots .$$

The random solution $\bar{x}(\omega)$ is unique in $S(x_0, r)$ if there exists an integer $m > 0$ such that (9) *holds with $n = m$ for all $y \in Y$ and $\Gamma_m(\omega, x)$ is onto, where $x \in S(x_0, r)$, with $Q_m(t, \omega) < t$ a.s. for $t > 0$.*

Interesting applications of the above results to some classes of random nonlinear integral equations are given in refs. [277], [278], [279], [281].

6.2.4. *On a General Convergence Theory for Solutions of Approximations*

H. Engl and W. Römisch [142] furnished a general theory for the convergence of solutions of approximations for random operator equations to solutions of the original equation, which is not necessarily uniquely solvable. Concepts like consistency, stability and compactness in sets of measurable functions are introduced and used. Their abstract results are illustrated by applications to convergence in probability, almost-sure and almost-uniform convergence. As a specific method for constructing approximate equations, they discussed discretization schemes, where the underlying probability space is discretized. Hence, the general concept based on Stummel's approach to approximate solutions of random equations ([365], [366]) is developed so as to allow the treatment of non-uniquely solvable equations involving random operators on stochastic domains.

Stummel's convergence concept is modified so as to include the special modes of convergence of random variables (for example, the conditions of [421] are not fulfilled for the almost-sure convergence).

Let $(\Omega, \mathscr{K}, P)$ be a complete probability space and let X, Y be Polish spaces. Let $\mathscr{V}_X^{\mathbb{N}}$ denote the set of all sequences in $\mathscr{V}_X$ and let c_X be a multivalued mapping from a subset $D(c_X) \subset \mathscr{V}_X^{\mathbb{N}}$ to $\mathscr{V}_X$.

DEFINITION 6.2.7. A multivalued mapping $c_X : D(c_X) \subset \mathscr{V}_X^{\mathbb{N}} \to \mathscr{V}_X$ will be called a *convergence on* $\mathscr{V}_X$ if it has the following properties :

(1) if $\{x_n\}_{n \in \mathbb{N}} = \{x, x, x, \ldots\}$ with $x \in \mathscr{V}_X$ is a constant sequence, then $\{x_n\}_{n \in \mathbb{N}} \in D(c_X)$ and $x = c_X\text{-}\lim\limits_{n \in \mathbb{N}} x_X$ (i.e., $x \in c_X((x_n)))$;

(2) if $\{x_n\}_{n \in \mathbb{N}} \in D(c_X)$ and $x = c_X\text{-}\lim\limits_{n \in \mathbb{N}} x_n$ and $\{x_{n_k}\}_{k \in \mathbb{N}}$ is a subsequence of $\{x_n\}_{n \in \mathbb{N}}$, then $\{x_{n_k}\}_{k \in \mathbb{N}} \in D(c_X)$ and $x = c_X\text{-}\lim\limits_{k \in \mathbb{N}} x_{n_k}$;

(3) let K, M be two disjoint infinite subsets of $\mathbb{N}$ with $K \cup M = \mathbb{N}$, $\{x_k\}_{k \in K} \in D(c_X)$, $\{\bar{x}_m\}_{m \in M} \in D(c_X)$, let $x = c_X\text{-}\lim\limits_{k \in K} x_k$ and $x = c_X\text{-}\lim\limits_{m \in M} \bar{x}_m$(i.e. $x \in c_X((x_k)) \cap c_X((\bar{x}_m)))$; for all $n \in \mathbb{N}$, let

$$\tilde{x}_n = \begin{cases} x_n & \text{if } n \in K \\ \bar{x}_n & \text{if } n \in M. \end{cases}$$

Then $(\tilde{x}_n) \in D(c_X)$ and $x = c_X\text{-}\lim\limits_{n \in \mathbb{N}} \tilde{x}_n$;

(4) if $x = c_X\text{-}\lim\limits_{n \in \mathbb{N}} x_n$, then $\bar{x} = c_X\text{-}\lim\limits_{n \in \mathbb{N}} x_n$ iff $x = \bar{x}$ a.s.; furthermore, if $\{\bar{x}_n\}_{n \in \mathbb{N}}$ is such that for all $n \in \mathbb{N}$, $\bar{x}_n = x_n$, then $(\bar{x}_n) \in D(c_X)$ and $x = c_X\text{-}\lim\limits_{n \in \mathbb{N}} \bar{x}_n$.

A sequence in $D(c_X)$ will be called c_X-*convergent*, and any element of $c_X(\{x_n\}_{n \in \mathbb{N}})$ will be called a c_X-*limit* of $\{x_n\}_{n \in \mathbb{N}}$.

Let us now consider the convergence of sets of random variables in the sense of Kuratowski.

DEFINITION 6.2.8. Let $\{Z_n\}_{n\in\mathbb{N}}$ be a sequence of nonempty subsets of $\mathscr{V}_X$.

(a) $c_X\text{-}\liminf_{n\in\mathbb{N}} Z_n = \bigcup_{\{x_n\}\in\{\{x_n\}\in\mathscr{V}_X^{\mathbb{N}}:\, x_n\in Z_n \text{ for all } n\in\mathbb{N}\}} c_X(\{x_n\})$

(b) $c_X\text{-}\limsup_{n\in\mathbb{N}} Z_n = \bigcup_{\{x_k\}\in\{\{x_k\}\in\mathscr{V}_X^{\mathbb{N}}:\, \exists\, n_1<n_2<\cdots,\ x_k\in Z_{n_k},\ \forall k\in\mathbb{N}\}} c_X(x_k)$

(c) $\{Z_n\}_{n\in\mathbb{N}}$ is called *c_X-convergent to $z\in\mathscr{V}_X$*, if $c_X\text{-}\limsup_{n\in\mathbb{N}} Z_n = c_X\text{-}\liminf_{n\in\mathbb{N}} Z_n = Z$. We then write $Z = c_X\text{-}\lim_{n\in\mathbb{N}} Z_n$;

(d) $\{Z_n\}_{n\in\mathbb{N}}$ is called *c_X-compact*, if for all $\{x_k\}\in\mathscr{V}_X^{\mathbb{N}}$ with $x_k\in Z_{n_k}$ for all $k\in\mathbb{N}$, where $n_1<n_2<n_3<\ldots$ are arbitrary integers, we have $c_X\text{-}\limsup_{k\in\mathbb{N}}\{x_k\}\neq\emptyset$.

Now, let C, C_n, $n\in\mathbb{N}$, be multivalued mappings from Ω into X and let $T: \mathrm{Gr}C\to Y$, $T_n: \mathrm{Gr}C_n\to Y$, $n\in\mathbb{N}$, be $\mathscr{K}\times\mathscr{B}_X$-measurable. Let

$$\mathscr{S}(C) = \{x:\Omega\to X: x \text{ measurable, } x(\omega)\in C(\omega) \text{ a.s.}\},$$

i.e., $\mathscr{S}(C)$ is the set of all random variables which are a.s. selectors of C. Then $T(\omega, x(\omega))$ need not be defined for all $\omega\in\Omega$, if $x\in\mathscr{S}(C)$. By $T(.,x(.))$ denote any $z\in\mathscr{V}_Y$ such that $z = T(.,x(.))$ a.s., where $\bar{x}=x$ a.s. and $\bar{x}(\omega)\in C(\omega)$ for all $\omega\in\Omega$; $T_n(.,x_n(.))$ with $x_n\in\mathscr{S}(C_n)$ are similarly defined. Thus, $T(.,x(.))$ is a set of random variables, any two of which are almost surely equal. If we use an expression like "$c_Y\text{-}\lim_{n\in\mathbb{N}} T_n(.,x_n(.)) = T(.,x(.))$" we mean "$c_Y\text{-}\lim_{n\in\mathbb{N}} Z_n = Z$" where Z_n and Z represent $T_n(.,x_n(.))$ and $T(.,x(.))$, respectively.

DEFINITION 6.2.9. (a) $\{T_n\}_{n\in\mathbb{N}}$ will be called "*(c_X, c_Y)-convergent to T*" if for all $\{x_n\}_{n\in\mathbb{N}}\subseteq\mathscr{V}_X^{\mathbb{N}}$ with $x_n\in\mathscr{S}(C_n)$ for all $n\in\mathbb{N}$ and all $x\in\mathscr{S}(C)$ with $x = c_X\text{-}\lim_{n\in\mathbb{N}} x_n$, we have $T(.,x(.)) = c_Y\text{-}\lim_{n\in\mathbb{N}} T_n(.,x_n(.))$.

(b) T and $\{T_n\}_{n\in\mathbb{N}}$ are called "*(c_X, c_Y)-consistent*" if for all $x\in\mathscr{S}(C)$ there exists a sequence $\{x_n\}_{n\in\mathbb{N}}$ with $x_n\in\mathscr{S}(C_n')$ for all $n\in\mathbb{N}$ such that $x = c_X\text{-}\lim_{n\in\mathbb{N}} x_n$ and $T(.,x(.)) = c_Y\text{-}\lim_{n\in\mathbb{N}} T_n(.,x_n(.))$.

(c) $\{T_n\}_{n\in\mathbb{N}}$ is called "*(c_X, c_Y)-stable*" if for all c_X-convergent sequences $\{x_n\}_{n\in\mathbb{N}}$, $\{\bar{x}_n\}_{n\in\mathbb{N}}$ with $x_n\in\mathscr{S}(C_n)$ and $\bar{x}_n\in\mathscr{S}(C_n)$ for all $n\in\mathbb{N}$, we have: If $\{T_n(.,x_n(.))\}_{n\in\mathbb{N}}$ is c_Y-convergent and $c_X\text{-}\lim_{n\in\mathbb{N}} x_n = c_X\text{-}\lim_{n\in\mathbb{N}} \bar{x}_n$ (i.e., $c_X(\{x_n\}) = c_X(\{\bar{x}_n\})$), then $\{T_n(\cdot, \bar{x}_n(\cdot))\}_{n\in\mathbb{N}}$ is c_Y-convergent and $c_Y\text{-}\lim_{n\in\mathbb{N}} T_n(\cdot, x_n(\cdot)) = c_Y\text{-}\lim_{n\in\mathbb{N}} T_n(\cdot, \bar{x}_n(\cdot))$.

(d) $\{T_n\}_{n\in\mathbb{N}}$ is called "(c_X, c_Y)-*inversely stable*" if, for all sequences $\{x_n\}_{n\in\mathbb{N}}$, $\{\bar{x}_n\}_{n\in\mathbb{N}}$ with $x_n \in \mathscr{S}(C_n)$ and $\bar{x}_n \in \mathscr{S}(C_n)$ for all $n \in \mathbb{N}$, we have: If $\{x_n\}_{n\in\mathbb{N}}$ is c_X-convergent and $\{T_n(\cdot, x_n(\cdot))\}_{n\in\mathbb{N}}$ and $\{T_n(\cdot, \bar{x}_n(\cdot))\}_{n\in\mathbb{N}}$ are c_Y-convergent to the same limit (set), then $\{\bar{x}_n\}_{n\in\mathbb{N}}$ is c_X-convergent and $c_X\text{-}\lim_{n\in\mathbb{N}} \bar{x}_n = c_X\text{-}\lim_{n\in\mathbb{N}} x_n$.

DEFINITION 6.2.10. (a) $\{T_n\}_{n\in\mathbb{N}}$ is called "c_Y-*compact*" if, for all $\{x_n\}_{n\in\mathbb{N}} \subset \mathscr{V}_X^{\mathbb{N}}$ with $x_n \in \mathscr{S}(C_n)$ for all $n \in \mathbb{N}$, $\{T_n(\cdot, x_n(\cdot))_{n\in\mathbb{N}}$ is compact in the sense of Definition 6.2.8 (d).

(b) T and $\{T_n\}_{n\in\mathbb{N}}$ are called "(c_X, c_Y)-*closed*" if, for all $n_1 < n_2 < \ldots$ and sequences $\{x_{n_k}\}_{k\in\mathbb{N}}$ with $x_{n_k} \in \mathscr{S}(C_{n_k})$ for all $k \in \mathbb{N}$, we have: If $x = c_X\text{-}\lim_{k\in\mathbb{N}} x_{n_k}$ and $y = c_Y\text{-}\lim_{k\in\mathbb{N}} T_{n_k}(\cdot, x_{n_k}(\cdot))$, then $x \in \mathscr{S}(C)$ and $T(\cdot, x(\cdot)) = y$ a.s.

(c) T and $\{T_n\}_{n\in\mathbb{N}}$ are called "(c_X, c_Y)-*regular*" if, for all $n_1 < n_2 < n_3 \ldots \in \mathbb{N}$ and all sequences $\{x_{n_k}\}_{k\in\mathbb{N}}$ with $x_{n_k} \in \mathscr{S}(C_{n_k})$ for all $k \in \mathbb{N}$, we have: If $y = c_Y\text{-}\lim_{k\in\mathbb{N}} T_{n_k}(\cdot, x_{n_k}(\cdot))$, then there exist $x \in \mathscr{S}(C)$ with $T(\cdot, x(\cdot)) = y$ a.s. and $x \in c_X\text{-}\limsup_{k\in\mathbb{N}} \{x_{n_k}\}$.

PROPOSITION 6.2.1. *Let* $\mathscr{S}(C) = c_X\text{-}\lim_{n\in\mathbb{N}} \mathscr{S}(C_n)$ *and let* $\{T_n\}_{n\in\mathbb{N}}$ *be* (c_X, c_Y)*-convergent to* T. *Then*

(a) T *and* $\{T_n\}_{n\in\mathbb{N}}$ *are* (c_X, c_Y)*-closed.*

(b) *If* $\{\mathscr{S}(C_n)\}_{n\in\mathbb{N}}$ *is* c_X*-compact, then* $\{T_n\}_{n\in\mathbb{N}}$ *is* c_Y*-compact.*

A convergence result for solutions of equations involving operators T and $\{T_n\}_{n\in\mathbb{N}}$, in addition to C and C_n, can also be obtained. Let $\mathscr{K} \times \mathscr{B}_X$-measurable mappings $U : \mathrm{Gr}C \to Y$ and $U_n : \mathrm{Gr}C_n \to Y$, $n \in \mathbb{N}$, be given and consider the random operator equations

(1) $T(\omega, x) = U(\omega, x)$

and

(1n) $T_n(\omega, x) = U_n(\omega, x)$.

Let

(2) $L = \{x \in \mathscr{S}(C) : T(., x(.)) = U(., x(.))\}$

and, for all $n \in \mathbb{N}$,

(2.n) $L_n = \{x_n \in \mathscr{S}(C_n) : T_n(., x_n(.)) = U_n(., x_n(.))\}$,

be the sets of all random solutions of (1) and (1n), respectively.

THEOREM 6.2.20. *Let L and L_n, $n \in \mathbb{N}$, be as in* (2) *and* (2.n), *respectively. Assume that U and $\{U_n\}_{n\in\mathbb{N}}$ are (c_X, c_Y)-regular, that T and $\{T_n\}_{n\in\mathbb{N}}$ are (c_X, c_Y)-closed and that $\{T_n\}_{n\in\mathbb{N}}$ is c_Y-compact. If, for all $n \in \mathbb{N}$,*

(3) $L_n \neq \emptyset$,

then

(4) $\emptyset \neq c_X\text{-}\lim\sup_{n\in\mathbb{N}} L_n \subseteq L$.

Let the random operator equation

(5) $T(\omega, x) = y(\omega)$

and C be a multivalued mapping from Ω into X, let $T : \mathrm{Gr}C \to Y$ be $\mathscr{K} \times \mathscr{B}_X$-measurable and let $y : \mathrm{Gr}C \to Y$ be measurable. Let $n \in \mathbb{N}$ be arbitrary, but fixed. Let $s_n \in \mathbb{N}$, $A_{n1}, \ldots, A_{ns_n} \in \mathscr{K}$ be pointwise disjoint with $\bigcup_{i=1}^{s_n} A_{ni} = \Omega$ and $P(A_{ni}) > 0$ for all $i \in \{1, \ldots, s_n\}$. Denote by $\mathscr{K}_n$ the σ-algebra on Ω generated by $\{A_{n1}, \ldots, A_{ns_n}\}$. For each $i \in \{1,2,\ldots, s_n\}$, let C_{ni} be a non-empty subset of X, $T_{ni} : C_{ni} \to Y$ and $y_{ni} \in Y$. Let also C_n be the multivalued mapping from Ω into X defined by

(6) $C_n(\omega) = C_{ni}$ if $\omega \in A_{ni}$;

let $y_n : \Omega \to Y$ be defined by

(7) $y_n(\omega) = y_{ni}$ if $\omega \in A_{ni}$;

and let $T_n : \mathrm{Gr}C_n \to Y$ be defined by

(8) $T_n(\omega, x) = T_{ni}(x)$ if $\omega \in A_{ni}$, $x \in C_{ni}$.

The collection $(\mathscr{K}_n, C_n, T_n, y_n)_{n\in\mathbb{N}}$ is called a "*discretization scheme*" (for (5)).

Note that $(\Omega, \mathscr{K}_n, P|_{\mathscr{K}_n})$ is a complete probability space and that the generating sets A_{ni}, $i \in \{1, \ldots, s_n\}$ are precisely the atoms of $\mathscr{K}_n$.

If such a discretization scheme is given, we use

(5.n) $T_n(\omega, x) = y_n(\omega)$

as an approximation for (5), where T_n and y_n are given by (8) and (7), respectively. Note that (5.n) reduces to the s_n deterministic equations

$$(9)\quad T_{ni}(x) = y_{ni} \text{ if } \omega \in A_{ni},\ i \in \{1, \ldots, s_n\}.$$

If all equations (9) are solvable, then one can construct a random solution of (5.n) in the following way : For $i \in \{1, \ldots, s_n\}$, let $x_{ni} \in C_{ni}$ be a solution of (9) and let $x : \Omega \to X$ be defined by

$$(10)\quad x(\omega) = x_{ni} \text{ if } \omega \in A_{ni}.$$

Then x solves (5.n) and x is $(\mathscr{K}_n, \mathscr{B}_X)$-measurable and, hence, also $(\mathscr{K}, \mathscr{B}_X)$-measurable. However, not all random solutions of (5.n) need be of form (10).

THEOREM 6.2.21. *Let $h : X \times [0, \infty) \to \mathbb{R}$ be such that $h(x,.)$ is continuous in 0 and $h(x, 0) = 0$ for all $x \in X$, and $\emptyset \neq K \subset C(X, Y)$. Let $T : \Omega \times X \to Y$ be a random operator such that, for all $\omega \in \Omega$:*

$$(11)\quad T(\omega,.) \in K$$

and

$$(12)\quad d(T(\omega, x), T(\omega, z)) \leqslant h(x, d(x, z)) \text{ for all } x, z \in X.$$

The following holds for each $n \in \mathbb{N}$:

There exist $s_n \in \mathbb{N}$, pairwise disjoint sets $A_{n1}, \ldots, A_{ns_n} \in \mathscr{K}$ with $\bigcup_{i=1}^{s_n} A_{ni} = \Omega$ and operators $T_{n1}, \ldots, T_{ns_n} \in \mathscr{K}$ such that with

$$(13)\quad T_n : \Omega \times X \to Y,\ (\omega, x) \to T_{ni}(X) \text{ if } \omega \in A_{ni},$$

we have : T_n is a random operator and (12) holds with T_n instead of T. Furthermore, for all $(\omega, x) \in \Omega \times X$, we have

$$(14)\quad \lim_{n \to \infty} T_n(\omega, x) = T(\omega, x).$$

T and $\{T_n\}_{n \in \mathbb{N}}$ are (c_X, c_Y)-consistent, where c_X is any convergence on $\mathscr{V}_X$ and c_Y denotes a.s.-convergence or convergence in probability. If there exists $\bar{h} : [0, \infty) \to \mathbb{R}$ continuous in 0, with $\bar{h}(0) = 0$ such that $h(x, t) \leqslant \bar{h}(t)$ for all $x \in X$, then $\{T_n\}_{n \in \mathbb{N}}$ is also (a.s., a.s.)-stable, and $\{T_n\}_{n \in \mathbb{N}}$ is (a.s.,a.s.)-convergent to T.

Some sufficient conditions for convergence and compactness of random operators and their domains with respect to a.s. and P-convergence are given in the last section of ref. [142].

6.3. NONLINEAR RANDOM EQUATIONS INVOLVING CERTAIN TYPES OF OPERATORS

Linear and nonlinear functional analysis is a basic tool for the study of random equations and random operators. Recently, Kannan and Salehi [251], [254], [255], Itoh [239] and Kravvaritis [267] proved the existence of solutions of random equations with single-valued monotone operators. The existence of random solutions of random differential equations associated with accretive operators was studied in ref. [240]. Similar problems for operator equations involving operators of type (M) were also studied [244]. Some results on the existence theorems for nonlinear random equations and inequalities containing multivalued monotone operators were given in ref. [268].

Some of the above-mentioned topics will be dealt with in this section.

Let us first introduce some preliminary concepts and results.

6.3.1. Given a real (or complex) Banach space X, let X^* be its dual, let $\|\,.\,\|$ be the norm and let $<.,.>$ denote the duality pairing between X and X^*. We use the convention that $(.,.) = \operatorname{Re} <.,.>$ if X is complex and $(.,.) = <.,.>$ if X is real. If X is a Hilbert space, then $<.,.>$ denotes the inner product on X and the same symbol $(.,.)$ as above will be used.

6.3.2. A nonlinear operator T from the Banach space X into the Banach space Y is said to be *bounded* if T takes bounded sets of X into bounded sets of Y.

Let D be a subset of the Banach space X. An operator $T: D \to \mathscr{P}(X^*)$ is said to be *quasi-bounded*, if, for each $M > 0$, there exists a $K(M) > 0$ such that if $\{(x, x^*)\} \in \operatorname{Gr} T$, $\|x\| \leqslant M$ and $(x^*, x) < M\|x\|$, then $\|x^*\| \leqslant K(M)$.

We use here the symbols $\to$ and $\rightharpoonup$ to denote convergence in the strong and weak topology, respectively. In the dual space X^*, one always denotes convergence in the weak* topology by $\rightharpoonup$.

6.3.3. A real Banach space X is said to be in *normal position* [255] with respect to a real Hilbert space S if

(i) $X^* \subseteq S \subseteq X$, algebraically and topologically,

(ii) $\langle u, v\rangle_s = (u, v)$, where $u \in X^*$, $v \in S$,

and

(iii) S is dense in X.

As an example of a space X in normal position, we may consider that $X = L_p(\Omega)$, $S = L_2(\Omega)$, $X^* = L_{p'}(\Omega)$, where Ω is a bounded measurable subset of Euclidean space, $1 \leqslant p \leqslant 2$ and

$p' = p/(p-1)$. Other examples could be the Sobolev spaces H_s, Orlicz spaces and Köthe spaces.

6.3.4. A nonlinear operator T from X into X^* is called *monotone* if $(Tx - Ty, x - y) \geqslant 0$ for all $x, y \in \mathscr{D}(T)$.

If T is an operator from X into $\mathscr{P}(X^*)$, T is said to be monotone if $(x^* - y^*, x - y) \geqslant 0$ for all $\{(x, x^*), (y,y^*) \in \mathrm{Gr}\,T$.

6.3.5. A monotone operator from X into X^* (from X into $\mathscr{P}(X^*)$) is said to be *maximal monotone* if it is monotone and if its graph is not properly contained in the graph of any other monotone operator T_1 from X into X^* (from X into $\mathscr{P}(X^*)$).

We note that any monotone operator which is continuous over the entire real Banach space X is maximal monotone.

6.3.6. An operator $T: X \to X^*$ is called *coercive* if there exists a function $c: \mathbb{R} \to \mathbb{R}$ such that $(Tx, x) \geqslant c(\|x\|)\|x\|$ and $c(r) \to \to \infty$ as $r \to \infty$.

6.3.7. An operator T from a subset D of a real Banach space X into X^* is said to be *hemicontinuous* if, for all $x \in D$, $y \in X$ with $x + t_n y \in D$, where $t_n \geqslant 0$, $t_n \to 0$, implies $T(x + t_n y) \rightharpoonup Tx$.

6.3.8. An operator T from a subset $D \subset X$ into X^* is said to be *demicontinuous* if $x_n \in D$ and $x_n \to x$ in X implies $Tx_n \rightharpoonup Tx$.

$T: X \to \mathscr{P}(X^*)$ is said to be demicontinuous if, for every sequence $\{x_n\}$ which converges to x_0 in X and for every choice $y_n \in T(x_n)$, there exists a subsequence $\{y_{n_k}\}$ which converges weakly to some $v_0 \in T(x_0)$.

6.3.9. Let X be a Banach space and $T: X \to X^*$. T is said to be *semimonotone* [239] if $Tx = S(x, x)$ for all $x \in X$, where S is a mapping of $X \times X$ into X^* with the following properties: (i) for each $y \in X$, $S(.,y)$ is hemicontinuous and monotone; (ii) for each $x \in X$, $S(x,.)$ is completely continuous (i.e., if $x_n \rightharpoonup x$ imply $Tx_n \to Tx$).

If X is a complex Banach space, a mapping $T: X \to X^*$ is said to be *complex-monotone* if, for every $N > 0$, there exists a strictly increasing continuous function $k_N: \mathbb{R}^+ \to \mathbb{R}^+$ such that $k_N(0) = 0$ and $|(Tx - Ty, x - y)| \geqslant k_N(\|x - y\|)$ if $x, y \in X$ with $\|x\| \leqslant N$, $\|y\| \leqslant N$. T is said to be *complex-semimonotone* if there exists a mapping $S: X \times X \to X^*$ with $Tx = S(x, x)$, $x \in X$, having the following properties: (i) for any $y \in X$, $S(.,y)$ is a complex-monotone mapping such that k_N is independent of y with $\|y\| \leqslant N$; (ii) for any bounded subset $B \subset X$, $S(x,.)$, $x \in B$ is completely continuous uniformly in x.

Let X be a real reflexive Banach space and $T: X \to \mathscr{P}(X^*)$. T is said to be generalized pseudomonotone if, for any sequence

$\{(x_n, x_n^*)\} \subset \operatorname{Gr} T$ such that $x_n \rightharpoonup x$ in X, $x_n^* \rightharpoonup x^*$ in X^* and $\limsup_{n\to\infty}(x_n^*, x_n - x) \leqslant 0$, we have (x, x^*) $\operatorname{Gr} T$ and $\lim_{n\to\infty}(x_n, x_n) = (x, x)$.

6.3.10. THEOREM I [254], [255]. *Let X be a reflexive Banach space and let $T: X \to X^*$ be monotone. Then T is maximal monotone if and only if $\mathscr{R}(I + T) = X^*$, i.e., $I + T$ is surjective.*

6.3.11. THEOREM II [78], [79]. *Let X be a reflexive Banach space and let $T: X \to X^*$ be maximal monotone and coercive. Then $\mathscr{R}(T) = X^*$.*

If T is hemicontinuous, monotone and coercive, then T is maximal monotone and $\mathscr{R}(T) = X^*$.

6.3.12. Two sequences $\{\lambda_n\}$ and $\{\theta_n\}$ of non-negative real numbers are said to be *acceptably paired* [81] if $\{\theta_n\}$ is decreasing, $\lim_{n\to\infty} \theta_n = 0$, and there exists a strictly increasing sequence $\{n(i)\}_{i\in\mathbb{N}}$ of positive integers such that

$$\liminf_i \theta_{n(i)} \sum_{j=n(i)}^{n(i+1)} \lambda_j > 0,$$

$$\lim_i [\theta_{n(i)} - \theta_{n(i+1)}] \sum_{j=n(i)}^{n(i+1)} \lambda_j = 0,$$

and

$$\lim_i \sum_{j=n(i)}^{n(i+1)} \lambda_j^2 = 0.$$

THEOREM III [81]. *Let T be a maximal monotone operator (on a Hilbert space H) with $0 \in T(y)$, and suppose U is defined and bounded on $S_r(y) = \{x : \|x - y\| \leqslant r\}$ for a certain $r > 0$. Then there exists a $\gamma > 0$ with the following property: whenever $\{\lambda_n\}$ and $\{\theta_n\}$ are acceptably paired and $\theta_1 \leqslant 1$, $\lambda_p/\theta_p \leqslant \gamma$ for all n, then for any $z \in S_{r/2}(y)$ and any initial guess $x_1 \in S_{r/2}(y)$, any sequence constructed from x_1 by*

$$x_{n+1} = x_n - \lambda_n(v_n + \theta_n(x_n - z)), \quad v_n \in T(x_n)$$

remains in $S_r(y)$ and converges strongly to the point $T^{-1}(0)$ closest to z.

6.3.13. THEOREM IV [254], [255]. *Let $T: \mathscr{D}(T) \subset X \to X^*$ be a monotone operator. If $\mathscr{D}(T)$ is open, then the hemicontinuity of T is equivalent to its demicontinuity.*

6.3.14. THEOREM V [255]. *Let $(\Omega, \mathscr{K}, P)$ be a complete probability space and let X and Y be separable Banach spaces. Assume* *that T is a random operator from X into Y such that, for each ω, $T(\omega)$ is demicontinuous. Then for any X-valued random variable x, the mapping $T(\omega,x(\omega))$ is a Y-valued random variable.*

Proof. Because T is demicontinuous for each $y^* \in Y^*$, the scalar-valued function $f(\omega, x) = (y^*, T(\omega, x))$ is continuous in X and measurable in ω. By a standard argument, $f(\omega,x(\omega)) = (y^*, T(\omega, x(\omega)))$ is a random variable for any X-valued random variable $x(\omega)$. Since this is true for every $y^* \in Y^*$, we conclude that the function $T(\omega,x(\omega))$ is a random variable.

6.3.15. A bounded linear monotone operator $T : X \to X^*$ is said to be *angle-bounded* if there exists a constant $c \geqslant 0$ such that $|(Tx, y) - (Ty, x)| \leqslant 2c[(Tx, x) \ (Ty, y)]^{1/2}$. It is clear that every symmetric monotone operator is angle-bounded with $c = 0$.

6.3.16. Let X be a Banach space and let $S : D \subset X \to X$ be a mapping. The mapping $S : [0, \infty) \times D \to D$ is said to be a *semigroup of type k* on D if :

(i) $S(t)S(s)x = S(t + s)x$ for $t,s \geqslant 0$ and $x \in D$;

(ii) $\|S(t)x - S(t)y\| \leqslant e^{kt}\|x - y\|$ for $t \geqslant 0$ and $x, y \in D$;

(iii) $\lim\limits_{t \downarrow 0} S(t)x = S(0)x = x$ for $x \in D$.

For each $x \in X$, denote $F(x) = \{x^* \in X^* ; (x^*, x) = \|x\|^2 = \|x^*\|^2\}$. Then $F(x)$ is nonempty and weakly* compact. For any $x, y \in X$, define $(x, y)_s = \sup\{(y^*, x), y^* \in F(y)\}$, $(y^*,x)_i = \inf\{(y^*, x), y \in F(y)\}$,

$$[x, y]_+ = \lim_{\lambda \downarrow 0} \{(\|x + \lambda y\| - \|x\|)/\lambda\},$$

and

$$[x, y]_- = \lim_{\lambda \uparrow 0} \{(\|x + \lambda y\| - \|x\|)/\lambda\},$$

respectively. A mapping $A : D \subset X \to X$ is said to be *accretive* if A satisfies one (all) of the following equivalent conditions :

(i) $[x - y, Ax - Ay]_+ \geqslant 0$ for $x, y \in D$;

(ii) $(Ax - Ay, \ x - y)_S \geqslant 0$ for $x, y \in D$;

(iii) For any $\lambda > 0$, $(I + \lambda A)^{-1}$ is a (single-valued) nonexpansive mapping from $\mathscr{R}(I + \lambda A)$ into D.

If X is a Hilbert space, then an accretive operator A is the same as a monotone operator.

6.3.17. Let X be a reflexive Banach space and let $T: X \to X^*$. T is said to be of type (M) if the following properties hold:

(i) if $x_n \rightharpoonup x$ in X, $T(x_n) \rightharpoonup g$ in X^* and $\limsup_n (Tx_n, x_n) \leqslant \leqslant (g, x)$, then $Tx = g$.

(ii) T is continuous from finite-dimensional subspaces of X into X^* equipped with weak* topology.

We say, as before, that a random operator $T: \Omega \times X^* \to X^*$ is of type (M) if $T(\omega)$ is of type (M) for all $\omega \in \Omega$.

It can be shown that every hemicontinuous monotone operator is of type (M). The sum of two operators of type (M) need not be of type (M), nor does a compact operator need be of type (M).

For $T: X \to \mathscr{P}(X^*)$, one says that T is of type (M) if, for any sequence $(x_n, x_n^*) \in \mathrm{Gr}\, T$ such that $x_n \rightharpoonup x$ in X, $x_n^* \rightharpoonup x^*$ in X^* and $\limsup_n (x_n^*, x_n - x) \leqslant 0$, one has $\{(x, x^*)\} \in \mathrm{Gr}\, T$.

6.3.18. THEOREM VI [244]. *Let X be a finite-dimensional Banach space and let $T: \Omega \times X \to X^*$ be a continuous random operator such that for all $\omega \in \Omega$*

$$(T(\omega)x, x) \geqslant 0 \text{ for } \|x\| = r.$$

Then the operator equation $T(\omega)x(\omega) = 0$ has a random solution $x(\omega) \in S_r (= \{x : \|x\| \leqslant r\})$.

Kannan and Salehi [255] considered the existence of random solutions for random nonlinear differential equations of the type

$$(1) \quad Lu + Nu = 0,$$

where L is a random linear self-adjoint positive differential operator and N is a random nonlinear operator. They utilized the existence results from the theory of monotone operators in the deterministic case and applied the theory of random operators in order to obtain random solutions.

In the equation of type (1) zero may or may not be an eigenvalue of L, whereas N is a monotone operator. When zero is not an eigenvalue of the associated linear problem of L, we rewrite (1) as a nonlinear Hemmerstein equation of the type $u + KNu = 0$ and then prove the measurability of solutions by applying the results on nonlinear Hammerstein equations in the deterministic case and the randomness properties of the operator $(I + KN)^{-1}$.

THEOREM 6.3.1 [255]. *Let X be a separable reflexive Banach space and let*

(i) $N: X^* \to X$ *be a hemicontinuous, monotone, bounded and random operator;*

(ii) $K: X \to X^*$ *be a random linear operator such that* $(Ku, u) \geqslant \mu \|Ku\|^2_{X^*}$, $\mu > 0$.

Then the random Hammerstein equation

$$u + KNu = a$$

has a unique random solution for each $a(\omega) \in X^*$.

Proof. Rewrite the Hammerstein equation as

$Tu = K^{-1}(u - a) + Nu \ni 0$.

Prove the existence of a unique solution for each fixed ω, as in ref. [79]. Since N is hemicontinuous and $\mathscr{D}(N) = X^*$, N is maximal monotone (cf. Th. II). K is linear, bounded and monotone and hence it is maximal monotone. Thus, K^{-1} is maximal monotone. Since $\mathscr{D}(N) = X^*$, one can apply some results for perturbation of maximal monotone operators (cf. Th. II) to conclude that $T: X^* \to X$ is maximal monotone. Also,

$$(Tu, u)/\|u\|_{X^*} = (K^{-1}(u - a) + Nu, u)/\|u\|_{X^*}.$$

From hypothesis (ii) and from the fact that N is monotone, it follows that $T: X^* \to X$ is coercive. Thus, Theorem II implies that $\mathscr{R}(T) = X$. If u_1, u_2 are any two solutions of the equation $u + KNu =$ a, for each ω, then $Tu_1 = Tu_2 = 0$. Hence, $0 = (Nu_1 - Nu_2, u_1 - u_2) + (K^{-1}(u_1 - a) - K^{-1}(u_2 - a), u_1 - u_2)$. Since N is monotone and since $(Ku, u) \geqslant \mu \|Ku\|^2$, it follows that $0 \geqslant \mu \|u_1 - u_2\|^2$. Thus, $u_1 = u_2$.

Hence the Hammerstein equation $u + KNu = a$ has a unique solution for every fixed ω.

Let us now show that the solution is random. We shall do so by proving that $(I + KN)^{-1}$ is random. Theorem 2.2.12 shows that this is true if $(I + KN)^{-1}$ is continuous for each ω. Since the operators N and K are random, from the theorem on the composition of operators it follows that KN is random too. As was shown above, $(I + KN)^{-1}$ is uniquely defined for each ω. Let us now prove the continuity of $(I + KN)^{-1}$ in two stages. First, show that $(I + KN)^{-1}$ is bounded.

Let $\|a\|_{X^*} \leqslant r$ and let $(I + KN)^{-1}a(\omega) = u(\omega)$. Then $u + KNu = a$ and then

$$\mu \|a - u\|^2_{X^*} = \mu \|KNu\|^2_{X^*} \leqslant (KNu, u) = (a - u, Nu) =$$
$$= -(a - u, Na - Nu) + (a - u, Na) \leqslant \|a - u\|_{X^*} \|Na\|_X,$$

since N is monotone. Thus, $\mu \|a - u\|_{X^*} \geqslant \|Na\|_X$. Since $\|a\|_{X^*} \leqslant r$ and since N maps bounded sets of X^* into bounded sets of X, it follows that $\|u\| \leqslant s(r)$, i.e., $(I + KN)^{-1}$ is bounded.

Let now $(I + KN)^{-1}\, a_i(\omega) = u_i(\omega)$, $i = 1,2$ and let $r > 0$ be such that $\|a_i\|_{X^*} \leqslant r$. Then $\|u_i\| \leqslant s(r)$. Likewise,

$$(u_1 - u_2) + (KNu_1 - KNu_2) = a_1 - a_2.$$

Thus,

$$(Nu_1 - Nu_2, a_1 - a_2) = (Nu_1 - Na_2, u_1 - u_2) +$$
$$+ (Nu_1 - Nu_2, KNu_1 - KNu_2) \geqslant \mu \|KNu_1 - KNu_2\|^2_{X^*} =$$
$$= \mu \|(a_1 - a_2) - (u_1 - u_2)\|^2_{X^*} \geqslant \frac{1}{2} \mu \|u_1 - u_2\|^2_{X^*} -$$
$$- \mu \|a_1 - a_2\|^2_{X^*}.$$

Hence,

$$\frac{1}{2} \mu \|u_1 - u_2\|^2_{X^*} \leqslant \mu \|a_1 - a_2\|_{X^*} + \|Nu_1 - Nu_2\|_X \|a_1 - a_2\|_{X^*}.$$

Since $\|Nu_1\|$, $\|Nu_2\| \leqslant s(r)$, it follows that $(I + KN)^{-1}$ is continuous. Using Theorem 2.2.12, we conclude that $(I + KN)^{-1}$ is a random operator. Hence, if $a(\omega)$ is measurable, it follows that $(I + KN)^{-1}a(\omega)$ is also so and the proof of the theorem is complete.

Another way of studying the measurability of solutions is by means of the *iterative approach*.

The random problem is rewritten in the form $T(\omega, u) = 0$, where T is the maximal monotone operator, and then one proves the randomness of $\{z_n\}_{n\in\mathbb{N}}$, where z_0 is arbitrary and random. The convergence of $\{z_n\}_{n\in\mathbb{N}}$ would establish the randomness of solution. Each z_{n+1} can be written in the form $(I + c_n\, T)^{-1}z_n$. Since T is maximal monotone and since $c_n > 0$, it follows that $T + (1/c_n)I$ is strictly monotone. Hence, $(T + (1/c_n)I)^{-1}$ is well-defined for each ω and also continuous. Hence, $T + (1/c_n)I$ is random and from Theorem 2.2.12 it follows that $(T + (1/c_n)I)^{-1}$ is random. The randomness of z_n implies the randomness of z_{n+1}. The randomness of T follows from the randomness of N and L.

In order to illustrate these methods, we consider :

THEOREM 6.3.2 [255, Th. 2]. *Let a separable reflexive Banach space X and a real Hilbert space S be such that (X, S, X^*) is in normal position. Let $N : X^* \to X$ be a random nonlinear hemicontinuous, bounded, monotone operator and let $K : X \to X^*$ be a random, linear symmetric continuous monotone operator. Then*

(a) *K can be written as J^*J, where J^* is a random linear operator from S to X^* and J is the dual of J^**

and

(b) *The sequence* $\{J^*v_n(\omega) + a(\omega)\}$, *where* $v_n(\omega)$ *is defined by*

$$v_{n+1} = (n/(n+1))v_n - (1/(n+1))JN[J^*v_n + a]$$

for any arbitrary measurable initial choice $v_1(\omega)$ *in* S, *converges to the unique random solution* $u(\omega)$ *of the Hammerstein equation*

$$u(\omega) + KNu(\omega) = a(\omega),$$

and, further,

$$\|J^*v_n(\omega) + a(\omega) - u(\omega)\| = 0(n^{-1/2}).$$

To illustrate the method of "Alternative Problems", we consider random differential equations of the type

$$(2) \quad Lu = Nu,$$

where L is a random differential operator and N is a random nonlinear operator. If L is such that $\mathcal{N}(L) = 0$, then the above equation could be rewritten as

$$u - (L^{-1})Nu = 0,$$

thus reducing to a Hammerstein type equation. For $-L^{-1}$ positive and suitable monotonicity hypotheses on N, we can apply the result of the above theorems to obtain the random solution of (2).

Let us now discuss the case when $\mathcal{N}(L) \neq \{0\}$. For the sake of simplicity, say that 0 is an eigenvalue of the associated linear problem. This corresponds to the case of resonance and the random differential equation (2) with a nonlinear right-hand side corresponds to the case of nonlinear perturbation at resonance. In the deterministic case, the essential idea is to split the differential equation into an equivalent system of two operator equations, one of which is over $\mathcal{N}(L)$ and the other over $\mathcal{N}(L)^\perp$. This method, which dates back to Lyapunov and Schmidt, was put in a functional analytic setting by Cesari [90] and is now widely referred to as the method of Alternative Problems. In the random case, $\mathcal{N}(L)$ changes with ω.

Let L be a random positive self-adjoint differential operator mapping a separable, reflexive Banach space X^* into a real Hilbert space S, where (X, S, X^*) is in normal position. Also, let N be a random nonlinear operator from X^* to S. We assume that the associated eigenvalue problem of L, i.e., $Lx + \lambda x = 0$, together with the boundary conditions, have a countable system of eigenvalues $\{\lambda_i(\omega)\}$, $\lambda_i(\omega) \geqslant 0$, $\lambda_i(\omega) \to \infty$ and a corresponding complete orthonormal system of eigenfunctions $\{\Phi_i(\omega)\}$ which form a basis for S. Let $S_0(\omega) = \mathcal{N}(L)$ (which is finite-dimensional) and let $P : S \to S_0(\omega)$ be the idempotent projection operator. Let also $H : S \to X^*$ be a linear operator such that

$$H(I - P)Lx = (I - P)x \text{ for all } x \in X^*.$$

The operator $L : X^* \subseteq S \to S$ is continuous and random. Hence, it is separable with respect to X^*. H is bounded and so one can conclude from Theorem 2.2.12 that H is a random operator. This implies that $\mathscr{R}(H)$ is measurable. Since $\mathscr{R}(H) = \mathscr{N}(L)^{\perp}$, it follows that $\mathscr{N}(L)$ is measurable. Thus, $H(I - P)$ is measurable. Then H satisfies $< -H(I - P)x, x \geqslant \lambda_{m+1}(\omega) \| - H(I - P)x\|^2$ for all $x \in X^*$. Since H is the generalized inverse of L, then Theorem 2.2.12 implies that $-H(I - P) : S \to X^*$ is a random operator.

Applying the operator $H(I - P)$ to the random differential equation $Lx = Nx$, one gets

$$(I - P)x = H(I - P)x$$

or

$$(3)\ x - H(I - P)Nx = Px.$$

Thus, every solution of the random differential equation is a solution of (3). If $x \in X^*$ is a solution of (3), then, applying L to both sides of (3),

$$Lx - (I - P)Nx = LPx = PLx.$$

Thus,

$$P(Lx - Nx) = Lx - Nx.$$

Hence, any solution of (3) is a solution of (2) if and only if

$$(4)\ P(Lx - Nx) = 0.$$

But $P : S \to \mathscr{N}(L)$ and thus (4) reduces to

$$(5)\ PNx = 0.$$

If, however, the random Hammerstein equation

$$(6)\ x - H(I - P)Nx = x^*(\omega)$$

has a unique solution $x(\omega)$ for any arbitrary $x^*(\omega) \in S_0$, then equation (5) can be rewritten as

$$(7)\ PN[I - H(I - P)N]^{-1}x^*(\omega) = 0.$$

Thus, we have reduced (under the above circumstances) the random nonlinear differential equation (2) to the equivalent system of operator equations (6) and (7). This method of splitting (2) is due to Cesari and Kannan [91]. Since L is symmetric it follows that $-H(I-P)$ is also so and, thus, (6) is of Hammerstein type (1). Applying the above theorems, we conclude that Eq. (6) has a unique solution for each $x^*(\omega) \in S_0$ if N is hemicontinuous random, bounded and monotone. As in Theorem 6.3.1, the operator $I - H(I-P)N^{-1}$ is continuous. Hence, the operator $I - H(I-P)N^{-1}$ is a random operator.

We now consider the solvability of equation (7) for each fixed ω. As may be seen in refs. [255], [254], Eq. (7) is of type

$$T_2 x^* = 0,$$

where T_2 is a nonlinear random operator from S_0 into itself and T_2 is maximal monotone. Furthermore, T_2 is coercive and, hence , (7) can be solved for x^*, for a fixed ω. From each solution x^* of (7), we can solve (6) from the theory of maximal monotone operators and Hammerstein integral equation. Each such solution x of (6) will then be a solution of the associated deterministic problem.

Let us now prove the randomness of the solution. First, show that if $x^*(\omega)$ is any random solution of Eq. (7), then the corresponding unique solution $x(\omega)$ of (6) is random. We know that $x - H(I-P)Nx = x^*$ or

$$x(\omega) = [I - H(I-P)N]^{-1} x^*(\omega),$$

N being bounded as in Theorem 6.3.1; we know that $[I - H(I-P)N]^{-1}$ is continuous. We have already proved that $[I - H(I-P)N]^{-1}$ is defined for each x^*. Hence, Theorem 2.2.12 implies that $[I - H(I-P)N]^{-1}$ is a random operator. If $x^*(\omega)$ is measurable, then the corresponding $x(\omega)$ is measurable. The bifurcation equation (7) may be written as

$$(8) \quad PN[I - H(I-P)N]^{-1} Px^* = 0.$$

From what was shown above, the operator

$$T_3 x = PN[I - H(I-P)N]^{-1} Px^*$$

is maximal monotone over S. Let us now apply a theorem due to Bruck [80, 81] and conclude that the boundedness of $[I - H(I-P)N]^{-1}$ implies the boundedness of N and P. We can apply Bruck's Theorem III [81] and conclude that there exist a $\gamma > 0$ and the sequences $\{\lambda_n\}$ and $\{\theta_n\}$ such that for any initial guess x_1^* belonging to a ball of radius γ about the origin, the sequence

$$x_{n+1}^* = x_n^* - \lambda_n [T_3 x_n^* + \theta_n x_n]$$

converges to the minimum norm solution of (7).

Thus, the operator T_3 over S, independent of ω, is maximal monotone and the sequence constructed above converges in S to the solution of minimum norm of $T_3x = 0$. However, as we change ω, and since the above sequence of random variables does not depend on ω, the randomness of T_3 and a choice of a random x_1^* ensures that the solution $x^*(\omega)$ of minimum norm of (7) is measurable. This implies that the corresponding $x(\omega)$ of Eq. (6) is random, and thus we have obtained a random solution of the random nonlinear differential equation (2). We thus have

THEOREM 6.3.3. *Let $L : X^* \to S$ be a self-adjoint random differential operator with eigenvalues $\{\lambda_i(\omega)\}$, $\lambda_i \geqslant 0$, $\lambda \to \infty$. Let $N : X^* \to S$ be a random, hemicontinuous monotone and coercive operator.*

Then the random nonlinear differential equation

$$Lu = Nu$$

has a random solution.

Another method is the "measurability graph method". This method is used when iterative techniques are not available, as in the case when one applies the Schauder fixed-point-theorem or the Leray-Schauder principle. The random problem is rewritten as $(I + T)u = 0$, where $u = (v, w) \in \mathscr{R}(L) \oplus \mathscr{N}(L)$ and where $T : (\Omega, (\mathscr{R}(L) \oplus \mathscr{N}(L))) \to (\mathscr{R}(L) \oplus \mathscr{N}(L))$ is random if L and N are random too. Assume that the hypotheses on N are such that, for each $\omega \in \Omega$, the deterministic problem is solvable. Thus, one can generate a correspondence $g : \omega \to \{u(\omega)\}$, such that, for every ω, $u(\omega)$ is the set of solutions of the deterministic problem. Let us now show that the graph of g belongs to the appropriate σ-algebra as in Theorem 2.2.1. From a selection theorem, one obtains a single-valued selection to the random problem.

For example, consider an alternative proof [254] to the above theorem using Theorem 2.1.4. Proceeding as in the earlier theorem, the random problem can be rewritten as the system (6) and (7). Since $H(I - P)$ is maximal monotone and since N is monotone hemicontinuous, the auxiliary equation is solvable for each $x^*(\omega) \in S_0(\omega)$. The bifurcation equation is then shown to be of the type $Tx^*_2 = 0$ with $T_2 : S_0(\omega) \to S_0(\omega)$ maximal monotone and coercive. Thus, $T_2x^* = 0$ is solvable for every ω. From the theory of maximal monotone operators, we know that the solution set is closed and convex for each ω.

The projection P is a random operator. Since $x^*(\omega) \in \mathscr{R}(P(\omega))$ for every ω, we can rewrite the bifurcation equation as (8).

Then the problem can be considered to be of the type

(9) $U(\omega, x) = 0$,

where $x \in S$ (a separable metric space). Since $[I - H(I - P)N]^{-1}$ is continuous, since N is defined on X^* and is hemicontinuous (hence demicontinuous) and since P is continuous, it follows that U is measurable in ω and continuous in x. Hence, Theorem 2.1.4 on joint measurability implies that U is jointly measurable. The set of couples $(\omega, x^*) \in \Omega \times \mathscr{R}(P)$ with $U(\omega, x^*) = 0$ is a measurable set. But this is precisely the graph of the function $\varphi : \Omega \to \mathscr{R}(P)$ defined by $\varphi(\omega) = x^*$, where x^* is a solution of the bifurcation equation. Hence, the possibly multivalued solution $x^*(\omega)$ is measurable in ω and thus the original random problem has a multivalued measurable solution. In particular, since the set of solutions of the bifurcation equation is closed and convex and since S is a real Hilbert space [363], we conclude that a singlevalued measurable selector exists. Since $[I + H(I - P)N]^{-1}$ is random, $x(\omega) = [I + H(I - P)N]^{-1}x^*$ is also measurable.

S. Itoh [239] considered some problems for which the uniqueness of the solution of the equation under consideration is not ensured. Using the theory of measurable selectors of multivalued mappings, one can obtain a random solution (not necessarily unique) of random equations. Some of their results are presented in the sequel.

THEOREM 6.3.4 [239, Th. 3.4]. *Let X be a separable reflexive Banach space and let $T : \Omega \times X \to X^*$ be a coercive semimonotone random operator. If $y \in B(\Omega, X^*) = \{X^*$-valued random variables z for which $\sup \|z(\omega)\| : \omega \in \Omega\} < \infty\}$, then there exists an $x \in B(\Omega, X)$ such that for every $\omega \in \Omega$, $T(\omega)x(\omega) = y(\omega)$.*

THEOREM 6.3.5 [239, Th. 4.3]. *Let X be a separable reflexive complex Banach space and let $T : \Omega \times X \to X^*$ be a bounded demicontinuous complex-semimonotone random operator. Suppose that there exists a function $c : (0, \infty) \to \mathbb{R}$ for which $|\langle Tx, x\rangle| \geqslant c(\|x\|)\|x\|(x \in X,\ \omega \in \Omega)$ and $c(r) \to \infty$ as $r \to \infty$. If $y \in B(\Omega, X^*)$, then there exists an $x \in B(\Omega, X)$ such that for every $\omega \in \Omega$, $T(\omega)x(\omega) = y(\omega)$.*

THEOREM 6.3.6 [239, Th. 6.1]. *Let x be a separable reflexive Banach space and let D be a dense linear subspace of X. Let $L : \Omega \times X \to X^*$ be a closed linear random operator, let $G : \Omega \times X \to X^*$ be a hemicontinuous bounded random operator and let $C : \Omega \times X \to X^*$ be a completely continuous random operator. Suppose that the mapping $T : \Omega \times D \to X^*$, defined by $T(\omega)x = L(\omega)x + G(\omega)x$, satisfies the following conditions: (i)* *for each* $\omega \in \Omega$, $T(\omega) + C(\omega)$ *is monotone*; (*ii*) *for each* $\omega \in \Omega$, $L(\omega)^*$ *is the closure of its restriction to* D; (*iii*) T *is coercive. If* $y \in B(\Omega, X^*)$, *then there exists an* $x \in B(\Omega, X)$ *such that* $T(\omega)x(\omega) = y(\omega)$ *for any* $\omega \in \Omega$.

An important result on the random Hammerstein equation is given by Theorem 7.1 [239].

The random analogue of nonlinear evolution equations involving monotonic nonlinearities can also be obtained:

THEOREM 6.3.7 [251]. *Let* $(\Omega, \mathscr{K}, P)$ *be a complete probability space and let* X *be a separable Banach space. Let* $T : \Omega \times X \to X^*$ *be maximal monotone with* Tx *nonempty for every* $x \in X$. *Further, let the equation* $0 \in Tx$ *have a solution for every* $\omega \in \Omega$. *Then there exists an* X*-valued random variable* $x(\omega)$ *such that*

$$0 \in Tx(\omega).$$

Some results on random differential equations associated with continuous nonlinear accretive operators in Banach or Hilbert spaces are given by S. Itoh in ref. [240]. Let us recall that $C([0, T], X)$ denotes the Banach space of all continuous mappings $v : [0, T] \to X$ with the norm $\|v\|_\infty = \max \{\|v(t)\|, 0 \leqslant t \leqslant T\}$. Derivatives are strong derivatives and integrals are Bochner integrals. $C_1([0, T], X)$ denotes the set of all continuously differentiable mappings on $[0, T]$. A mapping $u : [0, T] \times \Omega \to X$ is said to satisfy *condition* (C, Ω) (or (C_1, Ω)) if, for each $t \in [0, T]$, $u(t,.)$ is measurable and, for each $\omega \in \Omega$, $u(.,\omega)$ is continuous (or continuously differentiable).

THEOREM 6.3.8 [240, Th. 2.2]. *Let* X *be a separable Banach space and let* $A : \Omega \times X \to X$ *be a continuous random operator. Suppose there exists a function* $k : \Omega \to \mathbb{R}$ *such that* $\sup\{k(\omega) : \omega \in \Omega\} = M < \infty$ *and for each* $\omega \in \Omega$, $A(\omega) + k(\omega)I$ *is accretive. Then the following hold:*

(i) *There exists a mapping* $S : [0, \infty) \times \Omega \times X \to X$ *such that, for every* $\omega \in \Omega$, $S(.,\omega)$ *is a semigroup of type* $k(\omega)$ *on* X, *for every* $t \geqslant 0$ *and* $x \in X$, $S(t.,)$ *is measurable*;

(ii) *If* $v : \Omega \to X$ *is measurable, then the mapping* $u : [0, \infty) \times \times \Omega \to X$ *defined by* $u(t, \omega) = S(t, \omega)v(\omega)$ $(t \geqslant 0, \omega \in \Omega)$ *satisfies condition* (C_1, Ω) *and thus* u *is the unique mapping such that, for any* $\omega \in \Omega$, $u(0, \omega) = v(\omega)$ *and*

$$\mathrm{d}u(t, \omega)/\mathrm{d}t + A(\omega)u(t, \omega) = 0, \; t \geqslant 0.$$

Proof. According to Martin [289, Th. 6] and to the above hypotheses, for sufficiently small $\lambda > 0$, the range of $I + \lambda A(\omega)$ is equal to X for all $\omega \in \Omega$. Hence, the mapping $J : \Omega \times X \to X$ defined by $J_\lambda(\omega)x = (I + \lambda A(\omega))^{-1}x$, $\omega \in \Omega$,

$x \in X$, is a continuous (Lipschitz) random operator (see Crandall and Liggett [118], Lemmas 1.2 and 1.1). Define $S : [0, \infty) \times \Omega \times X \to X$ as

$$S(t, \omega)x = \lim_{n\to\infty} J_{t/n}(\omega)^n x \text{ for } t \geqslant 0,\ \omega \in \Omega \text{ and } x \in X.$$

Then, for every $\omega \in \Omega$, $S(.,\omega)$ is a semigroup of type $k(\omega)$ on X (see Crandall and Liggett [118], Theorem I) and for every $t \geqslant 0$ and $x \in X$, $S(t,.)x$ is measurable. Thus, (i) holds. For $\omega \in \Omega$ and $x \in X$, $S(.,\omega)x$ is continuously differentiable and

$$\mathrm{d}S(t, \omega)x/\mathrm{d}t = A(\omega)/S(t, \omega)x = 0,\ t \geqslant 0$$

(cf. Crandall and Liggett [118], p. 282—283). Moreover, if $v : \Omega \to X$ is measurable, then for each $t \geqslant 0$, $S(t,.)v(.)$ is measurable (cf. Remark 2.2.4) and from [118], Theorem II, there follows (ii).

An extension of (ii) is given by

THEOREM 6.3.9. *Let X be a separable Banach space, let $A : \Omega \times X \to X$ be a continuous random operator and let $f : [0, T] \times \Omega \to X$ be a mapping satisfying condition (C, Ω). Suppose there exists a function $k : \Omega \to R$ such that* $\sup\{k(\omega) : \omega \in \Omega\} \leqslant M < \infty$ *and, for any $\omega \in \Omega$, $A(\omega) + k(\omega)I$ is accretive. Then for any measurable mapping $v : \Omega \to X$, there exists a unique mapping $u : [0, T] \times \Omega \to X$ satisfying condition (C_1, Ω) such that $u(0, \omega) = v(\omega)$ and*

$$\mathrm{d}u(t, \omega)/\mathrm{d}t + A(\omega)u(t, \omega) = f(t, \omega),\ 0 \leqslant t \leqslant T, \textit{ for any } \omega \in \Omega.$$

The existence of a solution for a random differential equation with time-dependent accretive in a Hilbert space is given by

THEOREM 6.3.10 [24, Th. 3.2]. *Let H be a separable Hilbert space and let $A : [0, T] \times \Omega \times H \to H$ be a mapping having the following properties:*

(i) *For every $\omega \in \Omega$, $A(.,\omega)(.)$ is jointly continuous;*

(ii) *For every bounded subset D of H, $\sup\{\|A(t, \omega)x\| : 0 \leqslant t \leqslant T, \omega \in \Omega, x \in D\} < \infty$;*

(iii) *For every $t \in [0, T]$ and $x \in H$, $A(t,.)x$ is measurable;*

(iv) *There exists a function $k : [0, T] \times \Omega \to \mathbb{R}$ such that* $\sup\{k(t, \omega) : 0 \leqslant t \leqslant T, \omega \in \Omega\} = M < \infty$ *and $A(t, \omega) + k(t, \omega)I$ is monotone for every $t \in [0, T]$ and $\omega \in \Omega$.*

Then for each measurable mapping $v : \Omega \to H$, there exists a unique mapping $u : [0, T] \times \Omega \to H$ satisfying condition (C_1, Ω) such that $u(0, \omega) = v(\omega)$ and

$$\mathrm{d}u(t, \omega)/\mathrm{d}t + A(t, \omega)u(t, \omega) = 0,\ 0 \leqslant t \leqslant T, \textit{ for any } \omega \in \Omega.$$

Using the above existence theorems, one can obtain periodic solutions for various random differential equations in Banach or Hilbert spaces [240].

THEOREM 6.3.11. *Let X be a separable Banach space, let $A : \Omega \times X \to X$ be a continuous random operator and let $f : [0, T] \times \Omega \to X$ be a mapping satisfying condition (C, Ω). Suppose there exists a function $k : \Omega \to (-\infty, 0)$ such that, for every $\omega \in \Omega$, $A(\omega) + k(\omega)I$ is accretive. Then there exists a unique mapping $u : [0, T] \times \Omega \to X$ satisfying condition (C_1, Ω) such that for every $\omega \in \Omega$, $u(0, \omega) = u(T, \omega)$ and*

$$\mathrm{d}u(t, \omega)/dt + A(\omega)u(t, \omega) = f(t, \omega), \quad 0 \leqslant t \leqslant T.$$

Proof. Theorem 6.3.9 implies that, for any $x \in X$, there exists a unique mapping $u_x : [0, T] \times \Omega \to X$ satisfying condition (C_1, Ω) such that for any $\omega \in \Omega$, $u_x(0, \omega) = x$ and

$$\mathrm{d}u_x(t, \omega)/\mathrm{d}t + A(\omega)u_x(t, \omega) = f(t, \omega), \quad 0 \leqslant t \leqslant T.$$

Let us fix $\omega \in \Omega$. For $x, y \in X$, $\|u_x(.,\omega) - u_y(.,\omega)\|$ is absolutely continuous and, hence, it is differentiable i.e., $t \in [0, T]$ and

$$\mathrm{d}(\|u_x(t, \omega) - u_y(t, \omega)\|)/\mathrm{d}t \leqslant k(\omega)\|u_x(t, \omega) - u_y(t, \omega)\|.$$

Thus, for any $t \in [0, T]$,

$$\|u_x(t, \omega) - u_y(t, \omega)\| \leqslant e^{k(\omega)}\|u_x(0, \omega) - u_y(0, \omega)\| =$$
$$= \mathrm{e}^{k(\omega)}\|x - y\|.$$

Define $g : \Omega \times X \to X$ as $g(\omega, x) = u_x(T, \omega)$. Then the above inequality implies that g is a contraction random operator. Theorem 4.1.2 implies that there exists a unique measurable mapping $z : \Omega \to X$ such that $g(\omega, z(\omega)) = z(\omega)$ for all $\omega \in \Omega$. Let $u : [0, T] \times \Omega \to X$ be a unique mapping satisfying condition (C_1, Ω) such that for any $\omega \in \Omega$, $u(0, \omega) = z(\omega)$ and

$$\mathrm{d}u(t, \omega)/\mathrm{d}t + A(\omega)u(t, \omega) = f(t, \omega), \quad 0 \leqslant t \leqslant T.$$

From Theorem 6.3.9, it follows that $u(T, \omega) = u_{z(\omega)}(T, \omega) = g(\omega, z(\omega)) = z(\omega) = u(0, \omega)$ for every $\omega \in \Omega$. If $v : [0, T] \times \Omega \to X$ is another periodic solution, then, for any $t \in [0, T]$, $\omega \in \Omega$,

$$\|u(t, \omega) - v(t, \omega)\| \leqslant \mathrm{e}^{k(\omega)}\|u(0, \omega) - v(0, \omega)\|,$$

and, in particular,

$$\|u(T, \omega) - v(T, \omega)\| \leqslant \mathrm{e}^{k(\omega)}\|u(0, \omega) - v(0, \omega)\|.$$

Since $k(\omega) < 0$ and $\|u(T, \omega) - v(T, \omega)\| = \|u(0, \omega) - v(0,\omega)\|$, the above inequality implies that $u(0, \omega) = v(0, \omega)$, $\omega \in \Omega$, hence $u(t, \omega) = v(t, \omega)$, $0 \leqslant t \leqslant T$, $\omega \in \Omega$.

THEOREM 6.3.12. *Let H be a separable Hilbert space and let $A : [0, T] \times \Omega \times H \to H$ be a mapping satisfying conditions (i)—(iii) of Theorem* 6.3.10 *as well as the following condition*:

(iv) *There exists a function $k : [0, T] \times \Omega \to (-\infty, 0)$ such that for any $t \in [0, T]$ and any $\omega \in \Omega$, $A(t, \omega) + k(t, \omega)I$ is monotone and for any $\omega \in \Omega$, $k(.,\omega)$ is measurable with $-\infty < M(\omega) < 0$, where*

$$M(\omega) = \int_0^T k(t, \omega)\mathrm{d}t.$$

Then there exists a unique mapping $u : [0, T] \times \Omega \to H$ satisfying condition (C_1, Ω) such that for any $\omega \in \Omega$, $u(0, \omega) = u(T, \omega)$ and

$$\mathrm{d}u(t, \omega)/\mathrm{d}t + A(t, \omega)u(t, \omega) = 0, \quad 0 \leqslant t \leqslant T.$$

M. Joshi [244] obtained the random solution of the operator equation involving operators of type (M), e.g. the random analogy of the main deterministic result due to Petryshyn and Fitzpatrick [340]. A random solution of Hammerstein operator equations is given also and Theorem 6.3.2 is derived as a corollary.

THEOREM 6.3.13 [244, Th. 3.1]. *Let X be a separable reflexive Banach space and let $T : \Omega \times X \to X$ be a demicontinuous random operator of type (M). Further, let the following condition be satisfied*:

(c) *If $\{x_n\}$ is a bounded sequence in X such that, for every $\omega \in \Omega$, $\{(T(\omega)x_n, x_n)\}$ is also bounded, then $\{T(\omega)x_n\}$ is bounded for almost all $\omega \in \Omega$.*

If there exists an $r > 0$ such that $(T(\omega)x, x) \geqslant 0$ for $\|x\| = r$ and for all $\omega \in \Omega$, then there exists a random solution $x(\omega)$ of the operator equation

$$(*) \quad T(\omega)x(\omega) = 0.$$

Proof. Let $\{u_n\}_{n\in\mathbb{N}}$ be a linearly independent subset of X such that its linear span W is dense in X. Let $X_n = \{u_1, \ldots, u_n\}$ and let $j_n : X_n \to X$ be the injection mapping with dual j_n^*. Define $T_n(\omega) = X_n \to X^*$ as $T_n(\omega)x = j_n^* T_n(\omega) j_n x$.

Since j_n, j_n^* are continuous and since $T(\omega)$ is a demicontinuous random operator, it follows from Theorem V that $T_n(\omega)$ is a demicontinuous random operator from X_n to X_n. For every $x \in B_r \cap X_n$ $(B_r = \{x : \|x\| \leqslant r\})$, we have $(T_n(\omega)x, x) = (T(\omega)x, x) \geqslant 0$ for all $\omega \in \Omega$. It follows by Theorem VI that there exists $x_n(\omega) \in B_r$ such that $T_n(\omega)x_n(\omega) = 0$ for all $\omega \in \Omega$.

Let us now use the theory of measurable selector to extract a measurable subsequence of $\{x_n(\omega)\}$. For every n and every $\omega \in \Omega$, define $G : \Omega \to WK(B_r)$ as $G_n = \text{wcl}\{x_i(\omega),\ i \geqslant n\}$. Then Theorem 2.2.5 implies that the mappings G_n are w-measurable and so is the mapping $G : \Omega \to WK(B_r)$ defined by $G(\omega) = \bigcap_{n \in \mathbb{N}} G_n(\omega)$ (Theorem 2.2.2). From Kuratowski and Ryll-Nardzewski's theorem, it follows that there exists a w-measurable selector $x(\omega) : \Omega \to B_r$ of G. Because the space is separable, it follows that $x(\omega)$ is also measurable.

We claim that this random variable $x(\omega)$ is the solution of the operator equation $(*)$. If it were not so, then there would exist a measurable subset M of Ω with $\mu\,(M) > 0$ and $\xi_0(\omega) \in X$ such that $(T(\omega)x(\omega),\ \xi_0(\omega)) < 0$ for every $\omega \in M$. Since W is dense in X and since $\xi_0(\omega) \in X$, it follows that for some $n_0 \in \mathbb{N}$ there exists $\xi(\omega) \in X_{n_0}$ such that

$$(**)\quad (T(\omega)x(\omega),\ \xi(\omega)) < 0 \quad \text{for each} \quad \omega \in M.$$

Let us now proceed to get a contradiction of $(**)$. Fix $\omega \in M$ and consider $x(\omega) \in G(\omega)$. The definition of $G(\omega)$ implies that there exists a subsequence $\{x_{n_k}(\omega)\}$ of $\{x_n\}$ converging weakly to $x(\omega)$ and, hence, $0 = (T_{n_k}(\omega)x_{n_k}(\omega),\ x_{n_k}(\omega)) = (T(\omega)x_{n_k}(\omega),\ x_{n_k}(\omega))$. Condition (c) now implies that $\{T(\omega)x_{n_k}(\omega)\}$ is bounded and by virtue of the reflexivity of X, we may assume that $T(\omega)x_{n_k}(\omega) \rightharpoonup g(\omega)$ for some $g(\omega) \in X^*$. Since W is dense in X and $x(\omega) \in X$, there exists a sequence $\xi_{n_k}(\omega) \in X_{n_k}$ such that $\xi_{n_k}(\omega) \to x(\omega)$. Hence,

$$\begin{aligned}\limsup_{k\to\infty}(T(\omega)x_{n_k}(\omega),\ x_{n_k}(\omega)) &\leqslant \limsup_{k\to\infty}(T_{n_k}(\omega)x_{n_k}(\omega),\ x_{n_k} - \\ &- \xi_{n_k}(\omega)) + \limsup_{k\to\infty}(T(\omega)x_{n_k}(\omega),\ \xi_{n_k}(\omega) - x(\omega)) + \\ &+ \limsup_{k\to\infty}(T(\omega)x_{n_k}(\omega),\ x(\omega)).\end{aligned}$$

Since $T_{n_k}(\omega)x_{n_k}(\omega) = 0$, $\xi_{n_k}(\omega) \to x(\omega)$ and $T(\omega)x_{n_k}(\omega) \rightharpoonup g(\omega)$, the above inequalities imply that $\limsup_{k\to\infty} (T_{n_k}(\omega), x_{n_k}(\omega),\ x_{n_k}(\omega)) \leqslant (g(\omega),\ x(\omega))$. By virtue of the fact that $T(\omega)$ is of type (M) and that $x_{n_k}(\omega) \rightharpoonup x(\omega)$, it follows that $T(\omega)\,x(\omega) = g(\omega)$.

Now, $(T(\omega)x_{n_k}(\omega),\ \xi(\omega)) = 0$ for $n_k \geqslant n_0$ and, therefore,

$$0 = \lim_{k\to\infty} (T(\omega)x_{n_k}(\omega),\ \xi(\omega)) = (g(\omega),\ \xi(\omega)) = (T(\omega)x(\omega),\ \xi(\omega)),$$

which contradicts $(**)$. Thus, the proof of the theorem is complete.

An abstract random Hammerstein operator equation is

(I) $x(\omega) + K(\omega)N(\omega)x(\omega) = 0,$

where $N : \Omega \times X \to X^*$ is a bounded, demicontinuous random operator of type (M) and $K : \Omega \times X \to X^*$ is a suitable continuous linear random operator.

Since X is reflexive, we shall alternatively consider the operator equation

(II) $K^*(\omega)x(\omega) + K(\omega)N(\omega)K^*(\omega)x(\omega) = 0.$

THEOREM 6.3.14. *Let X be a separable reflexive Banach space and let $K : \Omega \times X \to X^*$ be a continuous linear random operator such that the functional $f(\omega) : X \to \mathbb{R}$ defined by $f(\omega)x = (K(\omega)x, x)$, $x \in X$, is weakly lower semicontinuous for all $\omega \in \Omega$. Suppose $N : \Omega \times X^* \to X$ is a bounded, demicontinuous random operator of type (M). If a number $r > 0$ exists such that*

$$(K^*(\omega)x, x) + (N(\omega)K^*(\omega)x, K^*(\omega)x) \geq 0 \text{ for } \|x\| = r$$

for all $\omega \in \Omega$, then there exists a random solution $x(\omega) \in X^*$ of (I).

One can weaken the above inequality by assuming that $K(\omega)$ is angle-bounded, i.e.

THEOREM 6.3.15. *Let X be a separable reflexive Banach space. Let $K : \Omega \times X \to X^*$ be a continuous, monotone, linear angle-bounded random operator with $\|K(\omega)\| \leq K_0$ and with a fixed constant c of angle-boundedness for all $\omega \in \Omega$. Let $N : \Omega \times X^* \to X$ be a bounded demicontinuous random operator of type (M) satisfying the inequality*

(α) $(N(\omega)x, x) \geq -k(\omega) + (N(\omega)0, x)$

for all $x \in X^*$ and for all $\omega \in \Omega$ ($k(\omega) \geq 0$ for all ω). If $k(\omega)(1 + c^2)K_0 < 1$, then (I) *has a random solution $x(\omega)$.*

Remark. Since every hemicontinuous monotone mapping is a demicontinuous mapping of type (M) satisfying inequality (α) for $k(\omega) = 0$, we get Theorem 6.3.2 of Kannan and Salehi.

Some generalizations of Itoh's [239] and Kravvaritis [267] results for multivalued operators of monotone type or of type (M) are given in ref. [268]:

THEOREM 6.3.16. *Let X be a separable reflexive Banach space and let D be a subset of X. Let $T : \Omega \times D \to \mathcal{P}(X^*)$ be a random operator of type (M), quasi-bounded and coercive. Suppose that* *there exists a dense linear subspace* X_0 *of* X *which is contained in* D *such that for every finite-dimensional subspace* H *of* X_0, *the random operator* $T : \Omega \times H \to \mathscr{P}(X^*)$ *is continuous, with* X^* *given as its weak topology, while for each* $\omega \in \Omega$ *and* $x \in X_0$, $T(\omega)x$ *is a bounded, closed convex subset of* X^*. *Then for every* $y \in B(\Omega, X^*)$ *there exists an* $x \in B(\Omega, X)$ *such that* $y(\omega) \in T(\omega)x(\omega)$ *for all* $\omega \in \Omega$.

Applying the above theorem, one can prove a theorem on the perturbation of a random maximal monotone operator:

THEOREM 6.3.17. *Let* X *be a separable reflexive Banach space and let* D_1, D_2 *be subsets of* X. *Let* $L : \Omega \times D_1 \to X^*$ *be a random operator, which is demicontinuous and maximal monotone with* $L(\omega)0 = 0$ *for every* $\omega \in \Omega$, *and let* $T : \Omega \times D_2 \to \mathscr{P}(X^*)$ *be a random operator, which is generalized pseudo-monotone, quasi-bounded and coercive. Suppose that there exists a dense linear subspace* X_0 *of* X *which is contained in* D_2 *such that, for every finite-dimensional subspace* H *of* X_0, *the random operator* $T : \Omega \times H \to \mathscr{P}(X^*)$ *is continuous,* X^* *being its given weak topology, while for every* $\omega \in \Omega$ *and* $x \in X_0$, $T(\omega)x$ *is a bounded, closed and convex subset of* X. *Then, for every* $y \in B(\Omega, X^*)$, *there exists an* $x \in B(\Omega, X)$ *such that* $x(\omega) \in L(\omega)x(\omega) + T(\omega)x(\omega)$ *for all* $\omega \in \Omega$.

Some nonlinear random inequalities involving multivalued operators of monotone type are proved in the last part of ref. [268].

Chapter 7

Some applications of random operators

Several applications of random operators have already been given in the previous chapters. We shall now deal with realistic examples of applications to wave propagation in random media, applications in turbulence theory, in the study of random vibration of elastic strings and bars, in stream temperature modelling, in stream pollution and in chemotherapy. Some applications of random operators in population growth problems and in telephone traffic theory are also given. Some considerations regarding stochastic systems, stochastic programming and stochastic optimal control with random models and operators will be dealt with in the last part of this chapter.

7.1. RANDOM EQUATIONS AND WAVE PROPAGATION IN RANDOM MEDIA

The mathematical equations used to describe a given physical phenomenon usually contain one or more coefficients or parameters, having some definite physical interpretation. For example, in the theory of wave propagation, the coefficients of the equation characterize the propagation medium, i.e., its refrative index $n(x)$, its density $\rho(x)$, its dielectric constant, etc. A random medium is a family of media, each characterized by a value of ω, and having its own index $n(n, \omega)$, density $\rho(x, \omega)$, etc. Wave propagation in a continuous random medium is governed by a linear differential equation with random coefficients, i.e., by random differential equations.

The problem of wave propagation in a random medium has attracted much research in recent years. It arises naturally in the study of acoustic wave propagation and scattering due to temperature inhomogeneities, magnetohydrodynamic waves in interplanetary media, penetration of temperature or elastic waves in solids, and many other important physical phenomena.

Insisting on the use of the probabilistic theory and thought in science and engineering, J. B. Keller [258] shows that the practical reason for studying stochastic equations is the belief that their solutions represent physical phenomena which could not be satisfactorily investigated in any other way. This is illustrated by the propagation of an electromagnetic wave through air. Because the wave travels so fast, it is adequate to assume that the air molecules do not move during the passage of the wave. If the locations of all molecules were known, one could in principle determine how they scatter the wave. However, the full details of the scattered wave would be too complex to be useful. Furthermore, it is out of the question to determine the locations of all molecules. Bearing these two difficulties in mind, it is hoped that the observable features of the wave scattering will be determined by the gross features of the molecular distribution. Therefore, the actual distribution of molecular positions is replaced by a random medium, i.e., a collection of molecular distributions with a probability associated with each of them. The mean particle density and some other statistics of the random medium are adjusted to equal the corresponding properties of the actual particle distribution. Then the mean scattered wave and other statistics of the random wave are presumed to equal the corresponding properties of the actual wave. Of course, the success of this procedure, like that of all other theories on natural phenomena, can only be judged by referring its predictions to observations.

7.1.1. *On Two Classes of Methods*

We illustrate below two techniques for solving random equations and investigating their statistical properties. We shall start from J. B. Keller's ideas [258] and his examples and from the subsequent developments due to K. K. Chen and T. T. Soong [102] regarding covariance properties of waves propagating in a random media. G. Adomian's methods rule out various limitations and allows the treatment of a wide class of physical applications which will be presented below.

According to Keller [258], various approximations methods for solving the random operator equation can be classified in two types: (1) "honest", and (2) "dishonest". It is known that in the first step of an honest method (obtain the solution $x(t, \omega)$ for each $\omega \in \Omega$), the role of randomness is minor. The second step in an honest method is to compute the expected value of $x(t, \omega)$ as well as higher moments from the set of explicit solutions obtained in the first step.

In a dishonest method, the probabilistic nature of the equation is taken in consideration before the solution $x(t, \omega)$ is obtained. For example, the random equation is "averaged" and some statistical assumptions on the averaged equation are introduced. The object of these assumptions is to obtain a deterministic problem from which some moments can be directly determined. If this approach is to be successful, the averaged equation should be simpler than the original one, and the statistical assumptions are partially chosen bearing this in mind. They usually involve assumptions of partial or total independence of various random quantities. Such assumptions are usually not justifiable in advance. Unfortunately, the solutions obtained using a dishonest method are at least suspect, and in many cases they are incorrect. Nevertheless, methods of this general type have long been used in statistical mechanics, in the study of turbulence [262], [103], as well as in the propagation of waves through random media [150], [66], [257], [258], [296], [259], [2]. Since an infinite system of equations must be considered for the simultaneous determination of all moments, some assumptions are needed. One method for obtaining finite sets of equations is to assume that some moment can be expressed in a particular way in terms of moments of lower order. Such a closure procedure is the basis for most of the dishonest methods [257], [258], [2], [361], [69] for analyzing stochastic equations. These may be called "hierarchy equations" and involve the use of some truncation procedure (closure approximation) to terminate the hierarchy. Hierarchy methods have been used to study random differential equations [361], [2] and random eigenvalue problems for random differential equations [69]. The hierarchy equations can be combined into a single functional differential equation for the generating functional of the moments of the solution and the random coefficients.

7.1.2. *Perturbation-theory Approach*

Consider a random linear operator equation

(1) $T(\omega)x = y$

in a separable Banach space X. Suppose first that the random variations are sufficiently small corrections to the deterministic solution and are of low order. Then the perturbation theory can be used to obtain the equation for the expected solution of the random operator equation (1), where $y = y(t)$ is a given deterministic function [258], [31], [2]. Assume that $T(\omega)$ depends on a small parameter ε and that, for $\varepsilon = 0$, the random operator $T(\omega, \varepsilon)$ reduces to a deterministic operator T. Expanding $T(\omega, \varepsilon)$ in powers of ε, we may write

(2) $T(\omega, \varepsilon) = T + \varepsilon T_1(\omega) + \varepsilon^2 T_2(\omega) + O(\varepsilon^3)$.

Thus, $T(\omega, \varepsilon)$ is the sum of a deterministic operator T and of two random operators $T_1(\omega)$ and $T_2(\omega)$ which represent random perturbations of T. Suppose, for simplicity, that the deterministic operator equation $Tx = y$ is solvable, i.e., there exists a T^{-1} and $x_0(t) = T^{-1}y(t)$. Using (2), one obtains

(3) $x(t, \omega) = x_0(t) - \varepsilon T^{-1}(T_1(\omega))x(t, \omega) -$
$- \varepsilon^2 T^{-1}(T_2(\omega))x(t, \omega) + O(\varepsilon^3)$.

In order to solve (3), one can use the iterative method or successive substitution

$$x(t, \omega) = x_0(t) + \varepsilon x_1(t, \omega) + \varepsilon^2 x_2(t, \omega) +$$

and, hence, (3) becomes

(3′) $x(t, \omega) = x_0(t) - T^{-1}(T_1(\omega))x_0(t) +$
$+ \varepsilon^2 T^{-1}[T_1(\omega)T^{-1}T_1(\omega) - T_2(\omega)]x_0(t) + O(\varepsilon^3)$.

Taking the expectation $(E(.))$ yields

(4) $E(x,(t, \omega)) = x_0(t) - \varepsilon T^{-1}E(T_1(\omega))x_0(t) +$
$+ \varepsilon^2 T^{-1}[E(T_1(\omega)T^{-1}T_1(\omega)) -$
$- E(T_2(\omega))]x_0(t) + O(\varepsilon^3)$.

To obtain an expression for the expected solution, eliminate the deterministic solution $x_0(t)$ from (4) by observing that

(5) $x_0(t) = E(x(t, \omega)) + \varepsilon T^{-1}E(T_1(\omega))x_0(t) + O(\varepsilon^2) =$
$= E(x(t, \omega)) + \varepsilon T^{-1}E(T_1(\omega))E(x(t, \omega)) + O(\varepsilon^2)$.

Substituting (5) in (4) yields

(6) $E(x(t, \omega)) = x_0(t) - T^{-1}E(T_1(\omega))E(x(t, \omega)) +$
$+ \varepsilon^2 T^{-1}[E(T_1(\omega)T^{-1}T_1(\omega)) -$
$- E(T_1(\omega))T^{-1}E(T_1(\omega)) -$
$- E(T_2(\omega))]E(x(t, \omega)) + O(\varepsilon^3)$.

Upon multiplication of (6) by T, one obtains

(7) $\{T + \varepsilon E(T_1(\omega)) + \varepsilon^2[E(T_1(\omega))T^{-1}E(T_1(\omega)) -$
$- E(T_1(\omega)T^{-1}T_1(\omega)) +$
$+ E(T_2(\omega))]\}E(x(t, \omega)) = y + O(\varepsilon^3)$.

If $E(T_1(\omega)) = E(T_2(\omega)) = \theta$ (the null operator), then one obtains an explicit equation for $E(x(t, \omega))$:

$$(8) \quad T - \varepsilon^2 E(T_1(\omega)T^{-1}T_1(\omega))E(x(t, \omega)) = y + O(\varepsilon^3),$$

where the term $O(\varepsilon^3)$ can be dropped off.

A dishonest approach yields the same result. Indeed, if one sets $T_2(\omega) = \theta$ and takes the expectation values

$$[T + \varepsilon T_1(\omega) + \varepsilon^2 T_2(\omega) + O(\varepsilon^3)]x = y(t),$$

then, assuming that $E(T_1(\omega)) = \theta$, one obtains

$$(9) \quad TE(x(t, \omega)) + \varepsilon E(T_1(\omega)x(t, \omega)) + O(\varepsilon^3) - y(t).$$

To determine $E(T_1(\omega)x(t, \omega))$, first multiply (3′) by $T_1(\omega)$ and then apply the expectation operator

$$(10) \quad E(T_1(\omega)x(t, \omega)) = -\varepsilon E(T_1(\omega)T^{-1}T_1(\omega)x(t, \omega)) + O(\varepsilon^3).$$

Assume now that

$$(11) \quad E(T_1(\omega)T^{-1}T_1(\omega)\, x(t, \omega)) = $$
$$= E(T_1(\omega)T^{-1}T_1(\omega))E(x(t, \omega))$$

and, hence, (10) becomes

$$E(T_1(\omega)x(t, \omega)) = -\varepsilon E(T_1(\omega)T^{-1}T_1(\omega))E(x(t, \omega)) + O(\varepsilon^3),$$

which is substituted in (9) to obtain (8). Relation (11) is called the *local-independence* assumption.

Note that it is not necessary for y to be determinate (see [2]), but it is necessary to assume that $T(\omega)$ and $y(t, \omega)$ are stochastically independent.

7.1.3. *Applications to Wave Propagation in Continuous Media*

The mathematical formulation of wave propagation in random media leads to linear partial differential equations whose coefficients are random functions of space and time. As was already mentioned, the wave motion is described mathematically by a function $u(x, t)$, where x denotes the vector space variables and t denotes the time. The transmission medium, say D, is characterized by a function $n(x, t)$ which enters the coefficients of the equation for $u(x, t)$. In the case of random media, the single nonhomogeneous medium D is replaced by a family of media $\{D(\omega), \omega \in \Omega\}$, where Ω is a probability space. Suppose that, for every fixed $\omega \in \Omega$, the properties of the transmission medium are characterized by a function $n(x, t, \omega)$ (for example, the index of refraction). Then $\{D(\omega)\}$ is said to be a random medium if $n(x, t, .)$ is a random variable for every fixed x and t [31], [216], [258], [150], [2].

For propagation in time-dependent media, the *random wave propagation equation* assumes the form

$$\text{(12)} \quad \frac{\partial u}{\partial t} = (A + B(\omega))u,$$

where the unknown (scalar or vectorial) function u is called the wave (or field), A is a deterministic linear partial differential operator and $B(\omega)$ is a linear partial differential operator with random coefficients. The radiation of harmonic time-dependent waves in time-independent random media leads to the random differential equation

$$\text{(13)} \quad (A + B(\omega))u = y,$$

where y is called the *source term* (given by a deterministic function or distribution) and A and B are time-independent.

We now confine our attention to the case of a medium having a random propagation speed $c[1 + \varepsilon\mu(x, t, \omega)]^{-1}$. Then the wave equation is

$$\text{(14)} \quad \Delta u - c^{-2}[1 + 2\varepsilon\mu + \varepsilon^2\mu^2]u_{tt} = 0.$$

Comparing equations (14) and (1) yields $y = 0$ while T, T_1 and T_2 are given by

$$\text{(15)} \quad T = \Delta - c^{-2}\partial_t^2,$$

$$\text{(16)} \quad T_1 = -2c^{-2}\mu\partial_t^2,$$

$$\text{(17)} \quad T_2 = -c^{-2}\mu^2\partial_t^2.$$

If $E(\mu(x, t, \omega)) = 0$, then $E(T_1(\omega)) = 0$ and (7) can be applied to the mean wave, which is also called the "coherent" wave. If x denotes a point in three-dimensional space, then T^{-1} appearing in (7) is given by

$$\text{(18)} \quad T^{-1}f(x, t) = -\frac{1}{4\pi}\int r^{-1}f(y, t - c^{-1}r)\mathrm{d}y, \quad r = |x - y|.$$

This choice of T^{-1} is such that $T^{-1}f \to 0$ as $t \to -\infty$ provided that f vanishes quickly enough at $t = -\infty$.

The term in (8) involving T^{-1} is

(19) $E(T_1(\omega)T^{-1}T_1(\omega))E(u(x, t, \omega))=$

$$= E\left\{- 2c^{-2}\mu(x, t, \omega)\partial_t^2 \frac{(-1)}{4}\int r^{-1}[- 2c^{-2}\mu(y, t - \right.$$

$$\left. - c^{-1}r, \omega)\partial_t^2 E(u(y, t - c^{-1}r, \omega))]\mathrm{d}y\right\} =$$

$$= \frac{-1}{\pi c^4} E\left\{\mu(x, t, \omega)\int r^{-1}\Big[\mu_{tt}(y, t - c^{-1}r, \omega)E(u(y, t - \right.$$

$$- c^{-1}r, \omega))_{tt} + 2\mu_t(y, t - c^{-1}r, \omega)E(u(y, t - c^{-1}r, \omega))_{ttt} +$$

$$\left. + \mu(y, t - c^{-1}r, \omega)E(u(y, t - c^{-1}r))_{ttt}\Big]\mathrm{d}y\right\}.$$

If the factor $\mu(x, t, \omega)$ is brought under the integral sign in (19), the expectation value of each term of the integrand can be expressed in terms of the correlation function R defined by

(20) $E(\mu(t, x, \omega)\, \mu(y, s, \omega)) = R(|x - y|, |t - s|).$

The dependence of R on $|x - y|$ and $|t - s|$ is similar, if the medium is statistically homogenous and isotropic in space and stationary in time. Using (20) and (19) in (7) with $E(T_1(\omega))= = 0$, we finally obtain

(21) $[\Delta - c^{-2}(1 + \varepsilon^2 E(\mu^2))\partial_t^2]E(u(x, t, \omega)) +$

$$+ \frac{2}{\pi c^4}\int r^{-1}[R_{\tau\tau}(r, r/c).\ E(u(y, t - c^{-1}r, \omega))_{tt} -$$

$$- 2R_\tau(r, r/c)E(u(y, t - c^{-1}r, \omega))_{ttt} +$$

$$+ R(c, r/c)E(u(y, t - c^{-1}, r\omega))_{tttt}]\mathrm{d}y = 0,$$

where R denotes the derivative of R with respect to its second argument whereas the term $O(\varepsilon^3)$ has been omitted. It should be noted that (20) implies that $E(\mu^2) = R(0, 0)$. If the medium is not statistically isotropic, then R depends on $x - y$ rather than on $|x - y|$ and (21) still holds.

The study of methods for determining the covariance properties of waves is useful in the quantitative determination of the covariance properties of waves.

Applying (21) to a coherent wave $E(u)$ (time-harmonic with angular frequency α) yields

$$E(u(x, t, \omega)) = \mathrm{e}^{-i\alpha t}v(x).$$

Introducing $k_0 = \alpha/c$ leads to the equation for $v(x)$:

$$(22)\quad [\Delta + k_0^2(1 + \varepsilon^2 E(\mu^2))]v(x) + \\ + \frac{\varepsilon^2 k_0^2}{\pi c^2}\int r^2 e^{ik_0 r}[\alpha^2 R(r, r/c) - i\alpha R_\tau(r, r/c) - \\ - R_{\tau\tau}(r, r/c)]v(y)\mathrm{d}y = 0.$$

An important consequence of (22) is the finding of an effective propagation constant k for a random medium. This can be done by considering a plane wave solution with propagation vector $\bar{k}$ with $k = |\bar{k}|$. Taking $v(x) = e^{ikx}$ one obtains the equation for k

$$k^2 = k_0^2(1 + \varepsilon^2 E(\mu^2)) + \frac{4\,\varepsilon^2 k_0^2}{k\varepsilon^2}\int_0^\infty e^{ik_0 r}\sin kr[\alpha^2 R(r, r/c) - \\ - i\alpha R_\tau(r, r/c) - R_{\tau\tau}(r, r/c)]\mathrm{d}r.$$

An interesting study of wave propagation in media of discrete scatterers, hierarchy equations and equations for generating functionals can be found in ref. [258].

The results of Sub-section 2.2 and the covariance properties of the waves allowed Chen and Soong [102] to obtain the following equation:

$$(23)\quad E(x(t_1, \omega))E(x(t_2, \omega)) = \\ = \{1 + \varepsilon^2 T^{-1}(t_1)T^{-1}(t_2)[E(T_1(t_1, \omega), T_1(t_2, \omega)) - \\ - E(T_1(t_1, \omega))E(T_1(t_2, \omega))]\}\, E(x(t_1, \omega))E(x(t_2, \omega)) + O(\varepsilon^3).$$

Combining this equation with (4) gives a direct evaluation of the correlation function.

A correlation equation for $x(t, \omega)$ independent of its mean value in the inhomogeneous part is

$$\{T(t_1)T(t_2) - \varepsilon^2[T(t_1)E(T_1(t_2, \omega)T^{-1}(t_2)T_1(t_2, \omega)) + \\ + T(t_2)E(T_1(t_1, \omega)T^{-1}(t_1)\cdot T^1(t_1, \omega)) + E(T_1(t_1, \omega)T_1(t_2, \omega)) - \\ - T(t_1)E(T_2(t_2, \omega)) - T(t_2)E(T_2(t_1, \omega)) + \\ + O(\varepsilon^3)]\}\, E(x(t_1, \omega), x(t_2, \omega)) = y(t_1)y(t_2).$$

The use of these equations is illustrated by an example in ref. [102].

The small-amplitude periodic classical motion of a lattice of particles about their equilibrium position is considered. The effect of random masses and random spring constants upon the coherent or mean motion can be treated by use of the equation for coherent motion obtained above (for example Eq. (7)).

7.2. A RANDOM INTEGRAL EQUATION IN TURBULENCE THEORY

We now present an application of random integral equations to the study of turbulent fluid motion, as given by C. P. Padgett and C. P. Tsokos [332]. The admissibility theory and Banach's fixed point theorem were used by these authors in their fundamental works on random integral equations [431], [432]. These results will also be used in the sequel in order to ensure the existence of a random solution to the stochastic integral equation

$$(1)\quad r(t, \omega) = \int_0^t u(s, r(s, \omega), \omega)\, ds,$$

which gives the position of a tagged material point in a continuous fluid in turbulent motion. Let the position at time $t > 0$ of a point which started at time zero at some reference point, say, the origin, be a vector-valued random variable $r(t, \omega)$, $\omega \in \Omega$, where Ω is the supporting set of the underlying probability space. At position $r(t, \omega)$, the velocity of the point is given by the Eulerian velocity field $u(r(t, \omega), t, \omega)$, which is a vector-valued random variable for each r and $t \geqslant 0$. The Eulerian approach for describing flow in fluid mechanics is that the velocity of a point in a continuous fluid moving by a fixed position at any time $t \geqslant 0$ can be given. However, if the flow is turbulent, then the velocity is random and the problem is to determine $r(t, \omega)$ if the statistical properties of $u(r(t, \omega), t, \omega)$ are known (a vector function whose components record the random rate of change of the coordinates of the tagged point at time t, and the length $|u(r, t, \omega)|$ records the random speed of the tagged point at time $t > 0$).

To examine some of the conditions under which equation (1) admits a random solution for each $t \geqslant 0$, we may assume that $r(t, \omega)$ is a continuous vector-valued function from $\mathbb{R}^+$ into $L_p(\Omega, \mathscr{K}, P)$, $1 \leqslant p \leqslant \infty$ and that the stochastic kernel $u(t, r(t, \omega))$ is a three-component vector-valued function which, for each $t \in \mathbb{R}^+$ and $\omega \in \Omega$, is continuous in r, and, for each $t \in \mathbb{R}^+$ and each vector $r(t, \omega)$, is in $L_p(\Omega, \mathscr{K}, P)$.

Let $C_p(\mathbb{R}^+, L_p(\Omega, \mathscr{K}, P)) = C_p$ be the space of all three-component vector-valued $\left(\text{i.e., } |r(t, \omega)| = \left(\sum_{k=1}^{3} |r_k(t, \omega)|^2\right)^{1/2}\right)$

continuous functions from $\mathbb{R}^+$ into $L_p(\Omega, \mathscr{K}, P)$ with the topology of uniform convergence on each compact interval $[0, m]$, $m > 0$ ($r_n(t, \omega) \to r(t, \omega) \in C_p$ if and only if $\left(\int_\Omega |r_n(t, \omega) - r(t, \omega)|^p\right)^{1/p} \to 0$ uniformly on every interval $[0, m]$).

Let D and E be a pair of Banach spaces such that $D, C \subset C_p$ and suppose that $T : C_p \to C_p$ is a linear operator such that $TE \subset D$. Then the pair (E, D) is said to be admissible with respect to T. If T is a continuous operator and the pair of Banach spaces D and E are stronger than C_p and if (E, D) is admissible with respect to T, then T is a continuous operator from E into D (i.e., there exists a constant $K_0 > 0$ such that $\|(Tr)(t, \omega)\|_D \leqslant K_0 \|r(t, \omega)\|_E$). The norm $\|r(t, \omega)\|_D$ is defined by $\sup_{t \geqslant 0} \|r(t, \omega)\|_p$.

THEOREM 7.2.1. *Let $D, E \subset C_p$ be an admissible pair of Banach spaces with respect to the operator*

$$(Tr)(t, \omega) = \int_0^t r(s, \omega)\,ds.$$

Let $u(t, ., \omega) : W \subset D \to E$ be a mapping from the set $W = \{r(t, \omega) : r(t, \omega) \in D, \|r(t, \omega)\|_D \leqslant \rho\}$ into the space E, for some $\rho \geqslant 0$, satisfying

$$\|u(t, r(t, \omega), \omega) - u(t, v(t, \omega), \omega)\|_E \leqslant b\|r(t, \omega) - v(t, \omega)\|_D$$

for $r(t, \omega)$, $v(t, \omega) \in W$ and $b \geqslant 0$ a constant.

If $\|u(t, r(t, \omega), \omega)\|_E \leqslant \rho/K$, where K is the norm of T and $bK < 1$, then there exists a unique random solution of (1).

Proof. Define the operator $U : W \to D$ by

$$(Ur)(t, \omega) = \int_0^t u(s, r(s, \omega))\,ds.$$

Since u is continuous in r, U is a continuous mapping from W into D.

Let us prove that $U(W) \subset W$ and that U is a contraction mapping on W. Indeed, if $r(t, \omega) \in W$, we have

$$\|(Ur)(t, \omega)\|_D = \left\| \int_0^t u(s, r(s, \omega), \omega)ds \right\|_D \leqslant$$

$$\leqslant K\|u(s, r(s, \omega), \omega)\|_E \leqslant \rho$$

so that $(Ur)(t, \omega) \in W$.

Let $r(t, \omega), v(t, \omega) \in W$ and, hence, $Ur - Us \in D$ and

$$\|(Ur)(t, \omega) - Uv(t, \omega)\|_D = \left\| \int_0^t u(s, r(s, \omega), \omega)ds - \right.$$

$$\left. - \int_0^t u(s, v(s, \omega), \omega)ds \right\|_D \leqslant$$

$$\leqslant K\|u(s, r(s, \omega), \omega) - u(s, v(s, \omega), \omega)\|_E \subset$$

$$\subset Kb\|r(t, \omega) - v(t, \omega)\|_D.$$

Thus, U is a contraction mapping on $W \subset D$ and Banach's fixed point theorem implies that there exists a unique element $r(t, \omega) \in W$ such that

$$(Ur)(t, \omega) = \int_0^t u(s, r(s, \omega), \omega)ds = r(t, \omega)$$

for every $t \in \mathbb{R}^+$.

The analytical study of dispersion of a suspended particle in turbulent fluid was initiated by Taylor (see [103, p. 28]). M. Kaç [245] introduced a random walk model. Let $v(t)$ denote the velocity of a marked particle moving along at time t, and let $|v(t)| = v_0$ be a constant. If the number of time reversals $N(t)$ is a Poisson process with the parameter a, the particle trajectory $\xi(t)$ satisfies the kinematic relation $d\xi/dt = \xi_0(-1)^{N(t)}$, where $\xi_0 = \pm v_0$ with equal probability. Let $\Phi(x)$ be a smooth real function on $\mathbb{R}$. If the expectational function

$$u(t, x) = E\{\Phi[\xi(t)]/\xi(0) = x\}$$

is defined, then u satisfies the dissipative wave equation

$$(2) \quad \partial_t^2 u + 2a\partial_t u = v_0^2 \partial_x^2 u$$

and the initial conditions

$$(2') \quad u(0, x) = \Phi(x), \quad \partial_t u(0, x) = 0.$$

If the observation time is long or if both the intensity and speed are high, then we have the turbulent diffusion approximation.

This probabilistic model was generalized to include the dispersion of particles with many speeds.

For a turbulent transport equation, one considers a marked Brownian particle drifting with the local turbulent fluid velocity $v(t, x, \omega)$. Then $\sigma(x)$ denotes the diffusion matrix of the Brownian motion $z(t)$ in $\mathbb{R}^3$. Then the trajectory $\xi(t, \omega)$ of the particle must obey the randomized stochastic equation

$$d_\xi(t) = v[t, \xi(t), \omega]dt + \sigma[\xi(t)]dz(t).$$

Let $\mathscr{F}_t$ denote the complete σ-field generated by the turbulent velocity $v(t,., \omega)$ up to time t. Suppose that the fluid is contained in a domain $D \subset \mathbb{R}^3$, and that, for every set $B \subset D$, the transition probability

$$P(s, y, t, B(\omega)) = E\{\xi(t) \in B | \mathscr{F}_t, \xi(s) = y\}$$

has a smooth probability density $p(s, y, t, x, \omega)$ as a function of $(t, x) \in D_T$ a.s. in ω. Then the particle concentration function

$$u(t, x, \omega) = \int_D \eta(y) p(0, y, t, x, \omega)\, dy$$

satisfies the stochastic Fokker-Planck equation

$$(3) \quad \partial_t u + v \cdot \nabla u = \frac{1}{2} \sigma[\sigma(x) \nabla \sigma(x)\, u]$$

and the initial-boundary conditions

$$(3') \quad u(0, x) = \eta(x), \quad \left.\frac{\partial u}{\partial x}\right|_{\partial D} = 0.$$

A detailed study of this problem was done by Chow [103].

7.3. RANDOM VIBRATION OF ELASTIC STRINGS AND BARS

The aim of this section is to present some problems of free vibrations of two simple elastic systems having random material properties: the transverse vibrations of a tightly stretched elastic string whose mass per unit length varies with position in a stationary random manner, and the longitudinal or transverse vibration of a bar of random cross-section. The results presented in this section are due to Boyce [67], [68], [69] Boyce and Goodwin [70] and Bharucha-Reid [31]. The boundary-value problem governing the free transverse vibrations of a tightly stretched string having both ends fixed can be written as

$$(1) \quad U'' + \lambda(1 + A)U = 0, \quad U(0) = U(1) = 0,$$

where U is the dimensionless transverse displacement, $A = A(x)$ is proportional to the random deviation of the linear density from its mean value at the point x, and λ is a parameter proportional to the square of the natural frequency. We assume that $A(x)$ is continuous and $1 + A > 0$. Then the eigenvalues λ_n (ordered in increasing size) are all real and positive.

Green's function $G(x, y)$ associated to the differential operator U'' subject to the given boundary condition is given by

$$(2) \quad G(x, y) = \begin{cases} x(1 - y), & 0 \leqslant x \leqslant y \leqslant 1, \\ y(1 - x), & 0 \leqslant y \leqslant x \leqslant 1. \end{cases}$$

Then the boundary-value problem (1) is converted into the integral equation

$$(3) \quad U(x) - \lambda \int_0^1 G(x, y)(1 + A(y))U(y)\mathrm{d}y = 0, \quad 0 \leqslant x \leqslant 1$$

or, for $u = (1 + A)^{1/2}U$,

$$(4) \quad u(x) - \lambda \int_0^1 K(x, y)u(y)\,\mathrm{d}y = 0, \quad 0 \leqslant x \leqslant 1,$$

i.e., a homogeneous Fredholm integral equation of the second kind with the positive symmetric kernel

$$K(x, y) = G(x, y)(1 + A(x))^{1/2}(1 + A(y))^{1/2}.$$

It is known [70], [163] that the eigenvalue λ_n of Eq. (4) are the same as those of Eq. (1) and that

$$(5) \quad \sum_{n=1}^{\infty} \lambda_n^{-m} = \int_0^1 K_m(x, x)\mathrm{d}x, \; m = 1, 2, \ldots$$

$$K_1(x, y) = K(x, y), \; K_m(x, y) = \int_0^1 K(x, z)K_{m-1}(z, y)\mathrm{d}z, \; m = 2, 3, \ldots.$$

If $A(x)$ is a random function, then (5) gives information concerning the statistical moments of λ_1^{-1} in terms of appropriate assumptions about the statistical properties of $A(x)$. Let us assume that $A(x)$ is a stationary random function and that

$$(6) \quad E(A(x)) = 0, \; 0 \leqslant x \leqslant 1$$

$$(7) \quad E(A(x)A(y)) = \mu^2 R(x - y), \quad 0 \leqslant x, y \leqslant 1,$$

where $\mu^2 = E(A^2)$ is constant, and the correlation function $R(.)$ depends only upon $x - y$, and not upon x and y separately.

As we know, in the deterministic case the following inequalities hold :

$$(8)\quad (1-\delta)(1/6-\pi^{-2}) \leqslant \int_0^1 K(x,x)\,dx - \lambda_1^{-1} \leqslant$$

$$\leqslant (1+\delta)(1/6-\pi^{-2})$$

$$(9)\quad (1-\delta)^2(1/90-\pi^{-4}) \leqslant \int_0^1 K_2(x,x)\,dx - \lambda_1^{-2} \leqslant$$

$$\leqslant (1+\gamma)^2(1/90-\pi^{-4}), \ldots,$$

$$(10)\quad \left|E\left(\int_0^1 K(x,x)dx\right) - (1/6-\pi^{-2}) -\right.$$

$$\left. - E(\lambda_1^{-1})\right| \leqslant \delta(1/6-\pi^{-2}),$$

$$(11)\quad E\left(\int_0^1 K_2(x,x)dx\right) - (1+\delta^2)(1/90-\pi^{-4}) -$$

$$- E(\lambda_1^{-2}) \leqslant 2\delta(1/90-\pi^{-4}).$$

From the definition of $K(x, x)$, it follows that

$$E\left(\int_0^1 K(x,x)\,dx\right) = \int_0^1 G(x,x)dx = 1/6$$

and that

$$E\left(\int_0^1 K_2(x,x)dx\right) =$$

$$= \int_0^1\int_0^1 G^2(x,y)(1+E(A(x)A(y)))\,dy\,dx = 1/90 +$$

$$+ \mu^2 \int_0^1\int_0^1 G^2(x,y)R(x-y)dx\,dy.$$

With the notation

$$J' = \int_0^1\int_0^1 G^2(x, y)R(x-y)\mathrm{d}y\,\mathrm{d}x$$

(10) and (11) become

(12) $|\pi^{-2} - E(\lambda_1^{-1})| < (1/6 - \pi^{-2})$;

(13) $|\mu^2 J' + \pi^{-4} - \delta^2(1/90 - \pi^{-4}) -$
$- E(\lambda_1^{-2})| \leqslant 2\delta(1/90 - \pi^{-4})$.

Similar estimates of $E(\lambda_1^{-m})$ are available in terms of higher-order correlation functions of $A(x)$. For fairly small values of δ, the accuracy should be adequate for many purposes (for instance, for $\delta = 0.1$, Eq. (12) gives the approximate value of $\pi^{-2} = 0.1013$ for $E(\lambda_1^{-1})$ with a maximum error of 0.0065).
Estimates for variance of λ_1^{-1} are given by

(14) $$\left|\mathrm{var}(\lambda_1^{-1}) - \mu^2 J' + \delta^2\left(\frac{7}{180} - \frac{1}{3\pi^2}\right)\right| \leqslant$$
$$\leqslant \delta\left(\frac{1}{45} + \frac{1}{3\pi^2} - \frac{4}{\pi^4}\right)$$

(15) $\mathrm{var}(\lambda_1^{-1}) = \mu^2 J' + 0(\delta)$

or

(16) $\mathrm{var}(\lambda_1^{-1}) = \mu^2 J'' + 0(\delta)$ where $J'' =$

$$= \int_0^1\int_0^1 G(x, x)G(y, y)R(x-y)\,\mathrm{d}y\,\mathrm{d}x$$

Thus, we have obtained upper and lower bounds for the statistical moments of the frequencies in terms of the corresponding moments and appropriate correlation functions for the random density.

Let us now consider the problem of analysing frequencies of transverse vibration of random elastic bars [70]. When the random element is the geometry of the cross-section of the bar, then the differential equation is of fourth order and its coefficients are random. The mathematical problem is then to estimate the effects of these random coefficients upon the natural frequencies of vibration, which appear as eigenvalues of the resulting boundary-value problem. Hence, we seek estimates for the statistical moments, in particular the mean and variance of the eigenvalues in terms of the corresponding parameters associated with the random coefficients occurring in the differential equation or boundary conditions. The differential equation governing the small transverse vibration of the elastic bar is

$$(17) \quad (EIu'')'' - m\,A\,l^4\omega^2 u = 0,$$

where $u(x)$ is the dimensionless time-independent transverse displacement, E is Young's modulus, I is the moment of inertia of the cross-section about an axis through the centroid perpendicular to the plane of vibration, m is the mass per unit length, A is the cross-sectional area, l is the length of the bar and ω is the frequency of transverse vibration. We assume that E, m, and l are determinate constants, but that the cross-section parameters A and I vary with position in a random manner.

If the area A and the moment inertia I are assumed to be stationary random functions with constant mean values A_0 and I_0, respectively, then

$$A = A_0(1 + B(x)), \quad I = I_0(1 + C(x)),$$

where $B(x)$ and $C(x)$ are stationary random functions with zero means. Equation (17) assumes the form

$$((1 + C)u'')'' - \lambda(1 + B)u = 0,$$

where

$$\lambda = \frac{mA_0 l^4\omega^2}{EI_0}.$$

If we assume that only the size and not the shape of the cross-section varies with position, then we can write

$$A(x) = k_1 r^2(x), \quad I(x) = k_2 r^4(x),$$

where k_1 and k_2 are constants and $r(x)$ denotes some typical length in the cross-section. If $r(x)$ is a stationary random function with mean value r_0, we have

$$r(x) = r_0(1 + D(x)),$$

where $D(x)$ is a stationary random function with mean value zero. Then equation (17) becomes

$$(17') \quad ((1 + D(x))^4 u'')u'' - \lambda(1 + D(x))^2 u = 0.$$

When $D(x)$ is small, a perturbation expansion about the uniform bar is natural and, for convenience, we introduce the slightly modified differential equation

$$(17'') \quad ((1 + \varepsilon D(x))^4 u'')u'' - \mu^4(1 + \varepsilon D(x))^2 u = 0$$

($\mu^4 = \lambda$) and consider μ as the eigenvalue rather than λ. As boundary condition, we take

$$(17''') \quad \begin{aligned} u(0) &= 0, \quad u''(0) - \delta_1 u'(0) = 0 \\ u(1) &= 0, \quad u''(1) - \delta_2 u'(1) = 0 \end{aligned}$$

corresponding to an elastic support at each end. As $\delta_1, \delta_2 \to 0$, the ends become hinged and as $\delta_1, \delta_2 \to \infty$, they become clamped. In order to make a comparison with other results, it is convenient to multiply (17'') by μ^{-4} and to assume that μ_n^{-1}, i.e. the reciprocal of the n-th eigenvalue, has the formal expansion

$$\mu_n^{-1} = \sum_{m=0}^{\infty} \mu_{n,m} \varepsilon^m.$$

Then one can obtain the following estimates :

$$E(\mu_n^{-1}) \cong \mu_{n,0} \quad E(\mu_n^{-2}) \cong \mu_{n,0}^2 \left(1 + E\left(\left(\frac{\mu_{n,1}}{\mu_{n,0}}\right)^2\right)\right)$$

$$\operatorname{var}(\mu_n^{-1}) = \mu_{n,0}^2 E\left(\left(\frac{\mu_{n,1}}{\mu_{n,0}}\right)^2\right).$$

In a similar way, one can describe the longitudinal vibrations of an elastic bar [69]. These vibrations are governed by the differential equation

$$(18) \quad \frac{\partial}{\partial s}\left(\alpha a \frac{\partial w}{\partial s}\right) = m a \frac{\partial^2 w}{\partial t^2},$$

where w represents the axial displacement in the rod, s is distance measured along the rod, α is Young's modulus, a the cross-sectional area, and m is the mass density per unit volume. Let α and m be constants and let a vary in a random manner along the length of the bar. If

$$w(s, t) = v(s)\, e^{i\lambda t}$$

and if

$$V = \frac{v}{l}, \quad x = \frac{s}{l},$$

where l is the length of the bar, then $V(x)$ satisfies

$$(18') \quad (aV')' + a(x) \frac{m l^2 \lambda^2}{\alpha} V = 0.$$

If the ends of the bar are held fixed, then the boundary conditions are

$$(18'') \quad V(0) = V(1) = 0.$$

If $E(a)$ = constant, then $a(x) = E(a)\,(1 + A(x))$, where $A(x)$ is a random function with zero mean value measuring the deviation of $a(x)$ from its value at the point x. Then, for $\gamma^2 = ml^2\lambda^2/\alpha$, the differential equation assumes the form

$$(18''') \quad ((1 + A)V')' + (1 + A)\gamma^2 V = 0.$$

Assuming expansions of the form

$$V_n = \sum_{m=0}^{\infty} V_{n,m}\,\varepsilon^m, \quad \gamma_n = \sum_{m=0}^{\infty} \gamma_{n,m}\,\varepsilon^m,$$

we can conclude [69] that

$$\gamma_{n,0} = n\pi, \quad \gamma_{n,1} = n\pi \int_0^1 A(x) \cos n\pi x \mathrm{d}x$$

$$E(\gamma_{n,1}) = 0, \quad E(\gamma_{n,1}^2) = n^2\pi^2 E(A^2) \int_0^1\int_0^1 R(x -$$

$$- y) \cos 2n\pi x \cos 2n\pi y \mathrm{d}x\mathrm{d}y$$

$$E(\gamma_n/n) = \pi, \quad E((\gamma_n/n)^2) = \pi^2 + E(\gamma_{n,1}^2)/n^2.$$

7.4. RANDOM METHODS IN STREAM TEMPERATURE MODELLING

Mathematical modelling of physical systems by means of stochastic systems has developed largely because of : (a) the scarcity or high cost of real-world data, (b) skepticism over the available data and (c) the reseachers' attempt to introduce more natural and general formulations to the respective problems. In a candid analysis of description of any physical system, at least one element of uncertainty should be included sooner or later. This is particularly true when considering the interaction of two or more thermodynamic systems. For example, the temperature of streams in contact with the atmosphere is influenced by many natural hydrodynamic and climatic uncertainties even neglecting those which are unnatural. These are only some of the arguments considered by W. L. Morse [295], [296] when proposing the random model which will be examined in the sequel

A quantity of matter of fixed mass and given identity is defined as a system. A system may change in shape, position and termodynamic properties, but it must always contain the same matter. The behaviour of the system is influenced by the reactions within the system and the interactions between the system and its surrounding media. This is the case of an unsteady stream of water moving through an essentially well-defined channel exposed to a variety of weather conditions. Water, as the moving media of open channel flow, is subject to gravitational forces, hydraulic pressure, viscous and frictional forces from the channel and the atmosphere.

Let $T(x, y, z, t)$ be the instantaneous temperature of the stream of water, where x, y, z are longitudinal, lateral and depth coordinates with the origin located on the projection of thalweg onto the water's surface. Let ρ be the mass density and let c_p be the isobaric specific heat, both of which are assumed to be constant. W. Morse [296] proved that the deterministic model of the one-dimensional form of the thermal-energy conservation principle is given by a quasi-linear partial differential equation :

$$(1)\quad \frac{\partial\theta}{\partial t} + V\frac{\partial\theta}{\partial t} = (\rho c_p z)^{-1}(A\theta^2 + B\theta + C),$$

where $\theta = T(x, t)$, $V = V_x(x, t)$, the nonhomogeneous term is the volumetric interphase transport of heat at the air-water interface, $z = z(x, t)$ is the stream depth and $H(x, t) = H(t) = H(t, T) = A(t)T^2 + B(t)T + C(t)$ is the approximated net heat flux for a fixed time unit ($\Delta t = 6$ or $\Delta t = 3$ [295]). Two methods of solution are presented [296] which lead to a unique solution and a stochastic differential equation with random forcing function from the meteorologic and forebay-depth analyses. Since the subsidiary differential (Lagrangian) system for (1) produces a Riccati ordinary diferential equation

$$\mathrm{d}\theta/\mathrm{d}t = \lambda(A\theta^2 + B\theta + C), \quad \lambda = \lambda(x, t) = [\rho c_p z(x, t)]^{-1},$$

one can give the stochastic formulation

$$(2)\quad \mathrm{d}\theta(t, \omega)/\mathrm{d}t = K(t, \theta(t, \omega), \omega), \quad \theta(0, \omega) = \theta_0(\omega),$$

where at some fixed site x_0 along the stream,

$$(2')\quad K(t, \theta, \omega) = \lambda(t, \omega)\, H(t, \theta, \omega)$$

$$H(t, \theta, \omega) = A(t, \omega)\theta^2(t, \omega) + B(t, \omega)\theta(t, \omega) + C(t, \omega) \text{ and } \lambda(t,\omega) = [\rho c_p z(t, \omega)]^{-1}.$$

Hence, a function $\theta(t, \omega) : [0, \infty) \times \Omega \to \mathbb{R}^n$ is a solution of (2) if, for all $t \in [0, \infty)$,

$$(3)\quad \theta(t, \omega) = \theta_0(\omega) + \int_{t_0}^{t} K(s, \theta(s, \omega), \omega)\, ds \text{ a.s.}$$

This stems from the fact that, for each fixed $\omega \in \Omega$, $K(t, \theta, \omega)$ is Lebesgue-integrable, and the sample functions $\theta(t, \omega)$ are absolutely continuous. Introducing the random integral operator $U(\omega)$ such that

$$(4)\quad U(\omega)(\theta) = \theta_0(\omega) + \int_{t_0}^{t} K(s, \theta(s, \omega), \omega) ds$$

and the function $\Phi(t, \omega)$ as a random fixed point of U, then $\Phi(t, \omega)$ is a random solution of (3) and (2).

The existence of this random solution Φ is ensured by any of the conditions given in Chapter 4 for the existence of random fixed points. For example, Morse [296] used the conditions of Theorem 4.1.2.

For physical reasons, the variances and covariances of the random variables in Eq. (3) are all finite, and so the stochastic process $\theta(t, \omega)$ is in the class of all square-summable $L_2(\Omega, \mathscr{K}, P)$. Thus, it can be shown that the assumptions of Theorem 4.1.2 are satisfied with

$$k(t', \omega) = \|\lambda(t', \omega)\| \, \|2A(t', \omega)\bar{\xi} + B(t', \omega)\|, \quad \bar{\xi} = (\xi_1 + \xi_2)/2,$$
$$t' \in [t_0, t].$$

The continuity of U follows from the integrability of K in (2) and (2′) and from the finiteness of the expectations of the meteorological and reciprocal-depth random variables.

W. Morse [296] mentions a dishonest method for solving the random equations (2) and (3). Worth mentioning is also the dynamic approach towards the stochastic-process solution. Even when we dispense with the assumption of independence or the knowledge on the distribution, Chebysheff's inequality would still permit a theoretical prediction for the range of values $\{\theta(t, \omega)\}$. Of great interest are Secs. 7.7. and 7.8 therein, which give concrete applications, tests and conclusions. From these, we retain the fact that during critical climateric or reduced flow conditions, the random model might aid water resource planners and reservoir-operating schedulers to make wise decisions regarding the preservation of valuable fishery or wildlife resources and still provide the electric energy required for everyday needs.

7.5. STOCHASTIC MODELS FOR STREAM POLLUTION

The pollution of natural water resources caused by the disposal of organic waste materials has become a major concern of the scientific community. In order to design any type of waste treatment plant, it is necessary to know the volume and concentration of a pollutant that may be discharged to the receiving stream such that the stipulated stream standards are not violated. This implies that there should be methods to predict the response of the receiving stream to waste loads placed therein. Many organic compounds discharged into lakes and streams are effectively degraded into inoffensive components by the action of bacteria in the water. As the organisms in a body of water degrade the pollution, they require oxygen, and hence use the dissolved oxygen (DO) in the water. Oxygen-consuming pollutants are measured in terms of the amount of oxygen required by the bacteria to stabilize them, called the biochemical oxygen demand (BOD). It is generally accepted that the BOD and DO, measured in parts per million (ppm), are the primary indicators for water quality in a stream or river. Dissolved oxygen is obtained directly from the atmosphere by aeration and, indirectly, through the photosynthetic process of aquatic plats.

Knowledge of the response of a stream DO concentration to the addition of organic wastes is essential to maintain sufficiently high DO concentrations to support auqatic life. Hence, this requires the development of realistic and workable mathematical models for the prediction of BOD and DO concentration at points downstream a waste source.

Several deterministic mathematical models and stochastic models for BOD and DO prediction are available [327], [328], [333], [431], [272], [119]. Thus, the Markov model, the birth-death-process approach, the random-walk model with a Monte Carlo simulating technique and the random differential equation approach are some of the instruments furnished by mathematics.

The general assumptions used for obtaining a deterministic of stochastic model are that there exist five major activities in the stream : (i) Due to pollution, (BOD) and (DO) are decreased by the action of bacteria. The rate of decrease is assumed to be proportional to the amount of pollution present with proportionality constant k_1 (deoxygenation coefficient) in units of dissolved oxygen per day (ppm), and there is always some oxygen present. (ii) The dissolved oxygen is increased by the process of re-aeration at a rate proportional to the dissolved-oxygen deficit DO (which is the DO saturation concentration minus the actual DO concentration), with proportionality constant k_2 (reaeration coefficient) in units of dissolved oxygen per day (ppm). (iii) The pollution is decreased by sedimentation and adsorption at a rate proportional to the amount of pollution present with proportionality constant k_3 in units of dissolved oxygen per day. (iv) The pollution is increased from small sources along the stretch of the stream with rate l_a in ppm per day, which is independent of the amount of pollution present.(v) The dissolved oxygen is decreased at a rate d_B in ppm per day. The variable d_B may have positive or negative values and represents the net change in dissolved oxygen due to the Benthal demand and respiration and photosynthesis of plants. It was also assumed that the stream flow was steady and uniform and that the conditions at every cross-section did not change with time.

Under the above assumptions, one obtains the following deterministic differential equations for BOD and DO [327], [328]:

$$-u\,\mathrm{d}l/\mathrm{d}t - (k_1 + k_3)l(t) + l_a = 0$$

(1)

$$-u\,\mathrm{d}c/\mathrm{d}t + k_2[c_s - c(t)] - k_1 l(t) - d_B = 0,$$

where t is the distance downstream from the pollution source, $l(t)$ is the (first stage) BOD in ppm at distance t, u is the average stream velocity along the stretch, c_s is the saturation concentration for dissolved oxygen, and $c(t)$ is the concentration of DO in ppm at distance t downstream. The initial conditions were $l(0) = l_0$ and $c(0) = c_0$.

Padgett [327] pointed out that the quantities k_1, k_2, k_3, u, l_a, c_s, d_B, l_0 and c_0 are more realistically considered to be random variables which have certain probability distributions rather than physical constants. This means that some of the uncertainty in modelling the BOD and DO processes may be expressed in the form of randomness in certain of the variables involved in the model. Therefore, a system of random differential equations given by

$$(2)\quad \mathrm{d}X/\mathrm{d}t = AX(t) + Y, \quad t \geqslant 0,$$

where

$$X(t) = \begin{pmatrix} l(t) \\ c(t) \end{pmatrix}, \quad Y = \begin{pmatrix} l_a/u \\ (k_2 c_s - d_B)/u \end{pmatrix},$$

$$A = \begin{pmatrix} -(k_1 + k_3)/u & 0 \\ -k_1/u & -k_2/u) \end{pmatrix},$$

and

$$X(0) = X_0 = \begin{pmatrix} l_0 \\ c_0 \end{pmatrix}$$

was proposed [327] as the stochastic model. Thus, the solutions of (2) are stochastic processes, $l(t)$ and $c(t)$, giving the BOD and DO concentrations, respectively, at each distance t downstream from the pollution source.

Assuming that l_a, d_B, l_0 and c_0 are random variables with finite second moments, the solution of system (2) is as given in ref. [327]. According to Soong [412], the vector-valued stochastic process $X(t)$, $t \geqslant 0$, is a *mean square solution* of (2) if $X(t)$ is a mean square continuous on $[0, \infty]$, that is $X(t+h) \xrightarrow{\text{m.s.}} X(t)$ as $h \to 0$ for each $t \geqslant 0$, $X(0) = X_0$ with probability one; and $AX(t) + Y$ is the mean square derivative of $X(t)$ on $[0, \infty)$. Using the techniques suggested by Soong [412, Chap. 7], the mean square solution of (2) may be represented as

$$(3) \quad X(t) = \Phi(t) + \int_0^t \Phi(t - s) Y \, ds, \quad t \geqslant 0,$$

where the integral is a mean square integral, and

$$\Phi(t) = \begin{pmatrix} \exp[-(k_1 + k_3)t/u] & 0 \\ -\dfrac{k_1}{k_1 - k_2 + k_3}\{\exp(-k_2 t/u) - \exp[-(k_1 + k_3)t/u]\} & \exp(-k_2 t/u) \end{pmatrix}$$

is the fundamental matrix solution for (2). The mean square solution of (2) is given by the stochastic process

$$(4) \quad \begin{pmatrix} l(t) \\ c(t) \end{pmatrix} =$$

$$=\begin{pmatrix} l_0\exp[-(k_1+k_3)t/u] \\ \dfrac{k_1 l_0}{k_1-k_2+k_3}\{\exp[-(k_1+k_3)t/u]-\exp[-k_2t/u]\}+c_0\exp[-k_2t/u] \end{pmatrix}+$$

$$+\begin{pmatrix} \dfrac{l_a}{k_1+k_3}\{1-\exp[-(k_1+k_3)t/u]\} \\ \dfrac{k_1 l_a}{k_1-k_2+k_3}\left\{\dfrac{1-\exp[-(k_1+k_3)t/u]}{k_1+k_3}-\dfrac{1-\exp[-k_2t/u]}{k_2}\right\}+ \\ +\dfrac{k_2c_s-d_B}{k_2}\qquad\{(1-\exp[-k_2t/u])\} \end{pmatrix},$$

$t \geqslant 0.$

Thus, under various assumptions about the distributions of the random variables l_a, d_B, l_0 and c_0, simulated trajectories of the BOD and DO processes given by (4) may be obtained. It is also possible to obtain the joint distribution of $l(t)$ and $c(t)$ at each value of $t \geqslant 0$ such that the statistical properties of the BOD and DO processes may be more completely determined.

Liouville's fundamental theorem in the theory of dynamic systems can be used to obtain the joint probability distribution of $X(t)$. To this aim, the random vector differential equation (2) is written in the form [333]

(5) $d\mathbb{X}(t)/dt = h(\mathbb{X}(t)\,;\,\mathbb{B}\,;\,t), \qquad \mathbb{C} = \mathbb{X}(0)$

where $\mathbb{B}$ denotes a random vector. Let the joint probability density function of $\mathbb{C}$ and $\mathbb{B}$ be $f_0(\mathbb{C}, l)$. The random solution of (5) has the form

$$\mathbb{X}(t) = g(\mathbb{C}, \mathbb{B}\,;\, t).$$

The problem of determining the joint probability density function $f(\mathbb{X}, \mathbf{b}, t)$ by use of the Liouville equation

$$(6)\quad \frac{\partial f(\mathbb{X}, \mathbf{b}, t)}{\partial t} + \sum_{j=1}^{n} \frac{\partial[fh_j]}{\partial x_j} = 0,$$

where h_j and x_j are the j-th components of $\mathbf{h}$ and $\mathbb{X}$ (assuming the existence of the mean square solution of system (5)) is an initial-value problem for first-order partial differential equations. the initial value being the joint probability density function of $\mathbb{C}$ and $\mathbb{B}$, $f_0(\mathbb{C}, \mathbf{b})$.

As developed by Soong [412], the solution is given by

$$(7)\ f(\mathbb{X}, \mathbf{b}; t) = f_0(\mathbb{C}, \mathbf{b}) \exp \left\{ \int_0^t \nabla \mathbf{h}[x = g(\mathbb{C}, \mathbf{b}; t), \mathbf{b}, s] \mathrm{d}s \right\} \Bigg|_{\mathbb{C}=g^{-1}(\mathbb{X}, \mathbf{b}; t)},$$

where $\nabla \mathbf{h}$ is the divergence of $\mathbf{h}$. Then the joint density function of $\mathbb{X}(t)$ can be found by integrating over $\mathbb{B}$ [333]

$$(8)\ f(\mathbb{X}; t) = \int_{\mathbb{R}^2} f(\mathbb{X}, \mathbf{b}, t) \mathrm{d}\mathbf{b}.$$

Simulated trajectories of $l(t)$ and $c(t)$ are obtained [333] from the mean-square solution and the mean functions and variance functions are obtained by numerical integration. W. J. Padgett considered in [328] that k_3, l_a and d_B have negligible effects on the BOD and DO trajectories and that k_1 and k_2 are dependent random variables. Then the stochastic model becomes

$$(9)\ -u\, \mathrm{d}l(t)/\mathrm{d}t - k_1 l(t) = 0$$
$$u\, \mathrm{d}d(t)/\mathrm{d}t - k_1 l(t) + k_2 \mathrm{d}(t) = 0$$

with $l(0) = l_0$ and $d(0) = d_0$, where $d(t) = c_s - C(t)$ is the random dissolved-oxygen deficit (OD) at each distance t downstream. The probability distribution of $l(t)$ and $d(t)$ at each t is obtained using a Liouville-type theorem. Computations and examples referred to earlier results may be found in ref. [328].

R. F. Curtain [119] used an infinite-dimensional model by appropriately modelling the Poisson-type noise as an infinite-dimensional stochastic process and the recent theory of estimation for linear systems disturbed by general noise processes to solve a river pollution problem.

Let us first recall the well-known model due to H. Kwakernaak [272] for the pollution of rivers and notice that he assumed that the numbers of deposits in a section of the river of infinitesimal length $\mathrm{d}x$ (x being the distance coordinate along the river) behave according to a Poisson process with rate parameter $\lambda(x)\,\mathrm{d}x$, where $\lambda(x)$ is a given function. The numbers of deposits in non-overlapping sections are independent processes, and the amounts of chemicals deposited each time at location x are independent stochastic variables with given distribution H_X and characteristic function φ_X. The time evolution of the concentration of chemicals at location x at time t,

$y(t, x)$ is supposed to be given by

$$(10)\quad \frac{\partial y(t, x)}{\partial t} = D\frac{\partial^2 y(t, x)}{\partial x^2} - V\frac{\partial y(t, x)}{\partial x} + \xi(t, x), \quad 0 \leqslant x \leqslant l$$

where D is the dispersion coefficient, V is the water velocity and $\xi(t, x)$ is the rate of increase in concentration at (t, x) as caused by the above-described deposits of chemical wastes.

A finite number of measuring stations placed along the river continuously measure the observed concentrations of the chemical. Kwakernaak solved the problem of reconstructing the times, location and amounts of deposits from the measured data by approximating (10) by a finite-dimensional model and applying the finite-dimensional filtering theory using a martingale approach.

R. F. Curtain [119] considered the problem of estimating the concentration of the chemical at any point along the river at any time in the past or future, based on continuous noisy measurements in time at a finite number of measuring stations along the river. He retained the infinite-dimensional character of the problem throughout and showed how the final equations can be solved by eigenfunction approximations. In order that (10) describes a stochastic differential equation, he defined $\xi(t, x)$ as some suitable distributed stochastic process. A model of the following type was tried:

$$(11)\quad \mathrm{d}y(t) = A_0 y(t)\mathrm{d}t + \mathrm{d}q(t), \qquad y(0) = y_0$$

where $y(t) \in H = L_2(0, l)$ for each $t \in [0, T]$ models the amount of pollutant dumped at (t, x) and A_0 is the diffusion operator defined by

$$(12)\quad A_0 y = D\frac{\partial^2 y}{\partial x^2} - V\frac{\partial y}{\partial x},\ \mathscr{D}(A_0) = \left\{y \in H, \frac{\partial y}{\partial x}, \frac{\partial^2 y}{\partial x^2} H \right.$$
$$\left. \text{and } y = 0, \text{ at } x = 0, 1\right\}.$$

In order to apply the theory of stochastic differential equations on a Hilbert space, it is first established that $q(t)$ is a compound Poisson process. Interesting results with numerical examples are presented in ref. [120].

7.6. A STOCHASTIC MODEL FOR CHEMOTHERAPY

Chemotherapy is the treatment of disease by injecting chemicals which kill certain types of cells while leaving the body relatively unharmed. The drug injected in the organism enters the extracellular and intracellular space of the various organs by diffusion, possibly reacts with some component of the intracellular space to form compounds, enters the various organs by some active transport process, circulates back to the heart which acts as a mixing chamber, travels back to the various organs, a.s.o. Mathematical models of drug distribution in a one-organ or two-organ biological system were developed deterministically by Bellman, Jacquez and Kalaba [18], [19], [20], [21]. A stochastic version of the above models was formulated and simulated by Padgett and Tsokos [329], [330], [431].

We shall now present some important aspects of these contributions in the stochastic modelling. For $t \geqslant 0$, we shall use the following notation : $u(s, t, \omega)$ is the concentration of drug in moles per unit volume at point s in the capillary at time t; $\omega \in \Omega$ is a random variable for each t; c is the constant volume flow rate of plasma in the capillary bed; and l is the mean length of capillary in the capillary bed. We assume that all capillaries in the capillary bed are lumped together into one capillary of length l, total volume flow rate c, and total surface area equal to that of all the capillaries combined.

For a one-organ system maintained by a simplified heart, assume that the heart is a mixing chamber of constant volume given by

$$V^* = V_e/\ln\ (1 + V_e/V_r),$$

where V_r is the residual volume of the heart and V_e is the ejection volume and let the heart have constant input and output flow rates c.

It is assumed that an initial injection is given at a constant rate right at the heart entrance, which gives concentration $u_I(t)$, $0 \leqslant t \leqslant t_I$, entering the heart, where t_I is the duration of the injection. Let the time required for the blood to flow from the heart exit to the organ entrance be $T > 0$ and let also T be the time required for blood to flow from the organ exit to the heart entrance. Thus, the plasma containing a concentration of drug reaches the organ after time T and, while flowing through the organ, the drug is diffused through the capillary walls into the organ tissue. Hence, after time T, the drug concentration in plasma at the heart entrance is a random variable $u_R(t, \omega)$ due to the diffusion process and recirculation of the blood, and is given by

$$(1)\quad u_R(t, \omega) = \begin{cases} 0, & t < 0, \\ u_I(t), & 0 \leqslant t < T, \\ u_I(t) + u(l, t - T, \omega), & t \geqslant T, \end{cases}$$

where $u_I(t) = 0$ for $t > t_I$, and $u(l, t, \omega)$ is the drug concentration in plasma leaving the organ a time t. The drug concentration in plasma leaving the heart, $u_L(t, \omega)$, satisfies the integral equation

$$(2)\quad u_L(t, \omega) = (c/V^*)\int_0^t [u_R(s, \omega) - u_L(s, \omega)]ds, \quad t \geqslant 0.$$

Then the drug concentration entering the organ at time t is given by

$$u(0, t, \omega) = \begin{cases} 0, & 0 \leqslant t < T, \\ u_L(t - T, \omega), & t \geqslant T. \end{cases}$$

Equation (2) is a semistochastic integral equation, since $u_R(t,\omega)$ is deterministic for $0 \leqslant t < T$ and is a random function for $t \geqslant T$.

Substituting (1) into (2) yields

$$(3)\quad u_L(t, \omega) = G(t) + \int_0^t K(s, u_L(s, \omega);\ \omega)ds,$$

where

$$G(t) = \int_0^{T(t)} (c/V^*)u_I(s)ds, \quad T(t) = \begin{cases} t, & 0 \leqslant t \leqslant t_I, \\ t_I, & t \geqslant t_I, \end{cases}$$

$$K(s, u_L(s, \omega), \omega) = (-c/V^*)[u_L(s, \omega) - u(l, s - T, \omega)],$$

and $u(l, s, \omega) = 0$ if $s < 0$. Let the initial concentration be given by

$$u_1(t) = \begin{cases} u^*, & 0 \leqslant t \leqslant t_I, \\ 0, & \text{otherwise.} \end{cases}$$

To prove that a semirandom solution of Eq. (3) exists, Padgett and Tsokos gave the following

THEOREM [329, Th. 2], [431, Th. 2.3.2]. *The random integral equation of the Volterra type of the form*

$$x(t, \omega) = y(t, \omega) + \int_0^t K(s, x(s, \omega), \omega)ds$$

with $t \in [0, M]$ for $0 < M < \infty$, has a random solution if the following conditions are satisfied:

(i) *The kernel* $K(t, x(t, \omega)\, \omega)$ *is a continuous function of* t *and* x, $t \in [0, \infty)$, $x \in \mathbb{R}$, *for each* $\omega \in \Omega$, *and* $K(t, x, \omega) \in L_p(\Omega, \mathscr{K}, P)$, $1 \leqslant p < \infty$, *for each* t *and* x.

(ii) *For all* $t \in t \in [0, M]$, $\sup_x |K(t, x, \omega)| \leqslant g(\omega) \in L_p(\Omega, \mathscr{K}, P)$, $1 \leqslant p < \infty$.

(iii) *The function* $y(t, \omega)$ *is a continuous function from* $[0, M]$ *into* $L_p(\Omega, \mathscr{K}, P)$ *and, for every* $t \in [0, M]$, $0 < M < \infty$, *there exists a constant* $N_M > 0$ *such that*

$$|y(t, \omega)| + \int_0^t \sup_x |K(s, x, \omega)|\, ds \leqslant N_M < \infty$$

for every $\omega \in \Omega$.

(iv) *The kernel satisfies the following weak continuity property: If* $x_n(\omega)$ *converges weakly to* $x(\omega)$ *in* $L_p(\Omega, \mathscr{K}, P)$, $1 \leqslant p < \infty$, *then* $K(t, x_n(\omega), \omega)$ *converges weakly to* $K(t, x(\omega), \omega)$ *in* $L_p(\Omega, \mathscr{K}, P)$, $1 \leqslant p < \infty$, *for each* $t \in [0, M]$.

The ordinary successive approximation method for the deterministic integral equation of the Volterra type, with $0 \leqslant t < T$, furnishes a deterministic solution

$$\xi(t) = G(t) + \sum_{m=1}^{\infty} (-c/V^*)^m \int_0^t K_m(t, y)\, G(y)dy$$

for the linear Volterra integral equation

$$\xi(t) = G(t) + \int_0^t (-c/V^*)\xi(s)ds$$

(the deterministic form obtained from (3) for $0 \leqslant t \leqslant T$), where

$$K_1(t, y) = \bar{K}(t, y) = \begin{cases} 1, & 0 \leqslant y \leqslant t, \\ 0, & \text{otherwise}, \end{cases}$$

$$K_m(t, y) = \int_y^t K(t, s)K_{m-1}(s, y)ds, \quad m \geqslant 1.$$

For $t \in [T, M]$, $T < M < \infty$, the concentration $u_L(t, \omega)$ and $u(l, t - T, \omega)$ can be considered as continuous functions of time, and hence $K(t, u_L(t, \omega), \omega)$ is continuous in t and u_L. For each t and u_L, $K(t, u_L, \omega)$ is a degenerate random variable (a constant) and, hence, for each t and u_L, $K(t, u_L, \omega) \in L_p(\Omega, \mathscr{K}, P)$, $1 \leqslant p < \infty$, and so condition (i) of the above theorem is satisfied.

By the nature of the physical problem, the maximum value that $u_L(t, \omega)$ or $u(l, t - T, \omega)$ can assume is u^*, and the minimum value is zero for all $t \in [T, M]$. Therefore,

$$\sup_{u_L} |K(t, u_L, \omega)| = \sup_{u_L} \{(c/V^*)| u_L(t, \omega) - u(l, t - T, \omega)|\} =$$

$$= c\, u^*/V^*.$$

Define the degenerate random variable $g(\omega) = cu^*/V^*$, $\omega \in \Omega$. Then $g(\omega) \in L_p(\Omega, \mathscr{K}, P)$ and condition (ii) of the above theorem is satisfied.

The stochastic free term $y(t, \omega)$ is the deterministic function $G(t)$ which is bounded since $|G(t)| \leqslant t_I cu^*/V^* < \infty$ for all $t \in [0, M]$. Hence, $y(t, \omega) = G(t)$ a.s. is in $L_p(\Omega, \mathscr{K}, P)$ and is a continuous function from $[0, M]$ into $L_p(\Omega, \mathscr{K}, P)$. Let $t \in [T, M] \subset$ $\subset [0, M]$, $t_I < M < \infty$. Then

$$|y(t, \omega)| + \int_0 \sup_{u_L} |K(t, u_L, \omega)| ds = |G(t)| + t(cu^*/V^*) \leqslant$$

$$\leqslant cu^*/V^*(t_I + M) < \infty,$$

i.e., condition (iii) of the above theorem is satisfied.

Suppose the sequence $u_{L,n}(t, \omega) \xrightarrow{\|\cdot\|L_p} u_L(t, \omega)$ for each $t \in [T, M]$. Then, for every $\varepsilon_1 > 0$, there exists an $N > 0$ such that whenever $n \in \mathbb{N}$,

$$\|u_{L,n}(t, \omega) - u_L(t, \omega)\|_{L_p} < \varepsilon_1.$$

Therefore, letting $\varepsilon = (c/V^*)\varepsilon_1 > 0$, we have

$$\|K(t, u_{L,n}(t, \omega), \omega) - K(t, u_L(t, \omega), \omega)\|_{L_p} =$$

$$= (c/V^*)\|u_{L,n}(t, \omega) - u_L(t, \omega)\|_{L_p} < (c/V^*)\varepsilon_1 = \varepsilon,$$

whenever $n \in \mathbb{N}$. Hence, condition (iv) of the above theorem is satisfied and a random solution of Eq. (3) exists for $t \in [T, M]$.

We have shown that the semistochastic integral equation (3) has a semirandom solution for $t \in [0, M]$, $0 < M < \infty$, $T < M < \infty$ and $t_I < M$. A computer simulation of the one-organ system was given in ref. [330].

An extension of the previous result to two organs is given in ref. [18]. A similar closed system with a simplified heart, two organs or capillary bends is assumed. With respect to the flow of blood, it is assumed as before that there is no mixing in the vessels. Assume that $u_j(s, t, \omega)$ is the random concentration of drug in organ j at point s in the capillary at time t, $j=1,2$; c_j is the constant-volume flow rate of blood in organ $j=1,2$ and $c = c_1 + c_2$ is the total constant volume flow rate of blood in the system. Thus, the drug enters each organ after a time T, and due to the diffusion process in the organs, the drug concentration entering the heart, $u_R(t, \omega)$, after time T is a random variable for each $t \geqslant T$ and is given by

$$u_R(t, \omega) =$$
$$= \begin{cases} 0, & t < 0 \\ u_I(t), & 0 \leqslant t, \\ u_I(t) + [c_1 u_1(l, t-T, \omega) + c_2 u_2(l, t-T, \omega)]/c, & T \leqslant t \leqslant M, \end{cases}$$

where $u_j(l, t, \omega)$ is the drug concentration in plasma leaving organ j (at the l end), at time t, $j = 1, 2$ and $u_I(t) = 0$ for $t \geqslant t_I$. Also, the drug concentration in plasma leaving the heart at time $t \geqslant 0$ is given by the semirandom solution $u_L(t, \omega)$ of Eq. (2). Then the drug concentration in plasma entering organ j at time t is given by

$$u_j(0, t, \omega) = \begin{cases} 0, & 0 \leqslant t < T, \\ u_L(t-T, \omega), & T \leqslant t \leqslant M \end{cases}, \quad j = 1, 2.$$

7.7. AN APPLICATION OF RANDOM CONTRACTORS IN POPULATION GROWTH PROBLEMS

A. C. Lee and W. J. Padgett considered [281] a random nonlinear Volterra integral equation which arises in the study of population growth in random environments. They assumed a growth rate depending only on population size and time and a probability of death depending only on age, i.e.,

$$(1)\quad x(t) = h(t) + \int_0^t g(x(t-s))l(s)\mathrm{d}s, \quad t \geqslant 0,$$

where $g(x)$ is the number of members added to the population in unit time when the population size is x. Then the growth rate of the population size in unit time per unit of population size is given by $g(x)/x$. The function $l(t)$ is the probability that a member of the population born at time zero is alive at time $t > 0$. Thus, the quantity $g(x(t-s))\ l(s)\Delta s$ approximates the number of members added to the population between the times $(t-s)$ and $(t-s+\Delta s)$ who survive to age s at time t. The forcing function $h(t)$ represents the number of members of the population who were already present at time zero and are still alive at time t.

Since in reality the size of a biological population is largely dependent on its ability to tolerate and compensate for random environmental variation such as changes in temperature, oxygen supply, nutrition, pressure, and the like, a stochastic model would be more realistic. This aim was set forth by Lee and Padgett [281] and will be presented in this section.

Let $(\Omega, \mathscr{K}, P)$ be a probability space and let $\omega \in \Omega$. Let $x(t, \omega)$ be a stochastic process which enumerates the conditionally expected number of members in the population at time $t > 0$ with a given realization of the environment. The term $g(t, x, \omega)$ is a stochastic process which enumerates at time t the number of members added to the population in unit time when the population size is x. The random process $l(t, \omega)$ represents the probability that a member of the population survives to age t. We assume that $l(0, \omega) = 1$ a.s. Let $h(t, \omega)$ be a stochastic process which is completely specified if the distribution of the population size $H(\omega)$ at time zero and the behaviour of the process $l(t, \omega)$ are known. For example, with no immigration from the outside, we have $h(t, \omega) = x(0, \omega)\, l(t, \omega)$. Therefore the random version of (1) is given by

$$(2)\quad x(t, \omega) = h(t, \omega) + \int_0^t g(t - s, x(t - s, \omega), \omega), l(s, \omega) ds, \quad t \geqslant 0.$$

In order to study the boundedness of the solution of Eq. (2) in an appropriate Banach space, Lee and Padgett introduced a bounded integral contractor of a random function. To prove the existence and uniqueness of the solution, some results on random contractors are also used.

In what follows, we assume that the stochastic processes in (2) are second-order, mean-square continuous processes on $\mathbb{R}^+ = [0, \infty)$. The space $C \equiv C(\mathbb{R}^+, L_2(\Omega, \mathscr{K}, P))$ is the set of all continuous functions from $\mathbb{R}^+$ into $L_2(\Omega, \mathscr{K}, P)$ with the topology of uniform convergence on compact sets. The space $\mathbb{C}_u = \mathbb{C}_u(\mathbb{R}^+, L_2(\Omega, \mathscr{K}, P))$ (where u is a nonnegative continuous function on $\mathbb{R}^+$) is the Banach space of all continuous functions $x(t, \omega)$ from $\mathbb{R}^+$ into $L_2(\Omega, \mathscr{K}, P)$ such that there exists a constant $M > 0$ with $\|x(t, \omega)\|_{L_2(\Omega, \mathscr{K}, P)} \leqslant Mu(t)$ for all $t \geqslant 0$ and with the norm in $\mathbb{C}_u$ defined by

$$\|x\|_{\mathbb{C}_u} = \sup_{t \geqslant 0} \{\|x(t, \omega)\|_{L_2(\Omega, \mathscr{K}, P)}/u(t)\}.$$

In addition to the assumption that $l(0, \omega) = 1$ a.s., we assume that the process $l(t, \omega)$ is P-essentially bounded for every

$t \geqslant 0$, i.e., $l(t, \omega) \in L_\infty(\Omega, \mathscr{K}, P) ||| (l(t, \omega) |||_{L_\infty(\Omega, \mathscr{K}, P)} = P\text{-ess sup}|l(t, \omega)|$.

Further, suppose that the mapping $t \to l(t, \omega)$ is continuous from $\mathbb{R}^+$ into $L_\infty(\Omega, \mathscr{K}, P)$. Then one can prove that the integral operator

$$(3)\ (Tx)(t, \omega) = \int_0^t l(t-s, \omega)x(s, \omega)\mathrm{d}s, \quad t \geqslant 0$$

is continuous from $\mathbb{C}$ into $\mathbb{C}$ when the Bochner integral exists. For B, D Banach spaces in $\mathbb{C}$ stronger than $\mathbb{C}$ and $T(B) \subset D$ (i.e., the pair (B, D) is *admissible* with respect to T), consider

DEFINITION 7.7.1. Let $f(t, x, \omega) : \mathbb{R}^+ \times \mathbb{R} \times \Omega \to \mathbb{R}$ such that $f(t, x(t, \omega), \omega) \in B$ whenever $x(t, \omega) \in D$. The function f is said to have a *bounded integral contractor* Γ with respect to (B, D) if :

(i) for every $t \in \mathbb{R}^+$ and $x \in \mathbb{R}$, there exists a bounded linear operator Γ from D to B such that $\|\Gamma(t, x)\|$ is continuous in (t, x) and $\|\Gamma(t, x)\| \leqslant Q$ for some positive constant Q; and

(ii) for every $t \in \mathbb{R}^+$ and $x, y \in D$, there exists an $\alpha > 0$ such that

$$\|f(t, x(t, \omega) + y(t, \omega) + [T\Gamma(t, x(t, \omega))y](t, \omega), \omega) -$$
$$-f(t, x(t, \omega), \omega) - [\Gamma(t, x(t, \omega))y](t, \omega)\|_B \leqslant \alpha \|y(t, \omega)\|_D.$$

The constant α will be called the *contractor constant.*

THEOREM 7.7.1. *Let g : $\mathbb{R}^+ \times \mathbb{R} \times \Omega \to \mathbb{R}$ be such that $x(t, \omega) \in D$ implies that $g(t, x(t, \omega), \omega) \in B$, where the pair of Banach spaces (B, D) in C is admissible with respect to the operator T given by* (3).

Let the random integral equation (2) *satisfy the condition that $h(t, \omega) \in D$ and that the function g has a bounded integral contractor Γ with contractor constant α. Then a random solution exists provided $K\alpha < 1$, where K is the norm of the operator T.*

Proof. Let $\{x_n\}_{n \in \mathbb{N}} \subset D$ be a sequence defined by

$$(4)\ x_{n+1}(t, \omega) = x_n(t, \omega) - \Big\{y_n(t, \omega) +$$
$$+\int_0^t [\Gamma(t-s, x_n(t-s, \omega))y_n(t-s, \omega)]l(s, \omega)\mathrm{d}s\Big\},$$

where

$$(4')\ y_n(t, \omega) = x_n(t, \omega) - h(t, \omega) -$$
$$-\int_0^t g(t-s, x_n(t-s, \omega), \omega)l(s, \omega)\mathrm{d}s$$

and $x_0 \in D$. It is obvious that $x_n, y_n \in D$ for $n = 1, 2, \ldots$. We shall show that $\lim_{n\to\infty} \|y_n\|_D = 0$. Indeed, (4) and (4′) imply that

$$y_{n+1}(t, \omega) = \int_0^t [g(t-s, x_n(t-s, \omega), \omega) - g(t-s, x_{n+1}(t-s, \omega), \omega) -$$

$$-\Gamma(t-s, x_n(t-s, \omega), \omega))y_n(t-s, \omega)]\, l(s, \omega)ds.$$

Then the condition that g has a bounded integral contractor implies that $\|y_{n+1}(t, \omega)\|_D \leqslant K\alpha\|y_n(t, \omega)\|_D$ and, since $K\alpha < 1$, it follows that $\lim_{n\to\infty} \|y_n(t, \omega)\|_D = 0$.

According to (4),

$$\|x_{n+1}(t, \omega) - x_n(t, \omega)\|_D \leqslant \|y_n(t, \omega)\|_D + K\|\Gamma(t, x_n(t, \omega))y_n(t, \omega)\|_D \leqslant$$
$$\leqslant (1 + KQ)(\alpha K)^{n-1}\|y_0(t, \omega)\|_D$$

and, hence, $\{x_n\}_{n\in\mathbb{N}}$ is a Cauchy sequence. Then there exists an $x \in D$ such that $\lim_{n\to\infty} x_n = x$.

Remark. The condition imposed on g in the above theorem is weaker than the usual Lipschitz condition. In general, this condition does not yield a unique solution.

To obtain an unique solution, one must impose in addition that there exists a solution in D for the random integral equation

$$(5)\ y(t, \omega) + \int_0^t \Gamma(t-s, x(t-s, \omega))y(t-s, \omega)l(s, \omega)ds = z(t, \omega)$$

where $x(t, \omega)$, $z(t, \omega) \in D$.

Indeed, if $x_1(t, \omega)$ and $x_2(t, \omega)$ are two random solutions in D for (2), then letting in (5) $x_1(t, \omega) - x_2(t, \omega) = z(t, \omega)$, one obtains

$$y(t, \omega) = \int_0^t [g(t-s, x_1(t-s, \omega)\omega) - g(t-s, x_2(t-s, \omega), \omega) -$$

$$-\Gamma(t-s, x_2(t-s, \omega))y(t-s, \omega)]l(s, \omega)ds.$$

Hence, $\|y(t, \omega)\|_D \leqslant \alpha K\|y(t, \omega)\|_D$ which implies that $y(t, \omega) = 0$, i.e., $x_1(t, \omega)$ and $x_2(t, \omega)$ are equivalent random solutions.

To assure the boundedness of the random solution of (2), we can choose the spaces B and D of the above theorem to be spaces of type $\mathbf{C}_u$.

THEOREM 7.7.2. *Let $u(t) > 0$ be a continuous function on $\mathbb{R}^+$ such that* $\int_0^t |||l(t-s,\omega)||| \, u(s)\,ds \leqslant MG(t)$ *for all* $t \in \mathbb{R}^+$, *where M is a positive constant and $G(t)$ is a positive continuous function on $\mathbb{R}^+$. Suppose that (2) satisfies the following conditions: (i) $h(t,\omega) \in \mathbb{C}_u$; (ii) $|g(t,x,\omega)| \leqslant u(t)\,[z(t,\omega) + \gamma(t)\,|x|]$ a.s., where $z(t,\omega)$ is a stochastic process in $\mathbb{C}_B$ and $\gamma(t)$ is a bounded continuous function defined for $t \geqslant 0$; and (iii) for each $t \in \mathbb{R}^+$ and $x, y \in \mathbb{C}_u$, there exists a bounded linear operator $\Gamma(t,x) \equiv \Gamma$ such that*

$$\Big\| g(t, x(t,\omega) + y(t,\omega) + \int_0^t (\Gamma y)(t-s,\omega)\, l(s,\omega)\,ds, \omega) -$$

$$- g(t, x(t,\omega),\omega) - (\Gamma y)(t,\omega)\Big\|_{L_2(\Omega,\mathcal{K},P)} \leqslant$$

$$\leqslant \alpha \, \| y(t,\omega)\|_{L_2(\Omega,\mathcal{K},P)}.$$

Then there exists a random solution $x(t,\omega)$ of (2) such that $E(|x(t,\omega)|^2) \leqslant A^2G^2(t)$, $t \geqslant 0$, where $A > 0$ is a constant, provided that $K\alpha < 1$.

Proof. First, we observe that $(\mathbb{C}_u, \mathbb{C}_G)$ is admissible with respect to T as given by (3), since for $x \in \mathbb{C}_u$ we have

$$\|(Tx)(t,\omega)\|_{L_2(\Omega,\mathcal{K},P)} \leqslant \|x(t,\omega)\|_{\mathbb{C}_u} \int_0^t |||l(t-s,\omega)|||\, u(s)\,ds \leqslant$$

$$\leqslant \|x(t,\omega)\|_{\mathbb{C}_u} M \cdot G(t), \quad t \geqslant 0.$$

Thus, $Tx \in \mathbb{C}_G$.

Condition (ii) implies that $x(t,\omega) \to g(t, x(t,\omega),\omega)$ maps from $\mathbb{C}_u$ into $\mathbb{C}_u$. Also (iii) implies the existence of a bounded integral contractor for the function g with contractor constant α. Therefore, by virtue of Theorem 7.7.1, there exists a random solution of (2) in $\mathbb{C}_G$.

The random contractor theory (see sub-section 6.2.3) can be used to obtain a unique random solution to (2).

THEOREM 7.7.3. *Consider the random integral equation (2) subject to the following conditions:*

(i) *$l(t,\omega) \leqslant \exp(-\mu t)$ a.s. for all $t \geqslant 0$, where $\mu > 0$ is a constant;*

(ii) $(\mathbb{C}_1, \mathbb{C}_1)$ $(\mathbb{C}_1 = \mathbb{C}_u$ *for* $u(t) = 1)$ *is admissible with respect to the random operator*

$$[T(\omega)x](t, \omega) = \int_0^t g(t - s, x(t - s, \omega), \omega)l(s, \omega)\mathrm{d}s;$$

(iii) $x(t, \omega) \to g(t, x(t, \omega), \omega)$ *is a continuous operator on* $S = \{x(t, \omega) \in \mathbb{C}_1 : \|x\|_{\mathbb{C}_1} \leqslant \gamma\}$ *with values in* $\mathbb{C}_1$ *satisfying*

$$\|g(t, x(t, \omega) + y(t, \omega), \omega) - g(t, x(t, \omega), \omega)\|_{L_2(\Omega, \mathscr{X}, P)} \leqslant$$
$$\leqslant \lambda \| y(t, \omega)\|_{L_2(\Omega, \mathscr{X}, P)}$$

for $x, y \in S$ *and* $t \geqslant 0$, $\lambda > 0$; *and*

(iv) $h(t, \omega) \in \mathbb{C}_1$.

Then there exists a unique random solution $x^*(t, \omega) \in \mathbb{C}_1$ *such that* $E(|x^*(t, \omega)|^2) < \gamma^2$ *for all* $t \geqslant 0$, *provided that* $\|h(t, \omega)\|_{\mathbb{C}_1} \leqslant \gamma(1 - \lambda/\mu)$ *and* $\lambda < \mu$.

To prove this theorem, it is easy to show that the identity random operator $I(\omega)$ is a random contractor for the random nonlinear operator on S defined by

$$[U(\omega)x](t, \omega) = x(t, \omega) - h(t, \omega) - [T(\omega)x](t, \omega).$$

Then the random operator equation $U(\omega)x = y(\omega)$ has a unique random solution [277, Th. 3.4].

Remark. The condition on $l(t, \omega)$ in Theorem 7.7.2 allows $|||l(t, \omega)|||$ to be constant or bounded by a constant for all t, letting $\int_0^t u(s)\mathrm{d}s = G(t)$. For example, if $u(t) = \exp(-\beta t)$ for, $\beta > 0$, then the variance function of a random solution converges exponentially to zero as $t \to \infty$. This condition is weaker than that on deterministic probability $P(t)$, which requires that $\int_0^\infty P(t)\mathrm{d}t < \infty$.

The above results are also applicable to the nonrandom problem represented by (1) and provide very general conditions for boundedness in that case.

If immigration from outside the population is allowed, then one can replace $h(t, \omega)$ in (2) by $h^*(t, \omega) = h(t, \omega) + i(t, \omega)$, where $i(t, \omega)$ is the number of immigrants having arrived from outside by time t, and the above results also hold.

7.8. A RANDOM EQUATION IN TELEPHONE TRAFFIC THEORY

In this section, we shall examine a stochastic integral equation arising in telephone engineering or in any system in which a number of servers provides services to customers in the absence of the formation of a queue. A related study in telephone engineering has been investigated by Fortet [148], Padgett and Tsokos [331], [431] and A. T. Bharucha-Reid [31].

Consider a telephone exchange with m (m finite and large) available trunks or channels. It is also assumed that any channel not being used may be utilized by an incoming call and that the holding time for a channel begins at the time instant that the call arrives at the exchange. This connection is realized if a channel is not busy at the time a call arrives. If all channels are busy at the time when a call arrives, then the call is lost. Suppose that the subscribers make calls at instants $t_1, t_2, \ldots, t_n, \ldots,$ where $0 < t_1 < t_2 < \ldots < t_n < \ldots < \infty$. These arrival times are considered to be random instants, the distribution of which on the time axis is a random distribution $G(t)$. We may consider G as being the sum of several random distributions G_i, $i = 1, \ldots, r$. The distributions of arrival times of each type of call are assumed to be independent. Various problems have been studied in this situation [26], [31], [331], [431]. We shall be concerned with the total number of conversations held at time t, $x(t, \omega)$, with $x(0, \omega) = 0$. Then $x(t, \omega)$ is a random process which may be described by a stochastic integral equation. Let $h(t, \omega)$ be the holding time for any call arriving at time t. Let $y(t, \omega)$ be a random function with value one if a call arrising at time $t > 0$ is not lost and value zero if the call is lost. Also, let

$$K(t, \tau, \omega) = \begin{cases} 1 & \text{if } t - \tau \in [0, h(\tau, \omega)] \\ 0 & \text{if } t - \tau \notin [0, h(\tau, \omega)] \end{cases}$$

such that $K(t, \tau, \omega)$ is equal to one if a conversation from a call arising at time τ is still being held at time $t \geqslant \tau$ and is equal to zero otherwise. Thus, we may write

$$(1)\quad x(t, \omega) = \int_0^t y(\tau, \omega) K(t, \tau, \omega)\, \mathrm{d}G(\tau).$$

Equation (1) may be interpreted as the total numbers of telephone calls arriving at time τ, $0 \leqslant \tau \leqslant t$, that were not lost such that the conversation is still being held at time t.

If $V(k)$ is any function such that

$$V(k) = \begin{cases} 1 & \text{if } k = 0, 1, \ldots, m-1, \\ 0 & \text{if } k \geqslant m \end{cases}$$

then, since $x(t, \omega) \leqslant m$ for all $t \in \mathbb{R}^+$ and $\omega \in \Omega$, we may write

$$V[x(t, \omega)] = \begin{cases} 1 & \text{if } x(t, \omega) = 0, 1, \ldots, m-1 \\ 0 & \text{otherwise.} \end{cases}$$

Then Eq. (1) may be written as the nonlinear stochastic integral equation

$$(2)\quad x(t, \omega) = \int_0^t K(t, \tau, \omega) V[x(t, \omega)] dG(\tau)$$

which is referred to [31], [431] as the Fortet integral equation.

If the distribution $G(t)$ has a density $a(t)$, then Eq. (2) reduces to a stochastic integral equation of the Volterra type

$$(3)\quad x(t, \omega) = \int_0^t K(t, \tau, \omega) f(\tau, x(\tau, \omega)) d\tau,$$

where

$$f(\tau, x(\tau, \omega)) = V[x(\tau, \omega)]a(\tau) = \begin{cases} a(\tau) & \text{if } x(\tau, \omega) = 0, 1, \ldots, m-1 \\ 0 & \text{otherwise} \end{cases}$$

and $K(t, \tau, \omega)$ is the stochastic kernel defined for $0 \leqslant \tau \leqslant t < \infty$, taking only the values 0 or 1.

In order to establish the existence of a solution to Eq. (2), Bharucha-Reid [31] used a method of successive approximations. Padgett and Tsokos [431], [331] established the existence and uniqueness of a solution of Eq. (2) in $L_2(\Omega, \mathscr{K}, P)$ using the following

THEOREM [331, Th. 3.1]. *Let us consider the random integral equation*

$$(4)\quad x(t, \omega) = y(t, \omega) + \int_0^t K(t, \tau, \omega) f(\tau, x(\tau, \omega)) d\tau$$

and assume that:

(i) *There exists a number* $M > 0$ *and a positive continuous function* $g(t)$ *on* $\mathbb{R}^+$ *such that*

$$\|K(t, \tau, \omega)\| = P\text{-ess}\sup_{\omega} |K(t, \tau, \omega)| \leqslant M, \quad 0 \leqslant \tau \leqslant t < \infty,$$

and

$$\int_0^\infty g(t)\mathrm{d}t < \infty;$$

(ii) $f(t, x)$ *is a continuous function for* $t \in \mathbb{R}^+$, $x \in \mathbb{R}$, *such that* $f(t, 0) \in C_g(\mathbb{R}^+, \mathbb{R})$ *and*

$$|f(t, x) - f(t, y)| < \lambda g(t) |x - y|,$$

for some constant $\lambda > 0$.

(iii) $y(t, \omega) \in C(\mathbb{R}^+, L_2(\Omega, \mathscr{K}, P))$

Then there exists a unique random solution of the above random integral equation, bounded on $\mathbb{R}^+$, *provided* $\|y(t, \omega)\|_C$, λ *and* $|f(t, 0)|$ *are sufficiently small.*

We now reproduce the remarks of Padgett and Tsokos that if we replace the words "telephone exchange" with "serving mechanism" and the words "channel", "call" and "conversation" with the words "server", "customer" and "service", respectively, then we are dealing with a general system in which the "customers" are being "served" by $m < \infty$ "servers". If we assume that a customer does not wait when he finds all m servers busy, then the random solution of (3) gives the total number of "services" being performed at time t.

On the other hand, the functions in (3) may be any functions which describe some physical situation. For example, the stochastic kernel may be of the form

$$K(t, \tau, \omega) = I_{X(\omega)}(t - \tau)\mathrm{e}^{-(t-\tau)},$$

where $I_{X(\omega)}$ is the indicator function of a random set $X(\omega)$, which means that solutions at earlier times $\tau \leqslant t$ have a decaying effect on the system.

7.9. STOCHASTIC OPERATORS AND STOCHASTIC SYSTEMS

In this paragraph we shall present some of G. Adomian's ideas and contributions to probabilistic analysis [7], [4], [5], [3].

A fundamental problem of science is to model phenomena of interest so as to allow the analysis of and insight into the essential relationships. The test of the model is given by how far the conclusions are in tune with observation. Every transformation or processing of an input into another form or output is a system. Most systems of interest are dynamical systems; they vary with time. The modelling of dynamical systems results in differential or integral equations which one seeks to solve to predict a behaviour or to control or optimize it. A stochastic system is one involving uncertainties or random fluctuations. The real systems are generally nonlinear. An important objective of applied mathematics is to find methods for the solution of nonlinear stochatic systems. However, since the solutions are stochastic, one can find various statistics of the solution process rather than explicit solutions.

The system of interest may be a large dynamical multi-input, multi-output system such as the national economy, a medium such as the atmosphere, the ocean, and the like, a plasma in which an electromagnetic or acoustic wave is propagating, or any device or measurement system which processes an input signal $x(t)$ into an output signal $y(t)$. If the system is represented by the operator $\mathscr{H}$ indicating the transformation which is applied to x by the system, we have the output $y = \mathscr{H}(x)$. The most interesting and general types of stochastic differential equations, linear or nonlinear, involve a stochastic operator by virtue of stochastic process coefficients in the differential operator. Some interesting and general approaches which allow for stochastic behaviour in the forcing as well as in the initial conditions whereas avoiding many of the common restrictions (white noise, small fluctuations, etc.) and assumptions (closure approximations, etc.) are given by G. Adomian [2], [7], [4], [5], [8], [6].

An iterative procedure was developed using the concept of stochastic Green's functions [1] and allowing the treatment of a wide class of problems. The method was shown [7] to be appropriate for solving partial differential equations and wave propagation in random varying media, without an assumption of monochromaticity, which ignores the spectral spread and without the closure approximations necessary in the hierarchy method.

On the other hand, the iterative procedure is independent of the presence of a small parameter, and it has been shown [2] that the results of the perturbation theory are obtained where perturbation theory is applicable. Extensive computer results have illustrated that the iterative method can be used to advantage.

A general expression was obtained for the stochastic Green function which yields the mean solution or the two point correlation of the solution. Relationships have been found between the various stochastic Green functions for different statistical measures and also between stochastic Green functions, random

Green's functions, and ordinary Green functions. The stochastic Green function is determined as a kernel of an integral expression yielding directly the desired statistical measure of the solution process (either the expectation of the solution or the correlation or covariance of the solution). The long-term objective is to develop methods for calculating the stochastic Green function for the desired statistical measure directly from knowledge of the given statistics (statistical measures) of the coefficients without the intermediate step of solving the equation.

The equation of interest $\mathscr{H}(y) = x$ is given by the case when the general operator $\mathscr{H}$ can be decomposed into a sum of purely linear stochastic operator $\mathscr{L}$ and a purely nonlinear stochastic operator $\mathscr{N}$, where $\mathscr{L}$ is an n-th order differential stochastic operator given by $\mathscr{L} = \sum_{k=0}^{n} a_k(t, \omega)\, \mathrm{d}^k/\mathrm{d}t^k$ with possibly stochastic coefficients $a_0, a_1, \ldots, a_{n-1}$, $t \in T$, $\omega \in (\Omega, \mathscr{K}, P)$, a probability space ($a_n > 0$). $\mathscr{N}$ is a nonlinear stochastic term given by $\mathscr{N} = \sum_{i=0}^{m} b_i(\omega, t)\, (y^{(i)})^{m_i}$, where $y^{(i)}$ is the i-th derivative, $m < n$, and $b_i(t, \omega)$, $t \in T$, $\omega \in \Omega$, are possibly stochastic processes for $i = 1, 2, \ldots, m$. The inhomogeneous term $x(t, \omega)$, $t \in T$, $\omega \in \Omega$, is a stochastic process statistically independent of a_k, b_i for all k, i allowed. The coefficient processes a_k, b_i are of class C^n on T for $\omega \in \Omega$ a.s. for the allowed ranges of k, i. In most of Adomian's papers, it is assumed that $\mathscr{L}$ can be decomposed into the sum $L + R$, where $L = E(\mathscr{L})$ is an invertible deterministic operator. We have $L = \sum_{k=0}^{n} E(a_k(t, \omega))\, \mathrm{d}^k/\mathrm{d}t^k$ and $R = \sum_{i=0}^{n-1} \alpha_k(t, \omega) \mathrm{d}^k/\mathrm{d}t^k$, where $E(a_k(t, \omega))$ exists and is continuous on T and α_k are zero-mean stochastic processes representing the fluctuating part of each coefficient. α_n is, of course, zero. For $\mathscr{N}$ too, one can write deterministic N and stochastic M parts such that $\mathscr{N} = N + M$. N and M are, of course, nonlinear. We have

$$Ly + Ry + \mathscr{N}(y, \dot{y}, \ldots) = x \text{ or } Ly = x - Ry - \mathscr{N}(y, \dot{y}, \ldots).$$

Since L is assumed to be invertible,

$$y = L^{-1} x + L^{-1} Ry - L^{-1} \mathscr{N}(y, \dot{y}, \ldots)$$

or, in terms of Green's function $l(t, \tau)$ for L,

$$y = \int_0^t l(t, \tau)x(\tau)\, d\tau - \int_0^t l(t, \tau)R[y(\tau)]d\tau -$$

$$- \int_0^t l(t, \tau)\mathcal{N}(y(\tau), \dot{y}(\tau), \ldots)d\tau.$$

Let $y = \sum_{i=0}^{\infty} (-1)^i y_i$. We get

$$y = \int_0^t l(t, \tau)x(\tau)d\tau - \int_0^t l(t, \tau)R[y_0(\tau) - y_1(\tau) + y_2(\tau) - \ldots]d\tau -$$

$$- \int_0^t l(t, \tau)\mathcal{N}(y_0(\tau) - y_1(\tau) + y_2(\tau) - \ldots)d\tau.$$

The term $L^{-1}Ry$ involves derivatives in $\mathbb{R}$. Consequently, since an iterative procedure is used to obtain y, it is desirable to integrate by parts or equivalently use the adjoint operator. Thus, $L^{-1}Ry$ becomes

$$L^{-1}Ry = \int_0^t l(t, \tau)Ry(\tau, \omega)d\tau =$$

$$= \sum_{k=0}^{n-1} (-1)^k \int_0^t [l(t, \tau)\alpha_k(\tau, \omega)]^{(k)} y(\tau, \omega)d\tau +$$

$$+ \sum_{k=0}^{n-1} \sum_{i=1}^{k-1} (-1)^i [l(t, \tau)\alpha_k(\tau, \omega)]^{(i)} y^{(k-1-i)}(\tau, \omega) \Bigg|_{\tau=0}^{\tau=t} =$$

$$= \int_0^t y(\tau, \omega)R^+[l(t, \tau)]d\tau + \sigma[y(\tau, \omega); l(t, \tau)] \Bigg|_0^t ,$$

where $\mathbb{R}^+[l(t, \tau)]$ is the adjoint operator and $\sigma[y(\tau, \omega);$

$l(t, \tau)]\Big|_0^t$ is the s.b.c. (stochastic biliniar concomitant). Hence,

$$\int_0^t [l(t, \tau)Ry(\tau, \omega) - y(\tau, \omega)R^+ l(t, \tau)]\, d\tau = \sigma[y(\tau, \omega); l(t, \tau)]\Big|_0^t$$

is a stochastic Green's formula.

The s.b.c. term is zero if the initial conditions are zero but does not vanish in the case of random initial conditions. Hence, the case of random initial conditions, which is essentially trivial in the case of a deterministic operator, becomes more difficult in the case of a stochastic operator since additional terms arise from the solution of the homogeneous equation and the value of the s.b.c. at $\tau = 0$. It does vanish at the upper limit because $(d^k/d\tau^k)l(t, \tau)|_{\tau=t=0}$ for $0 \leqslant k < n-1$, a known property of Green's function $l(t, \tau)$ most easily seem by differentiation applied to $l(t, \tau)$ (written in terms of determinants involving Φ, the solutions of $Ly = 0$) and noting that the determinants vanish if $k < n-1$ if $\tau = t$.

Writing $F(t)$ for $L^{-1}x$ and for any solution to the homogeneous equation $Ly = 0$, we get (for the case $\mathcal{N} = 0$)

$$y(t, \omega) = F(t, \omega) - L^{-1}Ry(t, \omega) =$$

$$= F(t, \omega) - \int_0^t R^+[l(t, \tau)]y(\tau, \omega)d\tau.$$

The iterative solution given by Adomian and Sibul [7], [8] reads as

$$y(t, \omega) = F(t, \omega) - \int_0^t \Gamma(t, \tau, \omega)F(\tau, \omega)d\tau,$$

$$\Gamma(t, \tau, \omega) = \sum_{m=0}^{\infty} K_m(t, \tau, \omega),$$

where

$$(*)\ K_m(t, \tau, \omega) = \int_0^t K(t, \tau_1)K_{m-1}(\tau_1, \tau)d\tau_1, K_1 = K = R^+ (l(t, \tau)).$$

We assume that the input process $x(t, \omega)$ is bounded a.s. on T and that $l(t, \tau)$ is continuous on T; thus, $|x(t, \omega)| < M_1$ or, equivalently, $|F(t, \omega)| < M_1'$, where M_1, M_1' are constants.

Further, α_k, $k = 0, 1, \ldots, n-1$, are bounded a.s. and $|(\partial^k/\partial t^k)\alpha_k(t)| < M_2$ (M_2 a constant), for $t \in T$, $\omega \in (\Omega, \mathscr{K}, P)$. From (*)

$$|K_m(t, \tau)| \leqslant \int_0^t |K(t, \tau_1)| \; |K_{m-1}(\tau_1, \tau)| \mathrm{d}\tau_1.$$

Since $(\partial^k/\partial t^k)l(t, \tau)$ is jointly continuous in t and τ on $T \times T$ for $0 \leqslant k \leqslant n-1$ and the derivatives of the α_k are bounded a.s., we can assume a bound directly on K_m. We observe that

if $|K_1| \leqslant M$, then $|K_2| \leqslant \int_0^t |K_1(t, \tau_1)| \, |K_1(\tau_1, \tau)| \mathrm{d}\tau_1 \leqslant M^2(t-\tau)$,

etc. Hence,

$$|K_n(t, \tau)| \leqslant M^m t^{m-1}/(m-1)!$$

and

$$\Gamma|(t, \tau, \omega)| = \sum_{m=1}^{\infty} (-1)^{m-1} |K_m(t, \tau, \omega)| \leqslant$$

$$\leqslant \sum_{m=1}^{\infty} M^m |t-\tau|^{m-1}/(m-1)! \leqslant$$

$$\leqslant M \sum_{m=0}^{\infty} M^m |t-\tau|^m m! = M \, \mathrm{e}^{M|t-\tau|}$$

which exists for a finite interval of observation $|t-\tau|$. The convergence of the series is ensured by the factorial in the denominator of the series terms due to the multiple integration $\int \ldots \int \mathrm{d}t$ with the bounded quantities taken outside the integrals.

Solutions of nonlinear stochastic equations in series form can be put in a convenient symmetrized form which is easily calculable [8]. For nonzero $\mathscr{N}$, we assume that the stochastic coefficients $b_k(t, \omega)$ are bounded a.s. on T for $\omega \in (\Omega, \mathscr{K}, P)$. The iteration now leads to extra-terms arising from $L^{-1}\mathscr{N}$. However, y_0 is bounded, as before, by hypothesis; y_1 differs from the y_1 for the linear case by the addition of the term $L^{-1}(\mathscr{N}(y_0))$; y_2 differs from the y_2 for the linear case by addition of $L^{-1}\mathscr{N}(y_0, y_1)$, etc. Thus,

$$y(t, \omega) = F(t, \omega) - \int_0^t \Gamma(t, \tau, \omega) \, F(\tau, \omega) \, \mathrm{d}\tau -$$

$$-\int_0^t l(t, \omega)\, \mathscr{N}(y, \dot{y}, \ldots)\mathrm{d}\tau =$$

$$= F(t, \omega) - \int_0^t \Gamma(t, \tau, 0)F(\tau, \omega)\mathrm{d}\tau -$$

$$- \int_0^t l(t, \tau) \sum_{k=0}^{m} b_k(\tau, \omega)\, (y^{(k)})^{m_k}\mathrm{d}\tau.$$

Adomian [7], [8], [4], [3], [5] investigated the symmetrized forms for polynomials nonlinearities, i.e., equations in which $\mathscr{N}(y, \dot{y}, \ldots) = by^2,\ by^3,\ by^m$ or e^y.

We shall assume, for $\mathscr{N}$ too, deterministic N and stochastic M parts such that $\mathscr{N} = N + M$. Note that N and M are, of course, nonlinear. Thus, we have

$$Ly + Ry + Ny + My = x$$

and since we supposed L^{-1} to be determinable, we have

$$(**)\ y = L^{-1}x - L^{-1}Ry - L^{-1}Ny - L^{-1}My.$$

Finally, we add a term γ to $L^{-1}x$ to include the effect of initial conditions — the solution of $Ly = 0$ — and possibly the terms resulting from a s.b.c. as well, if an adjoint operation is necessary on the Ry term because of the presence of derivatives in R. The solution to the stochastic equation is given by a decomposition of y, or, equivalently, $\mathscr{H}^{-1}x$, into components y_n or $\mathscr{H}_n^{-1}x$. The first term y_0 or $\mathscr{H}_0^{-1}\, x$ is assumed to be the known $L^{-1}x$ + terms added as was shown above. A parametrized decomposition $y = \sum_{n=0}^{\infty} \lambda^n y_n$, or $\sum_{n=0}^{\infty} \lambda^n \mathscr{H}_n^{-1}x$, may be convenient (as it permits a more convenient grouping of terms than in other works, yielding highly convenient forms for computation), where λ is later set equal to unity. In this case, (**) is also parametrized as

$$(***)\ y = y_0 - \lambda L^{-1}Ry - \lambda L^{-1}Ny - \lambda L^{-1}My$$

or

$$\sum_{n=0}^{\infty} \lambda^n \mathscr{H}_n^{-1}x = y_0 - \lambda L^{-1}R \sum_{n=0}^{\infty} \mathscr{H}_n^{-1}x - \lambda L^{-1}My,$$

which is known [4] as the inverse operator method.

The last two terms are dealt with as follows. Since Ny is a function, say $f(y)$ (and My is also a function $g(y)$ differing only inasmuch as it has stochastic coefficients) and since $y = y(\lambda)$, we assume $Ny = \sum_{n=0}^{\infty} A_n\lambda^n$ and $My = \sum_{n=0}^{\infty} B_n\lambda^n$ are respectively analytic and stochastically analytic forms for which the coefficients [2] are determinable by implicit differentiation [2], [7]:

$$A_0 = A_0(y_0) = f(y(\lambda))|_{\lambda=0} = f(y_0)$$

$$A_1 = A_1(y_0, y_1) = f'(y(\lambda))|_{\lambda=0} = (\mathrm{d}f/\mathrm{d}y)\,(\mathrm{d}y/\mathrm{d}\lambda)|_{\lambda=0}$$

$$A_2 = A_2(y_0, y_1, y_2) = \frac{1}{2}\{(\mathrm{d}^2f/\mathrm{d}y^2)(\mathrm{d}y/\mathrm{d}\lambda)^2 + (\mathrm{d}f/\mathrm{d}y)\,(\mathrm{d}^2y/\mathrm{d}\lambda^2)\}|_{\lambda=0}$$

$$A_3 = A_3(y_0, y_1, y_2, y_3) = \frac{1}{6}\{(\mathrm{d}^3f/\mathrm{d}y^3)\,(\mathrm{d}y/\mathrm{d}\lambda)^3 +$$

$$+ 2(\mathrm{d}^2f/\mathrm{d}y^2)\,(\mathrm{d}y/\mathrm{d}\lambda)\,(\mathrm{d}^2y/\mathrm{d}\lambda^2) +$$

$$+ (\mathrm{d}^2f/\mathrm{d}y^2)\,(\mathrm{d}^2y/\mathrm{d}\lambda^2)\,(\mathrm{d}y/\mathrm{d}\lambda) + (\mathrm{d}f/\mathrm{d}y)\,(\mathrm{d}^3y/\mathrm{d}\lambda^3)\}|_{\lambda=0}$$

The B_n are similarly determined for the function $My = g(y) = g(y(\lambda))$ and differ only inasmuch as they involve a stochastic process, e.g., in a term such as $My = b(t, \omega)y^3$, $b(t, \omega)e^y$, etc.

Equation (***) is now written as

$$\sum_{n=0}^{\infty} \lambda^n \mathscr{H}_n^{-1}x = y_0 - \lambda L^{-1}R \sum_{n=0}^{\infty} \lambda^n \mathscr{H}_n^{-1}x -$$

$$- \lambda L^{-1} \sum_{n=0}^{\infty} A_n\lambda^n - \lambda L^{-1} \sum_{n=0}^{\infty} B_n\lambda^n.$$

For $n = 0, 1, 2, \ldots,$ we find

$$y_1 = -L^{-1}R\mathscr{H}_0^{-1}x - L^{-1}A_0 - L^{-1}B_0$$

$$y_2 = -L^{-1}R\mathscr{H}_1^{-1}x - L^{-1}A_1 - L^{-1}B_1$$

— — — — — — — — — — — — .

Since y_0 is known and since every term is calculable in terms of the preceding term, and, hence, eventually in terms of y_0, no question of statistical separability will arise in the system and the input signals are statistically independent and no closure approximations are necessary.

If, for example, $f(y) = by^2$, we write $y = \sum_{n=0}^{\infty} (-1)^n \lambda_n y_m$ and have

$$y = L^{-1}x - \lambda L^{-1}Ry - \lambda L^{-1}by^2 = L^{-1}x - \lambda L^{-1}R -$$
$$- \lambda L^{-1}by^2 = L^{-1}x - \lambda L^{-1}R \sum_{n=0}^{\infty} (-1)^n \lambda^n y_n -$$
$$- L^{-1}b \sum_{i=0}^{\infty} \sum_{j=0}^{\infty} (-1)^{i+j} \lambda^i \lambda^j y_i y_j =$$
$$= y_0 - \lambda L^{-1}R[y_0 - \lambda y_1 + \lambda^2 y_2 - \ldots] -$$
$$- \lambda L^{-1}b[y_0 - \lambda y_1 + \lambda^2 y_2 + \ldots][L^{-1}y_0 -$$
$$- \lambda y_1 + \lambda^2 y_2 - \ldots] = y_0 - \lambda L^{-1}[Ry_0 + by_0^2] +$$
$$+ \lambda^2 L^{-1}[Ry_1 + 2by_0y_1] - \lambda^3 L^{-1}[Ry_1 + by_1^2 + 2by_0y_2].$$

Letting $\lambda = 1$ and using the initial conditions, we obtain

$$y_0 = L^{-1}x$$
$$y_1 = L^{-1}[Ry_0 + by_0{}^2]$$
$$y_2 = L^{-1}[Ry_1 + b(y_0y_1 + y_1y_0)]$$
$$y_3 = L^{-1}[Ry_2 + b(y_1^2 + y_0y_2 + y_2y_0)].$$

These can be re-written as

$$y_0 = L^{-1}x$$
$$y_1 = L^{-1}[Ry_0 + by_0y_0]$$
$$y_2 = L^{-1}[Ry_1 + b(y_0y_1 + y_1y_0)]$$
$$y_3 = L^{-1}[Ry_2 + b(y_1y_2 + y_1y_1 + y_2y_0)]$$
$$\ldots\ldots\ldots\ldots\ldots\ldots\ldots\ldots$$
$$y_{n+1} = L^{-1}[Ry_n + b(y_0y_n + y_1y_{n-1} +$$
$$+ y_2y_{n-2} + \ldots + y_ny_0)] = L^{-1}K(y_0),$$

where $K_n(y_0) = Ry_n + b(y_0y_n + y_1y_{n-1} + \ldots + y_ny_0)$ and $y = y_0 - L^{-1}\Gamma(y_0)$ with $\Gamma(y_0) = \sum_{n=0}^{\infty} (-1)^n K_n(y_0)$. The same letters Γ and K as in the linear case were used but are identical only if $\mathcal{N} = 0$ and $\sigma = 0$.

Consequently, we can easily write the general term in the above symmetric form and can program it on a computer. That convergence still holds in the nonlinear case can be seen from the form of $K_n(y_0)$. If one examines the bracketed term $y_0y_n + \ldots + y_ny_0$ and replaces backward each y_i by y_{i-1} until y_0 is reached, then each term yields n integrations or an $n!$ in the denominator as in the linear case. Now, we have n such terms in the general term yielding $1/(n-1)!$ and the convergence follows.

Possible extensions of these results to linear or nonlinear stochastic partial differential equations are given in refs. [4], [7].

It is of interest to notice that though the method was developed for stochastic problems to achieve statistical separability, it may be used for deterministic equations as well. The above iterative methods may be advantageous over the well-known Picard method for successive approximations since any stage of approximation is obtained by one more integration of a transformation of the preceding term whereas in the Picard series, the successive approximations become increasingly cumbersome. Further, the method appears to converge quicklier and the computation is relatively simpler.

With regard to the commonly used hierarchy equations or averaging method and its closure approximations, we now have the means at hand to evaluate the error [6] in the solution for closure approximation at any level of the hierarchy and it is quite clear that the iterative and inverse methods are preferable.

Itô equation methods are widely used in many fields of application but are limited to physical processes satisfying unrealistic mathematical assumptions and are less general than the proposed methods.

7.10. STOCHASTIC NONLINEAR PROGRAMMING AND STOCHASTIC OPTIMAL CONTROL PROBLEMS

Interesting applications of random fixed points theorems, of measurable solutions of random problems or of measurable selection theorems to stochastic nonlinear programming and stochastic optimal control problems, for the so-called "distribution problem" of stochastic optimization or to random convex programming were given by H. W. Engl [139], W. Römisch [368], A. Nowak [317], [316] and others. In what follows, we shall present some of these ideas and results.

Recently, some results have been proved on a certain class of stochastic linear programs which possesses an optimal solution depending on a random parameter in a measurable way, and with the optimal value measurable. In stochastic optimization, this is called "distribution problem". In the theory of operator equations, the basic problem mentioned above is treated under the name of random fixed point theory.

The model considered next is the stochastic analog of an older model [139]:

(A) $T(x, u) = 0$, $x \in \theta_x$, $y \in Q_y$, $S(x, u)$ max, where Q_x and Q_y are subsets of the Banach spaces X and U respectively, $T : Q_x \times Q_u \to Y$ (Banach spaces), $S : Q_x \times Q_u \to \mathbb{R}$. This model includes optimal control problems, where the state and the control variables may be constrained, as well as nonlinear optimization problems.

In order to increase the applicability, let T be multivalued and replace $T(x, u) = 0$ by $0 \in T(x, u)$. This allows us to treat inequality constraints in an easy way (cf. [139, Th. 14]).

We consider X, U, Y to be normed linear spaces, $(\Omega, \mathscr{K}, P)$ to be a complete σ-finite measure space, $C : \Omega \to 2^X$, $K : \Omega \to 2^U$, $T : \mathrm{Gr}(C \times K) \to CB(Y)$, $S : \mathrm{Gr}(C \times K) \to \mathbb{R}$, where T is a multivalued random operator with stochastic domain $C \times K$ (resp. S a random functional with the same stochastic domain) with C and K separable. T (resp. S) as above is called "continuous" if, for all $\omega \in \Omega$, $T(\omega,.,.)$ (resp. $S(\omega,.,.)$) is jointly continuous on $C(\omega) + K(\omega)$ with respect to the Hausdorff-metric (resp. the usual topology on $\mathbb{R}$). Hence, T and S are defined on $\{(\omega, x, u) \in \Omega \times X \times U : x \in C(\omega) \wedge u \in K(\omega)\}$, and, hence, T and S are families of operators (functionals) depending on $\omega \in \Omega$ such that the domain of the realization for a specific ω is just $C(\omega) \times K(\omega)$. For every fixed $(x, u) \in X \times U$, $T(.,x, u)$ and $(S.,x, u)$ depend in a measurable way on ω.

The stochastic version of model (A) is

$$(\mathrm{A}(\omega))\ (\omega, x, u) \in \mathrm{Gr}(Q_x \times Q_u),\ 0 \in T(\omega, x, u),\ S(\omega, x, u) \to \max,$$

where X, U, Y are separable Banach spaces, $Q_x : \Omega \to 2^X$, $Q_u : \Omega \to 2^U$, and $T : \mathrm{Gr}(Q_x \times Q_u) \to CB(Y)$ is a continuous multivalued random operator with stochastic domain $Q_x \times Q_u$ whereas $S : \mathrm{Gr}(Q_x \times Q_u) \to \mathbb{R}$ is a continuous random functional with stochastic domain $Q_x \times Q_u$.

DEFINITION 7.10.1. Model $A(\omega)$ is called "stochastically solvable" if there exist two measurable functions $\bar{x} : \Omega \to X$, $\bar{u} : \Omega \to U$ such that for all $\omega \in \Omega$, $\bar{x}(\omega) \in Q_x(\omega)$, $\bar{u}(\omega) \in Q_u(\omega)$, $0 \in T(\omega, \bar{x}(\omega), \bar{u}(\omega))$ and $S(\omega, \bar{x}(\omega), \bar{u}(\omega)) = v(\omega) = \sup\{S(\omega, x, u) : x \in Q_x(\omega),\ u \in Q_u(\omega),\ 0 \in T(\omega, x, u)\}$ and if $v : \Omega \to \mathbb{R}$ is measurable. The pair $(\bar{x}, \bar{u})$ is called "*random optimal solution*".

Model $A(\omega)$ is called "*deterministically solvable*" if for all $\omega_0 \in \Omega$ there exist $x_0 \in Q_x(\omega_0)$ and $u_0 \in Q_u(\omega_0)$ such that $0 \in T(\omega_0, x_0, u_0)$ and $S(\omega_0, x_0, u_0) = \sup\{S(\omega_0, x, u) : x \in Q_x(\omega_0) \wedge u \in Q_u(\omega_0) \wedge 0 \in T(\omega_0, x, u)\}$.

THEOREM 7.10.1 [139]. *Model $A(\omega)$ is stochastically solvable if and only if it is deterministically solvable.*

Proof. Let $A(\omega)$ be a deterministically solvable model and let D be the Hausdorff metric on $CB(Y)$. Let $N : \mathrm{Gr}(Q_x \times Q_u) \to \to \mathbb{R}$ be defined by $N(\omega, x, u) = d(0, T(\omega, x, u))$. Let us choose $\omega \in \Omega$, (x_1, u_1) and $(x_2, u_2) \in (Q_x \times Q_u)(\omega)$ arbitrary but fixed. Then

$$|N(\omega, x_1, u_1) - N(\omega, x_2, u_2)| \leqslant$$
$$\leqslant D(T(\omega, x_1, u_1), T(\omega, x_2, u_2)).$$

From the continuity of T, it follows that $N(\omega,.,.)$ is continuous on $Q_x(\omega) \times Q_u(\omega)$. Let now $(x, u) \in X \times U$ and let $r \geqslant 0$ be chosen arbitrarily but fixed. Then

$$(1) \quad \{\omega \in \Omega : x \in Q_x(\omega) \wedge u \in Q_u(\omega) \wedge N(\omega, x, u) < r\} =$$
$$= \{\omega \in \Omega : (x, u) \in (Q_x \times Q_u)(\omega) \wedge T(\omega, x, u) \cap$$
$$\cap \{y \in Y : \|y\| < r\} \neq \Phi\} \in \mathscr{K}$$

since T is a multivalued random operator.

Since $Q_x \times Q_u$ is separable (cf. Lemma 2.2.4), there exists a separator $Z \subset X \times U$. From the definition of Q_x and Q_u, Lemma 2.2.4 and Theorem 2.2.1, it follows that $\mathrm{Gr}(Q_x \times Q_u) \in \in \mathscr{K} \times \mathscr{B}_{X \times U}$. This, together with the continuity of each $N(\omega,.,.)$ and (1), imply that

$$\{(\omega, x, u) \in \mathrm{Gr}(Q_x \times Q_u) : N(\omega, x, u) \leqslant r\} =$$
$$= \{(\omega, x, u) \in \Omega \times X \times U : (x, u) \in (Q_x \times Q_u)(\omega) \wedge$$
$$\wedge N(\omega, x, u) \leqslant r\} = \bigcap_{n \in \mathbb{N}} \bigcup_{z \in \mathbb{Z}} [\{(\omega \in \Omega : z \in (Q_x \times Q_u)(\omega) \wedge$$
$$\wedge N(\omega, z) < r + 1/n\} \times \{(x, u) \in X \times U : \|(x, u) - z\|_{X \times U} <$$
$$< 1/n\}) \cap \mathrm{Gr}(Q_x \times Q_u)] \in \mathscr{K} \times \mathscr{B}_{X \times U}.$$

Thus, N is a measurable function from the measurable space $(\mathrm{Gr}(Q_x \times Q_u), \mathrm{Gr}(Q_x \times Q_u) \cap (\mathscr{K} \times \mathscr{B}_{X \times U}))$ into $\mathbb{R}$.

Let us now define $E : \Omega \to \{A : A \subseteq X \times U\}$ by

$$E(\omega) = \{(x, u) \in Q_x(\omega) \times Q_u(\omega) : 0 \in T(\omega, u, u)\}.$$

Because of the deterministic solvability of $A(\omega)$ each $F(\omega) \neq \neq \emptyset$. Let ω be an arbitrary but fixed point of Ω and let $(x_k, u_k) \in E(\omega)^{\mathbb{N}}$ be such that $(x_k, u_k) \to (x, u) \in X \times U$. Then

$$(x, u) \in (Q_x \times Q_u)(\omega) \text{ and } d(0, T(x, \omega, u)) \leqslant d(0, T(\omega, x, u)) +$$
$$+ D(T(\omega, x_k, u_k), T(\omega, x, u)) \text{ for all } k \in \mathbb{N}.$$

As all $(x_k, u_k,) \in E(\omega)$, $d(0, T(\omega, x_k, u_k)) = 0$.

The continuity of T and the closedness of $T(\omega, x, u)$ imply that $0 \in T(\omega, x, u)$, which implies that $(x, u) \in E(\omega)$. Hence, every $E(\omega)$ is closed.

The graph of E is

$$\mathrm{Gr}E = \{(\omega, x, u) \in \Omega \times X \times U : (x, u) \in E(\omega)\} =$$

$$= \{(\omega, x, u) \in \mathrm{Gr}N : N(\omega, x, u) = 0\} \in \mathscr{K} \times \mathscr{B}_{X \times U}$$

since N has joint measurability. This, together with Theorem 2.2.1, implies that $E : \Omega \to 2^{X \times U}$ is measurable.

From Theorem 2.2.5, it follows that there exists a countable set of measurable functions $f_1, f_2, \ldots : \Omega \to X \times U$ such that for all $\omega \in \Omega$, $E(\omega) = \mathrm{cl}\{f_1(\omega), f_2(\omega), \ldots\}$.

Now, let $v : \Omega \to \mathbb{R}$ be the function given in the above definition. Obviously, $v(\omega) = \sup\{S(\omega, x, u) : (x, u) \in E(\omega)\}$. By virtue of the continuity of S, we have

$$v(\omega) = \sup_{n \in \mathbb{N}} S(\omega, f_n)(\omega)).$$

In a similar way, one can prove that S is jointly measurable on $(\mathrm{Gr}(Q_x \times Q_u), \mathrm{Gr}(Q_x \times Q_u) \cap (\mathscr{K} \times \mathscr{B}_{X \times U}))$. Due to the measurability of the f_n's and to the fact that $\mathrm{Gr}(Q_x \times Q_u) \in \in \mathscr{K} \times \mathscr{B}_{X \times U}$, all $\omega \to (\omega, f_n(\omega))$ are measurable and, therefore, v is measurable.

Let us now define, for all $\omega \in \Omega$,

$$M(\omega) = \{(x, u) \in E(\omega) S(\omega, x, u) = v(\omega)\}.$$

For all $\omega \in \Omega$, $M(\omega)$ is nonvoid due to the deterministic solvability and is closed due to the continuity of $S(\omega, ., .)$ and the closedness of $E(\omega)$. Then

$$\mathrm{Gr}M = \{(\omega, x, u) \in \mathrm{Gr}E : S(\omega, x, u) = v(\omega)\} \in \mathscr{K} \times \mathscr{B}_{X \times U}$$

due to the joint measurability of S, the measurability of v and the fact that $\mathrm{Gr}E \in \mathscr{K} \times \mathscr{B}_{X \times U}$. Theorem 2.2.1 implies that $M : \Omega \to 2^{X \times U}$ is measurable. We apply the Kuratowski-Ryll-Nardzewski Selection Theorem (Theorem 2.2.4) to M and conclude on the existence of a measurable function $m : \Omega \to X \times U$ such that for all $\omega \in \Omega$, $m(\omega) \in M(\omega)$. Let p_1, p_2 be the natural projections from $X \times U$ onto X, U, let $\bar{x} : \Omega \to X$ and $\bar{u} : \Omega \to U$ be defined by $\bar{x} = p_1 \circ m$, $\bar{u} = p_2 \circ m$. Then $\bar{x}$ and $\bar{u}$ are measurable and $\bar{x}(\omega) \in Q_x(\omega)$, $\bar{u}(\omega) \in Q_u(\omega)$ for all $\omega \in \Omega$. Also, $0 \in T(\omega, \bar{x}(\omega), \bar{u}(\omega))$, and $S(\omega, \bar{x}(\omega), \bar{u}(\omega)) = = v(\omega)$. This proves that $(A(\omega))$ is stochastically solvable.

Since the converse implication is trivial, the proof is complete.

Remark. (1) An example can be given [139] to show that deterministic probability does not always imply stochastic solvability if the continuity assumption in $(A(\omega))$ is relaxed (2) If in the last part of the proof we use Theorem 2.2.5 instead of the Kuratowski-Ryll-Nardzewski theorem, we obtain the existence of a countable set of measurable functions $\bar{x}_n : \Omega \to \to X$, $\bar{u}_n : \Omega \to U$ with the property that, for all $\omega \in \Omega$, $(\bar{x}_n(\omega_0), \bar{u}_n(\omega_0))_{n \in \mathbb{N}}$ is a dense subset of the set of the optimal solutions of the realization of $(A(\omega))$ for ω_0.

As an application of the above general result to nonlinear programming and optimal control, consider

THEOREM 7.10.2 [139]. *Let* n, $m \in \mathbb{N}$, $f : \Omega \times \mathbb{R}^n \to \mathbb{R}$, $g : \Omega \times \mathbb{R}^n \to \mathbb{R}^m$, $c : \Omega \to \mathbb{R}^m$. *Assume that for all* $\omega \in \Omega$, $f(\omega,.)$ *and* $g(\omega,.)$ *are continuous and that* c *and for all* $x \in \mathbb{R}^n$, $f(.,x)$ *and* $g(.,x)$ *are measurable. Furthermore, assume that for all* $\omega_0 \in \Omega$, *there exists an* $x_0 \in \mathbb{R}$ *such that* $g(\omega_0, x_0) \leqslant c(\omega_0)$ *(componentwise) and* $f(\omega_0, x_0) = v(\omega_0) = \sup\{f(\omega_0, x) : x \in \mathbb{R}^n \wedge g(\omega_0, x) \leqslant \leqslant c(\omega_0)\}$ *("deterministic solvability"). Then* v *is measurable and there exists a measurable function* $\bar{x} : \Omega \to \mathbb{R}^n$ *such that for all* $\omega \in \Omega$, $g(\omega, \bar{x}(x)) \leqslant c(\omega)$ *and* $f(\omega, \bar{x}(\omega)) = v(\omega)$ *("stochastic solvability").*

Proof. The above problem can be regarded as a special case of the model $(A(\omega))$. Since there exists no control variable, we shall choose T and S independent of the third argument which, for the sake of simplicity, is omitted.

Let $X = \mathbb{R}^n$, $Y = \mathbb{R}^m$; denote by g_i and c_i the components of g and c. Let $T : \Omega \times X \to 2^Y$ be defined as $T(\omega, x) = = \times_{i=1}^{m} T_i(\omega, x)$, where $T_i(\omega, x) = [g_i(\omega, x) - c_i(\omega), |g_i(\omega, x) - - c_i(\omega, x)|]$. Then $0 \in T(\omega, x) \Leftrightarrow g(\omega, x) \leqslant c(\omega)$.

First, show that for every $x \in X$ and $i \in \{1, \ldots, m\}$, $T_i(.,x)$ is measurable as a multivalued mapping from Ω into $2^{\mathbb{R}}$. Let $x \in X$, $i \in \{1, 2, \ldots, m\}$, $y \in \mathbb{R}$, be arbitrary but fixed. Then the mapping $\omega \to d(y, T_i(\omega, x))$ is measurable since $d(y, T_i(x, \omega)) = \max\{0, |y - m(\omega)| - r(\omega)\}$ with

$$m(\omega) = \frac{|g_i(\omega, x) - c_i(\omega)| + g_i(\omega, x) - c_i(\omega)}{2} \quad \text{and}$$

$$r(\omega) = \frac{|g_i(\omega, x) - c_i(\omega)| + c_i(\omega) - g_i(\omega ,x)}{2}.$$

From Theorem 2.2.1 the measurability of $T_i(.,\omega)$ is immediate.

Let $G \subseteq Y$ be an open set of the form $G = \times_{i=1}^{n} G_i$ with $G_i \subseteq$

$\subseteq \mathbb{R}$ open. Then

$$\{\omega \in \Omega : T(\omega, x) \cap G \neq \emptyset\} =$$
$$= \bigcap_{i=1}^{m} \{\omega \in \Omega : T(\omega, x) \cap G_i \neq \emptyset\} \in \mathscr{K}$$

because of the measurability of all $T_i(\omega, x)$ and the measurability of $T(.,x)$ is immediate.

To prove the continuity of $T(\omega.,)$ with respect to the Hausdorff metric D on Y, we observe that $D([a, b], [d, e]) = \max\{|a-d|, |b-e|\}$ for the real intervals $[a, b]$, $[d, e]$. Together with the continuity of the $g_i(\omega.,)$, this implies that, for all $i \in \{1, \ldots, m\}$, $T_i(\omega.,)$ is continuous. Now, let $(x_j) \in X^{\mathbb{N}}$ with $(x_j) \to x \in X$. As for all $i \in \{1, \ldots, m\}$, $(T_i(\omega, x_j)) \to T_i(\omega, x)$, the definition of T implies that $(T(\omega, x_j)) \to T(\omega, x)$. So, $T(\omega.,)$ is continuous.

For the functional S in model $A(\omega)$, we take of course f, for $Q_x(\omega) = X$ for all $\omega \in \Omega$ and the proof follows by Theorem 7.10.1.

An optimal control problem, where the governing differential equation, the initial state, the objective functional and the set of feasible controls are random, is given by

THEOREM 7.10.3 [139]. *Let* $n, m, r \in \mathbb{N}$, $f : \Omega \times \mathbb{R}^n \times \mathbb{R}^m \to \mathbb{R}^n$, $g : \Omega \times \mathbb{R}^n \times \mathbb{R}^m \to \mathbb{R}$, $G : \mathbb{R}^n \to \mathbb{R}^r$, $x_0 : \Omega \to \mathbb{R}^n$, $u_0 : \Omega \times [0,1] \to \mathbb{R}^m$, $R : \Omega \to (0, +\infty)$. *We assume: G and for all $\omega \in \Omega$, $f(\omega,.,.)$, $g(\omega,.,.)$ and $u_0(\omega,.)$ are continuous; for all $(a, b) \in \mathbb{R}^n \times \mathbb{R}^m$, $f(,.a, b)$ and $g(.,a, b)$ are measurable; for all $t \in [0,1]$, $u_0(.,t)$ is measurable; R and x_0 are measurable. Furthermore, we assume that for all $\omega_0 \in \Omega$ there exist a continuously differentiable function $x_1 : [0,1] \to \mathbb{R}^n$ and a continuous function $u_1 : [0, 1] \to \mathbb{R}^m$ such that for all $t \in [0, 1]$, $x_1'(t) = f(\omega_0, x_1(t), u_1(t))$ and $\| u_1(t) = -u_0(\omega_0, t)\| \leqslant R(\omega_0)$ and that*
$$x_1(0) = x_0(\omega_0),\ G(x_1(1)) = 0 \ \textit{and} \int_0^t g(\omega_0, x_1(s), u_1(s))\, \mathrm{d}s = v(\omega_0) =$$
$$= \sup\left\{\int_0^1 g(\omega_0, x(s), u(s))\, \mathrm{d}s : x \textit{ and } u \textit{ fulfill the same conditions as } x_1 \textit{ and } u_1\right\}$$
("deterministic solvability").

Then there exist $\bar{x} : \Omega \times [0,1] \to \mathbb{R}^n$ and $\bar{u} : \Omega \times [0,1] \to \mathbb{R}^m$ with the following properties:

(1) *For all* $t \in [0, 1]$, $\bar{x}(.,t)$ *and* $\bar{u}(.,t)$ *are measurable.*

(2) *For all* $\omega \in \Omega$, $\bar{u}(\omega,.)$ *is continuous and* $\bar{x}(\omega,.)$ *is continuously differentiable.*

(3) *For all* $t \in [0, 1]$ *and* $\omega \in \Omega$,

$$\frac{\mathrm{d}\bar{x}(\omega, t)}{\mathrm{d}t} = f(\omega, \bar{x}(\omega, t), \bar{u}(\omega, t)), \quad \|\bar{u}(\omega, t) - u_0(\omega, t)\| \leqslant R(\omega)$$

(4) *For all* $\omega \in \Omega$, $\bar{x}(\omega, 0) = x_0(\omega)$ *and* $G(\bar{x}(\omega, 1)) = 0$,

(5) *For all* $\omega \in \Omega$, $\int_0^1 g(\omega, \bar{x}(\omega, s), \bar{u}(\omega, s))\, \mathrm{d}s = v(\omega)$.

Furthermore, v is measurable.

Proof. To define the above problem within the framework of model $A(\omega)$, we shall choose $X = Y = C_n[0, 1] = \{x : [0,1] \to \mathbb{R}^n,\ x \text{ continuous}\}$, $U = C_m[0,1]$. For all $\omega \in \Omega$, let $Q_x(\omega) = \{x \in X : G(x(1)) = 0\}$ and $Q_u(\omega) = \{u \in U : \|u - u_0(\omega,.)\| \leqslant R(\omega)\}$. According to the assumptions, all $Q_x(\omega)$ and $Q_u(\omega)$ are nonvoid and closed. Also Q_x and Q_u are separable.

We define $T : \mathrm{Gr}(Q_x \times Q_u) \to X$ by

$$T(\omega, x, u) = \left(t \to x(t) - x_0(\omega) - \int_0^t f(\omega, x(s), u(s))\, \mathrm{d}s\right)$$

and $S : \mathrm{Gr}(Q_x \times Q_u) \to \mathbb{R}$ by $S(\omega, x, u) = \int_0^1 g(\omega, x(s), u(s))\, \mathrm{d}s$.

From Theorem 2.2.1 and Lemma 2.2.4, $\mathrm{Gr}(Q_x \times Q_u) \in \mathscr{K} \times \mathscr{B}_{X \times U}$.

Then (cf. Lemma 2.2.6 and Theorem 2.2.1), we conclude that T and S are a random operator and a random functional with stochastic domain $Q_x \times Q_u$ respectively. For all $\omega \in \Omega$, $T(\omega,.,.)$ and $S(\omega,.,.)$ are continuous. From Theorem 7.10.1 and Lemma 2.2.6, the conclusion follows easily.

On the other hand, the convergence of measurable multivalued mappings and their measurable selection play an essential role for the design and study of approximation schemes in stochastic analysis and stochastic optimization. The first results on the convergence of measurable selections were given by Salinetti and Wets [376] and these results have prompted some more recent research.

Let X be a Polish space, let C, C_n, $n \in \mathbb{N}$, be multivalued mappings from Ω into X and let $\mathscr{S}(C)$ be the set of all measurable selectors of C. For later use, we define the following multivalued mappings from Ω into X for all $n \in \mathbb{N}$ and all $\varepsilon > 0$:

$$\Delta^+_{\varepsilon, n}(\omega) = C(\omega) - \varepsilon C_n(\omega), \text{ for all } \omega \in \Omega$$

$$\Delta_{\varepsilon, n}(\varepsilon) = \Delta^+_{\varepsilon, n}(\omega) \cup (C_n(\omega) - \varepsilon C(\omega)), \text{ for all } \omega \in \Omega,$$

where $\varepsilon A = \{x \in X : d(x, A) \leqslant \varepsilon\}$ for any subset $A \subseteq X$. Note that the multivalued mappings $\Delta^+_{\varepsilon, n}$ and $\Delta_{\varepsilon, n}$, $n \in \mathbb{N}$, $\varepsilon > 0$, are Gr-measurable [368].

DEFINITION 7.10.2. $(C_n)_{n \in \mathbb{N}}$ is said to converge in probability to C if, for all $\varepsilon > 0$ and any compact subset $K \subseteq X$, $\lim_{n \to \infty} P(\Delta^{-1}_{\varepsilon, n}(K)) = 0$.

Let $f : \Omega \times X \to \mathbb{R}$ be $\mathscr{K} \times \mathscr{B}_X$-measurable and let $C : \Omega \to \mathscr{P}(X)$ be a Gr-measurable multivalued mapping. Let us introduce the following notation:

$$\varphi : \Omega \to \overline{\mathbb{R}}, \ \varphi(\omega) = \inf\{f(\omega, x) : x \in C(\omega)\} \text{ for } \omega \in \Omega,$$

ψ is the multivalued mapping from Ω into X, defined by $\psi(\omega) = \{x \in C(\omega) : f(\omega, x) = \varphi(\omega)\}$ for $\omega \in \Omega$.

It is known [368, P.4.1] that φ is measurable and ψ is Gr-measurable and the question of the distribution or some of its characteristic values, of the "optimal value" φ and (or) of an "optimal solution" (i.e., a selection of ψ) is the so-called "distribution problem" of stochastic optimization.

Let us study approximations for the problem

$$\text{(I)} \quad f(\omega, x) = \text{Min! } s.t. x \in C(\omega), \ \omega \in \Omega.$$

Additionally, let $f_n : \Omega \times X \to \mathbb{R}$, $n \in \mathbb{N}$, be $\mathscr{K} \times \mathscr{B}_X$-measurable and let $C_n : \Omega \to \mathscr{P}(X)$, $n \in \mathbb{N}$, be Gr-measurable multivalued mappings.

Consider a sequence of stochastic optimization problems

$$\text{(II)} \quad f_n(\omega, x) \to \text{Min! } s.t. \ x \in C_n(\omega), \ \omega \in \Omega, \ n \in \mathbb{N},$$

and define, for all $n \in \mathbb{N}$ and $\omega \in \Omega$,

$$\varphi_n(\omega) = \inf\{f_n(\omega, x) : x \in C_n(\omega)\}$$

$$\psi_n(\omega) = \{x \in C_n(\omega) : f_n(\omega, x) = \varphi_n(\omega)\}.$$

A result on the convergence of the sequence of "optimal solution sets" $\{S(\psi_n)\}_{n \in \mathbb{N}}$ is given by

THEOREM 7.10.4. *Let f, f_n, $n \in \mathbb{N}$, be $\mathscr{K} \times \mathscr{B}_X$-measurable mappings from $\Omega \times X$ into $\mathbb{R}$ and let C, C_n, $n \in \mathbb{N}$, be Gr-measurable multivalued mappings from Ω into X. Assume that*

(1) C is closed-valued and $\{C_n\}_{n \in \mathbb{N}}$ converge in probability to C;

(2) for all $x \in \mathscr{S}(C)$ and $x_n \in \mathscr{S}(C_n)$, $n \in \mathbb{N}$, such that $x = P\text{-}\lim_{n\to\infty} x_n$ we have $f(.,x(.)) = P\text{-}\lim_{n\to\infty} f_n(.,x_n(.))$;

(3) ψ_n is defined as in (II) and is a multivalued mapping from Ω into X, for all $n \in \mathbb{N}$; ψ is as above.

Then $P\text{-}\lim_{n\to\infty} \sup \mathscr{S}(\psi_n) \subseteq \mathscr{S}(\psi)$.

Proof. It is known [368, Th. 3.4] that (1) implies that

$$(*) \quad \mathscr{S}(C) = P\text{-}\lim_{n\to\infty} \mathscr{S}(C_n).$$

Let $x \in P\text{-}\lim_{n\to\infty} \sup \mathscr{S}(\psi_n)$, i.e., $x = P\text{-}\lim_{k\to\infty} x_k$, $x_k \in \mathscr{S}(\psi_{n_k})$, for all $k \in \mathbb{N}$. Then $x_k \in \mathscr{S}(C_{n_k})$, $k \in \mathbb{N}$, and from $(*)$ it follows that $x \in \mathscr{S}(C)$.

By $(*)$ there exists a sequence $\bar{x}_n \in \mathscr{S}(C_n)$, $n \in \mathbb{N}$, such that $x = P\text{-}\lim_{n\to\infty} \bar{x}_n$. For all $n \in \mathbb{N}$, let

$$\tilde{x}_n = \begin{cases} x_n & \text{if } n \in \{n_1, n_2, \ldots\} \\ \bar{x}_n & \text{otherwise.} \end{cases}$$

Then $x = P\text{-}\lim_{n\to\infty} \tilde{x}_n$ and (2) implies that

$$P\text{-}\lim_{n\to\infty} f_n(.,\tilde{x}_n(.)) = P\text{-}\lim_{k\to\infty} f_{n_k}(.,x_k(.)) =$$

$$= P\text{-}\lim_{k\to\infty} \varphi_{n_k}(.) = f(.,x(.)).$$

It remains to show that $f(\omega, x(\omega)) \leqslant \varphi(\omega)$ a.s.

Let $\varepsilon > 0$ be arbitrary but fixed and consider the multivalued mapping $M : \Omega \to \mathscr{P}(X)$ defined by $M(\omega) = \{x \in C(\omega) : f(\omega, x) \leqslant r_\varepsilon(\omega)\}$, for all $\omega \in \Omega$, where

$$r_\varepsilon(\omega) = \begin{cases} \varphi(\omega) + \varepsilon, & \text{if } \varphi(\omega) > -\infty \\ -1/\varepsilon & \text{otherwise.} \end{cases}$$

It is clear that M is a Gr-measurable multivalued mapping. Let $\hat{x} \in \mathscr{S}(M) \subseteq \mathscr{S}(C)$. By $(*)$, there exists a sequence $\hat{x}_n \in \mathscr{S}(C_n)$, $n \in \mathbb{N}$, with $\hat{x} = P\text{-}\lim_{n\to\infty} \hat{x}_n$, and (2) implies that

$$P\text{-}\lim_{n\to\infty} f_n(.,\hat{x}_n(.)) = f(.,\hat{x}(.)).$$

Consequently, there exists a subsequence of $\{n_k\}_{k\in\mathbb{N}}$, say $\{n_{k_j}\}_{j\in\mathbb{N}}$, such that $(\varphi_{n_{k_j}})_{j\in\mathbb{N}}$ and $(f_{n_{k_j}}(.,\hat{x}_{n_{k_j}}(.)))_{j\in\mathbb{N}}$ converge a.s. Thus, we have

$$f(\omega, x(\omega)) = \lim_{j\to\infty} \varphi_{n_{k_j}}(\omega) \leqslant f(\omega, \hat{x}(\omega)) \leqslant r_\varepsilon(\omega) \text{ a.s.}$$

Since $\varepsilon > 0$ was arbitrary, this means that $f(\omega, x(\omega)) \leqslant \varphi(\omega)$ a.s. and thus, $x \in \mathcal{S}(\psi)$.

The study of approximation schemes (II) for the original stochastic optimization problem (I) is motivated by an approach to solve (I) *via* "discretizing" the random variables involved in (I).

Let us conclude this paragraph with a general approach to random problems due to A. Nowak [317] and based on the measurable selection theorem. *A random problem* Π is understood to mean a family of deterministic problems $\{\Pi_\omega\}_{\omega\in\Omega}$ depending on a random parameter. Let $\psi(\omega)$ be the set of all solutions of $\Pi\omega$ in a metric space X. We assume that $\psi(\omega) \neq \emptyset$ a.s. A *random solution* of Π is a measurable function $\eta:\Omega\to X$ such that $\eta(\omega)\in\psi(\omega)$ a.s. (i.e., η is a measurable a.s. selector for ψ).

In order to illustrate this method for a family of convex programming problems depending on a random parameter, we consider two real Banach spaces X and Y, E being a closed convex subset of X, and $C \subset Y$ denoting a closed convex cone with nonempty interior. The cone C induces the ordering on $Y: y \geqslant z$ for any $y, z \in Y$ if $y - z \in C$. Let Y^* be the dual space and let C^* be the dual cone, i.e. $C^* = \{p \in Y^* : \langle p, y\rangle \geqslant 0$ for all $y \in C\}$.

A *random convex programming problem* is understood to mean a family of optimization problems

$$(Z_\omega)\ \min f(\omega, x);\ g(\omega, x) \leqslant 0;\ x \in \varphi(\omega),$$

where $\varphi:\Omega \to 2^E$ is a multivalued mapping with nonempty closed convex values, and the functions $f:\Omega\times E\to\mathbb{R}$, $g:\Omega\times E\to Y$ are measurable in ω and continuous and convex in x. Additionally, we shall assume that the multifunction φ is measurable.

If E is separable and (Z_ω) a.s. has an optimal solution, then there exists a random optimal solution as in Theorem 7.10.1. A random version of the Kuhn-Tucker theorem is given by

THEOREM 7.10.5. *Let* ξ *be a random solution of the random convex programming problem* $\{Z_\omega\}_{\omega \in \Omega}$. *Suppose that* Y^* *is separable and each* Z_ω *satisfies the feasibility condition: for any* $p \in C^*$ *and any* $\omega \in \Omega$, *there exists an* $x \in \varphi(\omega)$ *such that* $\langle p, g(\omega, x)\rangle < 0$. *Then there exists a measurable function* $\eta : \Omega \to C^*$ *such that*

$$f(\omega, x) + \langle \eta(\omega), g(\omega, x)\rangle \geqslant f(\omega, \xi(\omega)) \text{ a.s.}$$

for all $x \in \varphi(\omega)$.

References

1. G. Adomian, Stochastic Green's function, *Proc. Symp. Appl. Math.*, **16** (1963), 1—39, Amer. Math. Soc., Providence, Rhode Island, 1964.
2. G. Adomian, *Random operator equations in mathematical physics*, I, Journ. Math. Phys., **11** (1970), 1069—1085 ; II, **12** (1971), 1944—1948 ; III, **12**(1971), 1948—1955.
3. G. Adomian, *Nonlinear stochastic differential equations*. J. Math. Anal. Appl., **55** (1976), 441—452.
4. G. Adomian, *Stochastic operators and dynamical systems*, CAM, Univ. of Georgia, No. 12, 1978.
5. G. Adomian and K. Malakian, Recent results in stochastic equations — the nonlinear case, in *Nonlinear Equations in Abstract Spaces*, V. Lakshnukantham, Academic Press, 1978, 3—23.
6. G. Adomian, *Closure Approximation Error in the Mean Solution of Stochastic Diff. Equations by the Hierarchy Method*, J. of Statist. Physics, **21** (1979), 181—189.
7. G. Adomian, Stochastic systems analysis, in *Applied Stoch. Processes* (G. Adomian), Academic Press, New York, 1980.
8. G. Adomian and L. H. Sibul, *Symmetrized solutions for nonlinear stochastic differential equations*, Internat. J. Math. and Math. Sci., **4** (1981), 529—542.
9. C. Alsina, *On countable product and algebraic convexifications of probabilistic metric spaces*, Pacific J. Math., **76** (1978), 291—300.
10. C. Alsina, On a family of functional inequalities, in *General Inequalities*, 2 (E. F. Beckenbach, Ed.), 1980, 419—427, Basel, Birkhäuser Verlag.
11. M. Altman, *Inverse differentiability contractors and equations in Banach spaces*, Studia Math., **46** (1974), 1—15.
12. A. Ambrosetti, *Un teorema di esistenza per le equazioni differenzialli negli spazi di Banach*, Rend. Sem. Mat. Univ. Padova, **39** (1967), 349—361.
13. G. F. Andrus and T. Nishiura, *Fixed points of random set-valued maps*, Nonlinear Analysis, **3**(1979), 65—72.
14. D. D. Angh and S. A. Thanh, *A probabilistic analogue of the Bohnenblust-Karlin fixed point*, Rev. Roum. Math. Pures. Appl., **19** (1974), 1177—1187.
15. N. A. Assad and W. A. Kirk, *Fixed point theorems for set-valued mappings of contractive type*, Pacific J. Math., **43** (1972), 553—562.

16. R. J. Aumann, *Integral of set-valued functions*, J. Math. Appl. (1965), 1–12.
17. S. Banach, *Théorie des opérations linéaires*, Warsaw, 1932.
18. R. Bellman, J. Jacquez and R. Kalaba, Mathematical models of chemotherapy, in *Proc. 4th Berkeley Symp. Math. Statist. and Prob.*, Vol. IV, pp. 55–66, Univ. of California Press, Berkeley, 1961.
19. R. Bellman, Some mathematical aspects of chemotherapy I: One-organ models, *Bull. Math. Biophys.*, **22** (1960), 181–198.
20. R. Bellman, Some mathematical aspects of chemotherapy II: The distribution of a drug in the body, *Bull. Math. Biophys.*, **22** (1960), 309–322.
21. R. Bellman, J. Jacquez, R. Kalaba and B. Kotkin, *A mathematical model of drug distribution in the body: Implications for cancer chemotherapy*, The RAND Corp., RM-3463-NIH, 1963.
22. C. Berge, *Espaces topologiques (fonctions multivoques)*, Dunod, Paris, 1959.
23. A. T. Bharucha-Reid, On random elements in Orlicz spaces, *Bull. Acad. Polon. Sci.*, **4** (1956), 655–657.
24. A. T. Bharucha-Reid, On random operator equations in Banach spaces, *Bull. Acad. Polon. Sci*, Sér. Sci. Math. Astronom. Phys., **7**(1959), 561–564.
25. A. T. Bharucha-Reid, On random solutions of integral equations in Banach spaces, *Trans. 2nd Prague Conf. on Information Theory. Statist. Decision Functions and Random Processes* (1959), Czechoslovak Acad. Sci., Prague, 1960; Acad. Press, New York, 1961, 27–28.
26. A. T. Bharucha-Reid, *Elements of the Theory of Markov Processes and Their Applications*, McGraw-Hill, N.Y., 1960.
27. A. T. Bharucha-Reid, On the theory of random equations, *Proc. Symp. Appl. Math.*, 1963, 40–69, Amer. Math. Soc., Providence, Rhode Island, 1964.
28. A. T. Bharucha-Reid, On random solutions of nonlinear integral equations, in *Nonlinear Problems in Engineering* (W. F. Ames, Ed.) 23–28, Academic Press, New York, 1964.
29. A. T. Bharucha-Reid, *Probabilistic methods in applied mathematics*, Vol. I, Academic Press, 1968.
30. A. T. Bharucha-Reid, *Probabilistic methods in applied mathematics*, Vol. II, Academic Press, 1970.
31. A. T. Bharucha-Reid, *Random integral equations*, Academic Press, New York, 1972.
32. A. T. Bharucha-Reid, Fixed point theorems in probabilistic analysis, *Bull. Amer. Math. Soc.*, **82** (1976), 641–657.
33. A. T. Bharucha-Reid, (Ed.). *Probabilistic analysis and related topics*, Vol. I, Academic Press, New York, 1978.
34. A. T. Bharucha-Reid (Ed.), *Probabilistic analysis and related topics*, Vol. II, Academic Press, New York, 1979.
35. A. T. Bharucha-Reid, (Ed.), *Approximate solutions of random equations*, North Holland, 1979.
36. A. T. Bharucha-Reid and R. Kannan, *Newton's method for random operator equations*, Nonlinear Analysis. Theory Methods and Appl., **4** (1980), 231–240.

37. A. T. Bharucha-Reid, Weak compactness of solutions measures associated with random equations, in *Probability in Banach spaces*, III (Ed. A. Beck), *Lecture Notes in Mathematics*, Vol. 860, Springer Verlag, Berlin, 1981, 25—34.
38. A. T. Bharucha-Reid, M. J. Christensen, Approximate solution of random integral equations : General Methods, in *Tenth IMACS Proceedings*, **4** (1982), 299—304.
39. A. T. Bharucha-Reid and W. Römisch, *Projective schemes for random operator equations I. Weak Compactness of approximate solution measures*, Journ. Integral Equations, **8** (1985), 95—111.
40. P. Billingsley, *Convergence of probability measures*, Wiley, New York, 1968.
41. G. D. Birkhoff and O. D. Kellogg, Invariant points in function spaces, *Trans. Amer. Math. Soc.*, **23** (1922), 96—115.
42. G. D. Birkhoff, *Lattice theory*, Amer. Math. Soc. Colloquium, Publ. 25, New York, 1948.
43. Gh. Bocşan, *Some applications of the function of Kuratowski*, Sem. on Probab. Metric Spaces, Univ. of Timişoara, No. 5, 1973.
44. Gh. Bocşan, *On the Kuratowski function in random normed spaces*, Sem. on Probab. Metric Spaces, Univ. of Timişoara, No. 8, 1974.
45. Gh. Boşcan, On some fixed point theorems in random normed spaces, *Proc. Fifth Conf. Probab. Theory*, Braşov—Romania, 1974, 153—155.
46. Gh. Boşcan, *Some remarks on a measure of noncompactness in probabilistic metric spaces*, Sem. on Probab. Metric Spaces, Univ. Timişoara, No. 18, 1974.
47. Gh. Bocşan, *A computational formula of Kuratowski function and applications to random normed spaces*, Sem. on Probab. Metric Spaces, Univ. of Timişoarà, No. 31, 1975.
48. Gh. Bocşan, *On some fixed point theorems in probabilistic metric spaces*, Math. Balkanica, **4**, 11(1974), 67—70.
49. Gh. Bocşan, *Măsuri aleatoare de necompacitate şi aplicaţii*, Doctoral thesis, Univ. of Timişoara, 1976.
50. Gh. Bocşan, *On random operators on separable Banach spaces*, Sem. on Probab. Metric Spaces, Univ. of Timişoara, Romania, No. 38, 1978.
51. Gh. Bocşan, *Some properties of continuous random functions on random domains with applications to continuous extensions and random fixed point theorems*, Sem. on Probab. Metric Spaces, Univ. of Timişoara, Romania, No. 45, 1979.
52. Gh. Bocşan, *On the existence of measurable solution of a differential equation with random parameters*, Sem. on Probab. Metric Spaces, Univ. of Timişoara, Romania, No. 48, 1979.
53. Gh. Bocşan, *A general random fixed point theorem and applications to random equations*, Rev. Roum. Math. Pures App., **26** (1981), 375—380.
54. Gh. Bocşan, On measurability of the inverse of a random mapping with random domain, *Proc. Seventh Conf. Probab. Theory*, Braşov, Romania, 1982, 399—401.
55. Gh. Bocşan, *The existence of fixed points for random densifying operators on Banach spaces*, Sem. on Probab. Theory Appl., Univ. of Timişoara, No. 65, 1983.

56. Gh. Bocşan and Gh. Constantin, The Kuratowski function and some applications to the probabilistic metric spaces, *Atti Accad. Naz. Lincei*, Vol. LV (1973), 236–240.
57. Gh. Bocşan, Gh. Constantin and V. Radu, *On a class of contractions and fixed point theorems*, Sem. on Probab. Metric Spaces, Univ. of Timişoara, No. 50, 1979.
58. Gh. Bocşan, Gh. Constantin and V. Radu, *On some asymptotic properties of the random nonexpansive operators on random subsets in a Hilbert space*, Sem. on Probab. Theory Appl., Univ. of Timişoara, No 52. 1980.
59. Gh. Bocşan, Gh. Constantin and V. Radu, *On the random extension property of a separable metric space*, Sem. on Probab. Theory Appl., Univ. of Timişoara, No. 53, 1980.
60. Gh. Bocşan, Gh. Constantin and V. Radu, *Some applications of the existence of random fixed points*, Sem. on Probab. Theory Appl., Univ. of Timişoara, No. 54, 1980.
61. Gh. Bocşan, Gh. Constantin and V. Radu, *On the existence and uniqueness of the solution for random differential equations in Banach spaces*, Rev. Roum. Math. Pures Appl., **26** (1981), 381–384.
62. Gh. Bocşan, Gh. Constantin and V. Radu, Some properties of random operators and applications to the existence theorems for random equations, *Proc. Seventh Conf. Probab. Theory*, Braşov, Romania, 1982, 403–408.
63. H. F. Bohnenblust and S. Karlin, On a theorem of Ville. *Contributions to the theory of games*, Vol. I, 155–160. *Annals of Math. Stud.*, No. 24, Princeton University Press, Princeton, 1950.
64. V. A. Bondarenko, On the existence of a universal measure of noncompactness. *Prob. Math. Anal. Comp. Syst.*, Vol. II. Voroenj, 1968, 18–21.
65. I. G. Borisovich and I. I. Sapronov, Topological theory of compact restriction mappings, *Trud. Sem. Funct-Anal. Voronej*, **12** (1969), 43–68.
66. R. C. Bourret, *Stochastically perturbed fields with applications to wave propagation in random media*, Nuovo Cimento, **26** (1962), 1–31.
67. W. E. Boyce, Random vibration of elastic springs and bars, *Proc. Fourth U.S. National Congress Appl. Mech.*, Berkeley, California, 1962, 77–85.
68. W. E. Boyce, *A dishonest approach to certain stochastic eigenvalue problems*, SIAM J. Appl. Math., **15** (1967), 143–152.
69. W. E. Boyce, *Random eigenvalue problems. Probabilistic Methods in Applied Mathematics* (Edited by A. T. Bharucha-Reid), Academic Press, New York, 1968, Vol. I, 1–73.
70. W. E. Boyce and B. E. Goodwin, *Random transverse vibrations of elastic beams*, J. SIAM, **12** (1964), 613–629.
71. D. W. Boyd and J. S. Wong, On nonlinear contractions, *Proc. Amer. Math. Soc.*, **20** (1969), 458–64.
72. M. S. Brodski and D. P. Milman, *On the center of a convex set*, Doklady Akad. Nauk SSSR (N.S.), **59** (1948), 837–840.
73. L. E. J. Brouwer, *Über Abbildungen von Mannigfaltigkeiten*, Math. Ann., **71** (1912), 320–327.

74. F. E. Browder, *On the fixed point index for continuous mappings of locally connected spaces,* Summa Brasil Math., **4** (1960), 253–293.

75. F. E. Browder, Fixed point theorems for noncompact mappings in Hilbert spaces, *Proc. Nat. Acad. Sci. USA,* **53** (1965), 1272–1276.

76. F. E. Browder, Semicontractive and semiaccretive nonlinear mappings in Banach spaces, *Bull. Amer. Math. Soc.,* **74** (1968), 660–665.

77. F. E. Browder, *On the convergence of successive approximations for nonlinear functional equations,* Indag. Math., **30** (1968), 27–35.

78. F. E. Browder, Nonlinear maximal monotone operators in Banach spaces, *Math. Ann.,* **75** (1968), 89–113.

79. F. E. Browder, Nonlinear functional analysis and nonlinear integral equations of Hammerstein and Urysohn type, *Contribution to Nonlinear Functional Analysis* (E. Zarantonello, Ed.), Academic Press, New York, 1971.

80. R. E. Bruck, The iterative solution of the equation $y \in x + Tx$ for a monotone operator T in Hilbert space, *Bull. Amer. Math. Soc.,* **79** (1973), 1258–1261.

81. R. E. Bruck, *A strongly convergent iterative solution of* $0 \in U(x)$ *for a maximal monotone operator* U *in Hilbert space,* J. Math. Anal. Appl., **48** (1974), 114–126.

82. G. Cain, *Fixed points of maps on uniform spaces with a contractive iterate at a point,* Math. Japonica, **30** (1985), 691–699.

83. G. L. Cain, *Fixed points and stability of multifunctions on random normed spaces* (in print).

84. G. L. Cain and R. H. Kasriel, *Fixed and periodic points of local contraction mappings on probabilistic metric spaces,* Math. Syst. Theory, **9** (1976), 289–297.

85. P. Calabrese, *The probability of the triangle inequality in probabilistic metric spaces,* Aequationes Math., **18** (1978), 187–205.

86. C. Castaing, *Sur les multi-applications mesurables,* Rev. Française d'Informatique et de Recherche Opérationnelle, **1** (1967), 91–126.

87. C. Castaing, *Version aléatoire du problème de râfle par un convexe variable,* C. R. Acad. Sci., Paris, 277(1973), 1057–1059, and Séminaire d'Analyse convexe, Montpellier, 1974, Exposé no. 1.

88. C. Castaing, *Râfle par un convexe aléatoire à variation continue à droite,* Séminaire d'Analyse Convexe Montpellier, 1975, Exp. 15.

89. C. Castaing and V. Valadier, *Convex Analysis and Measurable Multifunctions, Lect. Notes,* no. 580, Springer Verlag, 1977.

90. L. Cesari, *Functional analysis and Galerkin's method,* Michigan Math. J., **11** (1964), 385–414.

91. L. Cesari and R. Kannan, Functional analysis and nonlinear differential equations, *Bull. A.M.S.,* **79** (1973), 1216–1219.

92. L. Cesari, *Alternative methods in nonlinear analysis,* International Conference in Differential Equations, Los Angeles, 1974.

93. P. S. Chandrasekharan and A. T. Bharucha-Reid, *Approximate solution of fixed point equations for random operators in Approximate Solution of Random Equations*, edited by A. T. Bharucha-Reid, North Holland, 1979, 27–36.

94. S-s. Chang, *Random fixed point theorems for commuting random operators in probabilistic functional analysis*, Applied Math. and Mech., Hong Kong, **3** (1982), 383–392.

95. S-s. Chang, *On some fixed point theorems in probabilistic metric spaces and its applications*, Z. Wahrsch. verw. Geb., **63** (1983), 463–474.

96. S-s. Chang, *Some random fixed point theorems for continuous random operators*, Pacific J. Math., **105**(1983), 21–31.

97. S-s. Chang, *Fixed point theorems of mappings on probabilistic metric spaces with applications*, Scienția Sinica, **26** (1983), 1144–1155.

98. S-s. Chang, *On the theory of probabilistic metric spaces with applications*, Z. Wahrsch. verw. Geb., **67** (1984), 85–94.

99. S-s. Chang and S-k. Kang, *Fixed point theorems in probabilistic inner product spaces*, Applied Math. and Mech., Shanghai, **6** (1985), 67–74.

100. S-s. Chang, *Fixed points theorems for contractive type mappings in probabilistic metric spaces*, Rev. Roum. Math. Pures Appl. (in print).

101. S-s. Chang and S-k. Kang, *On a generalization of Istrățescu's fixed point theorem*, Rev. Roum. Math. Pures Appl. (in print).

102. K. K. Chen and T. T. Soong, *Covariance properties of waves propagating in a random medium*, Journ. Acoustic Soc. Amer., **49** (1971), 1639–1642.

103. P. L. Chow, Stochastic partial differential equations in turbulence related problems, in *Probabilistic Analysis and Related Topics* (edited by A. T. Bharucha-Reid), Academic Press, 1978, 1–43.

104. L. B. Cirić, A generalization of Banach's contraction principle, *Proc. Amer. Math. Soc.*, **45** (1974), 267–273.

105. L. B. Cirić, *On fixed points of generalized contractions on probabilistic metric spaces*, Publ. Inst. Math., **18** (32), (1975), 71–78.

106. Gh. Constantin, *Spectral and structural properties for some classes of linear operators on normed spaces*, Doctoral thesis, Univ. of Bucharest, 1972.

107. Gh. Constantin, Some spectral properties for the local α-contraction operators, *Boll. Un. Math. Ital.*, **4** (1972), 323–330.

108. Gh. Constantin, *On some applications of random norms in the information theory*, Comm. Sci. Ses., Univ. of Timișoara, 1977.

109. Gh. Constantin, *Some remarks on the nearly invariant points for mappings on random normed spaces*, Sem. on Probab. Metric Spaces, Univ. of Timișoara, No. 40, 1979.

110. Gh. Constantin, *On some classes of contraction mappings in Menger Spaces*, Sem. on Probab. Theory Appl., No. 76, Univ. of Timișoara, 1985.

111. Gh. Constantin and V. Radu, *Random-proximal spaces*, Sem. on Probab. Metric Spaces, Univ. of Timișoara, No. 7, 1974.

112. Gh. Constantin and Gh. Bocşan, On some measures of noncompactness in probabilistic metric spaces, *Proc. Fifth Conf. Probab. Theory*, Braşov, Romania (1974), 163—168.

113. Gh. Constantin, D. Hărăguş, *On the Darboux's problem for a partial differential equation with random parameters*, Sem. on Probab. Theory Appl., Univ. of Timişoara, No. 55, 1980.

114. Gh. Constantin and V. Radu, *On probabilistic δ-continuity and proximate fixed points for multivalued functions in PM-spaces*, Rev. Roum. Math. Pures Appl., **26** (1981), 393—397.

115. Gh. Constantin and I. Istrăţescu, *Elemente de Analiză Probabilistă şi Aplicaţii*, Ed. Acad., Bucureşti, 1981.

116. Gh. Constantin and I. Istrăţescu, Probability measures on indefinite inner product spaces, in *Studies in Probability and related topics*, Papers in Honour to Octav Onicescu on his 90th Birthday Anniversary, 95—100, Nagard Publisher, 1983.

117. Gh. Constantin and I. Istrăţescu, *On a random measure of noncompactness in PM-spaces and applications*, Sem. on Probab. Theory Appl., Univ. of Timişoara, No. 81, 1986.

118. M. G. Crandal and T. M. Liggett, *Generation of semigroups of nonlinear transformations on general Banach spaces*, Amer. J. Math., **93** (1971), 265—298.

119. R. F. Curtain, *Infinite-dimensional estimation theory applied to water pollution problem* (in print).

120. R. F. Curtain, *Modelling, filtering and control for water pollution* (in print).

121. J. Daneš, *Some fixed point theorems*, Comm. Math. Univ. Carolinae, **9** (1968), 223—235.

122. G. Darbo, *Punti uniti in transformazioni a codominio non compatto*, Rend. Sem. Math. Univ. Padova, **24** (1955), 84—92.

123. G. Debreu, Integration of correspondences, in: *Proceedings of the Fifth Berkeley Symposium on Math. Statist. and Probab.*, Vol. II, part I, 351—372, University of California Press, Berkeley, 1967.

124. J. B. Diaz and B. Margolis, A fixed point theorem of the alternative for contractions on a generalized complete metric space, *Bull. Amer. Math. Soc.*, **74** (1968), 305—309.

125. M. Driml and O. Hanš, Continuous stochastic approximations, Trans. of the Second Prague Conf., 1960, 113—122.

126. M. Driml and J. Nedoma, Stochastic approximations for continuous random processes, *Trans. Second Prague Conf. Inf. Theory*, Prague, 1960, 145—158.

127. N. Dunford and J. T. Schwartz, *Linear Operators*, Part I, Interscience, New York, 1958.

128. M. Edelstein, (i) *On fixed and periodic points under contractive mapping*, J. London Math. Soc., **37** (1962), 74—79; (ii) *An extension of Banach's contraction principle*, Proc. Amer. Math. Soc., **12** (1961), 7—10.

129. R. E. Edwards, *Functional analysis*, Holt, Rinehart and Winston, 1965.

130. R. J. Egbert, *Products and quotients of probabilistic metric spaces*, Pacific J. Math., **24** (1968), 437—455.

131. W. Engl, *Schwache Konvergenz asymptotisch regulärer Iterationsverfahren bei Fixpunktgleichungen nichtexpansiver Funktionen*, ZAMM, **57** (1977), T272—T273.

132. W. Engl, *Fixed point theorems for random operators on stochastic domains*, Dissertation zur Erlangung des akademischen Grades eines Doktors der technischen Wissenschaften, 1977, Eingreicht am Institut für Mathematic Johannes-Kepler-Universität Linz, 1—48.

133. W. Engl, *Weak convergence of asymptotically regular sequences for nonexpansive mappings and connections with certain Chebyshef centers*, Nonlinear Analysis, **1** (1977), 495—501.

134. W. Engl, *Weak convergence of Mann iteration for nonexpansive mappings without convexity assumptions*, Boll. U.M.I., **5**, **14**—A (1977), 471—475.

135. W. Engl, *Some random fixed point theorems for strict contractions and nonexpansive mappings*, Nonlinear Analysis, Theory Methods and Appl., **2** (1978), 619—626.

136. W. Engl, *A general stochastic fixed point theorem for continuous random operators on stochastic domains*, J. Math. Anal. Appl., **66** (1978), 220—231.

137. W. Engl, *Random fixed point theorems for multivalued mappings*, Pacific J. Math., **76** (1978), 351—360.

138. W. Engl, Random fixed point theorems, in *Proceedings on a Symposium on Nonlinear Equations in Abstract Spaces* (Edited by V. Lakshmikantham), Academic Press, 1978, 67—80.

139. W. Engl, *Existence of measurable optima in stochastic nonlinear programming and control*, Appl. Math. Optim., **5** (1979), 271—281.

140. W. Engl and M. Z. Nashed, *Stochastic projectional schemes for random linear operator equations of the first and second kinds*, Numer. Funct. Anal. and Optimiz., **1** (1979), 451—473.

141. W. Engl and M. Z. Nashed, *Generalized inverse of random linear operators in Banach spaces*, J. Math. Anal. Appl., **83** (1981), 582—610.

142. W. Engl and W. Römisch, *Convergence of approximate solutions of nonlinear random operator equations with non-unique solutions*, Stochastic Anal. Appl., **1** (1983), 239—298.

143. W. Engl and W. Römisch, *Approximate solutions of nonlinear random operator equations: Convergence in distribution*, Pacific J. Math., **120** (1985), 55—77.

144. W. Engl and W. Römisch, *Weak convergence of approximate solutions of stochastic equations with applications to random differential and integral equations*, Berichte des Instituts für Mathematik, Universität Linz, Report 287, 1985.

145. M. P. Ershov, *Extension of measures and stochastic equations*, Theor. Probab. Appl., **10** (1974), 431—444.

146. M. P. Ershov, Causal and non-anticipating solutions of stochastic equations, in : M. Kohlmann, W. Vogel (Eds.), *Stochastic Control Theory and Stochastic Diff. Systems, Lectures Notes in Control and Information Sciences*, Vol. 16, Springer Verlag, Berlin, 1979, 605—615.

147. Ky Fan, Fixed point and minimax theorems in locally convex topological linear spaces, *Proc. Nat. Acad. Sci. U.S.A.*, **38** (1952), 121—126.

148. R. M. Fortet, Random distributions with an application to telephone engineering, *Proc. Third. Berkeley Symp. Math. Statist. Prob.*, Univ. California Press, Berkeley, 1956, 81—88.

149. J. Ma. Fortuny, *Contribucio a l'estudi dels espais amb producte interior probabilistic,* Dissert., Palma de Mallorca, 1984.

150. U. Frisch, Wave propagation in random media, in *Probabilistic methods in applied mathematics*, Vol. I (edited by A. T. Bharucha-Reid), Academic Press, 1968, 75—191.

151. R. Fritsche, *Topologies for probabilistic metric spaces,* Fund. Math., **72** (1971), 7—16.

152. M. Furi and A. Vignoli, A fixed point theorem in complete metric spaces, *Boll. U.M.I.*, **4** (1969), 505—509.

153. M. Furi and A. Vignoli, *Fixed points for densifying mappings,* Atti Acad. Naz. Lincei, **47** (1969), 465—467.

154. M. Furi and A. Vignoli, *On α-nonexpansive mappings and fixed points,* Atti Accad. Naz. Lincei, **48** (1970), 195—198.

155. M. Furi, M. Martelli, A. Vignoli, On minimum problems for families of functionals, *Annali di Mat. Pura Appl.*, **86**(1970), 181—187.

156. B. G. Gabdulkhaev, *Solutions of the operator equations specifying,* Izvestia Vuzov Matem., **144** (1974), 680.

157. L. A. Gardner, Stochastic approximation and its application to problems of prediction and control synthesis, *Internat. Symp. Nonlin. Diff. Eqns. and Nonlin. Mechanics* (edited by J. P. LaSalle and S. Lefschetz), Academic Press, New York, 1963, 241—258.

158. V. M. Gerstein, *On theory of dissipative differential equations in Banach spaces,* Funct. Anal., **4** (1970), 99—100.

159. I. I. Ghihman and A. V. Skorohod, *Stochastic differential equations and their applications* (in Russian), Naukowa Dumka, Kiew, 1982.

160. I. L. Glicksberg, A further generalization of the Kakutani fixed point theorem, with applications to Nash equilibrium points, *Proc. Amer. Math. Soc.*, **3** (1952), 170—174.

161. L. S. Goldenstein, I. Ts. Gokhberg, A. S. Markus, *Investigations on some properties of bounded linear operators in connection with their q-norms,* Uch. Zap. Kishinev Gos. Univ., **22** (1957), 29—35.

162. L. S. Goldenstein and A. S. Markus, On a measure of noncompactness of bounded sets and linear operators, in *Studies in Algebra and Mathematical Analysis*, 45—54, Izdat. Karta Moldovenjaski, Kishinev, 1965.

163. B. E. Goodwin and W. E. Boyce, *The vibrations of a random elastic string: the method of integral equations,* Quarterly Appl. Math., **22** (1964), 261—266.

164. K. Göebel and W. Rzymowski, An existence theorem for the equations in Banach spaces, *Bull. Acad. Polon. Sci., Ser. Sci. Math. Astron. and Phys.*, **18**(1970), 367–370.

165. A. Granas, *The theory of compact vector fields and some of its applications to topology of functional spaces, I.* Dissertationes Math. Rozprawy Mat., **30**(1962), 1–93.

166. O. Hadžić, *A generalization of Krasnosel'skii's fixed point theorem in probabilistic locally convex spaces*, Matem. Vesnik, **1**(14), (29), (1977), 173–177.

167. O. Hadžić, *A probabilistic proof of a fixed point theorem in K-convex topological vector spaces*, Review Research Univ. Novi Sad, **7** (1977), 3–8.

168. O. Hadžić, *A fixed point theorem for mappings with sequentially compact iteration in probabilistic locally convex spaces*, Publ. Inst. Math., **22** (36) (1977), 71–76.

169. O. Hadžić, *A fixed point theorem in random normed spaces*, Review Research Univ. Novi Sad, **7**(1977). 23–27.

170. O. Hadžić, *A fixed point theorem for mappings in probabilistic locally convex spaces*, Rev. Roum. Math. Pures et Appl., **23** (1978), 735–744.

171. O. Hadžić, Fixed points for mappings on probabilistic locally convex spaces, *Bull. Math. Soc. Sci. Math. R. S. Roumanie*, **22** (70), (1978), 287–292.

172. O. Hadžić, *A remark on nonarchimedean Menger spaces*, Review of Research, Univ. Novi Sad, **8** (1978), 9–12.

173. O. Hadžić, *On the ε,λ,-topology of probabilistic locally convex spaces*, Glas. Mat., **13** (33), (1978), 293–297.

174. O. Hadžić, *A generalization of a fixed point theorem in probabilistic locally convex spaces*, Matem. Vesnik, **2** (15) (30), (1978), 203–207.

175. O. Hadžić, Fixed point theorems for multivalued mappings in some classes of topological vector spaces, *Colloq. Math., Soc. J. Bolyai*, **23**, *Topology*, Budapest, 1978, 569–573.

176. O. Hadžić, *Fundations of the fixed point theory*, Novi Sad (in Serbo-Croatian), 1978.

177. O. Hadžić, *Fixed point theorems for multivalued mappings in probabilistic metric spaces*, Mat. Vestnik, **3** (16)(31), (1979), 125–135.

178. O. Hadžić, *A fixed point theorem in Menger spaces*, Publ. Inst. Math., **20** (40), 1979, 107–112.

179. O. Hadžić, *A random fixed point theorem for multivalued mappings of Ciric's type*, Mat. Vestnik, **3** (16)(31), (1979), 397–401.

180. O. Hadžić, Fixed point theorems in probabilistic metric and random normed spaces, *Math. Sem. Notes*, Kobe University, **7** (1979), 261–270.

181. O. Hadžić, *A fixed point theorem for multivalued mappings in random normed spaces*, L'Analyse Numérique et la Théorie de l'Approximation, **8** (1979), 49–52.

182. O. Hadžić, A note on Istrăţescu's fixed point theorem in nonarchimedean Menger spaces, *Bull. Math. Soc., Sci. Math. R. S. Roumanie*, **24** (72), (1980), 359–362.

183. O. Hadžić, *A generalization of the contraction principle in probabilistic metric spaces*, Review Research Univ. Novi Sad., **10** (1980), 13–21.
184. O. Hadžić, *On a common fixed point in Banach and random normed spaces*, Review Research Univ. Novi Sad, **11** (1981), 11–18.
185. O. Hadžić, *Some fixed point and almost fixed point theorems for multivalued mappings in topological vector spaces*, Nonlin. Anal. Theory Math. and Appl., **9**(1981), 1009–1019.
186. O. Hadžić, *Some theorems on the fixed points in probabilistic metric and random normed spaces*, Boll. U.M.I., **68** (1982), 381–391.
187. O. Hadžić, *A common fixed point in probabilistic metric spaces*, Math. Sem. Notes, Kobe Univ., **10** (1982), 31–39.
188. O. Hadžić, *On almost continuous selection property*, Math. Sem. Notes, Kobe Univ , **10** (1982), 41–47.
189. O. Hadžić, Fixed point theorems for multivalued mappings in uniform spaces and their applications to *PM*-spaces, *Anal. Univ. Timişoara*, **21** (1983), 45–57.
190. O. Hadžić, *On common fixed points in metric and probabilistic metric spaces with convex structure*, Review Research Univ. Novi Sad, **14**(1984), 13–25.
191. O. Hadžić, *Fixed point theory in topological vector spaces*, Novi Sad, 1984.
192. O. Hadžić and O. Paunić, *Existence theorems for the system $x = H(x, y)$, $y = K(x, y)$ in probabilistic locally convex spaces*, Review Research Univ. Novi Sad, **8**(1978), 21–27.
193. O. Hadžić and M. Stojaković, *Two random fixed point theorems*, Review Research Univ. Novi Sad. **9** (1978), 49–52.
194. O. Hadžić, M. Budinčević, A fixed point theorem in probabilistic metric spaces, *Colloq. Math. Soc. J. Bolyai*, **23**, *Topology*, Budapest (1978), 569–579.
195. O. Hadžić, M. Stojaković, *On the existence of a solution of the system $x = H(x, y)$, $y = K(x, y)$ in random normed spaces*, Review Research Univ. Novi Sad, **9** (1979), 43–48.
196. O. Hadžić, N. Despotović, Some fixed point theorems in random normed spaces, *Anal. Univ. Timişoara*, **17** (1979), 39–47.
197. O. Hadžić, M. Budinčević, *A class of t-norms in the fixed point theory on probabilistic metric spaces*, Review of Research Univ. Novi Sad, **9**(1079) 37–41.
198. O. Hadžić, N. Dospotović, *Fixed point theorems of Krasnosel'skii's type in probabilistic locally convex spaces*, Review Research Univ. Novi Sad, **9** (1979), 23–28.
199. O. Hadžić, M. Stojaković, *Some applications of Bocşan's fixed point theorem*, Review Research Univ. Novi Sad, **10**(1980), 37–47.
200. O. Hadžić, R. Dedeić, A generalization of contraction principle for multivalued mappings in probabilistic metric spaces, III, *Conf. Appl. Math.*, Novi Sad, (1982), 13–22.
201. S. Hahn and F. Pötter, *Über Fixpunkte kompakter Abbildungen in topologischen Vectorräumen*, Studia Math., **L** (1974), 1–16.

202. J. K. Halle, *Continuous dependence of fixed points of condensing maps*, J. Math. and Appl., **46** (1974), 388–394.

203. O. Hanš, *Reduzierende zufällige Transformationen*, Czechoslovak Math. J., **7** (1957) 82, 154–158.

204. O. Hanš, Generalized random variables, *Trans. First Prague Conf. on Inform. Theory, Statist. Decision Functions and Random Processes* (Liblice, 1956), Czechoslovak Akad. Sci., Prague, 1957, 61–103.

205. O. Hanš, Random fixed point theorems, *Trans. First Prague Conf. on Inform. Theory, Statist. Decision Functions and Random Processes* (Liblice, 1956), Czechoslovak Akad. Sci., Prague, 1957, 105–125.

206. O. Hanš, Inverse and adjoint transforms of linear bounded random transforms, *Trans. First Prague Conf. on Inform. Theory, Statist. Decision Functions and Random Processes* (Liblice, 1956), Czechoslovak Akad. Sci., Prague, 1957, 127–133.

207. O. Hanš, Random operator equations, *Proc. 4th Berkeley Symp. on Math. Statist. and Probab.* (1960), Vol. II, Univ. California Press, Berkeley Calif., 1961, 185–202.

208. O. Hanš and A. Špaček, Random fixed point approximation by differentiable trajectories, *Trans. 2nd Prague Conf. Inform. Theory, Statist. Decision Functions and Random Processes* (1959), Czechoslovak Acad. Sci., Prague, 1960; Academic Press, New York, 1961, 203–213.

209. S. Heikkilä and S. Seikkala, *On generalized contractions in probabilistic metric spaces*, Rev. Roum. Math. Pures et Appl., **28** (1983), 297–306.

210. T. L. Hicks, *Fixed point theory in probabilistic metric spaces*, Review of Research Univ. Novi Sad, **13** (1983), 64–65.

211. C. J. Himmelberg, J. R. Porter and F. S. Van Vleck, Fixed point theorems for condensing multifunctions, *Proc. Amer. Math. Soc.*, **23** (1969), 635–641.

212. C. Himmelberg and F. S. Van Vleck, *Some selection theorems for measurable functions*, Can. J. Math., **21** (1969), 394–399.

213. C. Himmelberg and F. S. Van Vleck, *Lipschitzian generalized differential equations*, Rend. Sem. Mat. Univ. Padova, **48** (1972), 159–169.

214. C. Himmelberg and F. S. Van Vleck, Multifunctions on abstract measurable spaces and applications to stochastic decision theory, *Anali Mat. Pura et Appl.*, **101** (1974), 229–236.

215. C. Himmelberg, *Measurable relations*, Fundamenta Math., **LXXXVII** (1975), 53–72.

216. W. C. Hoffman, Wave propagation in a general random continuous medium, *Proc. Symp. Appl. Math.*, **16** (1963), 117–144, Amer. Math. Soc., Providence, Rhode Island, 1964.

217. I. Istrăţescu, *On generalized complete probabilistic metric spaces*, Sem. Probab. Metric Spaces, Univ. of Timişoara, No. 25, 1974.

218. I. Istrăţescu, *On some fixed point theorems with applications to the nonarchimedean Menger spaces*, Atti Accad. Naz. Lincei, **LVII** (1975), 374–378.

219. I. Istrăţescu, On some fixed point theorems in generalized metric spaces, *Bol. Univ. Mat. Ital.*, **5**, 13-A, 1976, 95–100.

220. I. Istrăţescu, *Some remarks on nonarchimedean probabilistic metric spaces*, Glasnik Math., **11** (31), 1976, 155–161.

221. I. Istrăţescu, *Fixed point theorems for some classes of contraction mappings*, Publ. Math. Debrecen, **25**(1978), 29–33.

222. I. Istrăţescu, *A fixed point theorem for mappings with a probabilistic contractive iterate*, Rev. Roum. Math. Pures et Appl., **26** (1981), 431–435.

223. I. Istrăţescu, On the existence of fixed points for mappings with contractive iterate, *Bul. St. Tehn. Inst. Polit. Timişoara*, **26** (40), 1981, 9–11.

224. I. Istrăţescu, On the convergence of the sequences of fixed points for probabilistic contractions, *Bul. St. Tehn. Inst. Polit. Timişoara*, 1981, 43–45.

225. I. Istrăţescu, *On some fixed point theorems for mappings on nonarchimedean probabilistic metric spaces*, Sem. on Probab. Theory and Appl., Univ. of Timişoara, No. 73, 1985.

226. I. Istrăţescu, *On some inequalities and fixed points problems on probabilistic metric spaces* (in print).

227. I. Istrăţescu and G. Palea, On nonarchimedean probabilistic metric spaces. *Anal. Univ. Timişoara*, **12** (1974), 115–118.

228. I. Istrăţescu and N. Crivăţ, *On some classes of nonarchimedean probabilistic metric spaces*, Sem. on Probab. Metric Spaces, Univ. of Timişoara, No. 12, 1974.

229. I. Istrăţescu and Gh. Babescu, *On the completion of nonarchimedean probabilistic metric spaces*, Sem. on Probab. Metric Spaces, Univ. of Timişoara, No. 25, 1974.

230. I. Istrăţescu and E. Roventa, On fixed point theorems for mappings on probabilistic metric spaces, *Bull. Math. Soc. Sci. R. S. Roumanie*, **19** (67), (1975), 67–69.

231. V. Istrăţescu, I. Văduva, *Products of statistical metric spaces*, Stud. Cerc. Mat., **12** (1961), 567–572.

232. V. I. Istrăţescu, *On generalized probabilistic metric spaces I*, Rev. Roum. Math. Pures et Appl., **14** (1969), 515–519.

233. V. I. Istrăţescu, On a measure of noncompactness, *Bull. Math. Soc. Sci. Math. R. S. Roumanie*, **16** (64), (1972), 195–197.

234. V. I. Istrăţescu, I. Săcuiu, *Fixed points theorems for contraction mappings on probabilistic metric spaces*, Rev. Roum. Math. Pures et Appl., **18** (1973), 1375–1380.

235. V. I. Istrăţescu, *Introducere în teoria punctelor fixe*, Editura Academiei R. S. România, 1973.

236. V. I. Istrăţescu, *Introducere în teoria spaţiilor metrice probabilistice cu aplicaţii*, Editura Tehnică, Bucureşti, 1974.

237. S. Itoh, *A random fixed point theorem for a multivalued contraction mapping*, Pacific J. Math., **68** (1977), 85–90.

238. S. Itoh and W. Takahashi, *Single-valued mappings, multivalued mappings and fixed-point theorems*, J. Math. Anal. Appl., **59** (1977), 514–521.

239. S. Itoh, *Nonlinear random equations with monotone operators in Banach paces*, Math. Ann., **236** (1978), 133–146.

240. S. Itoh, *Random differential equations associated with accretive operators*, J. Diff. Eq., **31** (1979), 139—154.
241. S. Itoh, *Measurable or condensing multivalued mappings and random fixed point theorems*, Kodai Math. J., **2** (1979), 293—299.
242. S. Itoh, *Random fixed point theorems with an application to random diff. equations in Banach spaces*, J. Math. Anal. Appl., **67** (1979), 261—273.
243. M. Joshi, *Nonlinear random equations with P-compact operators in Banach spaces*, Indian J. Pure Appl. Math., **11** (1980), 791—799.
244. M. Joshi, *Nonlinear random equations involving operators of type (M)*, J. Math. Anal. Appl., **94** (1983), 460—469.
245. M. Kac, Some stochastic problems in physics and mathematics, Magnolio Petroleum Co., *Lectures in Pure and Appl. Sciences*, No. 2, 1956.
246. S. Kakutani, *A generalization of Brouwer's fixed point theorem*, Duke Math. J., **8**(1941), 457—459.
247. R. Kannan, *Random operator equations, Dynamical Systems*, Conf. held in Gainesville, 113—137, Academic Press, New York, 1976.
248. R. Kannan, *Tikhonov regularization and nonlinear problems at resonance : deterministic and random*, Nonlinear Analysis, Academic Press, 1978, 109—124.
249. R. Kannan, *Une version stochastique du théorème de Kakutani*, C.R. Acad. Sci., Paris, **287** (1978), 551—552.
250. R. Kannan, Stochastic Approximation and nonlinear operator equations, in *Approximate solution of random equations* (edited by A. T. Bharucha-Reid), North Holland, New York, Oxford, 1979, 87—105.
251. R. Kannan, *Random correspondences and nonlinear equations*, J. Multiv. Anal., **11** (1981), 230—243.
252. R. Kannan et H. Salehi, *Mesurabilité du point fixe d'une transformation aléatoire séparable*, C. R. Acad. Sci., Paris, **281** (1975), Ser. A, 663—664.
253. R. Kannan and H. Salehi, Random solutions of nonlinear differential equations, *Boll. Un. Mat. Ital.*, **12** (1975), 209—213.
254. R. Kannan and H. Salehi, Measurability of solutions of nonlinear equations, *Proc. Conf. Nonlin. Funct. Anal.*, Dekker, New York, 1976.
255. R. Kannan and H. Salehi, *Random nonlinear equations and monotonic nonlinearities*, J. Math. Anal. Appl., **57** (1977), 234—256.
256. F. C. Karal and J. B. Keller, *Elastic, electromagnetic and other waves in a random medium*, J. Math. Phys., **5** (1964), 537—547.
257. J. B. Keller, Wave propagation in random media, *Proc. Symp. Appl. Math.*, **13**(1960), 227—246, Amer. Math. Soc., Providence, Rhode Island, 1962.
258. J. B. Keller, Stochastic equations and wave propagation in random media, *Proc. Symp. Appl. Math.*. **16** (1963), 145—170, Amer. Math. Soc., Providence, 1964.
259. J. B. Keller, G. C. Papanicolaou and J. Weilenmann, *Heat conduction in a one-dimensional random medium*, Comm. Pure Appl. Math., **32** (1978), 583—592.
260. P. V. Kirillov, *Random equations*, Kishineff, 1982 (in Russian).

261. V. Klee, *Stability of the fixed point property*, Colloq. Math., **81**(1961), 43–46.
262. R. H. Kraichnal, The closure problem in turbulence theory, *Proc. Symp. Appl. Math.*, New York., **13** (1960), 199–225, Amer. Math. Soc., Providence, 1962.
263. M. A. Krasnosel'skiĭ, *On a fixed point principle for completely continuous operators in functional spaces*, Doklady Akad. Nauk SSSR (N.S.), **73** (1950), 13–15.
264. M. A. Krasnosel'skiĭ, *Topological methods in the theory of nonlinear integral equations*, MacMillan, New York, 1964.
265. M. A. Krasnosel'skiĭ, G. M. Vainikko, P. O. Zabreiko, Ja. R. Rutickii and V. Ja. Stecenko, *Approximate solution of operator equations*, Nauka, Moscow, 1969.
266. G. Krause, *A strengthened form of Ling's theorem on associative functions*, Ph. Thesis, Illinois Inst. Techn., 1981.
267. D. Kravvaritis, *Nonlinear random operators of monotone type in Banach spaces* (in print).
268. D. Kravvaritis, *Existence theorems for nonlinear random equations and inequalities*, J. Math. Anal. Appl., **88** (1982), 61–75.
269. K. Kuratowski, *Sur les espaces complètes*, Fund. Math., **15** (1930), 301–309.
270. K. Kuratowski, *Topology*, Vol. I, Academic Press, New York and London, 1966.
271. K. Kuratowski and C. Ryll-Nardzewski, A general theorem on selectors, *Bull. Acad. Polon. Sci., Ser. Sci. Math. Astronom. Phys.*, **13** (1965), 397–403.
272. H. Kwakernaak, Filtering for systems excited by Poisson white noise, *Int. Symp. Control Theory, Numerical Methods and Computer Systems Modelling* (1974), Lecture Notes in Economics and Math. Systems, 107, Springer Verlag.
273. M. Kwapisz, General inequalities and fixed point problems, in: *General Inequalities. 2, Proc. Second Int. General Ineq.*, Oberwolfach, 1970.
274. E. Lami Dozo, Multivalued nonexpansive mappings and Opial's condition, *Proc. Amer. Math. Soc.*, **38** (1973), 286–292.
275. A. Lebow and M. Schechter, *Semigroups of operators and measures of noncompactness*, J. Funct. Anal., **7** (1971), 1–26.
276. A. C. Lee and W. J. Padgett, *On random nonlinear contractions*, Math. Syst. Theory, **11** (1977), 77–84.
277. A. C. Lee and W. J. Padgett, *Random contractors and the solution of random nonlinear equations*, Nonlin. Analysis. Theory Math. and Appl., (1977), 175–185.
278. A. C. Lee and W. J. Padgett, *Random contractors with random nonlinear majorant functions*, Nonlinear Analys. Theory Math. and Appl., **3** (1979), 707–715.
279. A. C. Lee and W. J. Padgett, *Solution of random operator equations by random step-contractors*, Nonlinear Analys. Theory Math. and Appl., **4** (1980), 145–151.

280. A. C. Lee and W. J. Padgett, *On a heavily nonlinear stochastic integral equation*, Utilitas Math., **9** (1976), 123–138.

281. A. C. Lee and W. J. Padgett, *A random nonlinear integral equation in population growth problems*, Journ. of Integral Eq., **2** (1980), 1–9.

282. S. Lefschetz, *Intersections and transformations of complexes and manifolds*, Trans. Amer. Math. Soc., **28** (1926), 1–49.

283. S. Lefschetz, *Algebraic topology*, Amer. Math. Soc., Colloq. Publ., Vol. 27, New York., 1942.

284. J. Leray and J. Schauder, Topologie et équations fonctionnelles, *Ann. Sci. École Norm. Sup.*, **51** (1934), 45–78.

285. C. H. Ling, *Representation of associative functions*, Publ. Math. Debrecen, **12** (1965), 189–212.

286. W. A. Luxemburg, *On the convergence of successive approximations in the theory of ordinary differential equations*, II, Nederl. Akad. Wetensch. Proc. Ser. A61, Indag. Math., **20** (1958), 540–546.

287. M. C. Maia, *Un osservazione sulle contrazione metriche*, Rend. Sem. Univ. Padova, **40** (1968), 139–143.

288. K. Margolis, On some fixed point theorems in generalized complete metric spaces, *Bull. Amer. Math. Soc.*, **74** (1968), 275–282.

289. R. H. Martin, *Differential equations on closed subsets of a Banach space*, Trans. Amer. Math. Soc., **179** (1973), 399–414.

290. K. Menger, Statistical metrics, *Proc. Nat. Acad. Sci. U.S.A.*, **28** (1942), 535–537.

291. K. Menger, Probabilistic geometry, *Proc. Nat. Acad. Sci. U.S.A.*, **37** (1951), 226–229.

292. K. Menger, Géometrie générale, *Mem. Sci. Math.*, **124** (1954), Chapter 7, Paris, Gauthier-Villars.

293. K. Menger, Mathematical implications of Mach's ideas: Positivistic geometry, the clarification of functional connections, in *Ernst Mach, Physicist and Philosopher (Boston Studies in Philosophy of Science*, edited by R. S. Cohen and R. J. Seeger, Vol. 6), 107–125, Dordrecht, Reidel.

294. B. Morrel and J. Nagata, *Statistical metric spaces as related to topological spaces*, General Topology Appl., **9** (1978), 233–237.

295. W. L. Morse, *Stream temperature prediction under reduced flow*, J. Hydraul. Div. Amer. Soc. Civil. Engn., **98** (HY6), (1972), 1031–1047.

296. W. L. Morse, *The dishonest method in stream temperature modeling*, Water Resources Research, **14** (1978), 45–51.

297. P. S. Mostert and A. I. Shields, On the structure of semigroups on a compact manifold with boundary, *Ann. of Math.*, **65** (1957), 117–143.

298. R. Moynihan, *Probabilistic metric spaces induced by Markov chains*, Z. Wahrsch. verw. Geb., **35** (1976), 177–187.

299. A. Mukherjea, *Random transformation on Banach spaces*, Ph. Dissertation, Wayne State Univ., Michigan, 1966.

300. A. Mukherjea, *Transformations aléatoires séparables: Théorème du point fixe aléatoire*, C.R. Acad. Sci., Paris, Sér. A–B, **263** (1966), A393–A395

301. A. Mukherjea, On the measurability of the resolvent of a random kernel, *Bull. Math. Soc. Sci. Math. R. S. Roumanie*, (N.S.), **14** (62), (1970), 55–59.
302. A. Mukherjea, *Contractions and completely continuous mappings*, Nonlin. Anal., **1** (1977), 235–247.
303. A. Mukherjea and A. T. Bharucha-Reid, *Separable random operators (I)*, Rev. Roum. Math. Pures et Appl., **14**(1969), 1553–1561.
304. D. H. Muštari, *On almost sure convergence in linear spaces of random variables*, Theory Prob. Appl., **15** (1970), 337–342.
305. S. B. Nadler, Jr., *Multivalued contraction mappings*, Pacific J. Math., **30** (1969), 475–488.
306. S. B. Nadler, Jr., *Some results on multivalued mapping in set-valued mapping, selections and topological properties of* q^X, edited by W. M. Fleischman, Springer Verlag, 1970.
307. M. Z. Nashed, *Steepest descent for singular linear operator equations*, SIAM J. Numer. Anal., **7** (1970), 358–363.
308. M. Z. Nashed and J. Wong, *Some variants of a fixed point theorem of Krasnoselskii and applications to nonlinear integral equations*, J. Math. Mech., **18** (1969), 767–777.
309. M. Z. Nashed and H. Salehi, *Measurability of generalized inverses of random linear operators*, SIAM J. Appl. Math., **25** (1973), 681–692.
310. M. Z. Nashed and H. W. Engl, Random generalized inverses and approximate solutions of random operator equations, in: *Approximate Solution of Random Equations* (edited by T. Bharucha-Reid), North Holland, New York, Oxford, 1979, 149–210.
311. V. Nemytskiĭ, *The methods of fixed points in analysis*, Uspehi Matem. Nauk, **1** (1936), 141–174.
312. E. Nishiura, *Constructive methods in probabilistic metric spaces*, Fund. Math., **67** (1970), 115–124.
313. A. Nowak, Random solutions of equations, *Trans. 8th Prague Conf. on Inform. Theory. Statist. Decision Functions, Random Processes*, Vol. B, 77–82, Acad. Prague, 1978.
314. A. Nowak, *A note on a random fixed point theorem*, Prace Mat., Katowice, **11** (1981), 33–35.
315. A. Nowak, *Random fixed points of multifunctions*, Prace Mat., Katowice, **11** (1981), 36–41.
316. A. Nowak, *Sequences of contractions and random fixed point theorems in dynamic programming*, Demonstratia Math., **14** (1981), 343–353.
317. A. Nowak, Random Kuhn-Tucker theorem, *Trans. 9th Prague Conf. on Inform. Theory. Statist. Decision Functions, Random Processes*, Vol. B, Acad. Prague, 95–97.
318. A. Nowak, *On generalized random differential equations*, Demonstratia Math., **16** (1983).
319. A. Nowak, *Applications of random point theorems in the theory of generalized random differential equations* (in print).

320. R. D. Nussbaum, The fixed point index and asymptotic fixed point theorems for K-set-contractions, *Bull. Amer. Math. Soc.*, **75** (1969), 490—495.
321. R. D. Nussbaum, *The radius of the essential spectrum*, Duke Math. J., **38** (1970), 473—478.
322. C. Olech, On the existence and uniqueness of solution of an ordinary differential equation in the case of Banach space, *Bull. Acad. Polon. Sci., Ser. Sci. Math. Astronom. Phys.*, **8** (1960), 667—673.
323. O. Onicescu, *Nombres et systèmes aléatoires*, Bucharest, Editions de l'Académie de la R. P. Roumaine, 1964.
324. Z. Opial, *Weak convergence cf the sequence of successive approximation for nonexpansive mappings*, Bull. Amer. Math. Soc., **73**(1967), 591—597.
325. K. G. Oza, *Identification problems and random contraction mappings*, Ph. D. Dissertation, Univ. of California, Berkeley, 1967.
326. K. G. Oza and E. I. Jury, *System identification and the principle of random contraction mapping*, SIAM J. Control, **6** (1968), 244—257.
327. W. J. Padgett, *A stochastic model for stream pollution*, Math. Biosciences, **25** (1975), 309—317.
328. W. J. Padgett, *A stream pollution model with random deoxygenation and reaeration coefficients*, Math. Biosciences, **42** (1978), 137—148.
329. W. J. Padgett and C. P. Tsokos, *On semi-stochastic model arising in a biological system*, Math. Biosciences, **9**(1970), 105—117.
330. W. J. Padgett and C. P. Tsokos, *A stochastic model for chemotherapy: computer simulation*, Math. Biosciences, **9** (1970), 119—133.
331. W. J. Padgett, and C. P. Tsokos, *On a stochastic integral equation of the, Volterra type in telephone traffic theory*, Journ. Appl. Probab., **8** (1971) 269—275.
332. W. J. Padgett and C. P. Tsokos, *Existence of a solution of a stochastic integral equation in turbulence theory*, J. Math. Phys., **12** (1971), 210—212.
333. W. J. Padgett, G. Schultz and C. P. Tsokos, *A random differential equation approach to the probability distribution of BOD and DO in streams*, SIAM J. Appl. Math., **32**(1977), 467—483.
334. W. J. Padgett and A.N.V. Rao, *Solution of a stochastic integral equation using integral contractors*, Information and Control, **41** (1979), 56—66.
335. W. J. Padgett, Approximate solution of a random nonlinear equation by random contractors, in : *Approximate solution of random equations* (edited by A. T. Bharucha-Reid), North Holland, New York, Oxford, 1979, 211—221.
336. N. S. Papageorgiou, *Random fixed point theorems for multifunctions*, Math. Japonica, **29** (1984), 93—106.
337. A. Pazy, *Asymptotic behavior of contractions in Hilbert spaces*, Israel J. Math., **9** (1971), 235—240.
338. M. V. Petryshyn, *On projectional solvalbility and Fredhom alternative for equations involving linear A-proper operators*, Arch. Rational Mech. Anal. **30**(1968), 270—284.
339. W. V. Petryshyn, *Remarks on condensing and K-set-contractive mappings*, J. Math. Anal. Appl., **39** (1972), 717—741.

340. W. V. Petryshyn and Fitzpatrick, *New existence theorems for nonlinear equations of Hammerstein type*, Trans. Amer. Math. Soc., **160** (1971), 39–63.
341. W. V. Petryshyn and Fitzpatrick, Degree theory for noncompact multivalued vector fields, *Bull. Amer. Math. Soc.*, **79** (1973), 609–613.
342. W. V. Petryshyn and Fitzpatrick, *Fixed point theorems for multivalued noncompact inward maps*, J. Math. Anal. Appl., **46** (1974), 756–767.
343. Phan Van Chuong, *On a smooth selection theorem and its application to multivalued integral equations* (in Russian), Mat. Sbornik, **105** (1978), 622–637.
344. H. Poincaré, *Sur les courbes définies par une équation différentielle, III*, J. Math. Pures Appl. (C. Jordan), **1** (1985), 167–244.
345. Yu. V. Prohorov, *Convergence of random processes and limit theorems in probability theory*, Theory Probab. Appl., **1** (1956), 157–214.
346. V. Radu, On linear operators on random normed spaces, *Bull. Math. Soc. Sci. Math. R. S. Roumanie*, **17** (1973), 217–220.
347. V. Radu, On some probabilistic structures on product spaces. *Proc. Fifth Conf. Probability Theory*, Brașov, 1974, 277–282, Bucharest.
348. V. Radu, *On minimal problem for functionals on probabilistic metric spaces*, Sem. on Probab. Metric Spaces, Univ. of Timișoara, No. 33(1975).
349. V. Radu, *Contribuții la studiul unor probleme de analiză funcțională pe spații normate aleatoare*, Doctoral Thesis, Univ. of Timișoara, 1976.
350. V. Radu, *On linear operators in random normed spaces over valued fields*, Sem. on Probab. Metric Spaces, Univ. of Timișoara, No. 46, 1979.
351. V. Radu, *On approximation method for random operator equations*, Rev. Roum. Math. Pures et Appl., **26** (1981), 469–473.
352. V. Radu, *A remark on contraction in Menger spaces*. Sem. on Probab. Theory Theory Appl., Univ. of Timișoara, No. 64, 1983.
353. V. Radu, *On the t-norms of Hadžić type and fixed points in probabilistic metric spaces*, Review of Research, Univ. Novi Sad, **13** (1983), 81–86.
354. V. Radu, *On the t-norms of Hadžić type and locally convex random normed spaces*, Sem. on Probab. Theory Appl., Univ. of Timișoara, No. 70, 1984.
355. V. Radu, *On the t-norms with the fixed point property*, Sem. on Probab. Appl., Univ. of Timișoara, No. 72, 1984.
356. V. Radu, On some contraction tpye mappings in Menger spaces, *Anal. Univ. Timișoara*, **23**(1985), 61–65.
357. V. Radu, *On some fixed point theorems in probabilistic metric spaces*, Sem. on Probab. Theory Appl., Univ. of Timișoara, No. 74, 1985.
358. V. Radu, *A family of deterministic metric on Menger spaces*, Sem. on Probab. Theory Appl., Univ. of Timișoara, No. 78, 1985.
359. J. Reinermann, *Fixpunktsätze von Krasnoselski-Typ*, Math. Zeitschrift., **119** (1971), 339–344.
360. J. Reinermann—E. Schönenberg, Some results in the fixed point theory of nonexpansive mappings and generalized contractions, in *Fixed point theory and its applications* (edited by S. Swaminathan), Academic Press, New York, 1976, 175–186.

361. J. M. Richardson, Application of truncated hierarchy techniques, *Proc-Symp. Appl. Math.*, **16** (1963), 290—302, Amer. Math. Soc. Providence, 1964.

362. H. Robbins and S. Monro, A stochastic approximation method, *Ann. Math. Stat*, **22** (1951), 400—407.

363. R. T. Rockafellar, *Measurable dependence of convex sets and functions on parameters*, J. Math. Anal. Appl., **28** (1969), 4—25.

364. W. Römisch und R. Schulze, *Kennwertmethoden für stochastische Volterrasche Integralgleichungen*, Wiss. Z. Humb. Univ. Berlin, **28** (1979), 523—533.

365. W. Römisch, *On the approximate solution of random operator equations*, Wiss. Z. Humb. Univ. Berlin, **30** (1981), 455—462.

366. W. Römisch, *Approximate solution of random operator equations*, Berichte des Instituts für Mathematik, Universität Linz, report 214, 1982.

367. W. Römisch, On an approximate method for random differential equations, in *Problems of Stochastic Analysis in Applications* (edited by J. von Scheidt), Zwickau, 1983, 327—337.

368. W. Römisch, *On the convergence of measurable selections and an application to approximations in stochastic optimization*, Humbold Universität zu Berlin, Sec. Math. Preprint. No. 71(1984).

369. I. A. Rus, *Generalized contractions*, Sem. on Fixed Point Theory, Univ. Babeş-Bolyai, Cluj-Napoca, No. 3, 1983, 1—130.

370. B. N. Sadovski, *On a fixed point principle*, Funct. Anal. Appl., **1** (1967), 74—76 (in Russian).

371. B. N. Sadovski, *On a measure of noncompactness and condensing operators*, Probab. Mat. Anal. Comp. Syst., Voronej, **2** (1968), 89—119.

372. B. N. Sadovski, *Ultimately compact and condensing mappings*, Usp. Mat. Nauk, **27**(1972), 81—146 (in Russian).

373. M. F. Sainte-Beuve, *On the extension of von Neumann-Aumann's theorems*, J. Funct. Analysis, **17** (1974), 112—129.

374. A. Saleski, *Entropy of self-homeomorphisms of statistical pseudometric spaces*, Pacific J. Math. **51** (1974), 537—542.

375. A. Saleski, *A conditional entropy for the space pseudo-Menger maps*, Pacific J. Math., **59** (1975), 525—533.

376. G. Salinetti and R. Wets, *On the convergence of closed-valued measurable multifunctions*, Trans. Amer. Math. Soc., **266** (1981), 275—289.

377. J. Schauder, *Zur Theorie stetiger Abbildungen in Funktionalräumen*, Math. Zeitschr., **26** (1927), 46—65; 417—431.

378. J. Schauder, *Der Fixpunktsatz in Funktionalräumen*, Studia Math., **2**(1930), 171—180.

379. B. Schweizer, *Probabilistic metric spaces, the first 25 years*, New York Statistician, **19** (1967), 3—6.

380. B. Schweizer, *Lectures on probabilistic metric spaces, Lectures Notes*, Univ. of Arizona, 1965.

381. B. Schweizer, *On the uniform continuity of the probabilistic distance*, Z. Wahrsch. verw. Geb., **5** (1966), 357—360.

382. B. Schweizer, *Multiplications on the space of probability distribution function*, Aequationes Math., **12** (1975), 156–183.
383. B. Schweizer, *Equivalence relations in probabilistic metric space*, Bul. Inst. Politech. Iaşi, **10** (1964), 67–70.
384. B. Schweizer and A. Sklar, *Espaces metriques aléatoires*, C.R. Acad. Sci., Paris, **247**(1958), 2092–2094.
385. B. Schweizer and A. Sklar, *Statistical metric spaces*, Pacific J. Math., **10** (1960), 313–334.
386. B. Schweizer and A. Sklar, *Associative functions and statistical triangle inequalities*, Publ. Math.. Debrecen, **8** (1961), 169–186.
387. B. Schweizer and A. Sklar, *Statistical metric spaces arising from sets of random variables in Euclidian n-spaces*, Teory Probab. Appl., **7** (1962), 447–456.
388. B. Schweizer and A. Sklar, *Associative functions and abstract semigroups*, Publ. Math. Debrecen, **10** (1963), 69–81.
389. B. Schweizer and A. Sklar, *Probabilistic metric spaces determined by measure-preserving transformation*, Z. Wahrsch. verw. Geb., **26** (1973), 235–239.
390. B. Schweizer and A. Sklar, *Operation on distribution functions not derivable from operations on random variables*, Studia Math., **52** (1974), 43–52.
391. B. Schweizer and A. Sklar, How to derive all L_p metrics from a single probabilistic metric, in *General Inequalities*, 2 (edited by E. F. Beckenbach), 429–434, Basel, Birkhäuser Verlag, 1980.
392. B. Schweizer, A. Sklar and E. Thorp, *The metrization of statistical metric spaces*, Pacific J. Math., **10**(1960), 673–675.
393. B. Schweizer and A. Sklar, *Probabilistic metric spaces*, North Holland, New York, Amsterdam, Oxford, 1983.
394. V. M. Sehgal, *Some fixed point theorems in functional analysis and probability*, Ph. D. Thesis, Wayne State University, 1966.
395. V. M. Sehgal, A fixed point theorem for mappings with a contractive iterate, *Proc. Amer Math. Soc.*, **23** (1969), 631–634.
396. V. M. Sehgal and A. T. Bharucha-Reid, *Fixed points of contraction mappings on probabilistic metric spaces*, Math. Systems Theory, **6** (1972), 92–102.
397. M. L. Senechal, *Approximate functional equations and probabilistic inner product spaces*, Ph. D. Thesis, Illinois Inst. of Techn., 1965.
398. A. N. Šerstnev, *Random normed spaces: problems of completeness*, Kazan Gos. Univ. Ucen. Zap., **122** (1962), 3–20.
399. A. N. Šerstnev, *On the notion of a random normed space*, Dokl. Akad. Nauk SSSR, **149** (1963), 280–283.
400. A. N. Šerstnev, *Best approximation problems in random normed spaces*, Dokl. Akad. Nauk SSSR, **149** (1963), 539–542.
401. A. N. Šerstnev, *On a probabilistic generalization of metric spaces*, Kazan Gos. Univ. Ucen. Zap., **124** (1964), 3–11.
402. A. N. Šerstnev, *Triangle inequalities for random metric spaces*, Kazan Gos. Univ. Ucen. Zap., **125** (1965), 90–93.

403. H. Sherwood, *Complete probabilistic metric spaces and random variables generated spaces*, Ph. D. Thesis, Univ. of Arizona, 1965.
404. H. Sherwood, *On the completion of probabilistic metric spaces*, Z. Wahrsch. verw. Geb., **6** (1966), 62–64.
405. H. Sherwood, *On E-spaces and their relation to other classes of probabilistic metric spaces*, J. London Math. Soc., **44** (1969), 441–448.
406. H. Sherwood., *Complete probabilistic metric spaces*, Z. Wahrsch. verw. Geb., **20** (1971), 117–128.
407. D. A. Sibley, *A metric for weak convergence of distribution functions*, Rocky Mountain J. Math., **1** (1971), 427–430.
408. G. Sâmboan and R. Theodorescu, *Statistical spaces*, Rev. Roum. Math. Pures et Appl., **7** (1962), 699–703.
409. A. V. Skorohod, Random operators in a Hilbert space, *Lectures Notes in Math.*, Vol. 330, Springer, Berlin, 1976, 562–591.
410. A. V. Skorohod, *Random linear operators*, Kieff, 1978.
411. D. S. Swart, *Fixed point theorems*, Cambridge University Press, Cambridge, 1974.
412. T. T. Soong, *Random differential equations in science and engineering*, Academic Press, 1973.
413. A. Špaček, *Zufällige Gleichungen*, Czechoslovak Math. J., **5** (80), (1955), 462–466.
414. A. Špaček, *Note on K. Menger's probabilistic geometry*, Czechoslovak Math. J., **6**(81), (1956), 72–74.
415. A. Špaček, Random metric spaces, *Trans. Second Prague Conf. Information Theory, Statist. Decision Funct. and Random Processes*, 1960, 627– 638.
416. R. R. Stevens, *Metrically generated probabilistic metric spaces*, Fund. Math., **61** (1968), 627–638.
417. M. Stojaković, *A common fixed point theorem of a family of mappings in probabilistic locally convex spaces*, Review of Research Univ. Novi Sad, **11** (1981), 77–87.
418. M. Stojaković, *A fixed point theorem in probabilistic metric spaces*, Review of Research Univ. Novi Sad, **12** (1982), 55–60.
419. M. Stojaković, *A common fixed point theorem in probabilistic metric spaces*, Z. Wahrsch. verw. Geb. (in print).
420. M. Stojaković, *Fixed point theorems in probabilistic metric spaces* (in print).
421. F. Stummel, Discrete convergence of mappings, in *Topics in Numerical Analysis*, edited by J. J. Miller, Academic Press, New York, 1973, 285–310.
422. S. Szufla, Some remarks on ordinary differential equations in Banach spaces, *Bull. Acad. Polon. Sci., Ser. Sci. Math. Astronom. et Phys.*, **16**(1968), 795–800.
423. D. H. Tan, *On probabilistic condensing mappings*, Rev. Roum. Math. Pures et Appl., **26**(1981), 1305–1317.
424. D. H. Tan, *A fixed point theorem for multivalued quasi-contractions in probabilistic metric spaces*, Review of Research, Univ. Novi Sad, **12** (1982), 43–54.

425. D. H. Tan, *A note on probabilistic measures of noncompactness*, Rev. Roum. Math. Pures et Appl., **28** (1983), 283—288.

426. D. H. Tan, *On the probabilistic inner measure of noncompactness*, Review of Research, Novi Sad, **13** (1983), 73—80.

427. D. H. Tan, *Some remarks on probabilisitc measures of noncompactness*, Rev. Roum. Math. Pures et Appl. (in print).

428. R. M. Tardiff, *Topologies for probabilistic metric spaces*, Pacific J. Math., **65** (1976), 233—251.

429. E. Thorp, *Generalized topologies for statistical metric spaces*, Fund. Math., **51** (1962), 9—21.

430. C. P. Tsokos and W. J. Padgett, Random integral equations with applications to stochastic systems, *Lecture Notes in Mathematics*, Vol. 233, Springer-Verlag, Berlin, New York, 1971.

431. C. P. Tsokos and W. J. Padgett, *Random integral equations with applications to life sciences and engineering*, Academic Press, New York, 1974.

432. A. Tychonoff, *Ein Fixpunktsatz*, Math. Ann., **111** (1935), 767—776.

433. D. H. Wagner, *Survey of measurable selection theorems*, SIAM J. Control Optimiz., **15**(1977), 859—903.

434. A. Wald, On statistical generalization of metric spaces, *Proc. Nat. Acad. Sci. U.S.A.*, **29** (1943), 196—197.

435. M. T. Wasan, *Stochastic approximation*, Cambridge Univ. Press, London, and New York, 1969.

436. A.F.S. Xavier, *On the product of probabilistic metric spaces*, Portugal Math., **27** (1968), 137—147.

437. P. P. Zabreico and J. B. Ledovskaia, *Existence theorems for equations in Banach spaces and mean principle problem*, Matem. Anal. Comp. Syst. Voronej, **3** (1968), 122—136.

Author Index

Subject Index

GPSR Compliance

The European Union's (EU) General Product Safety Regulation (GPSR) is a set of rules that requires consumer products to be safe and our obligations to ensure this.

If you have any concerns about our products, you can contact us on ProductSafety@springernature.com

In case Publisher is established outside the EU, the EU authorized representative is:

Springer Nature Customer Service Center GmbH
Europaplatz 3
69115 Heidelberg, Germany

Batch number: 10614741

Printed by Printforce, the Netherlands